Experimental Hydraulics

Methods, Instrumentation, Data Processing and Management

Volume I: Fundamentals and Methods

IAHR Monograph

Series editors

Peter A. Davies
Department of Civil Engineering,
The University of Dundee,
Dundee,
United Kingdom

Robert Ettema
Department of Civil and Environmental Engineering,
Colorado State University,
Fort Collins,
USA

The International Association for Hydro-Environment Engineering and Research (**IAHR**), founded in 1935, is a worldwide independent organisation of engineers and water specialists working in fields related to hydraulics and its practical application. Activities range from river and maritime hydraulics to water resources development and eco-hydraulics, through to ice engineering, hydroinformatics and continuing education and training. IAHR stimulates and promotes both research and its application, and, by doing so, strives to contribute to sustainable development, the optimisation of world water resources management and industrial flow processes. IAHR accomplishes its goals by a wide variety of member activities including: the establishment of working groups, congresses, specialty conferences, workshops, short courses; the commissioning and publication of journals, monographs and edited conference proceedings; involvement in international programmes such as UNESCO, WMO, IDNDR, GWP, ICSU, The World Water Forum; and by co-operation with other water-related (inter) national organisations. www.iahr.org

Supported by
Spain Water
and IWHR, China

Experimental Hydraulics

Methods, Instrumentation, Data Processing and Management

Volume I: Fundamentals and Methods

Editors

Marian Muste (Editor-in-Chief)
IIHR – Hydroscience & Engineering, University of Iowa, Iowa City, IA, USA

Dennis A. Lyn
Lyles School of Civil Engineering, Purdue University, West Lafayette, IN, USA

David M. Admiraal
Civil Engineering Department, University of Nebraska, Lincoln, NE, USA

Robert Ettema
Department of Civil and Environmental Engineering, Colorado State University, Fort Collins, CO, USA

Vladimir Nikora
School of Engineering, University of Aberdeen, Aberdeen, UK

Marcelo H. Garcia
Department of Civil and Environmental Engineering, University of Illinois at Urbana-Champaign, IL, USA

CRC Press
Taylor & Francis Group
Boca Raton   London   New York   Leiden

CRC Press is an imprint of the
Taylor & Francis Group, an **informa** business

A BALKEMA BOOK

Cover Illustration: Graphics assembled by David P. Herwaldt (IIHR – Hydroscience & Engineering) and Marian Muste with photographs courtesy of: Vlad Nikora, Jérôme Le Coz, and IIHR – Hydroscience & Engineering.

Applied for

Published by: CRC Press/Balkema
　　　　　　Schipholweg 107C, 2316 XC Leiden, The Netherlands
　　　　　　e-mail: Pub.NL@taylorandfrancis.com
　　　　　　www.crcpress.com – www.taylorandfrancis.com

First issued in paperback 2020

Visit the Taylor & Francis Web site at
http://www.taylorandfrancis.com

and the CRC Press Web site at
http://www.crcpress.com

Typeset by Integra Software Services Private Ltd
Although all care is taken to ensure integrity and the quality of this publication and the information herein, no responsibility is assumed by the publishers nor the author for any damage to the property or persons as a result of operation or use of this publication and/or the information contained herein.

Library of Congress Cataloging-in-Publication Data

ISBN 13: 978-1-138-02753-4 (hbk)(set of 2 vol)
ISBN 13: 978-0-367-57335-5 (pbk)(vol 1)
ISBN 13: 978-1-138-03816-5 (hbk)(vol 1)
ISBN 13: 978-1-138-03815-8 (hbk)(vol 2)

About the IAHR Book Series

An important function of any large international organisation representing the research, educational and practical components of its wide and varied membership is to disseminate the best elements of its discipline through learned works, specialized research publications and timely reviews. IAHR is particularly well-served in this regard by its flagship journals and by the extensive and wide body of substantive historical and reflective books that have been published through its auspices over the years. The IAHR Book Series is an initiative of IAHR, in partnership with CRC Press/ Balkema – Taylor & Francis Group, aimed at presenting the state-of-the-art in themes relating to all areas of hydro-environment engineering and research.

The Book Series will assist researchers and professionals by advancing and transferring contemporary knowledge needed for research, education and engineering practice. The series includes Design Manuals and Monographs. The Design Manuals, usually prepared by multiple authors, guide the application of theory and research findings to engineering practice; and, the Monographs give state-of-the-art coverage of various significant topics in water engineering.

The first and highly successful IAHR book was *"Turbulence Models and their Application in Hydraulics"* by W. Rodi, published in 1984 by Balkema. *"Turbulence in Open Channel Flows"* by I. Nezu and H. Nakagawa, also published by Balkema (in 1993), had an important impact on the field and, during the period 2000–2010, further authoritative texts (published directly by IAHR) included *Fluvial Hydraulics* by S. Yalin and A. Da Silva and *Hydraulicians in Europe* by W. Hager. All of these publications continue to strengthen the reach of IAHR and to serve as important intellectual reference points for the Association.

Since 2011, the Book Series is once again a partnership between CRC Press/Balkema – Taylor & Francis Group and the Technical Committees of IAHR. The present book is an exciting further contribution to IAHR's Book Series, substantially aiding water engineering research, education and practice, and showcasing the expertise IAHR fosters.

Series editors
Peter A. Davies
Robert Ettema

Editors Biography Volume I

Marian Muste
Editor-in-Chief

Marian Muste is Research Engineer with the IIHR-Hydroscience & Engineering, and Adjunct Professor in the Department of Civil & Environmental Engineering at the University of Iowa. Dr. Muste's areas of research include experimental river mechanics (laboratory and field investigations), instrumentation development and implementation (especially image-, acoustic-, and laser-based), uncertainty analysis, and hydroinformatics. He contributed to 185 peer-reviewed papers and 75 technical reports. He was chair for the International Association of Hydro-Environmental Engineering & Research's (IAHR) Experimental Methods and Instrumentation Technical Committee from 2000 to 2007.

Dennis A. Lyn

Dennis A. Lyn is Professor of Civil Engineering in the Lyles School of Civil Engineering at Purdue University. In addition to a general interest in statistical thinking and analysis of data, Dr. Lyn is specifically interested in turbulence, open-channel flows, sediment transport and scour phenomena, with applications to river engineering. Experimental studies in which he participated have appeared in Journal of Hydraulic Engineering and Journal of Fluid Mechanics. Dr. Lyn served on the editorial boards of several hydraulics-focused journals and was Editor of the American Society of Civil Engineers' Journal of Hydraulic Engineering from 2006 to 2009.

David M. Admiraal

David M. Admiraal is an Associate Professor in the Civil Engineering Department at the University of Nebraska–Lincoln. His research expertise includes laboratory and field investigations of sediment transport, river hydraulics, and hydraulic structures. Most of his work incorporates experimental modeling. Dr. Admiraal has over 50 journal and conference publications in a wide variety of hydraulic engineering topics. He is actively involved in American Society of Civil Engineers' Technical Committee on Hydraulic Measurements and Experimentation and has served as the chair of the committee. He has also acted as co-chair for two conferences on Hydraulic Measurements and Experimental Methods.

Robert Ettema

Robert Ettema is the Harold H. Short Professor at Colorado State University's Department of Civil and Environmental Engineering. He actively conducts laboratory and field experiments in several areas of hydraulic engineering, including hydraulic structures, alluvial-channel hydraulics and cold-regions hydraulics. Dr. Ettema chaired the committee that prepared the monograph *Hydraulic Modeling: Concepts and Practice*, Manual 97, published by the American Society of Civil Engineers (ASCE) and was Chief Editor for ASCE' Journal of Hydraulic Engineering. His prior positions were as a professor at the University of Wyoming and the University of Iowa, where he has been a member of IIHR—Hydroscience & Engineering.

Vladimir Nikora

Vladimir Nikora is the Sixth Century Chair in Environmental Fluid Mechanics at the School of Engineering, University of Aberdeen, UK. His main research areas relate to turbulent flows, sediment dynamics, hydraulic resistance, flow-biota interactions, and experimental methods. Dr. Nikora has served as Editor of IAHR Journal of Hydraulic Research and Associate Editor for AGU Water Resources Research and ASCE Journal of Hydraulic Engineering. He has been an active contributor to IAHR Instrumentation Section as a member, secretary and Section Chair. Dr. Nikora is Fellow of the Royal Society of Edinburgh.

Marcelo H. Garcia

Marcelo H. Garcia is the M.T. Geoffrey Yeh Chair in Civil Engineering and Director of the "Ven Te Chow" Hydrosystems Laboratory at the University of Illinois-Urbana. His research interests are in rivers, sediment transport, environmental fluid mechanics and experimental techniques. His scholarship includes 4 books, 19 book chapters, and more than 300 peer-reviewed papers and technical reports covering a wide range of fundamental and practical problems in hydraulic engineering. Dr. Garcia was Editor-in-chief of the ASCE Sedimentation Engineering Manual of Practice 110, served as Editor of the IAHR Journal of Hydraulic Research (2001–2006), and is recipient of numerous professional awards.

Galileo Galilei (1564 – 1642)

"Measure what is measurable, and make measurable what is not so."

"Measure what is measurable, and make measurable
what is not so."

Galileo Galilei (1564 – 1642)

Contents Volume I and II

Volume I

Volume II

Preface and Contributions

Hydraulics is the branch of engineering and science associated with all forms of water movement in natural and constructed channels and conduits. Physical interactions between many types of gaseous, liquid, and solid materials with water flows are also relevant to the field of hydraulics. Notable examples include sediment transport, ice flows, entrained air flows, and flows of water of varying salinity or turbidity. In addition, conventional hydraulics also addresses the transport of energy, heat, and dissolved chemicals by water. Related topics include water waves, thermal plumes in lakes and streams, and transport of contaminants and nutrients by rivers. Most problems in hydraulics are much too complex to be solved by direct mathematical formulation, making laboratory and field experiments crucial for understanding and advancing our knowledge about various flows.

This two-volume book is a comprehensive guide to designing, conducting, and interpreting laboratory and field experiments on a broad range of topics associated with modern hydraulic studies. The book provides guidance on the selection of appropriate laboratory facilities, instrumentation and experimental techniques. Advice is also given on performing experiments. Furthermore, methods are described for analysing data, assessing uncertainties in measurements, and presenting results derived from observations and experiments. The book considers experiments performed under a range of conditions, from well-equipped, well-staffed experimental facilities to environments where accessibility to advanced instrumentation and expertise is limited. Although the book focuses primarily on laboratory experiments, including hydraulic physical modelling, it also applies to field experiments of varying complexity and accessibility.

A comprehensive guide to experimental hydraulics is much needed. Most readily available guides are already several decades old, and do not reflect the remarkable developments that have occurred in contemporary hydraulics, especially regarding digital instrumentation and handling of large, high-density experimental datasets. The experts comprising the book's editorial team, aware of this major gap in literature on hydraulics, decided to embark on a concerted effort to write an authoritative and up-to-date reference text on hydraulic experiments. The book is the result of the first substantial effort in the community of hydraulic engineering to assemble in one place descriptions of all the components of hydraulic experiments along with a concise outline of essential hydraulic theory.

Aiming to ensure readability and accessibility to information, this book is organized into two volumes with content that follows a uniform, multi-layered approach (i.e., body text, summary tables, examples, appendices) across the chapters. The first volume focuses on fundamental hydraulics and on the phases of conducting and analysing the results of hydraulic experiments. The second volume concentrates on flow visualization and measurement methods and comprehensively describes the measurement instrumentation available for hydraulic experiments. The coverage of established experimental hydraulics subjects (such as conventional measurement instruments) is briefer, while descriptions of contemporary topics in both experimental methods and instrumentation are more extensive.

Target readers are college faculty, researchers, practitioners, and students involved in various aspects of experimental hydraulics. The book's primary purpose is to provide readers a clear and readily accessible guide to hydraulic experiments, the necessary steps when conducting them, and the extensive information about the principles, capabilities, and limitations of available instrumentation (particularly contemporary technologies). It is assumed that readers understand elementary fluid mechanics and statistics and have a general technical background. Ample references are given to complementary books and articles on engineering hydraulics and experimentation, enabling the interested reader to delve further into the topics covered.

This book was written by a team of more than 45 experts who have been at the forefront of research, teaching or practice in experimental hydraulics. Most of the writers have also been active members of the International Association for Hydro-Environment Engineering and Research's (IAHR) Technical Committee on Experimental Methods and Instrumentation in the last three decades. The need for this book was conceived, discussed, and developed during multiple meetings held at various conferences and through extensive correspondence. During these discussions, an editorial group was established to handle critical editorial duties. Most of the credit in creating the book, though, must be given to the dedicated team of authors. Their desire to advance experimental hydraulics, and their consistently good-natured collaboration, led to the book's realization. The authors contributing to chapters in this volume are named in the table below. Additionally, the affiliations for the contributing authors for this volume are subsequently listed.

The editorial group also thanks Professor Peter Davies, the former IAHR Book Editor, and Janjaap Blom, Senior Publisher at Taylor & Francis, for their guidance and support, as well as patience. Many other people also should be acknowledged and thanked for contributing figures and background material used in preparing the book and for reviewing portions of the book. The summary table below recognizes the volunteers who provided assistance to the authors.

Everybody involved in preparing this book trusts that it achieves the vision of an authoritative and accessible reference resource for experimentation in hydraulics. Readers are invited to contact the editorial team regarding various omissions, misprints, and suggestions for improvements to this edition of the book. Future editions will address such items and include on-going developments in instrumentation and experimental techniques

Marian Muste
Editor-in-Chief

List of authors contributing to chapters and sections in Volume I

Chapter/ Section	Title	Contributing Author(s)
Chapter 1. Introduction		**Coordinators: Muste M., Ettema R.**
1.1	Book overview	Muste M.
1.2	Role of hydraulic experiments	Ettema R.
1.3	Approach	Muste M., Ettema R.
1.4	Structure of Volume I	Muste M.
Chapter 2. Hydraulic flows: Overview		**Coordinators: Garcia M.H., Nikora V.**
2.1	Introduction	Nikora V.
2.2	Turbulent flows in hydraulic engineering	Nikora V.
2.3	Turbulence mechanics: concepts and descriptive frameworks	Nikora V.
2.4	Open-channel flows	Garcia M.H.
2.5	Complex flows	Garcia M.H.
2.5.2	Flows in vegetated channels	Garcia M.H., Fytanidis D.K.
2.5.3	Aerated flows	Chang K.A., Socolofsky S.A.
2.5.4	Ice-laden flows	Ettema R.
Chapter 3. Similitude		**Coordinator: Ettema R.**
3.1	Introduction	Ettema R.
3.2	Basics	Ettema R.
3.3	Dynamics similitude from flow equations	Ettema R.
3.4	Water flow	Ettema R.
3.5	Multi-phase flow and transport processes	Ettema R.
3.6	Addressing similitude shortcomings	Heller V.
Chapter 4. Selection and design of the experiment		**Coordinators: Muste M., Admiraal D.M.**
4.1	The experimental process	Muste M.
4.1.2	Experiment phases	Muste M., Admiraal D.M.
4.1.3	Safety considerations	Admiraal D.M.
4.2	Experimental setup components	Muste M.
4.2.3	Measurement systems	Admiraal D.M., Muste M.
4.3	Laboratory facilities	Muste M., Uijttewaal W.S.J.
4.4	Instrument selection	Muste M.
4.4.3	Practical relationships for estimating turbulent flow scales	Muste M., Garcia C.M.
4.5	From signals to data	Muste M.
Chapter 5. Experiment execution		**Coordinators: Muste, M., Admiraal D.M.**
5.1	Instrument-flow and facility-flow interactions	Muste M.
5.1.2	Types of interactions	Admiraal D.M.
5.1.4	Influence of turbulence levels	Admiraal D.M., Muste M.
5.2	Conducting the experiment	Muste M.
5.2.1	Activities	Admiraal D.M.
5.2.2	Instrument calibration	Admiraal D.M.
5.2.3	Establishing the measurement procedure	Admiraal D.M.
5.2.6	Flow control	Uijttewaal W.S.J., Muste, M.
5.2.8	Record keeping	Admiraal D.M.
5.3	Field experiments	Aberle J., Rüther N.
5.4	Complex experiments	Muste M.
5.4.1	Sediment transport	Uijttewaal W.S.J

ACKNOWLEDGEMENTS

Some of the chapters and sections benefitted from contributions made by colleagues who are not included in the authorship. The acknowledged colleagues are:

Chapter 2: Dimitrios K. Fytanidis (editing tasks)

Chapter 6. David M. Admiraal, Gary Parker, Tony Wahl, and Cary Troy (data for examples included in the chapter)

Chapter 7: Paul Pilon, Janice Fulford, Tom Yorke (editing tasks)

Chapter 8: Venkatesh Merwade (editing tasks)

All chapters: Heng-Wei Tsai, Mary Windsor, Carl Smith, – students affiliated with IIHR- Hydroscience & Engineering, The University of Iowa (editing and compilation tasks)

SAMPLE CITATIONS FOR BOOK CHAPTERS AND SECTIONS:

Chang K.A. & Socolofsky S.A. (2017). Aerated flows, Section 2.5.3 in Experimental Hydraulics, Volume I; M. Muste, D.A. Lyn, D.M. Admiraal, R. Ettema, V. Nikora, and M.H. Garcia (Editors); Taylor & Francis, New York, NY.

Moore, S.A. & Lemmin, U. (2017). Acoustic backscattering, Section 5.3.3.1 in Experimental Hydraulics, Volume II; J. Aberle, C.D. Rennie, D.M. Admiraal, M. Muste (Editors); Taylor & Francis, New York, NY.

List of Contributing Authors Vol. 1

Jorge Abad   Department of Environmental Engineering
                    Universidad de Ingeniería y Tecnología
                    Barranco-Lima, Peru
                    Email: jabadc@utec.edu.pe

Jochen Aberle   Leichtweiß-Institute for Hydraulic Engineering and Water Resources
                    Technische Universität Braunschweig
                    D-38106 Braunschweig, Germany

                    Department of Civil and Environmental Engineering
                    Norwegian University of Science and Technology
                    Trondheim 7491, Norway
                    Emails: j.aberle@tu-braunschweig.de; jochen.aberle@ntnu.no

David Admiraal   Civil Engineering Department
                    University of Nebraska – Lincoln
                    Lincoln, Nebraska 68588, U.S.A
                    Email: dadmiraal@unl.edu

Ram Balachandar   Civil and Environmental Engineering
                    University of Windsor
                    Windsor, Ontario N9B 3P4, Canada
                    Email: rambala@uwindsor.ca

Kuang-An Chang   Zachry Department of Civil Engineering
                    Texas A&M University
                    College Station, Texas 77843, U.S.A
                    Email: kchang@civil.tamu.edu

George   Department of Civil and Environmental Engineering, IIHR – Hydroscience
Constantinescu   & Engineering
                    University of Iowa
                    Iowa City, Iowa 52242, U.S.A
                    Email: sconstan@engineering.uiowa.edu

Ibrahim Demir   Department Civil and Environmental Engineering, IIHR – Hydroscience &
                    Engineering,
                    University of Iowa
                    Iowa City, Iowa 52242, U.S.A
                    Email: ibrahim-demir@uiowa.edu

Robert Ettema   Civil & Environmental Engineering Department
                    Colorado State University
                    Fort Collins, Colorado 80523, U.S.A
                    Email: Robert.Ettema@colostate.edu

Dimitrios K.
Fytanidis
Ven Te Chow Hydrosystems Laboratory
University of Illinois at Urbana-Champaign
Urbana, Illinois 61801, U.S.A
Email: fytanid2@illinois.edu

Carlos M. Garcia
School of Exact, Physical and Natural Sciences
Córdoba National University
Institute for Advanced Studies for Engineering and Technology
(IDIT0CONICET-UNC)
Córdoba, Argentina
Email: carlos.marcelo.garcia@unc.edu.ar

Marcelo H. Garcia
Ven Te Chow Hydrosystems Laboratory, Civil and Environmental
Engineering Department
University of Illinois at Urbana-Champaign
Urbana, Illinois 61801, U.S.A
Email: mhgarcia@illinois.edu

Valentin Heller
Department of Civil Engineering
University of Nottingham
Nottingham NG7 2RD, UK
Email: valentin.heller@nottingham.ac.uk

Dongsu Kim
Civil & Environmental Engineering Department
Dankook University,
Yongin, Gyeonggi 31116, South Korea
Email: dongsu-kim@dankook.ac.kr

Jérôme Le Coz
Waters – River Hydraulics
IRSTEA
Lyon-Villeurbanne 69616, France
Email: jerome.lecoz@irstea.fr

Kyutae Lee
Streamgauging Unit
South Florida Water Management District
West Palm Beach, FL, 33406, U.S.A
Email: alkene78@gmail.com

Danxun Li
Department of Hydraulic Engineering,
Tsinghua University, Beijing, China
Email: lidx@mail.tsinghua.edu.cn

Dennis A. Lyn
Lyles School of Civil Engineering,
Purdue University,
West Lafayette, Indiana 47907, U.S.A
Email: lyn@purdue.edu

Marian Muste
IIHR – Hydroscience & Engineering, Civil & Environmental Engineering
Department
University of Iowa
Iowa City, Iowa 52242, U.S.A
Email: marian-muste@uiowa.edu

Vladimir Nikora
School of Engineering
University of Aberdeen
Aberdeen AB24 3UE, United Kingdom
Email: v.nikora@abdn.ac.uk

Colin Rennie
Civil Engineering Department
University of Ottawa
Ottawa, Ontario K1N 6N5, Canada
Email: Colin.Rennie@uottawa.ca

Nils Ruther

Department of Civil and Environmental Engineering
The Norwegian University of Science and Technology
Trondheim 7491, Norway
Email: nils.ruther@ntnu.no

Scott A.
Socolofsky

Zachry Department of Civil Engineering
Texas A&M University
College Station, Texas 77843, U.S.A
Email: socolofs@tamu.edu

Shivam Tripathi

Department of Civil Engineering,
Indian Institute of Technology
Kanpur, Uttar Pradesh 208016, India
Email: shiva@iitk.ac.in

Wim S.J.
Uijttewaal

Hydraulics Engineering – Civil Engineering and Geosciences
Delft University of Technology
Delft, 2628CN, The Netherlands
Email: w.s.j.uijttewaal@tudelft.nl

Soheil G. Zare

Water Power Department
Hatch
Calgary, Alberta T2P 3G2, Canada
Email: soheil.zare@hatch.com

Chapter 1

Introduction

1.1 BOOK OVERVIEW

The purpose of this book is to aid contemporary laboratory and field experimentation in civil engineering hydraulics. In so doing, it presents current and foreseeable practice in the design and conduct of experiments in hydraulics, including guidelines for selecting and operating instrumentation, for analyzing measurements, and for the use of consistent methods for quantifying and documenting the quality of measurements.

Experiment design and conduct are commonly influenced by the availability of instrumentation and facilities and the expertise of the experimenter. Accordingly, as instrumentation and facilities have evolved, the level of sophistication of hydraulic experiments has increased, as has the requisite level of expertise. Consequently the range of experiment capabilities has expanded – experimenters today can measure, record and observe far more than ever before. Additionally, improvements in methods for quantifying measurement reliability provide a more solid technical foundation for experimental findings.

This book is published in two volumes to make it easily accessible without eliminating important information related to hydraulic experimentation. The two volumes have the following emphases:

Volume I: Fundamentals and Methods –guidelines for the design and conduct of
  hydraulic experiments and the analysis and presentation of measured data
Volume II: Instrumentation and Measurement Techniques – information about the
  principles, capabilities, and limitations of hydraulic instruments as well as guidance
  regarding selection of instrumentation and techniques for measurement and flow
  visualization

The ensuing sections of this chapter introduce the structure of Volume I and present the topics associated with it, including the role of experiments and the proper approach to experiment design and conduct. It is important to note that Volume I begins by recapping fundamental aspects of theoretical hydraulics before proceeding with guidelines of experiment methodology. The authors hold that a sound grasp of fundamentals of hydraulics is requisite for sound experimentation. Theoretical hydraulics is based on fluid mechanics, especially as applied to the turbulent flow of water and the transport of various constituents (*e.g.*, sediment, air, ice) by water.

It is useful to note that hydraulic experiments benefit from advances in other areas of fluid mechanics, especially areas involving sophisticated aspects of fluid mechanics, such as ship hydrodynamics and aircraft aerodynamics. An important aspect of contemporary experimentation in hydraulics is that it is increasingly cross-disciplinary. In this context, the goals of the investigations exceed the needs of typical hydraulics-focused studies (*e.g.*, acquiring velocity and stress distributions) to include aspects related to water flows that are drivers or consequences of riverine natural processes (*e.g.*, water bio-geo-chemistry, aquatic habitat health). The replication of these processes in the laboratory or the direct observations carried out in the field increases the complexity of the experiments (see Sections 2.5 and 5.4). Further, experiments in hydraulics often benefit from being part of a broader problem-solving approach by interacting with numerical simulation. The broader approach is commonly termed "hybrid modeling," and uses the attributes of numerical simulation and physical experimentation. The authors also hold that, for the conceivable future, hydraulic experiments remain at the cutting edge of new information discovery.

1.2 THE ROLE OF EXPERIMENTS IN HYDRAULICS

Well-conducted laboratory and field experiments play a crucial role in hydraulic research, development and design because many aspects of water flow elude analytical formulation or, for the moment at least, are not readily amenable to numerical simulation. Additionally, experiments often offer a practical and economical way to better understand hydraulic processes and solve hydraulic problems. As with all branches of science and engineering practice, laboratory or field experiments are commonly used to provide the data and observations needed to better understand a phenomenon or corroborate a formulation or numerical simulation. In line with substantial advances in the sophistication of instrumentation and experimental methods, experiments have gained the capacity to measure more flow features and produce more (and more detailed) data.

The ability to conduct experiments and evaluate findings from experiments is an essential theme in science and engineering education. Accordingly, experiments have a significant role in hydraulics education. They handily demonstrate, confirm or reveal flow processes, and usefully show the extents to which selected variables influence flow processes. Moreover, students of hydraulics benefit greatly from knowing how selected characteristics of flows can be measured, appreciating the limitations of measuring flow properties, and having the experience to judge quality or reliability of data and observations obtained from experiments. Science and engineering, after all, rely on sound data and observations acquired from laboratory or field experiments for strengthening the understanding of cause-and-effect relationships, and sharpening critical thinking.

Hydraulic experiments feature prominently in the progress of hydraulic engineering. The exact origins of hydraulics experimentation are rather blurred, dating to antiquity, but their significance clearly emerges in the seventeenth, eighteenth and nineteenth centuries. Contemporary hydraulic experimentation proliferated in the early twentieth century with the emergence of numerous formal hydraulic laboratories. They typically were associated with universities and engineering departments of governments and some large companies. Rehbock's River-Hydraulics Laboratory at the Technical University of Karlsruhe is credited as being one of the first and best laboratories (*e.g.*, Rouse & Ince, 1975). Founded in 1901, it quickly demonstrated that experiments utilizing model similitude concepts could significantly advance hydraulic engineering through the practical application of fluid mechanics. Dedicated experimental facilities and advanced measurement techniques enhanced the capabilities of such laboratories. The early twentieth century also saw the establishment of laboratories for other areas of fluid mechanics research, including ship hydrodynamics, hydro-machinery and aerodynamics. Various books and journal articles indicate the momentous influence of laboratory experiments on the progress of hydraulic engineering; *e.g.*, Rouse and Ince (1957), and Hager (2003, 2009, 2013).

Four general considerations and related problems associated with water flow prompt the on-going need for laboratory experiments: spatial complexity; certain aspects of complexity in time; turbulence; and, transport processes. These considerations are at the heart of most, if not all, hydraulic experiments, and they often occur together. Most real world flows are spatially complex, hampering direct analysis. Spatial complexity is often associated with variability of the flow boundary geometry and surface roughness, both of which can significantly impede estimation of flow and its distribution. Complexity in time, especially regarding flow separation in conditions of unsteady flow, is also common, both locally (*e.g.*, flow around individual flexible plants) and more globally (*e.g.*, flow around a barge or ship hull moving into a shallowing channel). Turbulence often embodies these spatial and temporal complexities. Yet turbulence is also rotational and comprises vorticity of fluid motion, a semi-random flow property that is highly significant in the fate of flow energy and in the transport of flow constituents like sediment and heat. Transport processes themselves typically introduce additional mechanics and associated variables that complicate analysis and numerical simulation. Classical examples in this regard include transport of bed sediment by water and adjustments in bed morphology due to hydrodynamic forces.

There are several types of experiments in hydraulics. The content of this book implicitly applies to all of them. The types can be grouped as follows:

1. Laboratory experiments focus on fundamental processes and usually do not relate to specific project sites or structures. Graduate thesis research commonly entails such experiments;
2. Hydraulic modeling usually involves a reduced-scale simulation of a specific project site or representative location. Experiments of this type usually aim at solving a particular problem or improve a design;
3. Specialty laboratories that focus on specific areas such as coastal engineering, stream rehabilitation, soil erosion, ice hydraulics, biomedical or other specific flows;
4. Educational experiments and demonstrations show noteworthy processes in engineering hydraulics. Undergraduate and graduate courses benefit when students see and touch flow processes, and experience the approaches used to measure flow properties;
5. Performance testing of hydraulic equipment such as flow meters, velocity meters, pressure meters, flow valves and gates. Most widely-used hydraulic equipment must be tested for sound performance and measurement accuracy. This category of experiments also includes the design and performance testing of turbines or pumps. The latter experiments normally require specially configured laboratories

(*e.g.*, rigid floor, accurate flow and pressure metering) to yield accurate discharge, flow head and power data;

6. Field measurements of river-related processes in their natural environment. Most contemporary natural-scale experiments are specifically designed to address open scientific questions that cannot be adequately addressed in laboratory experiments (*e.g.*, flow through vegetation, processes involving interaction between water-borne abiotic and biotic factors, etc.); and,

7. Field measurements for monitoring purposes. These measurements are required for many aspects of water resources management (*e.g.*, water use, flood forecasting and mitigation). They are increasingly used in conjunction with the construction and validation of numerical models for hydrology and hydraulics. Most of these measurements are administered by specialized agencies, but researchers are increasingly involved in complementing the available data with supplementary datasets used for validation of theories and hypotheses.

It is evident from these types that hydraulic experiments are conducted for a range of purposes, varying from basic research to education to commercial testing. Whatever their purpose, experiments share the same need for reliable instrumentation and methods.

Irrespective of their type, experiments are effective means for illuminating and developing many aspects of engineering hydraulics. The range and sophistication of their use has grown substantially in recent decades, as contemporary instrumentation and computer-based techniques for data-acquisition and analysis and visualization have increasingly been applied to problems in engineering hydraulics. This growth facilitates experiments that until now have been considered impractical. Experiments involving complex flows and requiring high degrees of temporal and spatial resolution are more and more common, both in the laboratory and in the field. However, there remains the continual need for these advances to be matched with practical guidance on conducting experiments in hydraulics. Furthermore, development of sophisticated experimentation equipment also demands better methods for evaluating measurement reliability and for handling and visualizing the massive amounts of data that experiments may produce.

1.3 APPROACH

Volume I of Experimental Hydraulics is an advanced guide for designing and conducting experiments in hydraulics. Its content delineates the contemporary frontier of knowledge and practice regarding experiments over the range of flows representative of engineering hydraulics. Assembling theoretical fundamentals and experimental methods in one place is a distinct strength of the book that, by and large, is not found in most textbooks on engineering hydraulics. The succession of chapters follows the usual sequence of the measurement process, from connecting the measurements with their governing equations to presentation of results. It first describes essential theoretical considerations for water flow and water-related transport processes, and then addresses key concerns in experiment design and the selection of appropriate facilities, instrumentation and measurement techniques. Further, it offers guidance on processing, presentation, quantifying uncertainties, and archiving data and observations from experiments.

Although this book could serve as a textbook for a course on experimental hydraulics, it was designed as a handbook intended to provide fast and convenient access to information. Therefore, its style is concise and makes frequent use of tables and figures to present information. The contributors have gone to great lengths to provide specific guidance on technologies and methods currently used in hydraulic experimentation. However, given that experimental hydraulics comprises a wide variety of topics and special applications, the depth of the descriptions varies from section to section, commensurate with the complexity, usage, and novelty of the technique or procedure under discussion. While abundant references are provided to allow the reader to explore beyond the covers of the book, most of the emerging and specialized methods are presented in great detail by assembling information from many references into an easily accessible format. Implementation examples are provided for new or emerging methods, especially for those presented in the final sections of this volume, to illustrate how theoretical considerations are transferred to practice.

The various hydraulics- and experimentation-related concepts and terms are defined as the experimental phases are described, but the level of detail of the definitions varies commensurate with the particular needs of each chapter or section. For example, statistical terminology appears in Chapters 2, 4, 5, 6 and 7. In the chapters preceding Chapter 6, the exploratory statistics suffice for those chapters whereby the characterization of experimental samples is the main interest. Common descriptive statistics (a.k.a. frequentist statistics) is used to determine flow scales in Chapters 2 and 4, to establish the necessary parameters for data acquisition in turbulent flows described in Chapter 5, and for evaluating standard uncertainties in Chapter 7. Chapter 6 utilizes more specialized statistical concepts and terminology to infer general population parameters from limited samples of measurements and to express them in probabilistic terms.

The consistency of terminology notation across chapters preoccupied the authors from the onset of the book development. The consistency challenge resides in the fact that the book makes use of variable names from several domains (*e.g.*, fluid dynamics, hydraulics, statistics, and data-driven inferences) that have developed stand-alone nomenclature and conventions that are often in conflict across domains. In response to this challenge, a two-prong approach was established for this handbook. First, the authors were requested to adhere to a skeleton nomenclature and conventions that are as close as possible to standard fluid mechanics conventions – specifically hydraulics conventions. Common notations in this category include symbols for variable types (vectors, tensors, or complex quantities), dimensionless numbers, operators, Greek symbols, subscripts and superscripts for variables. For example, acommon convention was used for the notation of Reynolds decomposition for describing turbulence. Second, important terminology is defined within relevant chapters, and a list of important notations is given at the end of each chapter if the terminology or notation is deemed central to the chapter. Deviations from the agreed-upon strategy are identified by individual authors.

The book uses Standard International (SI) units to accommodate a convention widely adopted by the international scientific and practice communities. Exceptions are tables or figures replicated in their original form from various literature sources.

Numbering of the figures, tables, equations, and examples is based on the numbering of the section in which they are found. For example, the first figure of Section 2.2 of Chapter 2 is identified as Figure 2.2.1. To improve accessibility, appendices immediately follow the sections in which they are mentioned.

1.4 STRUCTURE OF VOLUME I

The two volumes of Experimental Hydraulics contain a total of sixteen chapters. The first eight chapters in Volume I deal with the theoretical foundation of hydraulics and experimental methods, including similitude analysis, an important tool for the design of experiments. The eight chapters in Volume II describe flow visualization, measurement instrumentation, and experimental techniques that lead to measurements. The chapters in Volume II group instruments based on the type of variable measured. While arranged in distinct volumes, experimental methods and instrumentation are both integral parts of experimental hydraulics, so there are many cross-references between the volumes.

Chapter 2 describes the fundamentals of hydraulic flows, which are grouped as wall-bounded shear and free-shear flows. Distinction is further made between canonical (*i.e.*, fully-developed, steady flows in simple geometries) and non-canonical (more complex flows with complex geometries and spatio-temporal variation) flows. Definitions and descriptions for all of these flows are provided at the introductory level and are intended to guide the design and execution of the experiments that are often flow specific. Only turbulent flows are considered since pure laminar flows are usually of low importance in most practical hydraulic engineering problems. For turbulent flows, concepts and descriptive frameworks are introduced to highlight the mechanics of the flows, their governing equations, and the intersections between theory and experiments. Starting with Section 2.4, the chapter focuses on turbulent open-channel flows that comprise a large majority of experimental investigations in hydraulics. The flows described in this latter section include clear-water flows, sediment-laden flows, flow through vegetation, aerated flows, and flows under ice cover. Analysis assumptions, flow structures, and governing equations for these complex flows are provided both to support experimental design and data processing as well as to substantiate the role of experiments in the formulation of semi-empirical relationships used to describe most open-channel flows.

Chapter 3 covers similitude analysis, which links theory and experiments for attaining substantial and practical insights into both simple and complex flows. This analysis precedes the design of experimental facilities (Chapter 4) and guides the selection of investigation targets based on the resources and facilities available to the user (Chapter 5). Similitude can also complement the interpretation of experimental data and observations. The focus of Chapter 3 is on core similitude analysis aspects that are common to experiments in hydraulics. The chapter details typical approaches for solving hydraulic problems through experiments by either scaling the equation for the flow and transport processes (Section 3.3) or by directly comparing ratios of variables that govern the behavior of the flow under investigation (Sections 3.4 and 3.5). Both approaches stem from dimensional analysis which is described in Appendix 3.A. Section 3.6 addresses the shortcomings of similitude results that stem from the inability to match all governing similitude parameters. Hydraulics-specific practices are then proposed to mitigate these short comings. These practices are synthesized in comprehensive summary tables to aid experimenters with the design of their experiments and for avoiding unwanted scale effects in physical modeling studies.

Chapter 4 introduces criteria for selecting facilities and instruments for common and special experiments. Section 4.1 highlights that experiments in hydraulics ensue a continuum of activities progressing in orderly manner through totally-controlled (laboratory) or partially-controlled (field) measurement protocols. The components of the experimental setup along with their characteristics and behavior are subsequently described in Section 4.2. Laboratory facilities are extensively described in Section 4.3, given that their configuration, scale, and sophistication level dictate the extent of the physical modeling scope, the degree of investigation details, and the quality of experiment outcomes. Section 4.4 guides users in the selection of measurement instruments with due consideration of the scope of the experiments, flow characteristics (e.g., spatial-temporal variability of turbulent flows), and the capabilities of available instruments. Section 4.5 provides insights on how signals travel within data acquisition systems (from sensor to instrument display), identifying processes and effects that can negatively impact the quality and accuracy of experimental data.

Chapter 5 outlines essential steps in experiment execution. Section 5.1 warns the reader of possible unwanted effects that might occur due to interaction between the flow and the measurement instruments as well as between the flow and the facility where the experiments take place. Section 5.2 describes the protocols to be followed for setting instrument operating parameters and for ensuring a good quality experiment (i.e., preliminary runs, flow control, quality control, and record keeping). With a good understanding of the execution of the laboratory experiments, considerations are extended in Section 5.3 to in-situ experiments with emphasis on the logistics needed to ensure the safety of personnel, the public and deployed equipment. A series of complex experimental situations is described in Section 5.4 to highlight additional concerns that arise when investigating flow-driven transport processes (e.g., sediment-laden flows, aerated flows) or flow through vegetation and under ice covers. Finally, Section 5.5 discusses an emerging trend in hydraulic investigations whereby experiments are increasingly coupled with numerical modeling (i.e., numerical simulations) to obtain the information needed for solving specific practical problems. Essentials of this investigative synergy, often called "hybrid modeling," are illustrated through examples.

Chapter 6 presents conventional and modern approaches for post-processing and analysis of data acquired through experiments. As most of these methods are of a statistical nature, the descriptions highlight the relation between the obtained parameters and the physical hydraulic context and the subtleties in interpretations that are often missing in purely statistical accounts. The readers are assumed to be familiar with elementary concepts of probability and statistics theory. Section 6.1 offers a brief review of basic statistical concepts and extended descriptions for the more specialized statistical concepts. Methods for characterization, exploration, and conditioning of the datasets are subsequently presented in Sections 6.2 and 6.3. Sections 6.4 and 6.5 detail the methods for quantification of the extent to which a limited number of measurements are representative of the whole data population for the variables under investigation.

As many experimental studies entail formulation of relationships between variables, Sections 6.6 and 6.7 focus on data regression analysis and methods for evaluation of their level of confidence. Bayesian interpretation of the probability associated with the confidence levels is also introduced at this point. Chapter 6 also comprises post-processing methods that have not yet been extensively used in hydraulic experimentation. Among them are inference methods based on data classification and machine-learning that uniquely enhance handling and analysis of large datasets and thus insight into physical processes. With the extensive use of contemporary measurement techniques and numerical simulations, hydraulics investigations collect and process unprecedented amounts of data, making these special analysis tools irreplaceable. Sections 6.9 and 6.10 elaborate on the inference methods and demonstrate their implementation through examples. Most of the presented methods have been widely used in related domains such as hydrology (*e.g.*, Singh, 2017), and the large number of examples presented in this chapter demonstrate the value of extending these methods to experimental hydraulics analyses.

Sections 6.11 and 6.12 present methods that reveal more sophisticated information than that derived from the descriptive statistics presented in the first part of the chapter. For example, statistical analysis accomplished through autocorrelation and the related power spectra may yield information that is not revealed by other methods, leading therefore to additional insights into the flow dynamics (*e.g.*, energy distribution in the turbulent flow structures). Furthermore, the spatial interpolation method presented in Section 6.13 elaborates on means of enhancement of measured data though visualization of the measured processes. Section 6.14 describes methods for extraction of flow features from high-density temporal and spatial datasets. The methods have been originally developed for analysis of numerical simulations output but they work as well with the voluminous, multi-dimensional datasets acquired with laser- and image-based measurement techniques. These methods enable efficient visualization of large datasets, drastically reducing the complexity that is often associated with understanding flow turbulence.

Chapter 7 addresses the issue of data quality. Some areas of hydraulic engineering require rigorous procedures (*i.e.*, standards) for quantitatively assessing the uncertainties of measurements. These uncertainties are determined by analytically propagating the impacts of quantifiable uncertainty sources to the uncertainty of a final result. Section 7.1 reviews these methods with emphasis on the formal procedures used in experimental mechanics and aerodynamics whereby the requirement for attaining high levels of experiment reliability is the norm. Following the standards review, the metrology standard "Guide to Expression of Uncertainty in Measurement" (GUM, 1993) is recommended for adoption, as it offers a suitable method for evaluating the uncertainties of a variety of hydraulic measurements. As GUM (1993) does not provide detailed guidelines for a specific area of measurement, Sections 7.2 and 7.3 give the terminology, classification of errors, and procedures suggested by this standard. The GUM protocol is subsequently applied to a hydraulic measurement (*i.e.*, discharge in natural streams) to illustrate its implementation. Section 7.4 describes the intercomparison-based method, a surrogate for inferring measurement uncertainty when the resources available for data quality assessment are limited. Section 7.5 suggests overall guidelines for ensuring the reliability and accuracy of measured quantities.

Chapter 8 introduces the topic of hydroinformatics, a fast growing area of hydraulic engineering that links advances in measurement instrumentation with dramatic progress in computer science and computational thinking. This cross-disciplinary approach originated in computational hydraulics during the early 1990s, driven by the need to efficiently manage and extract information and knowledge from multi-disciplinary data (*e.g.*, hydraulics, hydrology, and environmental engineering) for problem-solving in water resources. Following the introduction of hydroinformatics, prototypes of information-centric systems are presented with the intent to illustrate their capabilities to equip experimenters with the means to integrate datasets acquired over large spatio-temporal scales by multiple data providers for subsequent analyses of complex water-related issues. A simple GIS-based data model is presented in the closing example of this chapter to illustrate the next generation of data warehousing in hydraulic experiments.

REFERENCES

GUM (1993) ISBN 92-67-10188-9. *Guide to the expression of uncertainty in measurement.* Geneva, Switzerland, International Organization for Standardization.

Hager, W.H. (2003) *Hydraulicians in Europe 1800–2000: A biographical Dictionary of Leaders in Hydraulic Engineering and Fluid Mechanics.* Volume 1. London, UK, IAHR, CRC Press.

Hager, W.H. (2009) *Hydraulicians in Europe 1800–2000: A biographical Dictionary of Leaders in Hydraulic Engineering and Fluid Mechanics.* Volume 2. London, UK, IAHR, CRC Press.

Hager, W.H. (2015) *Hydraulicians in the USA 1800–2000 A biographical Dictionary of Leaders in Hydraulic Engineering and Fluid Mechanics.* London, UK, IAHR CRC Press.

Singh, V.P. (2017) *Handbook of Applied Hydrology.* New York, NY, McGraw Hill.

Rouse, H. & Ince, S. (1957) *History of Hydraulics.* Iowa City, IA, Iowa Institute of Hydraulic Research, The University of Iowa.

Chapter 2

Hydraulic Flows: Overview

2.1 INTRODUCTION

Modern hydraulics encompasses a wide range of water flow phenomena that can be categorized into two fundamental classes: wall-bounded shear flows and free-shear flows. Each class can be further subdivided into canonical and non-canonical flow types and patterns. Canonical (or 'benchmark') flows include flows of relatively simple geometries. They have been extensively studied over past decades and therefore considerable knowledge on such flows is available. Examples of canonical flows include: two-dimensional boundary layers; fully developed pipe, conduit and open-channel flows; jets; wakes; and mixing layers. Non-canonical flows are typically either not fully-developed (or not in equilibrium) or highly three-dimensional, unsteady, and/or exhibit unusual geometrical features. Three-dimensional boundary layers (*e.g.*, Dwyer, 1996; Schlichting & Gersten, 2003; Wallace, 2013), complex flow separations (*e.g.*, Williams, 1996), flows in ice-covered rivers (*e.g.*, Ettema *et al.*, 2008), and vegetated flows (*e.g.*, Nepf, 2012) may serve as examples of non-canonical flows. Most of these hydraulic flows are fully turbulent. Moreover, they usually combine several canonical flows.

To varying degrees, hydraulic flows can be approximated using canonical flow types or their combinations. This attribute explains the long-lasting attraction in studying such flows, many features of which remain poorly understood. Non-canonical flow types often attract even more attention due to their practical significance for specific situations, although theoretical analyses of such flows are highly challenging. Section 2.2 of this volume provides brief definitions of ordinary (canonical) hydraulic flows and describes their main features. Section 2.3 subsequently focuses on key turbulence concepts and descriptive frameworks equally important and applicable to all types of hydraulic flows. Sections 2.4 and 2.5 of this volume are devoted to open-channel and complex flows and cover governing equations, flow structure, sediment transport, and vegetation effects, among other issues. This information is intended to guide considerations of the experimental methods and techniques that are often flow-specific. The treatment below is introductory, as in-depth consideration of hydraulic flows is beyond the scope of this book and can be found elsewhere (*e.g.*, Monin & Yaglom, 1971, 1975; Bernard & Wallace, 2002; Schlichting & Gersten, 2003; Pope, 2005). Laminar flows are not discussed in this chapter as they usually have low practical relevance to hydraulic research and engineering. Interested readers can find comprehensive treatments of laminar flows in Shames (2003), Janna (2009) and other texts.

2.2 TURBULENT FLOWS IN HYDRAULIC ENGINEERING

2.2.1 Wall-bounded shear flows

Canonical wall-bounded shear flows include turbulent boundary layers over flat plates, open-channel (free-surface) flows, pipe flows and conduit/channel flows (Figure 2.2.1).

A *Boundary Layer* (BL) is the flow domain where velocity changes from the wall velocity (zero for immobile wall surfaces) to the mainstream (external) velocity U_m away from the wall surface (Figure 2.2.1a). Three types of boundary layers occur adjacent to flat boundaries: (i) smooth-wall BLs ($\Delta << \delta_v$); (ii) transitional BLs ($\Delta \sim \delta_v$), and (iii) rough-wall BLs ($\Delta >> \delta_v$), where $\delta_v \sim 5(v/u_*)$ is the thickness of the viscous sublayer, Δ is the roughness height, $u_* = (\tau_0/\rho)^{0.5}$ is the local shear velocity, τ_o is the local wall (or bed) shear stress, v is kinematic fluid viscosity, and ρ is fluid density (*e.g.*, Monin & Yaglom, 1975; Schlichting & Gersten, 2003). BL type can be identified using the roughness Reynolds number $Re_* = u_*\Delta/v$ which represents a ratio of the roughness height to the thickness of the viscous sublayer, *i.e.*, $Re_* = u_*\Delta/v = \Delta/(v/u_*) \propto \Delta/\delta_v$. Based on the accepted thickness of the viscous sublayer ($\delta_v = 5(v/u_*)$, *e.g.*, Monin & Yaglom, 1971; Nezu & Nakagawa, 1993), roughness elements protrude beyond the viscous sublayer if $\Delta > 5(v/u_*)$ (*i.e.*, if $Re_* > 5$). Thus, depending on flow conditions, the same wall can be *dynamically smooth* ($Re_* \lesssim 5$) in the case of a smooth-wall BL, *dynamically transitionally rough*

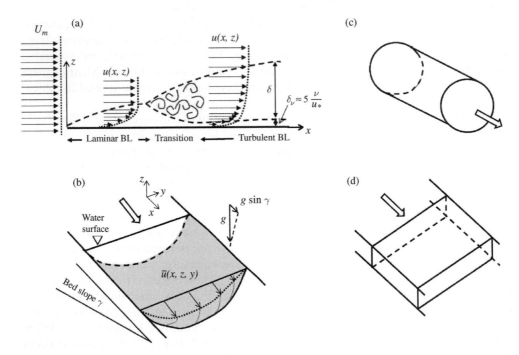

Figure 2.2.1 Sketches of turbulent wall-bounded flows: (a) smooth-wall boundary layer (rough-wall-boundary layer exhibits a similar structure); (b) open-channel flow; (c) pipe flow; and (d) conduit flow (also known as channel flow).

$(5 < \mathrm{Re}_* < 70)$ for a transitional BL, or *dynamically fully rough* $(\mathrm{Re}_* \geq 70)$ for a rough-wall BL. It should be noted that the given values of Re_*, separating wall roughness regimes, are not strictly defined and may vary slightly among different texts (*e.g.*, Schlichting & Gersten, 2003).

The most relevant bulk measures of *smooth-wall* BLs involve three forms of the Reynolds number: $\mathrm{Re}_x = U_m x/\nu$, $\mathrm{Re}_\delta = U_m \delta/\nu$ and $\mathrm{Re}_L = U_m L/\nu$, where x is the distance from the leading edge (*e.g.*, upstream edge of a plate), δ is the BL thickness which increases with x (*i.e.*, with $\mathrm{Re}_x = U_m x/\nu$), and L is the length of the plate. Occasionally, a local shear velocity u_* is also used as a characteristic velocity instead of U_m. In addition to its traditional interpretation as the ratio of inertial forces to viscous forces, the Reynolds number can also be interpreted as a ratio of length scales (this was illustrated above with Re_*). Indeed, $\mathrm{Re}_x = U_m x/\nu = x/(\nu/U_m) \propto x/\delta_v$, as the thickness of the viscous sublayer is $\delta_v \propto \nu/u_* \propto \nu/U_m$. Thus, a high value of the Reynolds number $\mathrm{Re}_x = U_m x/\nu$ means that the viscous sublayer is very thin compared to the distance from the leading edge of the BL. Similarly, $\mathrm{Re}_\delta = U_m \delta/\nu$ can be viewed as a ratio of the BL thickness δ to the thickness of the viscous sublayer δ_v: $\mathrm{Re}_\delta = U_m \delta/\nu = \delta/(\nu/U_m) \propto \delta/\delta_v$. This interpretation considers $\mathrm{Re}_\delta = U_m \delta/\nu$ as a measure of the relative 'submergence' of the viscous sublayer by the turbulent region.

For *rough-wall* BLs, the three forms of the Reynolds number mentioned above should be supplemented with a fourth measure δ/Δ, which is the relative submergence of roughness elements. An analogy exists between $\mathrm{Re}_\delta \propto \delta/\delta_v$ for *smooth-wall* BLs and δ/Δ for *rough-wall* BLs: both represent a ratio of the key characteristic scales featuring in these BLs.

Since the introduction of Prandtl's BL concept in 1904, there have been numerous experimental, numerical, and theoretical studies of turbulent boundary layers over smooth and rough walls. Their focus has traditionally been on obtaining predictive relationships for drag coefficients and other BL parameters such as velocity distributions, boundary layer thicknesses, and wall shear stresses. Although Table 2.2.1 provides some conventional examples of such relationships, the reader is encouraged to consult specialized texts for more extensive information (*e.g.*, Monin & Yaglom, 1971; Schlichting & Gersten, 2003; Wallace, 2013).

Theoretical considerations of canonical boundary layers often involve assumptions not strictly valid for many real-life flows encountered in nature and engineering. Very high Reynolds number and/or very high relative submergence (to justify the occurrence of the universal logarithmic velocity distribution) are typical among such assumptions. As a result, analysis of non-canonical BL flows often involves relaxation of BL assumptions. Such flows include flows with small relative submergence (*e.g.*, gravel-bed or sand-dune rivers), flows over highly porous spatially-extended walls/beds (*e.g.*, vegetated rivers), roughness transitions along and/or across the flow, and internal boundary layers formed by roughness patches. Although some standard BL assumptions are violated for such flows, the conventional results obtained for canonical BLs are a useful starting point and may even provide a sufficient approximation. For instance, many (if not most) rough-wall hydraulic flows exhibit relative submergence δ/Δ much less than 40, the minimum value required for the genuine logarithmic velocity profile with (quasi-) universal constants to occur (*e.g.*, Jimenez, 2004). However, the attempts to approximate measured velocity distributions with the logarithmic formula (Table 2.2.1) for flows with $\delta/\Delta < 40$ are numerous. The measured data suggest that with decreasing δ/Δ the

Table 2.2.1 Representative relationships characterizing canonical turbulent boundary layer flows (e.g., Shames, 2003; Schlichting & Gersten, 2003; Janna, 2009)

Assumed velocity profile	Wall shear stress τ_0	BL thickness δ	Drag coefficient $C_D(L) = \frac{F_D}{0.5\rho U_m S}$	$Re_x = \frac{U_m x}{\nu}$
Smooth-wall boundary layer				
$\dfrac{\overline{u}}{U_m} = \left(\dfrac{z}{\delta}\right)^{1/7}$	$\dfrac{0.0225\rho U_m^2}{Re_\delta^{1/4}}$	$\dfrac{0.37}{Re_x^{1/5}}x$	$\dfrac{0.074}{Re_L^{1/5}}$	5×10^5 to 5×10^7
$\dfrac{\overline{u}}{U_m} = \left(\dfrac{z}{\delta}\right)^{1/10}$	$\dfrac{0.0100\rho U_m^2}{Re_\delta^{1/6}}$	$\dfrac{0.201}{Re_x^{1/7}}x$	$\dfrac{0.0305}{Re_L^{1/7}}$	2.9×10^7 to 5×10^8
$\dfrac{\overline{u}}{u_*} = \dfrac{1}{\kappa}\ln\dfrac{u_* z}{\nu} + A$	–	–	$\dfrac{0.455}{(\log Re_L)^{2.58}}$	$>10^7$
Rough-wall boundary layer				
$\dfrac{\overline{u}}{u_*} = \dfrac{1}{\kappa}\ln\dfrac{z-d}{\Delta} + B$	–	–	$[1.89 + 1.62\log\frac{L}{\Delta}]^{-2.5}$	$>10^7$

Symbols: $\overline{u}$ is mean streamwise velocity (overbar defines ensemble or time-averaging), z is distance from the wall, d is the zero-plane displacement (i.e., it defines the origin of the vertical coordinate for the log profile), $\kappa \approx 0.40$ is the von Kármán constant, $A \approx 5.5$ and $B \approx 8.5$ are empirical constants assumed to be universal (although the data show that κ, A, and B may vary depending on flow conditions; e.g., Marusic et al., 2010).

approximations for κ and B deviate from their 'universal' counterparts obtained for flows with very large relative submergence (i.e., $k = 40$ and $B = 8.5$). Researchers typically account for these deviations by developing additional relationships linking parameters κ and B with roughness parameters and flow submergence δ/Δ. The proposed relationships, however, are empirical and reflect trends that are likely flow-specific and thus may not be general. In addition, the time-averaged velocity profiles of flows with low submergence δ/Δ often exhibit high spatial heterogeneity due to the influence of the local geometry of roughness elements. This heterogeneity can be eliminated by spatial averaging of the velocity profiles, but then a new formulation of the logarithmic law is required to accommodate spatial averaging (see Section 2.3.2 in Volume I). Similar to the rough-wall logarithmic velocity distribution in low-submergence flows, the smooth-wall logarithmic velocity distribution is often used at moderate Re_δ (i.e., at small relative submergence of the viscous sublayer) by adjusting κ and A.

Open-channel flow (OCF) is free-surface flow driven by gravity and contained within channel boundaries (Figure 2.2.1b). The canonical OCF is a steady uniform turbulent flow in a straight channel of rectangular cross-section. The key differences between canonical OCF and BL are: (1) unlike BL thickness which increases along the flow, OCF depth does not change; (2) corners at the intersections of the channel walls, bed, and free-surface of OCFs generate turbulence anisotropy that induce secondary currents (in very wide channels the secondary currents may weaken or even die out away from the channel walls); and (3) the free surface acts as a mobile moving boundary and may have significant impact on flow dynamics. Altogether, these features make OCF a special class of wall-bounded turbulent flows distinctly different from BLs as well as from conduits and pipes where the free surface is absent. Nevertheless, there are many similarities between the vertical distributions of mean velocities, turbulence structure, mixing behavior, and friction laws of OCFs and BLs. In fact, many experimental and fundamental results obtained for BLs have direct relevance for other canonical flows. Non-canonical open-cannel flows include rivers and artificial channels with complex (non-straight) plan forms, irregular cross-sections, and mobile sedimentary boundaries. These features induce additional dynamic effects that may not be easily accommodated with simple parameterizations. Since OCF is of great importance for most branches of hydraulic engineering it is further considered in Section 2.4 in Volume I. Readers interested in comprehensive treatment of OCF are encouraged to consult specialized texts (*e.g.*, Nezu & Nakagawa, 1993; Chanson, 2004; Sturm, 2009).

Pipe flow (PF) is a class of internal flows driven by a pressure gradient and fully confined within a closed channel of circular cross-section (Figure 2.2.1c). The canonical form of PF is a flow in a straight circular pipe with wall roughness that is homogeneous around and along the flow. Due to the circular geometry of the pipe cross-section, the distributions of flow properties (*e.g.*, mean velocity or turbulent energy) are axially-symmetric, making generation of turbulence-induced secondary currents impossible (although they can develop in curved pipes). Canonical PF has been extensively studied over several decades and has served as a test ground for many theoretical and phenomenological developments (*e.g.*, Kim, 2012).

Conduit flow (CF) is a flow type that is very similar to pipe flow: it is also driven by a pressure gradient within a closed channel. However, there is an important difference between PF and CF that relates to the non-circular geometry of the CF cross-section (Figure 2.2.1d). Non-circular geometry introduces turbulence anisotropy that induces secondary currents in CFs distinguishing them from the canonical pipe flow.

There are many similarities between turbulent boundary layers and flows in open-channels, pipes and conduits, particularly related to the turbulence structure in the near-wall region and its manifestation in terms of wall-normal distributions of mean velocities, turbulent energy, and higher-order velocity moments (*e.g.*, Nezu, 2005; Kim, 2012; Wallace, 2013). However, even in the near-wall region, subtle differences between canonical flows have been discovered in recent studies (*e.g.*, Marusic *et al.*, 2010; Marusic & Adrian, 2013), and it is not surprising that intensive theoretical, numerical and experimental studies of wall-bounded flows continue.

2.2.2 Free-shear flows

Canonical free-shear flows include wakes, jets, and mixing layers (Figure 2.2.2). Although free-shear flows are free from the effects of solid boundaries by definition, in most cases they originate near boundaries. For instance, wakes form due to the deficit of momentum immediately downstream of solid bodies, and mixing layers typically develop downstream of a solid boundary separating two flows of varying momentum.

In free-shear flows, as in most wall-bounded flows, cross-flow variation of flow properties is much stronger than in the along-flow direction. Therefore, the appropriate velocity and length scales for such flows are the variation $\Delta \bar{u}$ of the mean streamwise velocity $\bar{u}$ across the flow and the half-width l of the velocity shear region, respectively. The scale l is defined as a cross-stream distance from the flow centerline to a location of a certain fraction of $\Delta \bar{u}$. In the longitudinal direction, the free-shear flows can be subdivided into three distinct regions: (1) the near-field region immediately downstream of the flow-forming solid body or structure; (2) the transitional region within which specific effects of the flow-forming body vanish; and (3) the far-field self-

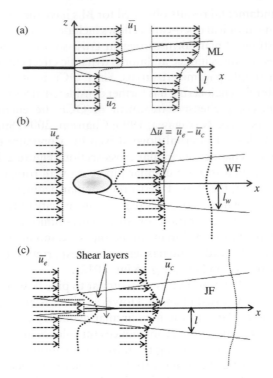

Figure 2.2.2 Sketches of free-shear flows: (a) mixing layer (ML); (b) wake flow (WF); and (c) jet flow (JF).

similarity (or self-preservation) region where the spatial distributions of the flow parameters normalized with local scales $\Delta \bar{u}$ and l do not change along the flow. The relative size and significance of each of the three regions depend on the flow type and forcing (*e.g.*, Brown & Roshko, 2012).

A *Mixing layer* (ML) is a shear layer that forms at the interface between two streams of different velocities ($\bar{u}_1$, $\bar{u}_2$) as a result of Kelvin-Hemholtz instability (KHI), which may occur if an inflection point is present in the velocity profile (Figure 2.2.2a). A classical example of a ML is the plane spatial shear layer which forms behind a splitter plate. In principle, MLs may be classified into spatially-growing (but stationary in time) MLs, temporally-growing (but spatially homogeneous) MLs, stationary and homogeneous MLs, stratified MLs, three-dimensional (deep-water) MLs, two-dimensional (shallow) MLs, and other types. Studies of MLs use a variety of characteristic scales, the most important being $\Delta \bar{u} = \bar{u}_2 - \bar{u}_1$ and $\bar{u}_o = 0.5(\bar{u}_1 + \bar{u}_2)$ as velocity scales; and $l_\omega = \Delta \bar{u}/(d\bar{u}/dz)_{\max}$ as a length scale (known as the vorticity thickness), where z is the distance from the ML centerline (Figure 2.2.2a). Other length scales, such as the momentum thickness, are also frequently used. A comprehensive examination of ML dynamics is beyond the scope of this book. Therefore only two ML characteristics, both supported by extensive experiments and numerical simulations, are presented herein. The first one is the relationship for the across-flow velocity distribution as described by Gortler's theoretical equation (*e.g.*, Bernard & Wallace, 2002):

$$\bar{u}(z) \approx \bar{u}_o \left[1 + \frac{\Delta \bar{u}}{2\bar{u}_o} erf(\psi) \right] \qquad (2.2.1)$$

where *erf* is the error function, $\psi = 0.5\text{Re}_{ML}[z/(x - x_o)]$ is a relative distance from the ML centerline, $\text{Re}_{ML} = \bar{u}_o l/v$, $l = \alpha(x - x_o)$, α is a constant, and x_o is a 'virtual' starting point (origin) of the ML. The second characteristic relates to the ML growth rate that was postulated to be constant, *i.e.*, $dl_\omega/dx = const$ (*e.g.*, Bernard & Wallace, 2002). Mixing layers are widely present in environmental and industrial applications and therefore their studies, particularly of non-canonical cases, continue to thrive. Furthermore, mixing layers are key elements of jets and wakes that are defined next.

Wake flow (WF) is a region of disturbed flow downstream of a body within which the momentum transport is reduced, *i.e.*, the wake is a region of momentum deficit (Figure 2.2.2b). Wake flows can be subdivided into three broad categories: (1) planar wakes formed behind two-dimensional bodies such as cylinders; (2) axisymmetric wakes behind bodies of circular shape in the across-flow plane; and (3) wakes behind complex three dimensional bodies.

The planar wake behind a cylinder is the most extensively studied wake flow (*e.g.*, Zdravkovich 1997, 2003). The relevant velocity scale of the planar wake is $\Delta\bar{u} = \bar{u}_e - \bar{u}_c$, where $\bar{u}_e$ is the external flow velocity and $\bar{u}_c$ is the flow velocity at the wake centerline. This velocity scale, together with the half-width length scale l_w, define the wake flow behavior, particularly in the far-field self-similarity region. A general feature of any wake is the development of strong shear layers separating the low-momentum fluid inside the wake from the surrounding flow. Another common feature of turbulent wakes is flow separation at the body surface (see Section 2.2.3 in Volume I).

The structure and organization of wake flow are controlled by the Reynolds number $\text{Re}_W = \bar{u}_e b/v$, where b is the body width (or diameter). For a circular cylinder, an initially laminar planar wake at small Re_W becomes unstable at around $\text{Re}_W \approx 60$, resulting in the formation, detachment, and convection of alternate two-dimensional vortices downstream of the cylinder. A well-defined sequence of these vortices forms when $60 \leq \text{Re}_W \leq 5000$; this organized sequence of vortices is known as a von Kármán vortex street. At $\text{Re}_W \approx 1300$, the shedding frequency f_s of von Kármán vortices reaches a maximum, corresponding to a Strouhal number $St = bf_s/\bar{u}_e \simeq 0.21$. At $\text{Re}_W \geq 5000$ the wake becomes fully turbulent, and the regular von Kármán vortex street disappears.

The WF dynamics in the near-field region is highly dependent on the body shape and surface roughness. The wake far-field (self-similarity) region starts at approximately $100b$ or further downstream of a body. In this region, the wake structure is dominated by large eddies formed, most likely, as a result of shear instabilities associated with the turbulent mean flow (*e.g.*, Bernard & Wallace, 2002). The analysis of the momentum equations shows that in the far-field region the momentum thickness θ of the wake flow is constant, *i.e.*, it does not change along the flow. Another result for this region, directly relevant to experimental work, is the relation between the drag force F_D (per unit length) and the momentum deficit, *i.e.*:

$$F_D = \rho \int_{-\infty}^{\infty} \bar{u}(\bar{u}_e - \bar{u})dz = \rho\bar{u}_e^2\theta \qquad (2.2.2)$$

The mean velocity distribution across a planar wake is Gaussian while $\Delta\bar{u} = \bar{u}_e - \bar{u}_c$ and l_w change along the flow as (*e.g.*, Schetz & Fuhs, 1996):

$$\frac{\Delta \overline{u}}{\overline{u}_e} \propto \left(\frac{C_D b}{x}\right)^{1/2}, \quad l_w \propto (C_D b x)^{1/2} \tag{2.2.3}$$

where C_D is the body drag coefficient. Similar relations for the axisymmetric wake (*e.g.*, behind a sphere) are:

$$\frac{\Delta \overline{u}}{\overline{u}_e} \propto \left(\frac{C_D A}{x^2}\right)^{1/3}, \quad l_w \propto (C_D A x)^{1/3} \tag{2.2.4}$$

where A is the projected area of the body (Schetz & Fuhs, 1996). More general wake problems involving bodies of irregular geometry embedded in complex flows do not permit simple self-similarity solutions. They are typically studied with dedicated laboratory experiments and numerical simulations. Examples of such wake flows include shallow wakes behind porous obstacles embedded in a boundary layer (*e.g.*, wakes behind vegetation patches in shallow rivers or estuaries; Nepf, 2012), interactions of wakes behind wind turbines, or momentumless wakes behind self-propelling objects.

Jet flow (JF) is defined as a flow pattern formed by passing one fluid into another, typically ambient fluid, with associated injection of either momentum, buoyancy, or both (Figure 2.2.2c). In principle, jets can be inertia-, friction-, or buoyancy-dominated. They can be subdivided into laminar and turbulent jets, round and plane jets, co-flow and cross-flow jets, and others. The simplest (canonical) configurations of a jet flow are planar or axisymmetric jets entering into still fluid occupying an infinite domain. Similar to a WF structure, the cores of jets are separated from motionless surrounding fluid by strong shear layers. The dynamics of these layers is largely responsible for the interaction between jets and the ambient environment in the near-field region. Immediately downstream from a jet entrance (*i.e.*, nozzle), the shear layers grow by entraining the surrounding fluid. At about $6b$ to $10b$ (where b is the nozzle width or diameter) the shear layers on opposite sides of the jet meet, delineating the near-field region. Further downstream of the nozzle, the jet develops into a single expanding structure with near-Gaussian cross-flow distributions of the mean streamwise velocity and turbulence parameters. At approximately $30b$ to $60b$ downstream of the nozzle, cross-flow distributions of velocity and turbulence properties become completely self-similar, marking the beginning of the far-field (self-similarity) region. The main scales of the far-field region are the centerline velocity $\overline{u}_c$ and jet half-width l (or radius for axially-symmetric jets). Therefore, the local jet Reynolds number is often defined as $\mathrm{Re}_J = \overline{u}_c l / v$.

The key self-similarity results for canonical planar and axially-symmetric jets include Gaussian cross-flow distributions of mean streamwise velocities and turbulence parameters, linear dependence of the jet width on the distance from the nozzle and approximately constant spreading angle. However, the decrease of the centerline velocity follows different relations: $\overline{u}_c \sim x^{-1/2}$ for planar jets and $\overline{u}_c \sim x^{-1}$ for circular jets. Although these conventional results are useful for guiding generic jet studies, there are many non-canonical jet configurations important for applications that require dedicated experimental and numerical studies.

Being fundamentally different, canonical free-shear flows (mixing layers, wakes, and jets) also share some similarities. First, mixing layers, wakes, and jets expand laterally along the flow. Second, they all exhibit self-similarity properties in the far-field regions, where appropriately normalized transverse distributions of mean velocities and turbulence characteristics do not change along the flow. Third, the typical feature of most free-shear flows is a high level of regularity of vortical (coherent) structures emerging through sequences of hydrodynamic instabilities (*e.g.*, Monin & Yaglom, 1971; Bernard & Wallace, 2002). A recent comprehensive review of free-shear flows can be found in Brown and Roshko (2012).

2.2.3 Flow near interfaces

The canonical flows outlined in the previous sections aid in understanding more complex flow configurations encountered in hydraulic applications. Examples include three-dimensional boundary layers, flow separations, vegetated flows, ice-covered flows, sediment-laden flows, and flows over mobile boundaries as in rivers with active bedload. Below we briefly consider two common flow configurations of broad practical significance: flow separations and hydrodynamic interfaces.

Flow separation (FS) occurs in locations where a boundary layer detaches from the surface on which it forms. Despite extensive studies of flow separation over many decades, it remains as one of the least understood complex fluid-mechanics phenomena. Flow separation was first described and defined by Prandtl (1952), who proposed that if an adverse pressure gradient is present in a boundary layer, near-wall fluid particles decelerate until they reach the separation point where they completely lose their momentum, causing the local wall shear stress to vanish. At the separation point the boundary layer detaches and a reverse flow replaces the detached BL. Subsequent theoretical and experimental studies have confirmed this initial conceptual picture, at least for two-dimensional flow separations. Practical interest in flow separation phenomena is motivated by the fact that FS is associated with the large pressure drops that result in pressure drag. Thus, in many applications drag control and drag reduction are tantamount to flow separation control.

In general, flow separations can be subdivided into two distinctly different classes: two-dimensional and three-dimensional flow separations. Each class can be further classified as either laminar or turbulent and steady or unsteady. A distinction is also made between separation from smooth surfaces and from corners (*e.g.*, Williams, 1996). The physical understanding and predictive capabilities for many separation scenarios are still limited. Nevertheless, flow separation is prevalent in hydraulics and has a profound impact on the fluid dynamics of important flows like the flows around bedforms, vegetation patches, rocks, ships, bridge piers and many other submerged objects.

A *Hydrodynamic interface* (HI) can be defined as a thin region between neighboring flow domains of distinctly different properties (*e.g.*, water flow-porous bed interface). The maxima of gradients of flow and concentration properties are often found at hydrodynamic interfaces, and they have consequently become the subject of many active theoretical and experimental studies, particularly during the last few decades. The close attention of researchers to interface regions is not surprising, as the interfaces often control the overall mixing, transport, and hydraulic resistance. The range of interfaces studied by modern fluid mechanics is vast, reflecting the expansion of fluid mechanics toward its borders with other disciplines where interfacial phenomena ranging from the nano-scale to the astronomical scale occur (*e.g.*, Stone, 2010). Ecologically-critical heterogeneous interfaces are a typical example (*e.g.*, Marion *et al.*, 2014). Considering the nature of interfacial phases, interface mobility, and interface dimensionality, hydrodynamic interfaces can be classified as: (1) gas-liquid, liquid-liquid, liquid-solid interfaces; (2) fixed or moving interfaces; and (3) two-dimensional or three-dimensional interfaces. This simplified classification yields twelve types of hydrodynamic interfaces, which in turn can be subdivided into more subtypes depending on factors such as scale or prevailing dynamics.

To study HIs researchers typically start with conventional ideas such as interfacial instabilities and concepts of boundary and mixing layers. However, the application of the conventional approaches for interpreting unconventional interfaces may not be straightforward and often presents major challenges. A good example is the interface between a free surface flow and subsurface flow in porous media. Analysis of such a system can be done at a range of scales from sub-particle (or void) scale to the whole system scale. Transition from small to large scales requires homogenization procedures so both 'phases' can be treated at large (macro-) scales as a continuum. For example, such a problem has been encountered in studies of vegetated flows, where an inflection point in mean velocity profiles is often observed at the vegetation-flow interface (*i.e.*, at the vegetation canopy top). This feature (and associated Kelvin-Helmholtz instability) have led to the development of the mixing layer analogy, as an expansion of a conventional mixing layer concept for the case of two domains of different nature: porous canopy domain and free flow above the canopy (Finnigan, 2000; Nepf, 2012).

2.3 TURBULENCE MECHANICS: CONCEPTS AND DESCRIPTIVE FRAMEWORKS

Starting with the pioneering work of Osborne Reynolds in the 1890s, the prevailing approach in turbulence research has been statistical, reflecting seemingly fully random fluctuations of flow velocities in turbulent flows. By the end of the 1930s, an alternative structural approach has emerged highlighting some important regularities in turbulent flows, largely overlooked in the statistical approach (*e.g.*, Adrian, 2007; Adrian & Marusic, 2012). Since then, these two fundamental approaches, statistical and structural, have dominated theoretical and experimental studies. Among the most influential outcomes of the statistical approach is the energy cascade concept, while the coherent structures concept is the 'backbone' of the structural approach (*e.g.*, Davidson, 2004).

Initially, the statistical and structural approaches were developed independently and only recently researchers have begun exploring the connections between them, recognizing that they represent two equally important facets of turbulent flow dynamics. Both approaches are underpinned by fundamental conservation equations for momentum, energy, vorticity, and substances. Various forms of these equations represent descriptive frameworks that serve as a basis for experimental design, numerical modeling, data interpretation, and predictions. This section therefore starts with a brief outline of key descriptive frameworks (Sections 2.3.1 and 2.3.2) followed by a summary on the statistical theory of turbulence (Section 2.3.3), coherent structures (Section 2.3.4), and turbulent mixing (Section 2.3.5). It concludes with an outline of key turbulence hypotheses pertinent for hydraulic experimentation (Section 2.3.6).

2.3.1 Reynolds-averaging Navier-Stokes (RANS) framework

Most flows in hydraulic applications are governed by the Navier-Stokes (or momentum) equations:

$$\frac{\partial u_i}{\partial t} + u_j \frac{\partial u_i}{\partial x_j} = g_i - \frac{1}{\rho}\frac{\partial p}{\partial x_i} + \frac{\partial}{\partial x_j}\left(v\frac{\partial u_i}{\partial x_j}\right) \qquad (2.3.1)$$

$$\underset{\substack{\text{local} \\ \text{accelerations}}}{} \quad \underset{\substack{\text{convective} \\ }}{} \quad \underset{\substack{\text{gravity} \\ \text{force}}}{} \quad \underset{\substack{\text{pressure} \\ \text{force}}}{} \quad \underset{\substack{\text{viscous shear} \\ \text{force}}}{}$$

and the continuity equation:

$$\frac{\partial u_i}{\partial x_i} = 0 \qquad (2.3.2)$$

where ρ is fluid density, v is kinematic viscosity, g_i is the ith component of the gravity acceleration, p is pressure, and u_i is the instantaneous velocity vector. Equations (2.3.1) and (2.3.2) and the equations that follow are written using Cartesian indicial notation (*e.g.*, Davidson, 2004), where the subscripts i and j can be replaced with the indices 1, 2, or 3, representing the three principal orthogonal coordinate directions. In many cases, the indices 1, 2, and 3 define the streamwise, transverse, and vertical (or orthogonal to the bed) directions, respectively. In other words, x_1, x_2, and x_3 represent x, y, and z; and u_1, u_2, and u_3 represent u, v, and w, respectively. By convention, terms that contain a repeated index (known as a dummy index) are summed over all possible values of the index. For example, $\partial u_i/\partial x_i = \partial u_1/\partial x_1 + \partial u_2/\partial x_2 + \partial u_3/\partial x_3$ since i appears twice in the term $\partial u_i/\partial x_i$. In Equations (2.3.1) and (2.3.2) it is assumed that the fluid (*e.g.*, water) is incompressible which means that ρ is constant and thus $\partial\rho/\partial t = 0$.

Although many equations in hydraulic research (Figure 2.3.1) stem from Equation (2.3.1), it is rarely employed in its original form. Indeed, Equation (2.3.1) operates with instantaneous variables which in turbulent flows are highly fluctuating and thus Equation (2.3.1) is impracticable. This inconvenience of Equation (2.3.1) for engineering applications has been highlighted by Reynolds (1895) who introduced two important procedures: (1) decomposition of hydrodynamic fields into slow (or mean) and fast (or turbulent) components (known as the Reynolds decomposition); and (2) averaging (or filtering) of instantaneous variables and corresponding hydrodynamic equations.

The first procedure can be interpreted as a *scale decomposition* or *separation of scales*. The second procedure can be formulated in many different ways among which time, ensemble, and area/volume averaging are most common. This second procedure can be viewed as a *scaling-up procedure* that changes the scale of consideration from one level in the time-space-probability domain to another. In this respect, scale is an inherent feature of any hydrodynamic equation, a feature not always recognized in engineering applications (see also Sections 2.3.3 and 4.4.4 in Volume I).

As first proposed by Reynolds (1895), the averaging of the Navier-Stokes Equation (2.3.1) and continuity Equation (2.3.2) produces a new set of equations widely known as the Reynolds Averaged Navier-Stokes equations or RANS equations (Figure 2.3.1):

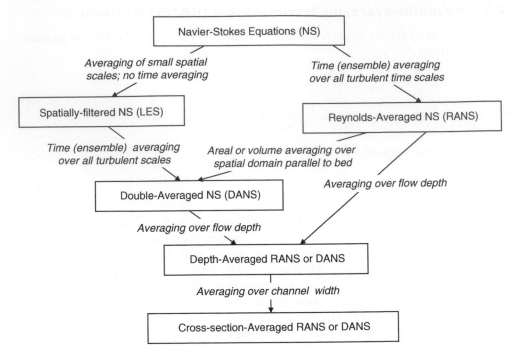

Figure 2.3.1 Interrelations between the key governing equations used in hydraulic applications.

$$\frac{\partial \overline{u}_i}{\partial t} + \overline{u}_j \frac{\partial \overline{u}_i}{\partial x_j} = g_i - \frac{1}{\rho}\frac{\partial \overline{p}}{\partial x_i} + \frac{1}{\rho}\frac{\partial}{\partial x_j}(-\rho\overline{u_i'u_j'}) + \frac{\partial}{\partial x_j}\left(v\frac{\partial \overline{u}_i}{\partial x_j}\right)$$ (2.3.3)

$$\quad\underset{\substack{local \\ accelerations}}{} \underset{convective}{} \underset{\substack{gravity \\ force}}{} \underset{\substack{pressure \\ force}}{} \underset{\substack{'turbulent' \\ force}}{} \underset{\substack{viscous\ shear \\ force}}{}$$

and

$$\frac{\partial \overline{u}_i}{\partial x_i} = 0$$ (2.3.4)

where overbars denote ensemble- (or time-) averaged values and primes denote deviations of the instantaneous value of each variable from its mean value. For example, the instantaneous pressure p can be decomposed into an average value $\overline{p}$ and a deviation p' such that $p = \overline{p} + p'$. Note that averaged variables $\overline{u}_i$ and $\overline{p}$ are often expressed in the literature (and in some chapters of this book) as U_i and P, respectively. Averaging of the non-linear term $u_j\partial u_i/\partial x_j$ in Equation (2.3.1) results in an additional term $(1/\rho)\partial(-\rho\overline{u_i'u_j'})/\partial x_j$ in the averaged Equation (2.3.3) where $-\rho\overline{u_i'u_j'}$ is known as the turbulent stress tensor or Reynolds stress tensor (this tensor is often represented as $-\overline{u_i'u_j'}$ if stress is normalized with fluid density). The tensor $-\rho\overline{u_i'u_j'}$ denotes the averaged transport of fluctuating momentum $\rho u_i'$ by the fluctuating velocity component u_j'. It is thus interpreted as an additional fluid stress due to turbulent momentum exchange. From Equation (2.3.3) it can be shown that the vertical distribution of the fluid stress for steady ($\partial \bar{u}_i / \partial t = 0$), uniform ($\partial \bar{u}_i / \partial x_1 = 0$), two-dimensional ($\partial \bar{u}_i / \partial x_2 = 0$) free-surface flow over a flat bed is linear; *i.e.*,

$$\frac{\tau(z)}{\rho} = -\overline{u'w'} + v\frac{d\bar{u}}{dz} = u_*^2 \left(1 - \frac{z}{H}\right) \tag{2.3.5}$$

where $u_* = (\tau_o / \rho)^{0.5}$ is shear (friction) velocity, H is the flow depth, $\tau_o = \rho g S H$ is wall (bed) shear stress, and S is bed slope. An equation similar to (2.3.5) can be written in cylindrical coordinates for a steady uniform pipe flow where the pipe radius should be used instead of H. For high-Reynolds number flows, the viscous stress in Equation (2.3.5) can be neglected as $-\overline{u'w'} \gg v\,d\bar{u}/dz$ and thus $-\overline{u'w'} = (\tau_o/\rho)(1 - z/H)$. Equation (2.3.5) is particularly important for hydraulic experimentation because it is often used as a diagnostic tool in tests of flow uniformity and two-dimensionality as well as for obtaining the wall shear stress from the turbulent stress measurements (see Section 6.5 in Volume II).

Other important RANS-type equations, also developed by Reynolds (1895), are the kinetic energy budgets for the mean flow (2.3.6, MKE) and turbulence (2.3.7, TKE):

$$\underbrace{\frac{\partial}{\partial t}\left(\frac{\bar{u}_i \bar{u}_i}{2}\right) + \bar{u}_j \frac{\partial}{\partial x_j}\left(\frac{\bar{u}_i \bar{u}_i}{2}\right)}_{\text{rate of change of MKE}} = \underbrace{g_i \bar{u}_i}_{\substack{\text{energy input} \\ \text{from gravity}}} - \frac{\partial}{\partial x_j}\left[\underbrace{\frac{1}{\rho}\overline{p}\bar{u}_j}_{\substack{\text{pressure} \\ \text{transport}}} + \underbrace{\bar{u}_i \overline{u'_i u'_j}}_{\substack{\text{turbulent} \\ \text{transport}}} - \underbrace{v\bar{u}\left(\frac{\partial \bar{u}_i}{\partial x_j} + \frac{\partial \bar{u}_j}{\partial x_i}\right)}_{\substack{\text{viscous} \\ \text{transport}}}\right] +$$

$$\underbrace{\overline{u'_i u'_j}\frac{\partial \bar{u}_i}{\partial x_j}}_{\substack{\text{energy transfer} \\ \text{between mean} \\ \text{flow and turbulence}}} - \underbrace{\frac{v}{2}\left(\frac{\partial \bar{u}_i}{\partial x_j} + \frac{\partial \bar{u}_j}{\partial x_i}\right)^2}_{\substack{\text{dissipation} \\ \text{of mean flow} \\ \text{energy}}} \tag{2.3.6}$$

and

$$\underbrace{\frac{\partial}{\partial t}\frac{\overline{u'_i u'_i}}{2} + \bar{u}_j \frac{\partial}{\partial x_j}\frac{\overline{u'_i u'_i}}{2}}_{\text{rate of change of TKE}} = \underbrace{-\overline{u'_i u'_j}\frac{\partial \bar{u}_i}{\partial x_j}}_{\substack{\text{energy transfer} \\ \text{between mean flow} \\ \text{and turbulence}}} - \frac{\partial}{\partial x_j}\left[\underbrace{\frac{1}{\rho}\overline{p'u'_j}}_{\substack{\text{pressure} \\ \text{transport}}} + \underbrace{\frac{\overline{u'_i u'_i u'_j}}{2}}_{\substack{\text{turbulent} \\ \text{transport}}} - \underbrace{vu'_i\left(\frac{\partial u'_i}{\partial x_j} + \frac{\partial u'_j}{\partial x_i}\right)}_{\substack{\text{viscous} \\ \text{transport}}}\right] -$$

$$\underbrace{\frac{v}{2}\overline{\left(\frac{\partial u'_i}{\partial x_j} + \frac{\partial u'_j}{\partial x_i}\right)^2}}_{\text{viscous dissipation of TKE}} \tag{2.3.7}$$

The MKE Equation (2.3.6) is obtained by multiplying Equation (2.3.3) by $\bar{u}_i$ and rearranging the result (*e.g.*, Monin & Yaglom, 1971). Equation (2.3.7) can be derived by first multiplying the Navier-Stokes Equation (2.3.1) with u_i, replacing u_i with $u_i =$

$\overline{u}_i + u'_i$ and p with $p = \overline{p} + p'$, and then averaging to obtain the full kinetic energy equation. Subtraction of Equation (2.3.6) for MKE from the full kinetic energy equation produces Equation (2.3.7) for TKE. Equations (2.3.6) and (2.3.7) show that in gravity-driven flows, gravity provides energy to the mean flow only. The mean flow feeds part of this energy to turbulence through the interaction between velocity shear and turbulent stresses, *i.e.*, via the production term $-\overline{u'_i u'_j} \partial \overline{u}_i / \partial x_j$, which is present with different signs in both Equations (2.3.6) and (2.3.7). For steady, uniform, two-dimensional flow regions where the energy production $P = -\overline{u'_i u'_j} \partial \overline{u}_i / \partial x_j$ is mainly balanced by the energy dissipation $\overline{\varepsilon}$ (the last term in 2.3.7), the TKE budget equation reduces to:

$$P = -\overline{u'w'} \frac{\partial \overline{u}}{\partial z} = \frac{v}{2} \overline{\left(\frac{\partial u'_i}{\partial x_j} + \frac{\partial u'_j}{\partial x_i} \right)^2} = \overline{\varepsilon} \tag{2.3.8}$$

Equation (2.3.8), together with a linear stress distribution, Equation (2.3.5), and the logarithmic relation for the velocity distribution (Table 2.2.1) leads to the simple relation for the vertical distribution of the energy dissipation:

$$\overline{\varepsilon} = -\overline{u'w'} \frac{\partial \overline{u}}{\partial z} = u_*^2 (1 - \frac{z}{H}) \frac{\partial \overline{u}}{\partial z} \approx u_*^2 (1 - \frac{z}{H}) \frac{u_*}{\kappa z} = \frac{u_*^3}{\kappa H} (\frac{H}{z} - 1) \tag{2.3.9}$$

where viscous stress in Equation (2.3.5) is neglected and $d\overline{u}/dz \approx u_*/(\kappa z)$ represents the logarithmic velocity distribution. Experimental studies show that Equation (2.3.9) may serve as a fairly good approximation even when the flow is not strictly two-dimensional and uniform (*e.g.*, Grinvald & Nikora, 1988). This equation has also been used to estimate values of shear velocity from turbulence measurements invoking wall similarity in open channel flows (Lopez & Garcia, 1999; Bigillon *et al.*, 2006).

Following an approach similar to that used to derive Equations (2.3.3), (2.3.6), and (2.3.7), equations for statistical moments $\overline{u'^m_i u'^n_j u'^l_k}$ of various orders of m, n, and l can also be derived, as first proposed by Keller and Fridman in the 1920s (*e.g.*, Monin & Yaglom, 1971). The equation for a two-point correlation moment is probably the most widely known from this family of equations as its simplified forms for homogeneous and isotropic turbulence produced a number of fundamental results in the statistical theory of turbulence (*e.g.*, Batchelor, 1970; Davidson, 2004; see Section 2.3.3 in Volume I). Another well-known example is the budget equation for turbulent stresses $-\overline{u'_i u'_j}$ that has been heavily used in algebraic turbulence models (*e.g.*, Bernard & Wallace, 2002; Pope, 2005).

In addition to the equations for velocity moments, the Navier-Stokes equations can be used to produce equations for other quantities important for flow characterization. Among them, the most widely used equation is the one for vorticity $\omega_k = (\partial u_i / \partial x_j - \partial u_j / \partial x_i)$:

$$\underbrace{\frac{\partial \omega_i}{\partial t} + u_j \frac{\partial \omega_i}{\partial x_j}}_{\substack{\text{vortex} \\ \text{convection}}} = \underbrace{\omega_j \frac{\partial u_i}{\partial x_j}}_{\substack{\text{vortex} \\ \text{stretching}}} + \underbrace{\frac{\partial}{\partial x_j} \left(v \frac{\partial \omega_i}{\partial x_j} \right)}_{\substack{\text{viscous} \\ \text{diffusion}}} \tag{2.3.10}$$

The vorticity Equation (2.3.10) is obtained by taking the curl of the momentum Equation (2.3.1) (or by its cross-differentiation). A counterpart of the continuity Equation (2.3.2) is:

$$\frac{\partial \omega_i}{\partial x_i} = 0 \qquad\qquad (2.3.11)$$

Equation (2.3.11) shows that vorticity has zero divergence, resulting from the combination of $\omega_k = (\partial u_i/\partial x_j - \partial u_j/\partial x_i)$ and continuity Equation (2.3.2). The mean vorticity equation can be derived from (2.3.10) by splitting u_i and ω_i into mean and fluctuating parts and then averaging (alternatively, it can be obtained by taking the curl of the mean momentum Equation (2.3.3)), i.e.:

$$\frac{\partial \overline{\omega}_i}{\partial t} + \overline{u}_j \frac{\partial \overline{\omega}_i}{\partial x_j} = \overline{\omega}_j \frac{\partial \overline{u}_i}{\partial x_j} \; - \; \frac{\partial}{\partial x_j}\left(\overline{\omega_i' u_j'} - \overline{\omega_j' u_i'}\right) \; + \; \frac{\partial}{\partial x_j}\left(\nu \frac{\partial \overline{\omega}_i}{\partial x_j}\right) \qquad (2.3.12)$$

$$\underset{\substack{\text{rate of change}\\\text{of vorticity}}}{} \qquad \underset{\substack{\text{mean vortex}\\\text{stretching}}}{} \underset{\substack{\text{turbulent}\\\text{transport}}}{} \underset{\substack{\text{turbulent}\\\text{stretching}}}{} \qquad \underset{\substack{\text{viscous}\\\text{diffusion}}}{}$$

where the Reynolds decomposition of velocity and vorticity is used, i.e., $u_i = \overline{u}_i + u_i'$ and $\omega_i = \overline{\omega}_i + \omega_i'$. The second term on the RHS of (2.3.12) represents effects of anisotropy and spatial heterogeneity of turbulent stresses expressed through correlations of fluctuating vorticity and velocity components. The derivative $\partial \overline{\omega_j' u_i'}/\partial x_j$ represents the gain (or loss) of mean vorticity due to stretching/tilting of the fluctuating vorticity by fluctuating strain rates, while the term $\partial \overline{\omega_i' u_j'}/\partial x_j$ represents vorticity transport in the x_j direction (e.g., Davidson, 2004). Various forms of Equation (2.3.12) have been used in hydraulic research, particularly in theoretical and experimental studies of secondary currents. For instance, Nezu (2005) used Equation (2.3.12) as a basis for the subdivision of secondary flows in open channels into Prandtl's first and second kinds (Prandtl, 1952). A review of applications of Equation (2.3.12) in rivers is given in Nikora and Roy (2012).

Similar to the mean energy and turbulent energy, the coupling between the mean and fluctuating vorticity can be expressed using equations for $\overline{\omega}_i^2/2$ and $\overline{\omega_i'\omega_i'}/2$, which represent two components of the mean product $\overline{\omega_i\omega_i}/2 = \overline{\omega}_i^2/2 + \overline{\omega_i'\omega_i'}/2$, where $\omega_i\omega_i/2$ is known as enstrophy. Although there are some analogies between the MKE and mean enstrophy $\overline{\omega}_i^2/2$ (ME), and between the TKE and the turbulent enstrophy $\overline{\omega_i'\omega_i'}/2$ (TE), their physical nature is different; i.e., enstrophy represents a measure of the density of the kinetic energy of helical motions rather than of all motions. As with Equation (2.3.6) for MKE, the mean enstrophy balance can be obtained by multiplying Equation (2.3.12) by $\overline{\omega}_i$; i.e.,

$$\frac{\partial}{\partial t}\left(\frac{\overline{\omega}_i\overline{\omega}_i}{2}\right) + \overline{u}_j \frac{\partial}{\partial x_j}\left(\frac{\overline{\omega}_i\overline{\omega}_i}{2}\right) = -\frac{\partial}{\partial x_j}\overline{\omega}_i\overline{u_j'\omega_i'} \; + \; \overline{\omega_i'u_j'}\frac{\partial \overline{\omega}_i}{\partial x_j} \; + \; \overline{\omega}_i\overline{\omega}_j S_{ij} \; + \; \overline{\omega}_i\overline{\omega_j'\frac{\partial u_i'}{\partial x_j}} \; +$$

$$\underset{\text{rate of change of ME}}{} \qquad\qquad \underset{\substack{\text{transport of ME}\\\text{by velocity-vorticity}\\\text{interactions}}}{} \quad \underset{\substack{\text{gradient}\\\text{production}\\\text{of ME}}}{} \quad \underset{\substack{\text{ME}\\\text{stretching}\\\text{by mean strain}}}{} \quad \underset{\substack{\text{ME stretching}\\\text{by turbulent}\\\text{strain rate}}}{}$$

$$\nu \frac{\partial^2}{\partial x_j \partial x_j}\left(\frac{\overline{\omega}_i\overline{\omega}_i}{2}\right) - \nu \frac{\partial \overline{\omega}_i}{\partial x_j}\frac{\partial \overline{\omega}_i}{\partial x_j}$$

$$\underset{\substack{\text{viscous}\\\text{transport}}}{} \qquad \underset{\substack{\text{dissipation of}\\\text{mean enstrophy}}}{}$$

$$(2.3.13)$$

where $S_{ij} = 0.5(\partial \bar{u}_i/\partial x_j + \partial \bar{u}_j/\partial x_i)$. The procedure for deriving the turbulent enstrophy balance is identical to that for the TKE balance (2.3.7). It involves multiplication of the equation for ω_i' by ω_i', and then subsequent time (ensemble) averaging (or, alternatively, subtraction of the mean enstrophy balance from the total enstrophy balance), *i.e.*:

$$\frac{\partial}{\partial t}\left(\frac{\overline{\omega_i'\omega_i'}}{2}\right) + \bar{u}_j \frac{\partial}{\partial x_j}\left(\frac{\overline{\omega_i'\omega_i'}}{2}\right) = -\overline{\omega_i'u_j'}\frac{\partial \bar{\omega}_i}{\partial x_j} - \frac{\partial}{\partial x_j}\frac{\overline{u_j'\omega_i'\omega_i'}}{2} + \overline{\omega_i'\omega_j'\frac{\partial u_i'}{\partial x_j}} + \overline{\omega_i'\omega_j'\frac{\partial \bar{u}_i}{\partial x_j}} + \bar{\omega}_j\overline{\omega_i'\frac{\partial u_i'}{\partial x_j}} +$$

| rate of change of TE | gradient production of TE | turbulent transport | TE production by turbulent stretching | change of TE by mean strain | 'mixed' production |

$$\nu \frac{\partial^2}{\partial x_j \partial x_j}\left(\frac{\overline{\omega_i'\omega_i'}}{2}\right) - \nu \overline{\frac{\partial \omega_i'}{\partial x_j}\frac{\partial \omega_i'}{\partial x_j}}$$

| viscous transport | viscous dissipation |

$$(2.3.14)$$

Equations (2.3.13) and (2.3.14) have been extensively studied in turbulence research with particular focus on their simplified versions for high-Reynolds number flows with homogeneous turbulence. There have been few, if any, attempts, to use these equations in hydraulic research (*e.g.*, for studies of secondary flows). The main reason for this is probably the absence of experimental assessments of the terms in Equations (2.3.13) and (2.3.14). However, with recent advances in laboratory and field instrumentation it is quite likely that such experimental data will soon appear. In addition, recent progress in numerical simulation techniques and computing capabilities increases the potential of Equations (2.3.13) and (2.3.14) for studies of hydraulic flows. The enstrophy balances presented above highlight a potentially fruitful theoretical framework for coupling mean and fluctuating vorticity fields, with the latter formed, most likely, by helical coherent structures (see also Section 6.12 in Volume I). There may be several coupling mechanisms between these fields, with the gradient production term $-\overline{\omega_i'u_j'}\partial \bar{\omega}_i/\partial x_j$ being the most plausible candidate as it is included in both (2.3.13) and (2.3.14), similar to the TKE production term $\overline{u_i'u_j'}\partial \bar{u}_i/\partial x_j$ in Equations (2.3.6) and (2.3.7).

2.3.2 Double-averaging Navier-Stokes (DANS) framework

Many wall-bounded hydraulic flows (both natural and artificial) can often be classified as low-submergence rough-bed flows with high levels of heterogeneity in time-averaged hydrodynamic fields due to the effects of roughness elements, especially in the near-bed region. These flows have been typically described using the Reynolds averaged Navier-Stokes equations (Figure 2.3.1), which deal with time- (or ensemble-) averaged variables and involve no spatial averaging (see Section 5.2.4 in Volume I). Such an approach, however, is usually inconvenient and impracticable due to the complex and often mobile boundary conditions that lead to the high flow heterogeneity and intermittency. As a result, 2-D approximations based on the RANS equations, as well as similarity considerations for time-averaged variables, are not possible for the near-bed region in rough-bed flows. Also, most applications deal with spatially-averaged roughness parameters that cannot be linked explicitly with local (point) flow properties provided by the RANS equations. Therefore, it is useful to spatially average the time-averaged hydrodynamic variables and equations within a thin slab defined by the longitudinal, transverse, and vertical length-scales of roughness elements. In other words, time (or ensemble) averaging should be combined with volume or area averaging.

The double-averaged (in time and space) equations relate to the time-(ensemble)-averaged equations as the time-averaged RANS equations relate to the Navier-Stokes equations for instantaneous hydrodynamic variables (Figure 2.3.1). The development of this methodology for rough-boundary turbulent flows was initiated by atmospheric scientists for describing air flows within and above terrestrial canopies such as forests or bushes (Wilson & Shaw, 1977; Raupach & Shaw, 1982; Finnigan, 2000), and later it was adopted in studies of water flows (e.g., Gimenez-Curto & Corniero Lera, 1996; Lopez & Garcia, 1998, 2001; Nikora et al., 2001, 2007a, 2007b, 2013; Nepf, 2012).

Applications of the double-averaging approach to a wide range of flows, from porous media flows to rough-bed open-channel flows to atmospheric boundary layers, are discussed in a special issue of *Acta Geophysica* (Nikora & Rowinski, 2008), which highlights the advantages of this methodology: (1) rigor and self-consistency; (2) refined definitions for rough-bed flows such as flow uniformity, two-dimensionality, and the bed shear stress; (3) a consistent link between spatially-averaged roughness parameters, bed shear stress, and double-averaged flow variables; (4) explicit accounting for the viscous drag, form drag and form-induced stresses and substance fluxes as a result of rigorous derivation rather than intuitive reasoning; (5) a framework for scaling considerations and parameterizations based on double-averaged variables; and (6) the possibility for the rigorous scale partitioning of the roughness parameters and flow properties. These advantages underpin the use of the double-averaged hydrodynamic equations in developing numerical models and associated closures for environmental rough-bed flows; designing laboratory, field, and numerical experiments; data analysis and interpretation; and guiding conceptual developments and parameterizations. The derivation and discussion of the general double-averaged equations for hydraulic flows over fixed and mobile rough beds are provided in Nikora et al. (2007a, 2007b, 2013). In principle, the appearance of the double-averaged hydrodynamic equations, as well as the physical meaning of the double-averaged quantities, may depend on the averaging approach, i.e., *simultaneous space-time averaging; consecutive time-space averaging;* or *consecutive space-time averaging* (Nikora et al., 2013). For consistency with RANS and to take advantage of existing data (which in most cases are collected within a RANS framework), the consecutive time-space averaging is adopted in the presentation below.

Derivation of the double-averaged hydrodynamic equations requires three key elements: (1) averaging operators; (2) equations that link double-averaged derivatives to derivatives of the double-averaged variables, known as the averaging theorems; and (3) modified Reynolds decomposition of instantaneous variables. A double-averaged quantity $\langle \overline{\theta} \rangle$ is obtained with a *consecutive time-space* averaging operator as:

$$\langle \overline{\theta} \rangle = \frac{1}{V_m} \int\limits_{V_o} \frac{1}{T_f} \int\limits_{T_o} \theta \gamma(x_i, t) dt dV = \frac{1}{V_m} \int\limits_{V_o} \overline{\theta} dV, \qquad (2.3.15)$$

where an overbar denotes time averaging and angular brackets denote spatial averaging; $\gamma(x_i, t)$ is a 'clipping' or 'distribution' function equal to 1 in the fluid and 0 otherwise; V_m is the part of the total averaging volume V_o that has been 'visited' by fluid, even briefly, within the averaging period T_o; and T_f is the total period of time (not necessarily continuous) when fluid passes through a fixed point x_i ($T_f = \int_{T_o} \gamma(x_i, t) dt$). Equation (2.3.15) represents so-called intrinsic averaging, over volume V_m and over time duration T_f (*i.e.*, when the space is occupied by fluid only). Another form of averaging used in derivations and presentation of hydrodynamic variables is superficial averaging, over the whole averaging domain of volume V_o and time interval T_o. The superficial averaging in the equations below is indicated by the index s.

Selection of the shape and dimensions of the spatial averaging domain and the averaging time depends on the roughness geometry and the turbulence structure (particularly on their characteristic scales, Section 2.3.3 in Volume I; Nikora *et al.*, 2007a, 2013). It is noteworthy that most hydraulic flows exhibit strong gradients in flow properties in the vertical direction, especially near the bed, and therefore the volume-averaging domain should be designed as a thin slab parallel to the average bed surface, being much thinner than the roughness elements. The dimensions of the averaging domain in the plane parallel to the average bed surface should be much larger than the dominant roughness scales, but much smaller than the large-scale features of the bed topography. For example, for gravel-bed rivers it should be much larger than gravel particles, but much smaller than dimensions of riffles or pools. The averaging time should well exceed the integral time scales of turbulence, still being much shorter than the duration of hydrological events such as floods (see also Section 5.2.4.2 in Volume I). The exact shape and dimensions of the averaging domain and averaging time vary with the flow problem considered.

The averaging theorems in the consecutive time-space form can be presented as:

$$\left\langle \phi_T \frac{\overline{\partial \theta}}{\partial t} \right\rangle = \frac{1}{\phi_{Vm}} \frac{\partial \phi_{Vm} \langle \phi_T \overline{\theta} \rangle}{\partial t} + \frac{1}{\phi_{Vm} V_o} \overline{\iint\limits_{S_{int}} \theta v_i n_i dS}^s$$

and

$$\left\langle \phi_T \frac{\overline{\partial \theta}}{\partial x_i} \right\rangle = \frac{1}{\phi_{Vm}} \frac{\partial \phi_{Vm} \langle \phi_T \overline{\theta} \rangle}{\partial x_i} - \frac{1}{\phi_{Vm} V_o} \overline{\iint\limits_{S_{int}} \theta n_i dS}^s \qquad (2.3.16)$$

where v_i is the velocity vector of the fluid at the fluid-boundary interface; n_i is the unit vector normal to the interface surface and directed into the fluid; S_{int} is the total area of the fluid-boundary interface within the averaging domain; index s defines superficial averaging (over the whole averaging domain) while averaged quantities shown without indices are intrinsic (*i.e.*, averaged over the sub-domain occupied by fluid only); $\phi_T(x_i, t)$ is the local time porosity defined as the ratio T_f/T_o (note that the quantity $\phi_T(x_i, t) = \overline{\gamma}^s = T_f/T_o$ involves no spatial averaging); and $\phi_{Vm} = V_m/V_o$ is the space porosity based on non-zero $\overline{\gamma}^s$. The combined space-time porosity is defined as

$\phi_{VT} = \phi_{Vm}\langle\phi_T\rangle$. The modified Reynolds decomposition for consecutive time-space averaging takes the form $\theta = \bar{\theta} + \theta' = \langle\bar{\theta}\rangle + \tilde{\bar{\theta}} + \theta'$, where a prime defines turbulent fluctuation (i.e., $\theta' = \theta - \bar{\theta}$) and an overbar with a tilde above it denotes the deviation of a time-averaged quantity from its double-averaged value (i.e., $\tilde{\bar{\theta}} = \bar{\theta} - \langle\bar{\theta}\rangle$).

Derivation of the double-averaged equations starts with applying an operation of superficial double-averaging to each individual term of the initial equations, and then transforming these terms using the double-averaging theorems (2.3.16) supplemented with the decomposition of the instantaneous variables as $\theta = \langle\bar{\theta}\rangle + \tilde{\bar{\theta}} + \theta'$, similar to how the Reynolds-averaged equations are obtained. The double-averaged counterpart of the continuity Equation (2.3.2) for mobile boundaries is obtained as:

$$\frac{\partial\phi_{Vm}\langle\phi_T\rangle}{\partial t} + \frac{\partial\phi_{Vm}\langle\phi_T\bar{u}_i\rangle}{\partial x_i} = 0 \tag{2.3.17}$$

For the fluid phase we assume incompressibility, which means $\rho = \bar{\rho} = \langle\bar{\rho}\rangle = const$ and thus $\partial\rho/\partial t = 0$. However, deriving Equation (2.3.17) requires considering the full continuity equation to account for an effect of changing ϕ_{Vm} and ϕ_T in the double-averaged continuity equation. In other words, variation of $\phi_{Vm}\langle\phi_T\rangle$ may be interpreted as 'compressibility' of the flow-mobile-rough-bed medium even though both components of this medium, fluid and roughness elements, are incompressible. For spatially uncorrelated local time porosity and flow velocities (i.e., $\langle\phi_T\bar{u}_i\rangle = \langle\phi_T\rangle\langle\bar{u}_i\rangle$), Equation (2.3.17) can be simplified, using a combined space-time porosity $\phi_{VT} = \phi_{Vm}\langle\phi_T\rangle$, as:

$$\frac{\partial\phi}{\partial t} + \frac{\partial\phi\langle\bar{u}_i\rangle}{\partial x_i} = 0 \tag{2.3.18}$$

where we use $\phi = \phi_{VT}$ for brevity. For fixed flow boundaries, Equation (2.3.18) remains the same, with ϕ serving as the conventional bed porosity, i.e., $\phi = V_f/V_o$ where V_f is the volume of fluid within the total averaging domain V_o. The double-averaged counterpart of the momentum Equation (2.3.1) for mobile-rough-bed conditions is:

$$\underbrace{\frac{\partial\phi_{Vm}\langle\phi_T\bar{u}_i\rangle}{\partial t}}_{1} + \underbrace{\frac{\partial\phi_{Vm}\langle\phi_T\rangle\langle\bar{u}_i\rangle\langle\bar{u}_j\rangle}{\partial x_j}}_{2} = \underbrace{\phi_{Vm}\langle\phi_T\rangle g_i}_{3} - \underbrace{\frac{1}{\rho}\frac{\partial\phi_{Vm}\langle\phi_T\bar{p}\rangle}{\partial x_i}}_{4} - \underbrace{\frac{\partial\phi_{Vm}\langle\phi_T\overline{u_i'u_j'}\rangle}{\partial x_j}}_{5}$$

$$- \underbrace{\frac{\partial\phi_{Vm}\langle\phi_T\tilde{\bar{u}}_i\tilde{\bar{u}}_j\rangle}{\partial x_j}}_{6} + \underbrace{\frac{\partial}{\partial x_j}\left(\phi_{Vm}\left\langle\phi_T\nu\frac{\overline{\partial u_i}}{\partial x_j}\right\rangle\right)}_{7} - \underbrace{\frac{\partial\phi_{Vm}\langle\phi_T\tilde{\bar{u}}_i\rangle\langle\bar{u}_j\rangle}{\partial x_j}}_{8} - \underbrace{\frac{\partial\phi_{Vm}\langle\phi_T\tilde{\bar{u}}_j\rangle\langle\bar{u}_i\rangle}{\partial x_j}}_{9}$$

$$+ \underbrace{\frac{1}{\rho}\frac{1}{V_o}\iint_{S_{int}}\overline{pn_i\,dS}^s}_{10} - \underbrace{\frac{1}{V_o}\iint_{S_{int}}\overline{\left(\nu\frac{\partial u_i}{\partial x_j}\right)n_j\,dS}^s}_{11}$$

$$\tag{2.3.19}$$

Terms 1 and 2 in Equation (2.3.19) represent local and convective accelerations, respectively. The third term is the gravity term; the fourth term is the pressure gradient; the fifth, sixth and seventh terms are contributions from turbulent $(-\langle \phi_T \overline{u'_i u'_j} \rangle)$, form-induced $(-\langle \phi_T \widetilde{\overline{u}}_i \widetilde{\overline{u}}_j \rangle)$, and viscous fluid stresses, respectively; the eighth and ninth terms represent momentum fluxes (stresses) due to potential spatial correlations between the local time porosity and time-averaged velocities; and the final two terms, *i.e.*, tenth and eleventh, are pressure and viscous drag terms. For the case when spatial correlations between the local time porosity and time-averaged flow parameters can be neglected, Equation (2.3.19) can be simplified, using the continuity Equation (2.3.18) and the space-time porosity $\phi = \phi_{VT} = \phi_{Vm} \langle \phi_T \rangle$, as:

$$\frac{\partial \langle \overline{u}_i \rangle}{\partial t} + \langle \overline{u}_j \rangle \frac{\partial \langle \overline{u}_i \rangle}{\partial x_j} = g_i - \frac{1}{\rho} \frac{1}{\phi} \frac{\partial \phi \langle \overline{p} \rangle}{\partial x_i} - \frac{1}{\phi} \frac{\partial \phi \langle \overline{u'_i u'_j} \rangle}{\partial x_j} - \frac{1}{\phi} \frac{\partial \phi \langle \widetilde{\overline{u}}_i \widetilde{\overline{u}}_j \rangle}{\partial x_j} +$$

$$\frac{1}{\phi} \frac{\partial}{\partial x_j} \left(\phi \left\langle \overline{\nu \frac{\partial \overline{u}_i}{\partial x_j}} \right\rangle \right) + \frac{1}{\rho} \frac{1}{\phi} \frac{1}{V_o} \overline{\iint\limits_{S_{int}} p n_i dS}^{\,s} - \frac{1}{\phi} \frac{1}{V_o} \overline{\iint\limits_{S_{int}} \left(\nu \frac{\partial u_i}{\partial x_j} \right) n_j dS}^{\,s} \qquad (2.3.20)$$

When required, the double-averaged hydrodynamic equations can also be formulated within *simultaneous space-time* and *consecutive space-time* averaging frameworks, which are analytically linked to Equations (2.3.17) to (2.3.20) obtained based on *consecutive time-space* averaging. The double-averaged equations for the fixed-bed conditions are equivalent to Equations (2.3.18) and (2.3.20), where $\phi = \phi_{VT} = \phi_{Vm} \langle \phi_T \rangle = V_f / V_o$, as $\langle \phi_T \rangle \equiv 1$ and thus $\phi_{Vm} = \phi$.

Double-averaged conservation equations provide a mathematical framework for studying turbulent mobile-rough-bed flows such as gravel-bed rivers during flood events or flows over vegetated beds. The data on such flows, especially within moving roughness elements, are currently very limited due to both measurement difficulties and the remaining uncertainty of what exactly to measure, interpret, and model. The measurement techniques (*e.g.*, refractive index matching Particle Image Velocimetry or those based on Magnetic Resonance Imaging) and modeling capabilities (*e.g.*, Large Eddy Simulation method) have been improved in recent years and it is likely that extensive data on hydrodynamic variables within mobile roughness elements will appear soon. Equations (2.3.17) to (2.3.20), and similar equations for higher-order moments, help in designing experiments to obtain, interpret and parameterize such data.

2.3.3 Statistical characterization of turbulence

This section summarizes the key definitions and basic concepts of the statistical approach, and then outlines the benchmark statistical turbulence theories. The presentation is at the introductory level. For a comprehensive treatment see specialized turbulence texts (*e.g.*, Batchelor, 1970; Monin & Yaglom, 1971, 1975; Davidson, 2004).

2.3.3.1 Definitions and basic concepts

Within the statistical approach, vector (*e.g.*, velocity) and scalar (*e.g.*, pressure) fields in turbulent flows are considered as random fields that can be characterized using a set of statistical measures. The most important among them are probability distributions and their moments, correlation functions, structure functions, and velocity spectra. Although these measures can be defined for any turbulent flow, they are mostly used in studies of homogeneous and isotropic turbulence.

Turbulence is defined as *(globally) homogeneous* if the statistical properties of hydrodynamic variables do not depend on the locations of measurement points within the flow. In other words, moving the coordinate system without changing its orientation in the field of homogeneous turbulence does not change the statistical characteristics. Homogeneous turbulence is *isotropic* if its statistical parameters are independent of the rotations and reflections of the coordinate system. The notions of the global homogeneity and isotropy typically apply to the 'primary' hydrodynamic variables such as velocities and pressure. In contrast, the notions of *local homogeneity* and *local isotropy*, although similar to their global counterparts, typically apply to the relative values of hydrodynamic variables; *i.e.*, their differences between the points spatially and temporally separated within the flow. In this sense, local homogeneity and isotropy are more general as they may occur even in globally inhomogeneous and anisotropic flows. The notions of global and local homogeneity may apply for a particular direction in the flow, *i.e.*, not necessarily along all coordinates (*e.g.*, a canonical pipe flow is homogeneous along the flow and inhomogeneous across the flow). If used in relation to the time series (*e.g.*, time records of the hydrodynamic variables at particular points), the notion of homogeneity is known as *stationarity* (see also Sections 6.2.1 and 6.3 in Volume I).

To define the turbulence statistics some sort of averaging is required. There are at least five forms of averaging used in theoretical analyses and in treatment of experimental data: (1) time averaging over a specified time period (typically much longer than a lifetime of largest eddies but much shorter than a time scale of external forcing); (2) spatial averaging over an extended domain with dimensions well exceeding the sizes of the largest eddies (this type of averaging was employed in the original derivations of Reynolds (1895), who used spherical averaging domain); (3) time-space averaging as specified by Equation (2.3.15); (4) spatial (temporal) filtering which is a general form of spatial (temporal) averaging as it employs a weighting function (or kernel of the filter) that typically changes within the averaging domain, in contrast to standard spatial and time averaging where the kernel is always equal to 1 throughout the domain; and (5) ensemble (or probabilistic) averaging over a large number of realizations, which is the most general averaging approach. At this point it is worth mentioning an important class of homogeneous and stationary random functions and fields known as ergodic. The property of *ergodicity* means that statistics obtained through ensemble averaging are equal to the statistics obtained from a single realization. Assuming this property, the ensemble averages may well be approximated with time and/or spatial averages. Ergodicity is a key assumption in most experimental studies where obtaining an ensemble of realizations is impossible. The generic averaging operation is defined in Section 2.3.3 in Volume I with an overbar.

2.3.3.2 Statistical moments

A theoretically complete description of a random function or field is given by the n-dimensional probability density function (PDF) when $n \to \infty$. In relation to turbulence analysis this n-dimensional measure is impracticable and researchers typically resort to much simpler quantities such as low-dimensional PDFs and their moments. The one-dimensional PDF can be defined as the probability that a velocity (or pressure) value will occur in the range $u_i, u_i + \Delta u_i$ (or $p, p + \Delta p$). The initial moments, defined for u_i, and the central moments, defined for $u'_i = u_i - \bar{u}_i$, of a single-point PDF $P(u_i)$ are:

$$\overline{u_i^k} = \int_{-\infty}^{\infty} u_i^k P(u_i)du_i \approx \frac{1}{N}\sum_{n=1}^{n=N} u_{in}^k \tag{2.3.21}$$

and

$$\overline{u_i'^k} = \int_{-\infty}^{\infty} u_i'^k P(u_i)du_i \approx \frac{1}{N}\sum_{n=1}^{n=N} u_{in}'^k \tag{2.3.22}$$

where N is the number of samples which should be sufficient to achieve appropriate statistical convergence (see also Section 6.2 in Volume I). Among the initial moments of Equation (2.3.21), the most widely used quantity is first-order moment ($k = 1$) which is the mean velocity $\bar{u}_i$. Among the central moments of Equation (2.3.22), the most relevant parameters are moments of the second- ($k = 2$), third- ($k = 3$) and fourth- ($k = 4$) orders. The second-order moment gives velocity variance and thus TKE, i.e.,:

$$\overline{u_i'^2} = \int_{-\infty}^{\infty} u_i'^2 P(u_i)du_i \approx \frac{1}{N}\sum_{n=1}^{n=N} u_{in}'^2, \quad TKE = k = 0.5(\overline{u_1'^2} + \overline{u_2'^2} + \overline{u_3'^2}) \tag{2.3.23}$$

The square root of variance is velocity standard deviation $\sigma_i = \left(\overline{u_i'^2}\right)^{1/2}$ known in turbulence research as the turbulence intensity. The third-order moment provides a normalized measure of the PDF asymmetry known as skewness Sk:

$$Sk = \overline{u_i'^3}/\overline{u_i'^2}^{3/2}, \quad \overline{u_i'^3} = \int_{-\infty}^{\infty} u_i'^3 P(u_i)du_i \approx \frac{1}{N}\sum_{n=1}^{n=N} u_{in}'^3 \tag{2.3.24}$$

The skewness is positive if large (but rare) positive fluctuations are present in the record. Similarly, it is negative when the negative fluctuations are strongest. For symmetrical PDFs, such as the Gaussian distribution, the skewness is zero. The fourth-order moment characterizes the PDF 'peakedness' and is typically expressed in normalized form as flatness F:

$$F = \overline{u_i'^4}/\overline{u_i'^2}^2, \quad \overline{u_i'^4} = \int_{-\infty}^{\infty} u_i'^4 P(u_i)du_i \approx \frac{1}{N}\sum_{n=1}^{n=N} u_{in}'^4, \tag{2.3.25}$$

For the Gaussian distribution, the flatness F equals 3 and therefore this measure is often presented as kurtosis $Ku = F - 3$, to keep similarity with skewness (which is zero for the Gaussian distribution). The flatness (or kurtosis) serves in turbulence research for assessments of intermittency. Indeed, it may be presented as a measure of variation of variance, i.e.,:

$$F = \overline{u_i'^4}/\overline{u_i'^2}^2 = \frac{\overline{u_i'^2 u_i'^2}}{\overline{u_i'^2}\ \overline{u_i'^2}} = \frac{\overline{u_i'^2}\ \overline{u_i'^2}}{\overline{u_i'^2}\ \overline{u_i'^2}} + \frac{\overline{(u_i'^2)'(u_i'^2)'}}{\overline{u_i'^2}\ \overline{u_i'^2}} = 1 + \frac{\overline{(u_i'^2)'^2}}{\overline{u_i'^2}^2} = 1 + \left(\frac{\overline{(u_i'^2)'^2}^{1/2}}{\overline{u_i'^2}}\right)^2 \tag{2.3.26}$$

where $(u_i'^2)' = u_i'^2 - \overline{u_i'^2}$ is the deviation of the squared fluctuating velocity from its mean value (which is variance). Equation (2.3.26) can be re-arranged as $\overline{(u_i'^2)'^2}^{1/2} / \overline{u_i'^2} = (F-1)^{1/2}$ to provide an explicit expression for the coefficient of variation of the squared fluctuating velocity.

Considering two or more velocity components at a point, the joint PDFs can also be defined with associated joint velocity moments. The most influential mixed second-order moment is the Reynolds stress tensor $\overline{u_i' u_j'}$:

$$\overline{u_i' u_j'} = \int_{-\infty}^{\infty} \int_{-\infty}^{\infty} u_i' u_j' P(u_i, u_j) du_i du_j \approx \frac{1}{N} \sum_{n=1}^{n=N} u_{in}' u_{jn}' \qquad (2.3.27)$$

When normalized on velocity standard deviations σ_i and σ_j it is known as the correlation coefficient $R_{ij} = \overline{u_i' u_j'} / (\sigma_i \sigma_j)$. Mixed third-order moments $\overline{u_i' u_j' u_l'}$ can be defined similarly to Equation (2.3.27). These triple correlations arise in the second-order RANS equations for Reynolds stresses and the TKE balance and are responsible for turbulent fluxes of TKE within the flow.

2.3.3.3 Correlation functions

Equation (2.3.27) is specified for single-point joint PDFs. When considering velocities in two different points within the flow and at different times, it can be extended for the moments of two-point joint PDFs, *i.e.*,

$$B_{ij}(\mathbf{x}, \mathbf{x}', t, t') = \overline{u_i'(\mathbf{x}, t) u_j'(\mathbf{x}', t')} = \int_{-\infty}^{\infty} \int_{-\infty}^{\infty} u_i'(\mathbf{x}, t) u_j'(\mathbf{x}', t') P(u_i, u_j) du_i du_j$$

$$(2.3.28)$$

which are known as *correlation functions* $B_{ij}(\mathbf{x}, \mathbf{x}', t, t')$ as they depend not only on the point coordinates (as in Equation (2.3.27) but also on the distance between the points in spatial and temporal domains; $\mathbf{x}, \mathbf{x}'$ are coordinate vectors of two points, and t, t' are different time moments. For velocity fields which are homogeneous in time and space, the correlation function, Equation (2.3.28), depends only on the relative positions $(\mathbf{r} = \mathbf{x} - \mathbf{x}')$ and $(\tau = t - t')$ of the points but not on the coordinates themselves.

Although fully dimensional correlation functions (2.3.28) are frequently used in theoretical considerations, in experimental studies they are employed in reduced forms depending on the data availability (typically, only one- or two- dimensional sections of $B_{ij}(\mathbf{x}, \mathbf{x}', t, t')$ can be measured). For example, for stationary but inhomogeneous flow (*e.g.*, conduit flow) it is convenient to use $B_{ij}(\mathbf{x} = \mathbf{x}_0, \mathbf{x}', \tau = 0)$, which provides correlations between a hydrodynamic parameter at a fixed point $\mathbf{x} = \mathbf{x}_0$ (*e.g.*, located in the near-bed region) with another parameter (or with itself) at surrounding points $\mathbf{x}'$. This correlation function is frequently used if multi-point data are available; *e.g.*, from sensor arrays or Particle Image Velocimetry (PIV). Examples of useful reduced forms of Equation (2.3.28) for stationary and homogeneous (at least in one direction) flows include the time autocorrelation function $B_{ii}(\mathbf{r} = 0, \tau)$ (*i.e.*, correlation with itself in the time domain), spatial auto-correlation function $B_{ii}(\mathbf{r}, \tau = 0)$, single-point cross-correlation function $B_{ij}(\mathbf{r} = 0, \tau)$, spatial cross-correlation function $B_{ij}(\mathbf{r}, \tau = 0)$, and streamwise space-time cross-correlation function

$B_{ij}(r_1, r_2 = 0, r_3 = 0, \tau)$. These functions help identify typical flow patterns, turbulence scales, and eddy convection velocities as discussed in Section 4.4.3, Volume I. Note that the single-point second-order moment (2.3.23) is a value of $B_{ij}(\mathbf{x}, \mathbf{x}', t, t')$ at $\mathbf{x} = \mathbf{x}', t = t'$ when $i = j$. Similarly, the second-order joint moment (Equation (2.3.27)) is a value of $B_{ij}(\mathbf{x}, \mathbf{x}', t, t')$ at $\mathbf{x} = \mathbf{x}', t = t'$ when $i \neq j$. Thus, the Reynolds stress tensor is a subset of the correlation function (2.3.28).

The mixed third-order three-point correlation functions $B_{ijl}(\mathbf{x}, \mathbf{x}', \mathbf{x}'', t) = \overline{u'_i(\mathbf{x})u'_j(\mathbf{x}')u'_l(\mathbf{x}'')}$ can be defined similarly as in Equation (2.3.28). At $\mathbf{x}' = \mathbf{x}''$ the correlation function $B_{ijl}(\mathbf{x}, \mathbf{x}', \mathbf{x}'', t)$ is interpreted as two-point third-order correlation function. Such triple correlation functions are fundamental in turbulence analyses as they provide information about turbulent energy exchanges between eddies of different scales (*e.g.*, Davidson, 2004). For isotropic turbulence, the correlation functions $B_{ij}(\mathbf{x}, \mathbf{x}', t, t') = \overline{u'_i(\mathbf{x}, t)u'_j(\mathbf{x}', t')}$ and $B_{ijl}(\mathbf{x}, \mathbf{x}', t, t') = \overline{u'_i(\mathbf{x})u'_j(\mathbf{x})u'_l(\mathbf{x}')}$ can be expressed through scalar functions such as longitudinal and transverse correlation functions.

2.3.3.4 Structure functions

Another important turbulence tool known as *structure functions* was introduced by Kolmogorov to study small-scale turbulence (Kolmogorov, 1941; Monin & Yaglom, 1975). As with the correlation functions, structure functions are also statistical moments, but now of velocity increments $\Delta u_i(\mathbf{r}, \tau) = u_i(\mathbf{x} + \mathbf{r}, t + \tau) - u_i(\mathbf{x}, t)$ rather than of the 'original' velocities $u_i(\mathbf{x}, t)$. The velocity increments can be defined for single points (*i.e.*, at different time moments) or for multiple points (so that velocity differences are taken between spatial points). For example, the two-point p-order structure function for velocity component u_i can be defined as:

$$D^p_{ii}(\mathbf{r}, \tau) = \overline{[u_i(\mathbf{x}, t) - u_i(\mathbf{x} + \mathbf{r}, t + \tau)]^p} \qquad (2.3.29)$$

where the meaning of the arguments is the same as in the definitions for the correlation functions in the preceding section. A wide variety of structure functions have been employed in studies of small-scale turbulence involving either scalar or vector fields or both (*e.g.*, Monin & Yaglom, 1975). The most frequently used structure functions in studies of locally-isotropic fields are the second-order longitudinal $D_{LL}(r, \tau = 0)$ and transverse $D_{NN}(r, \tau = 0)$ structure functions:

$$D^{p=2}_{LL}(r, \tau = 0) = D_{LL}(r, \tau = 0) = \overline{[u_L(\mathbf{x}) - u_L(\mathbf{x} + r)]^2} \qquad (2.3.30)$$

$$D^{p=2}_{NN}(r, \tau = 0) = D_{NN}(r, \tau = 0) = \overline{[u_N(\mathbf{x}) - u_N(\mathbf{x} + r)]^2} \qquad (2.3.31)$$

where $u_L(\mathbf{x})$ is the projection of the velocity vector on the direction of r while $u_N(\mathbf{x})$ is the projection of the velocity vector on any direction orthogonal to the r direction.

The structure function brings two key benefits for turbulence analyses: (1) it filters out the large-scale time and spatial heterogeneities due to the external flow geometry; and (2) it serves as a measure of the mean energy of an 'eddy' of size $\leq r$, as becomes clear when considering the second order-structure function $\Delta u_i^2(r)$. If the velocity field is not only locally isotropic but also globally isotropic then the second-order structure functions (2.3.30) and (2.3.31) are connected with corresponding correlation functions as:

$$D_{LL}(r) = 2[B(r=0) - B_{LL}(r)], \quad D_{NN}(r) = 2[B(r=0) - B_{NN}(r)] \tag{2.3.32}$$

where $B(r=0)$ is velocity variance (the same in all directions and for all three velocity components), $B_{LL}(r) = \overline{u_L(\mathbf{x})u_L(\mathbf{x}+r)}$ and $B_{NN}(r) = \overline{u_N(\mathbf{x})u_N(\mathbf{x}+r)}$ are two-point longitudinal and transverse correlation functions, respectively. The higher-order structure functions are also extensively used in theoretical considerations, particularly in studies of turbulence intermittency in high-Reynolds number turbulent flows.

2.3.3.5 Spectra of hydrodynamic variables

The next fundamental tool in the turbulence analyses is *velocity spectra* that characterize the distribution of the total variance across the scales and represent the Fourier transforms of the correlation functions and the second-order structure functions. For a one-dimensional scalar variable, such as pressure at a fixed point, the frequency spectrum $S_p(\omega)$ is related to the correlation function $B_p(\tau)$ (and vice versa) as:

$$S_p(\omega) = \frac{1}{2\pi}\int_{-\infty}^{\infty} B_p(\tau)e^{-i\omega\tau}d\tau, \quad B_p(\tau) = \int_{-\infty}^{\infty} S_p(\omega)e^{i\omega\tau}d\omega \tag{2.3.33}$$

where $i = \sqrt{-1}$, $\omega = 2\pi/T$ is the angular frequency, and T is the time period of a spectral component. In physical terms, the spectral density $S_p(\omega)$ characterizes the density of the 'local' variance $p'^2(\omega)$ of a scalar variable at a frequency ω, i.e., $S_p(\omega) \propto p'^2(\omega)/\Delta\omega \propto p'^2(\omega)/\omega$. Considering a spatial record (e.g., of a single velocity component u_i) the above relations will take the following forms:

$$S_{u_i}(k) = \frac{1}{2\pi}\int_{-\infty}^{\infty} B_{u_i}(r)e^{-ikr}dr \tag{2.3.34a}$$

and

$$B_{u_i}(r) = \int_{-\infty}^{\infty} S_{u_i}(k)e^{ikr}dk \tag{2.3.34b}$$

where $k = 2\pi/\lambda$ is the wavenumber, and λ is the wavelength (or a spatial scale). Similar to $S_p(\omega)$, the spectrum $S_{u_i}(k)$ is a measure of the density of the 'local' velocity variance $u_i'^2(k)$ at a wavenumber k, i.e., $S_{u_i}(k) \propto u_i'^2(k)/\Delta k \propto u_i'^2(k)/k$. In other words, $u_i'^2(k) \propto kS_{u_i}(k)$ can be interpreted as the energy of an 'eddy' of size λ. The relationship $u_i'^2(k) \propto kS_{u_i}(k)$ is a key reason for the frequent use of the so-called pre-multiplied spectra $kS_{u_i}(k)$ in turbulence research instead of the spectral densities $S_{u_i}(k)$. Consistent with the above consideration, Equation (2.3.34b) shows that the total energy of the ith velocity component is:

$$B_{u_i}(0) = \overline{u_i'^2} = \int_{-\infty}^{\infty} S_{u_i}(k)dk = \int_{-\infty}^{\infty} kS_{u_i}(k)d\ln k \tag{2.3.35}$$

The spectral densities, defined above, are two-sided, *i.e.*, they relate to the whole frequency (wavenumber) range from negative values to positive values. One-sided spectra, obtained by doubling two-sided spectra, are also in frequent use.

Moving to vector hydrodynamic quantities, we can now generalize velocity spectra (2.3.34a) as:

$$\Phi_{ij}(k_1, k_2, k_3) = \frac{1}{8\pi^3} \iiint_{\mathbf{r}} B_{ij}(r_1, r_2, r_3) e^{-i\mathbf{kr}} dr_1 dr_2 dr_3 \qquad (2.3.36)$$

where $B_{ij}(r_1, r_2, r_3) = \overline{u'_i(x_1, x_2, x_3) u'_j(x_1 + r_1, x_2 + r_2, x_3 + r_3)}$ is the velocity correlation function for a homogeneous velocity field at $\tau = 0$ that follows from (2.3.28). The quantity $\Phi_{ij}(k_1, k_2, k_3)$ is a tensor and thus it includes 9 components, each of which depends on three wavenumbers (k_1, k_2, k_3) that represent components of the wavenumber vector $\mathbf{k}$ for the longitudinal, transverse, and vertical directions, respectively. Projecting the three-dimensional spectra $\Phi_{ij}(k_1, k_2, k_3)$ on the (k_k, k_l) planes reduces their dimensionality from 3 to 2 making them more approachable in experiments. However, most frequently researchers use one-dimensional spectra that can be directly (or indirectly) measured with available instruments. The one-dimensional spectra relate to the correlation functions and fully-dimensional spectra as:

$$S_{ij}(k_1) = \frac{1}{2\pi} \int_{r_1} B_{ij}(r_1, 0, 0) e^{-ik_1 r_1} dr_1 = \iint_{k_2 k_3} \Phi_{ij}(k_1, k_2, k_3) dk_2 dk_3 \qquad (2.3.37)$$

The most commonly used components of $S_{ij}(k_1)$ are the longitudinal $S_{11}(k_1)$, transverse $S_{22}(k_1)$, and vertical $S_{33}(k_1)$ auto-spectra that characterize the energy distribution across the scales. The spectrum $S_{u_i}(k)$ in Equation (2.3.34a) is thus equivalent to $S_{ii}(k_1)$ if we invoke the vectorial nature of the velocity field. The cross-spectrum $S_{13}(k_1)$, although less frequently used, also provides important information related to contributions of eddies of different sizes to the primary Reynolds stress $\overline{u'_1 u'_3}$. Unlike real-valued even functions of auto-spectra, the cross-spectrum $S_{13}(k_1)$ is a complex-valued function $S_{13}(k_1) = C_{13}(k_1) + i Q_{13}(k_1)$ where $C_{13}(k_1)$ is the coincident spectral density (co-spectrum) and $Q_{13}(k_1)$ is the quadrature spectral density (quad-spectrum). The co-spectrum $C_{13}(k_1)$ shows the distribution of 'joint' energy of u'_1 and u'_3 across the scales and thus $\overline{u'_1 u'_3} = \int_{-\infty}^{\infty} C_{13}(k_1) dk_1$, which is similar to (2.3.35) for auto-spectra. Two other important measures that follow from $S_{13}(k_1) = C_{13}(k_1) + i Q_{13}(k_1)$ are the coherence function and the phase spectrum. The coherence function $Co_{13}(k_1) = [C_{13}^2(k_1) + Q_{13}^2(k_1)]/[S_{11}(k_1) S_{33}(k_1)]$ can be interpreted as a squared spectral 'correlation coefficient' that characterizes the degree of correlation between u'_1 and u'_3 at a wavenumber k_1. The phase spectrum $\theta_{13}(k_1) = \tan^{-1}[Q_{13}(k_1)/C_{13}(k_1)]$ gives a phase shift between fluctuating u'_1 and u'_3 at a wavenumber k_1 (or, equivalently, spatial lag $r_{13}(k_1) = \theta_{13}(k_1)/k_1$).

Although some PIV setups already allow estimation of the one-dimensional wavenumber spectra, the majority of data sets in hydraulic engineering are still obtained as velocity time series at fixed points. Therefore, one-dimensional wavenumber spectra $S_{ij}(k_1)$ are often computed from one-dimensional frequency spectra $S_{ij}(\omega)$ using Taylor's 'frozen' turbulence hypothesis, *i.e.*, $S_{ij}(k_1 = \omega/\overline{u}_1) = \overline{u}_1 S_{ij}(\omega)$ (see sections 2.3.6 and 4.4.3 in Volume I for more details).

Another useful form of the three-dimensional spectrum $\Phi_{ij}(k_1, k_2, k_3)$ can be obtained by its integration over surface A of a sphere of radius $k = |\mathbf{k}| = (k_1^2 + k_2^2 + k_3^2)^{0.5}$:

$$E_{ij}(k) = \int_A \Phi_{ij}(k_1, k_2, k_3) dA(k) \tag{2.3.38}$$

In physical terms, $E_{ij}(k)dk$ is the contribution to the tensor $\overline{u_i' u_j'}$ from wavenumbers whose magnitudes lie between k and $k+dk$. The half-sum of the diagonal components of the tensor $E_{ij}(k)$ gives the *energy spectrum function*:

$$E(k) = \frac{1}{2}[E_{11}(k) + E_{22}(k) + E_{33}(k)] \tag{2.3.39}$$

which is a key quantity in most spectral theories of turbulence. From Equation (2.3.39) it follows that:

$$\frac{1}{2}\overline{u_i' u_i'} = \frac{1}{2}\left(\overline{u_1' u_1'} + \overline{u_2' u_2'} + \overline{u_3' u_3'}\right) = \int_0^\infty E(k)dk$$

$$= \frac{1}{2}\iiint_\mathbf{k} \Phi_{ii}(k_1, k_2, k_3) dk_1 dk_2 dk_3 \tag{2.3.40}$$

For isotropic turbulence, one-dimensional spectra $S_{11}(k_1)$ and $S_{22}(k_1) = S_{33}(k_1)$, 'measurable' in experiments, relate to the energy spectrum function $E(k)$ as:

$$S_{11}(k_1) = \frac{1}{2}\int_{k_1}^\infty \left(1 - \frac{k_1^2}{k^2}\right)\frac{E(k)}{k}dk \quad \text{and} \quad E(k) = k^3 \frac{d}{dk}\left[\frac{1}{k}\frac{dS_{11}(k)}{dk}\right] \tag{2.3.41}$$

$$S_{22}(k_1) = S_{33}(k_1) = \frac{1}{4}\int_{k_1}^\infty \left(1 + \frac{k_1^2}{k^2}\right)\frac{E(k)}{k}dk = \frac{1}{2}S_{11}(k_1) - \frac{1}{2}k_1\frac{dS_{11}(k_1)}{dk_1} \tag{2.3.42}$$

Equations (2.3.41) and (2.3.42) show that special care should be taken when experimental results for $S_{11}(k_1), S_{22}(k_1)$, and $S_{33}(k_1)$ are compared with theoretical predictions, which are typically derived in terms of $E(k)$. The second-order correlation functions, structure functions and spectra contain the same information about random fields, which is presented either in the time/space domains (as in correlation and structure functions) or in frequency/wavenumber domains (as in spectra).

2.3.3.6 Turbulence scales

Turbulent flows involve a wide range of spatial and temporal velocity fluctuations, which can be characterized using the statistical measures summarized in the preceding sections. The physical roles of velocity fluctuations (or 'eddies') in most real-life flows change with the eddy size. This feature suggests that there may be a few special scales representative of the whole scale spectrum. These special scales are known as characteristic turbulence scales. Marusic and Adrian (2013) subdivide the characteristic scales into two distinctly different groups: extrinsic and intrinsic. The extrinsic scales are controlled by the boundary conditions, initial conditions, external forcing, and fluid properties such as density and viscosity. The intrinsic scales are determined by the flow response to the extrinsic conditions. The notion of characteristic scales is among the most important in turbulence research as they are widely used in simplified analytical considerations, developing predictive relationships, and scaling turbulence statistics. The key turbulence scales are velocity, length, and time scales. These scales are often supplemented with problem-specific scales (*e.g.*, if heat transfer or transport of substances are involved).

There are at least four characteristic length scales important in turbulence analyses.

(1) The *external turbulence scale* $\mathcal{L}$ represents the largest turbulent eddies comparable to the whole flow dimension(s) (*e.g.*, pipe diameter or river flow depth or width). In most turbulent flows, the energy supply from the mean flow to turbulence occurs through eddies of this scale. The external scale $\mathcal{L}$ is an example of an extrinsic scale.

(2) The *integral turbulence scale* L_{ii} can be defined in a number of ways, such as:

$$L_{ii} = \frac{1}{\overline{u_i' u_i'}} \int_0^{\infty} B_{ii}(r_1, \tau = 0) dr_1 \tag{2.3.43}$$

where r_1 is the spatial lag along the homogeneous flow direction (*e.g.*, along the straight pipe flow). The values of L_{ii} for different i^{th} velocity components may be different and be specific to the turbulent flow type (note that no summation on repeated indices applies in Equation (2.3.43)). It can be seen from Equation (2.3.43) that L_{ii} is a correlation length scale, *i.e.*, a measure of the distance within which velocity correlation between two points along x_1 vanishes. In wall-bounded flows L_{ii} is proportional to the distance from the wall. In free-shear flows it is proportional to a cross-stream distance l from the flow centerline to a location of a certain fraction of $\Delta \overline{u}$ (Figure 2.2.2). The integral turbulence scale L_{ii} is an example of an intrinsic scale. The scales L_{ii} and $\mathcal{L}$ typically relate to each other as $L_{ii} < \mathcal{L}$.

(3) The *Taylor microscale* λ_T characterizes eddies of intermediate sizes within the full scale range. It can be introduced considering the correlation function (2.3.28) for the case of homogeneous flow (*e.g.*, in the x_1 direction). At small r_1, the normalized correlation function $R_{ii}(r_1) = B_{ii}(r_1, \tau = 0)/\overline{u_i' u_i'}$ can be expanded in a Taylor series as:

$$R_{ii}(r_1) \approx 1 + r_1 \frac{dR_{ii}}{dr_1}\bigg|_{r_1=0} + \frac{r_1^2}{2} \frac{d^2 R_{ii}}{dr_1^2}\bigg|_{r_1=0} \tag{2.3.44}$$

The Taylor microscale λ_T can be defined as the intersection of the parabolic approximation (2.3.44) with the r_1 axis (*i.e.*, where $R_{ii}(r_1) = 0$), and thus can be obtained from the following Equation:

$$0 \approx 1 + \lambda_T \frac{dR_{ii}}{dr_1}\bigg|_{r_1=0} + \frac{\lambda_T^2}{2} \frac{d^2 R_{ii}}{dr_1^2}\bigg|_{r_1=0} \tag{2.3.45}$$

Similar to the integral turbulence scales, L_{ii}, the Taylor scale is an intrinsic scale.

(4) The *dissipative scale* η represents the smallest (dissipative) eddies which are influenced by fluid viscosity and thus where the turbulent kinetic energy is transformed into heat. The size of dissipative eddies is controlled by the energy flux across the scales and thus it is an intrinsic scale. The scale η is also known as the *internal scale* as together

with the *external scale* $\mathcal{L}$ they bound the whole range of turbulent fluctuations in the flow.

The four spatial scales, defined above, have their temporal counterparts that can be defined in similar ways as (2.3.43) to (2.3.45). Turbulence analyses also require characteristic velocity scales, which are appropriate for scaling turbulence quantities. As mentioned in Section 2.2.1, Volume I, the shear velocity u_* and external (or maximum) velocity U_m are typical characteristic velocity scales for wall-bounded flows. The appropriate velocity scale for free-shear flows is the variation $\Delta \bar{u}$ of the mean velocity across the flow (Figure 2.2.2, Section 2.2.2 in Volume I).

2.3.3.7 Statistical theories

This section briefly reviews some of the theories that attempt to conceptualize the turbulence physics using statistical means. Given that much of the statistical data is obtained from experimentation it is relevant for the present context to highlight the type of measurements that are needed from experiments to advance the understanding of turbulence. The statistical approach, initiated by O. Reynolds at the end of the 19th century, has led to a set of RANS-based equations for key turbulence properties outlined in section 2.3.1, Volume I. The next major step after Reynolds' (1895) pioneering work was a theory proposed by L.V. Keller and A.A. Fridman who in the 1920s introduced space-time correlations for turbulence description as well as proposed the general method for deriving hydrodynamic equations for multipoint correlation functions (Monin & Yaglom, 1971, 1975). The balance equations for the Reynolds stresses, currently popular in turbulence modeling, are the simplest case of the Keller-Fridman equations written for single-point velocity correlations.

In the middle of the 1930s, the correlation functions were re-invented by G.I. Taylor who significantly expanded the turbulence tools by adding velocity spectra and notions of turbulence homogeneity and isotropy (Batchelor, 1970). The theoretical background for isotropic turbulence, created by G.I. Taylor, a few years later was used by T. von Kármán and L. Howarth who derived an equation for the correlation functions in isotropic turbulence (Monin & Yaglom, 1975). Various forms of this equation (*e.g.*, in spectral form or written in terms of structure functions), have been among the most influential equations underlying modern statistical theories of turbulence.

The conceptual advancements of the 1930s underpinned the work of A.N. Kolmogorov at the beginning of the 1940s who introduced notions of local homogeneity, local isotropy, and self-similarity (Kolmogorov, 1941, 1962). By applying Taylor's global homogeneity and isotropy in consideration of small-scale fluctuations in non-homogeneous flows, Kolmogorov extended the applicability of these notions to turbulent flows that may not be homogeneous and isotropic at large scales. Based on these notions, Kolmogorov proposed a theory of locally isotropic turbulence that included a set of the characteristic scales and testable relationships for the second-order and third-order structure functions (Batchelor, 1970; Monin & Yaglom, 1975; Frisch, 1995). Kolmogorov's turbulence theory and its follow-up modifications remain the most influential part and inspiration of the statistical direction in modern turbulence research (*e.g.*, Frisch, 1995; Davidson, 2004).

It is useful to briefly outline Kolmogorov's theory as it combines testable statistical hypotheses with an insightful account of key physical mechanisms involved in turbulence dynamics. According to Kolmogorov (1941), a high-Reynolds number flow represents a combination of multi-scale eddies superimposed with the mean flow (in the Reynolds sense). The largest eddies are formed as a result of hydrodynamic instability of the mean flow and thus they receive energy directly from the mean flow. These eddies are scaled with the flow dimensions and thus can be represented with the external scale $\mathcal{L}$. The largest eddies, in turn, are also unstable and form smaller eddies of size r and characteristic velocity u'_r. This process of creation of smaller and smaller eddies continues until the eddy Reynolds number $u'_r r/v$ becomes so small that further generation of finer eddies is blocked by viscosity. These smallest eddies, represented by the dissipative scale η, serve as the mechanism for the energy transfer from turbulent fluctuations into heat. Thus, Kolmogorov's concept assumes that the energy supply to turbulence occurs through largest eddies of scale $\mathcal{L}$, then the energy 'cascades' from larger to smaller eddies without dissipation or production until the eddy size becomes small enough for viscosity to overcome the inertia. This energy cascade concept provides a clear picture of energy fate in turbulence dynamics.

The shape and kinematics of the largest eddies are strongly influenced by the geometry of external flow boundaries and therefore at scale $\mathcal{L}$ the turbulent flow is inhomogeneous and anisotropic. However, the memory of the external conditions is gradually weakened with decreasing scale and sufficiently small eddies (but still larger than dissipative eddies) may be viewed as locally homogeneous and isotropic. The range of these eddies expands with increase in the scale separation $\mathcal{L}/\eta$, which in turn increases with the Reynolds number $Re_\mathcal{L} = u'_\mathcal{L} \mathcal{L}/v$ as $\mathcal{L}/\eta \propto Re_\mathcal{L}^{3/4}$. To study locally-homogeneous and locally-isotropic small scale turbulence, Kolmogorov proposed to consider velocity differences between two separated points and their statistical moments (2.3.29) known as structure functions (see Section 2.3.3.4 in Volume I). First introduced by Richardson (1922) at a qualitative level, the turbulent cascade concept was independently developed by Kolmogorov as a platform for his statistical self-similarity hypotheses. These are briefly summarized below.

Kolmogorov's first hypothesis assumes that at very high Reynolds number, all statistical properties of the relative velocities $\Delta u(r, \tau) = u(x_o + r, t_o + \tau) - u(x_o, t_o)$, at r and τ where turbulence is locally isotropic, are uniquely determined by the fluid viscosity v and mean energy dissipation $\bar{\varepsilon} = 0.5 v \overline{(\partial u'_i/\partial x_j + \partial u'_j/\partial x_i)^2}$. The range of scales where the first hypothesis is valid is called the *quasi-equilibrium range* (Figure 2.3.2). The following characteristic scales can be deduced from this hypothesis: *length scale* $\eta = (v^3/\bar{\varepsilon})^{1/4}$, *velocity scale* $v_\eta = (v\bar{\varepsilon})^{1/4}$, and *time-scale* $\tau_\eta = (v/\bar{\varepsilon})^{1/2}$. According to Kolmogorov's cascade picture, the mean energy dissipation $\bar{\varepsilon}$ at smallest scales should be equal to the energy supply $P \propto u'^3_\mathcal{L}/\mathcal{L}$ at the largest scales, *i.e.*, $P = \bar{\varepsilon} \propto u'^3_\mathcal{L}/\mathcal{L}$ (as it is assumed that there is no energy dissipation or production within the range between $\mathcal{L}$ and η). Combining this equality with $\eta = (v^3/\bar{\varepsilon})^{1/4}$ one may derive $\mathcal{L}/\eta \propto Re_\mathcal{L}^{3/4}$ already mentioned above.

Kolmogorov's second hypothesis states that, at very high Reynolds number, all statistical properties of the relative velocities $\Delta u(r, \tau)$ for $\eta = (v^3/\bar{\varepsilon})^{1/4} << r << \mathcal{L}$ and $\tau_\eta = (v/\bar{\varepsilon})^{1/2} << r << \mathcal{L}/u'_\mathcal{L}$ are unambiguously determined by the mean energy dissipation $\bar{\varepsilon}$ and are independent of the fluid viscosity v. The subrange of scales within

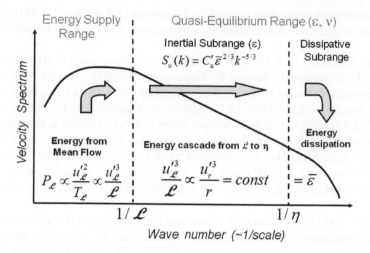

Figure 2.3.2 A sketch of Kolmogorov's spectrum showing the energy supply range and the quasi-equilibrium range, that includes an inertial subrange and a dissipative subrange.

the *quasi-equilibrium range* where the second hypothesis is valid is called the *inertial subrange* (Figure 2.3.2). Note that Taylor's microscale λ_T, defined in Section 2.3.3.6, Volume I, occupies an intermediate position between the dissipative scale η and the external scale $\mathcal{L}$ (*e.g.*, Davidson, 2004).

Based on his hypotheses, Kolmogorov expressed the tensor of the second moment $\overline{\Delta u_{ij}^2}$ in terms of two scalar functions, the longitudinal $D_{LL}(r)$ and transverse $D_{NN}(r)$ structure functions, defined by (2.3.30) and (2.3.31). For $r \ll \eta$, he obtained:

$$D_{LL}(r) = Ar^2, \quad D_{NN}(r) = A'r^2, \quad D_{NN}(r) = 2D_{LL}(r) \tag{2.3.46}$$

where A and A' are constants independent of r but dependent on $\bar{\varepsilon}$ and v. The result $D_{NN}(r) \propto D_{LL}(r) \propto r^2$ reflects 'smoothness' of the velocity field at very small r where $\Delta u \propto r$. For the inertial subrange $\eta \ll r \ll \mathcal{L}$, Kolmogorov derived:

$$D_{LL}(r) = C\bar{\varepsilon}^{2/3}r^{2/3}, \quad D_{NN}(r) = C'\bar{\varepsilon}^{2/3}r^{2/3}, \quad D_{NN}(r) = \frac{4}{3}D_{LL}(r) \tag{2.3.47}$$

where $C' - (4/3)C$. Equations (2.3.47) are known in turbulence theory as Kolmogorov's two-thirds law.

The expressions for the third-order longitudinal structure function $D_{LLL}(r) = \overline{[u_L(\mathbf{x},t) - u_L(\mathbf{x}+r)]^3}$ are given as:

$$D_{LLL}(r) = Br^3 \quad \text{for } r \ll \eta \tag{2.3.48}$$

and

$$D_{LLL}(r) = -\frac{4}{5}\bar{\varepsilon}r \quad \text{for } \eta \ll r \ll \mathcal{L} \tag{2.3.49}$$

where B is a constant dependent on $\bar{\varepsilon}$ and v. Equations (2.3.46) to (2.3.48) are based on the assumptions outlined in Kolmogorov's hypotheses and thus are phenomenological in nature.

However, Equation (2.3.49) for the third-order longitudinal structure function $D_{LLL}(r)$ has been rigorously derived by Kolmogorov from the Navier-Stokes equation through a procedure similar to that first proposed by von Kármán and Howarth for the correlation functions (Batchelor, 1970; Monin & Yaglom, 1975). The minus sign in Equation (2.3.49) reflects energy flux from larger scales to smaller scales thus supporting Kolmogorov's picture of the energy cascade. Equation (2.3.49), known as Kolmogorov's four-fifths law, is considered to be the most significant result to date in turbulence theory as it is both exact and non-trivial. It is regarded as a "boundary condition" for turbulence theories: to be acceptable the theory must satisfy the "4/5" law.

Equation (2.3.49) is also crucially important for experimental studies as it contains no empirical constants and therefore can be directly used for obtaining energy dissipation $\bar{\varepsilon}$ through measurements of $D_{LLL}(r) = \overline{[u_L(\mathbf{x}) - u_L(\mathbf{x} + r)]^3}$. Since $D_{LLL}(r)$ is an even-order moment, the measurement errors are essentially cancelled out through the averaging procedure making Equation (2.3.49) ideal for experimental studies (although the requirements of high Reynolds number and statistical convergence should be kept in mind). Equation (2.3.49) has therefore significant advantages over frequently used relations like Equation (2.3.47) that contain (semi-)empirical constants and involve the measurement errors that are difficult to avoid when dealing with second-order statistics.

Spectral equivalents of the second-order structure functions (2.3.47) are:

$$S_{LL}(k) = C_S \bar{\varepsilon}^{2/3} k^{-5/3}, \quad S_{NN}(k) = C'_S \bar{\varepsilon}^{2/3} k^{-5/3}, \quad S_{NN}(k) = \frac{4}{3} S_{LL}(k) \qquad (2.3.50)$$

where $C'_S = (4/3)C_S$ (e.g., Monin & Yaglom, 1975). Equations (2.3.50) are consistent with Equations (2.3.47) as the second-order structure functions and spectra are connected by Fourier transforms. Relations similar to those shown in Equation (2.3.50) have been independently obtained by A.M. Obukhov, L. Onsager, W. Heisenberg, and von C.F. Weizsäcker (e.g., Batchelor, 1970; Monin & Yaglom, 1975; Frisch, 1995). Kolmogorov's theory is summarized for the spectral domain in a sketch in Figure 2.3.2.

The theory outlined above is known in the turbulence community as K41 (since it was published in 1941). The experimental data related to the low-order statistics (2.3.46)–(2.3.50) are fairly consistent with K41 theory. However, the high-order statistics show some deviations from K41 predictions that may be due to more complicated mechanisms than those assumed in K41. In 1962, Kolmogorov himself proposed a refined theory where he replaced $\bar{\varepsilon}$ with a quantity ε_r, which is the energy dissipation averaged over a sphere of radius r (Kolmogorov, 1962). This replacement allowed Kolmogorov to account for the fluctuations of the energy supply to large-scale turbulence leading to intermittency of the energy dissipation at smallest scales. The follow up studies, which remain active, have explored a variety of mechanisms that could explain the deviations of the measured high-order statistics from K41 predictions. Frisch (1995), Davidson (2004) and Tsinober (2009) among others give detailed reviews of the post-Kolmogorov theories. However, experiments aimed at confirming the newly-emerging theories have been inconclusive as they have not produced adequate data in terms of resolution and statistical convergence.

2.3.4 Structural characterization of turbulence

Although the RANS and DANS equations and statistical theories outlined in the previous sections provide useful insights into turbulence mechanics, they also have significant limitations. Indeed, the averaging procedures can hide important details on key mechanisms and patterns that contribute to the 'time/space-averaged' structure, limiting the capacity of statistical approaches. A strength of the structural characterization approach, with its concept of *coherent structures*, is that it seeks to recognize organized mechanisms and patterns in turbulent flows.

A coherent structure (or motion) can be broadly defined as a persistent three-dimensional flow region over which at least one fundamental flow variable exhibits significant correlation with itself or with another variable over a range of space and/or time (*e.g.*, Robinson, 1991; Adrian, 2007). Coherent structures or their parts can be rotational or irrotational, in contrast to 'eddies' that are always rotational (Adrian & Marusic, 2012; Marusic & Adrian, 2013). There are also alternative and more specific definitions of coherent structures that account for particular topological or kinematic features of organized motions (*e.g.*, Townsend, 1975). A great variety of coherent structures have been identified depending on the flow type and Reynolds number (*e.g.*, Brown & Roshko, 2012; Kim, 2012; Wallace, 2012). It has been shown that the coherent structures play a significant role in mass and momentum transfer in free-shear flows, wall-bounded flows, and in various complex flows.

Indirect indications of the existence of organized motions in turbulent flows have been reported since the beginning of the 1900s (*e.g.*, Rumelin, 1913, who also proposed a quantitative theory for large eddies). However, the concept of coherent structures found a proper recognition only in the 1960s–1970s, after high-quality visualization experiments with wall-bounded flows (*e.g.*, Kline *et al.*, 1967) and mixing layers (*e.g.*, Brown & Roshko, 1974) were completed. Section 6.14 in Volume I presents post-processing means to identify and track the dynamics of turbulent organized motions (structures) from high-density, spatio-temporal datasets. These data have been typically provided by numerical simulations, but with the advancement of the newer non-intrusive technologies similar data can be obtained from experiments.

Based on extensive experimental studies, the following classification has been proposed for coherent structures in hydraulically-smooth wall-bounded flows (Adrian, 2007; Marusic *et al.*, 2010a; Smits *et al.*, 2011; Marusic & Adrian, 2013): (1) *quasi-streamwise vortices*, residing within the viscous and buffer sublayers, up to $100(v/u_*)$ in diameter, $1000(v/u_*)$ in length, and with $\sim 100(v/u_*)$ transverse spacing (Figure 2.3.3); (2) *hairpin vortices*, 'growing' from the solid surface and scaled with the distance from the surface (Figure 2.3.4); these vortices are representative of the influential concept of 'attached eddies' first introduced by Townsend (1976); (3) *large-scale motions* which are epitomized by hairpin packets reaching 2 to 4 flow depths (Figure 2.3.4); and (4) *very-large scale motions* (or *superstructures*), the diameter of which is comparable to the flow thickness H (*e.g.*, boundary layer thickness or pipe radius) while their length may reach up to $50H$ or even more (Figure 2.3.5). A similar classification is also applicable for rough-bed flows, with the exception of the smallest structures which

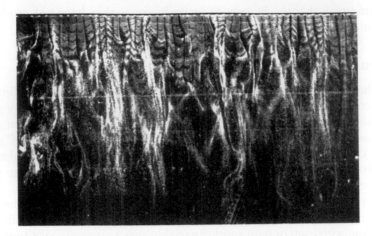

Figure 2.3.3 The structure of a flat plate turbulent boundary layer at $z = 9.6(v/u_*)$, as visualized with the hydrogen bubble technique (the flow is from top to bottom, Kline *et al.*, 1967).

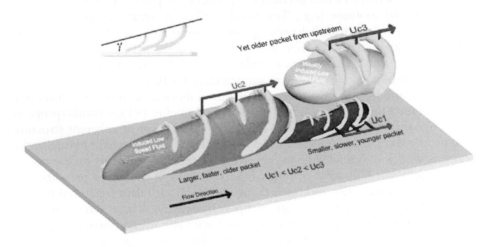

Figure 2.3.4 Conceptual picture of the packets of the hairpin structures growing from the wall and aligned in the streamwise direction (Adrian *et al.*, 2000). Note how the coherent alignment of vortices induces low momentum regions inside the packets.

are associated with *wake eddies* behind roughness elements rather than with *quasi-streamwise vortices* as in smooth-bed flows.

Visualization experiments with free shear layers revealed even more spectacular organized motions than seen in wall-bounded flows (*e.g.*, Brown & Roshko, 1974, 2012). As can be observed in Figure 2.3.6, free shear layers are dominated by large-scale spanwise structures. These quasi-two-dimensional structures control the growth of the mixing layer through a repeated pairing process in which two smaller structures form a bigger single structure. Similar to mixing layers, wake flows also display a high

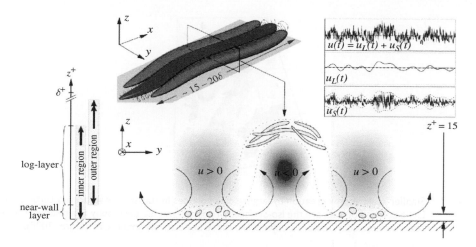

Figure 2.3.5 Very-large scale motions (or superstructures) and their interactions across the turbulent boundary layer. Note how superstructures extend from the log region down toward the wall, modulating the near-wall region. The sample time series of the streamwise velocity u highlight the modulation effect of the large scales on the small scale; the near-wall location corresponds to the peak turbulence intensity. The features shown in gray indicate elongated filamentary vortex structures and their conjectured alignment with the superstructure (From Marusic et al. 2010b. Reprinted with permission from AAAS).

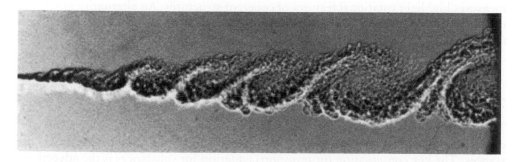

Figure 2.3.6 A photo of a mixing layer between helium (upper) and nitrogen (lower) reported in Brown and Roshko (1974) as evidence of large-scale organized structures that control the overall flow dynamics.

level of organization, with the well-known 'von Kármán vortex street' behind a cylinder as a typical example.

The statistical and structural approaches have been traditionally considered as contraposing each other (e.g., Lumley, 1989). However, this perception has changed in recent years and researchers started viewing statistical and structural features of turbulent flows as different facets of the same phenomenon (e.g., Tsinober, 2009). It is likely that multi-scale coherent structures play key roles in the energy cascade discussed in the previous section. Indeed, the Direct Numerical Simulation presented by Goto

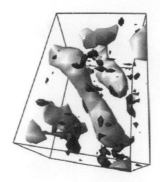

Figure 2.3.7 Smaller-scale vortex creation by larger-scale vortices in statistically stationary turbulence. Light-colored objects are the iso-surfaces of coarse-grained enstrophy at a relatively large scale, whereas dark-colored objects are those at relatively small scale (Goto, 2008).

(2008) shows that the energy cascade may occur as a result of the successive creation of smaller-scale tubular vortices in the larger-scale straining regions existing between pairs of larger-scale structures, as illustrated in Figure 2.3.7.

Although the importance of coherent structures for turbulence dynamics is widely recognized, the issues of defining, detecting and quantifying coherent structures remain an active research topic. Most current methods for detecting coherent structures are based on a variety of measures derived from the velocity gradient tensor. These measures highlight rotational features of coherent structures. An alternative approach is based on the Proper Orthogonal Decomposition (known as POD) that defines the most energetic modes of motion. A brief review of these methods with examples is given in Section 6.14, Volume I.

2.3.5 Mixing, diffusion, and dispersion

Turbulence knowledge is required for addressing multiple problems in hydraulic engineering. One of most significant among them is mixing and transport of substances. The key conservation (or transport) equation for substances is the advection-diffusion equation:

$$\underbrace{\frac{\partial C}{\partial t}}_{\substack{local \\ change}} + \underbrace{\frac{\partial C u_j}{\partial x_j}}_{advection} = \underbrace{\frac{\partial}{\partial x_j}\left(\chi_m \frac{\partial C}{\partial x_j}\right)}_{\substack{molecular \\ diffusion}} + \underbrace{F}_{source} \qquad (2.3.51)$$

where C is the instantaneous substance concentration, *i.e.*, the ratio of substance mass (or volume in some cases) to the total sampling volume; χ_m is the molecular diffusion coefficient; and F is source/sink of substance C (in other words, the rate of homogeneous reaction within the fluid volume). The concentration C at a spatial point can be also interpreted as a probability to find a substance particle, released from an upstream source, in the vicinity of this point. Note that in Equation (2.3.51) the substance flux due to molecular diffusion is expressed as $\chi_m \partial C / \partial x_j$. This linear dependence of the flux of substance on the concentration gradient is known as Fick's first law.

Similar to RANS, the Reynolds-averaged advection-diffusion equation is more suitable for practical applications than Equation (2.3.51). After decomposition of instantaneous concentration, source term, and velocity in Equation (2.3.51) as $u_i = \overline{u}_i + u'_i$, $F = \overline{F} + F'$ and $C = \overline{C} + C'$ and then averaging, it follows:

$$\underbrace{\frac{\partial \overline{C}}{\partial t}}_{\substack{local \\ change}} + \underbrace{\frac{\partial \overline{C}\overline{u}_j}{\partial x_j}}_{\substack{mean \\ advection}} = \underbrace{\frac{\partial}{\partial x_j}\left(\chi_m \frac{\partial \overline{C}}{\partial x_j}\right)}_{\substack{molecular \\ diffusion}} + \underbrace{\frac{\partial(-\overline{C'u'_j})}{\partial x_j}}_{\substack{turbulent \\ transport}} + \underbrace{\overline{F}}_{source} \tag{2.3.52}$$

As in the case of the Reynolds-averaged momentum equation, averaging the advection term in Equation (2.3.51) leads to the appearance of an additional term $\partial(-\overline{C'u'_j})/\partial x_j$ responsible for the substance mixing and transport by turbulence. Unlike Equation (2.3.51), Equation (2.3.52) contains three unknowns: turbulent fluxes $-\overline{C'u'}$, $-\overline{C'v'}$, and $-\overline{C'w'}$. Thus, there is a need for additional equations, which should link the fluxes $-\overline{C'u'_j}$ with the averaged concentration $\overline{C}$. Taylor (1915) proposed that similar to the molecular flux, the turbulent flux $-\overline{C'u'_j}$ is proportional to the concentration gradient. For the tree-dimensional turbulent flow, this assumption can be expressed as:

$$-\overline{C'u'_j} = \chi_{ji}^T \frac{\partial \overline{C}}{\partial x_i} \tag{2.3.53}$$

where χ_{ji}^T is the tensor of turbulent diffusion coefficient (superscript T stands for turbulent). In general, χ_{ji}^T contains nine components dependent on the spatial coordinates and time. For the simplest case of homogeneous and isotropic mixing, the turbulent diffusivity χ_{ji}^T is reduced to a scalar parameter χ^T. However, in most practical cases it is assumed to be a tensor with zero off-diagonal components and thus the turbulent diffusivity is viewed as a vector χ_{ii}^T (Rutherford, 1994). This assumption is equivalent to the assumption that the flow coordinate axes coincide with the principal axes of the tensor χ_{ji}^T, which in practice rarely happens. Nevertheless, it is widely used as the experience shows that contributions of the off-diagonal components of χ_{ji}^T are often insignificant. Thus, Equation (2.3.52) can be simplified as:

$$\frac{\partial \overline{C}}{\partial t} + \frac{\partial \overline{C}\overline{u}_j}{\partial x_j} = \frac{\partial}{\partial x_j}\left(\chi_m \frac{\partial \overline{C}}{\partial x_j}\right) + \frac{\partial}{\partial x_j}\left(\chi_{ji}^T \frac{\partial \overline{C}}{\partial x_j}\right) + \overline{F} \tag{2.3.54}$$

where $\chi_{ji}^T = 0$ when $i \neq j$. Equation (2.3.54) serves as a starting point in many hydraulic applications where transport of substances is the focus. Its success largely depends on how accurately the turbulent diffusivities χ_{ii}^T are presented. The theoretical analysis shows that $\chi_{ii}^T = \overline{u'^2_i} T_i$ where $\overline{u'^2_i}$ is variance of the ith velocity component and T_i is the Lagrangian integral turbulence scale (*i.e.*, obtained as an integral of the Lagrangian correlation function). In practical applications, it is often expressed as $\chi_{ii}^T = \overline{u'^2_i} T_i = a_i u_* L_{di}$ where L_{di} is a relevant length scale such as flow depth, width or a characteristic scale of channel forms; relevant coefficients α_i are taken from laboratory and field experiments (Fischer *et al.*, 1979; Rutherford, 1994).

In the case of rough-bed flows, the conventional Equations (2.3.52) and (2.3.54) are not applicable due to local roughness effects on velocity and concentration fields. Similar to Equation (2.3.19), the advection-diffusion Equation (2.3.51) can be presented in the double-averaged form to provide a basis for description of substance transport in mobile rough-bed flows such as vegetated flows or flows over sand dunes (Nikora *et al.*, 2013). Using the advection-diffusion Equation (2.3.51) for instantaneous variables, the following double-averaged advection-diffusion equation can be derived, similarly to Equation (2.3.19):

$$\underbrace{\frac{\partial \phi_{Vm}\langle \phi_T \overline{C}\rangle}{\partial t}}_{1} + \underbrace{\frac{\partial \phi_{Vm}\langle \phi_T\rangle\langle \overline{C}\rangle\langle \overline{u}_j\rangle}{\partial x_j}}_{2} = \underbrace{\frac{\partial}{\partial x_j}\left(\phi_{Vm}\left\langle \phi_T \chi_m \frac{\partial \overline{C}}{\partial x_j}\right\rangle\right)}_{3} - \underbrace{\frac{\partial \phi_{Vm}\left\langle \phi_T \overline{C'u'_j}\right\rangle}{\partial x_j}}_{4}$$

$$-\underbrace{\frac{\partial \phi_{Vm}\left\langle \phi_T \widetilde{\overline{C}}\widetilde{\overline{u}}_j\right\rangle}{\partial x_j}}_{5} - \underbrace{\frac{\partial \phi_{Vm}\left\langle \phi_T \widetilde{\overline{C}}\right\rangle\langle \overline{u}_j\rangle}{\partial x_j}}_{6} - \underbrace{\frac{\partial \phi_{Vm}\left\langle \phi_T \widetilde{\overline{u}}_j\right\rangle\langle \overline{C}\rangle}{\partial x_j}}_{7} \qquad (2.3.55)$$

$$-\underbrace{\frac{1}{V_o}\iint_{S_{int}}\overline{\left(\chi_m \frac{\partial C}{\partial x_j}\right)^s}n_j dS}_{8} + \underbrace{\phi_{Vm}\left\langle \phi_T \overline{F}\right\rangle}_{9}$$

Terms 1 and 2 in Equation (2.3.55) represent local change of concentration and convective transport. The third, fourth, and fifth terms are due to the molecular diffusion, turbulent transport ($\langle \phi_T \overline{C'u'_j}\rangle$), and form-induced transport ($\langle \phi_T \widetilde{\overline{C}}\widetilde{\overline{u}}_j\rangle$), respectively; the sixth and seventh terms represent substance fluxes due to potential spatial correlations between the local time porosity and time-averaged velocities and concentrations; the final two terms, *i.e.*, eighth and ninth, are an interfacial flux term (*i.e.*, heterogeneous reaction rate) and a homogeneous reaction rate, respectively. For the case when spatial correlations between the local time porosity and time-averaged velocities and concentrations can be neglected, Equation (2.3.55) can be simplified, using the continuity Equation (2.3.18) and the space-time porosity $\phi = \phi_{VT} = \phi_{Vm}\langle \phi_T\rangle$, as:

$$\frac{\partial \langle \overline{C}\rangle}{\partial t} + \langle \overline{u}_j\rangle \frac{\partial \langle \overline{C}\rangle}{\partial x_j} = \frac{1}{\phi}\frac{\partial}{\partial x_j}\left(\phi\langle \chi_m \frac{\partial \overline{C}}{\partial x_j}\rangle\right) - \frac{1}{\phi}\frac{\partial \phi\langle \overline{C'u'_j}\rangle}{\partial x_j} - \frac{1}{\phi}\frac{\partial \phi\langle \widetilde{\overline{C}}\widetilde{\overline{u}}_j\rangle}{\partial x_j}$$

$$-\frac{1}{\phi}\frac{1}{V_o}\iint_{S_{int}}\overline{\left(\chi_m \frac{\partial C}{\partial x_j}\right)^s}n_j dS + \langle \overline{F}\rangle \qquad (2.3.56)$$

Advection-diffusion equations similar to Equations (2.3.55) and (2.3.56) can also be derived for fine suspended sediments at low concentrations at which the advection-diffusion approximation is appropriate. Compared to Equation (2.3.52), Equations

(2.3.55) and (2.3.56) contain three more unknown fluxes (*i.e.*, form-induced fluxes $\langle \widetilde{C}\widetilde{u}_j \rangle$) and thus additional closure equations are needed. Their development requires extensive laboratory and field experiments underpinned by physical considerations.

Although currently available data on form induced fluxes $\langle \widetilde{C}\widetilde{u}_j \rangle$ are scarce, there are indications suggesting that their contributions may be significant (Lopez & Garcia, 1998; McLean *et al.*, 2007). Section 2.5 in Volume I provides more details on this approach in relation to sediment transport and vegetated flows.

Three-dimensional Equations (2.3.54) to (2.3.56) can be averaged over depth or width leading to two-dimensional equations more suitable for applications. Cross-sectional averaging produces a one-dimensional dispersion equation (Taylor, 1954; Fischer *et al.*, 1979; Rutherford, 1994):

$$\frac{\partial \overline{\overline{C}}}{\partial t} + \overline{\overline{u}}_1 \frac{\partial \overline{\overline{C}}_j}{\partial x_1} = \frac{1}{A} \frac{\partial}{\partial x_1} \left(\frac{AD \partial \overline{\overline{C}}}{\partial x_1} \right) + \overline{\overline{F}} \tag{2.3.57}$$

where double overbar defines cross-section averaging; A is the cross-section area; and D is the dispersion coefficient that combines effects of streamwise, vertical and lateral mixing. There are analytical methods for obtaining D based on the known velocity fields and diffusion coefficients. However, in most cases the dispersion coefficient is obtained based on the tracer experiments. The data from such experiments are often used for parameterizations of the dispersion coefficient, which typically involves flow depth or width (or both) and cross-sectionally-averaged velocity.

In some applications it may be important to consider evolution of the substance variance. Second-order equations similar to Equations (2.3.6) for MKE and (2.3.7) for TKE can also be derived for substance transport by turbulent flows. These second-order equations provide a basis for data interpretation in experiments where both fluctuating velocities and concentrations are measured (*e.g.*, using methods based on the laser-induced fluorescence). A comprehensive treatment of the second-order equations for substances can be found, *e.g.*, in Bernard and Wallace (2002) and Davidson (2004).

Finally, it should be noted that Equations (2.3.53), (2.3.54), and (2.3.57) belong to the class of so-called Fickian-type equations where diffusive fluxes are presented as linearly proportional to mean concentration gradients, following an analog of Fick's first law, *i.e.*, $-\overline{C'u'_j} = \chi^T_{ij} \partial \overline{C}/\partial x_j$. However, many studies highlighted some deviations from the Fickian diffusion, which are due to non-Gaussian effects in turbulent flows. One of the mechanisms that can lead to non-Fickian diffusion relates to effects of coherent structures, particularly to superstructures, as highlighted by Adrian and Marusic (2012). Thus, the role of experimental work in clarifying the nature of turbulent diffusion and the most appropriate theoretical frameworks for its description and modeling remains significant.

2.3.6 Key working hypotheses pertinent for experimentation

Although constantly growing computational capabilities and advancements in direct numerical simulations (DNS) promise significant step changes in turbulence understanding, the role of physical experiments in studies of turbulent flows will always remain critically important. Tsinober (2009, pp. 68, 69) highlights a number of important factors that make the experiments in turbulence research irreplaceable:

"Though the scale resolution problem is serious, both in laboratory and numerical experiments, there is an essential difference between the two: inadequate resolution in numerical experiments leads usually to erroneous results of the whole output, whereas in laboratory/field experiments one has the *true* flow and correct results for the scales resolved even when some range of scales is not resolved. Second, even at small Reynolds numbers, it is not always possible to use the numerical approach either. An important example is represented by flows in complex geometries such as flows with rough walls and flows in plant canopies. Similar problems arise in handling flows involving sediment transport and other additives (various particles, bubbles), and especially polymers and surfactants for which even no adequate equations seem to be known. This is a partial list only of the limitations of numerical approaches and the reasons for the importance of physical experiments. One more reason is that it is the physical experiment (dealing with real turbulent flows) that provides the final verdict to the results of both numerical simulations and theories."

Even high-precision and high-resolution data produced in the experiments are always limited in terms of uncertainties, resolution, and dimensionality. The data interpretation therefore often requires some hypotheses and assumptions, particularly when statistical measures and theories are involved. Although many such hypotheses are flow-dependent, some of them are fairly general in turbulence research, particularly assumptions of stationarity, homogeneity, and ergodicity (see informal definitions in Section 2.3.3, Volume I). All three assumptions are often essential for estimating most statistical measures outlined in Section (2.3.3), Volume I.

Strictly speaking, the stationarity condition requires that one-dimensional probability distributions of all flow quantities to be independent of time while their multi-dimensional distributions should depend only on the time separation between observations. Similarly, the spatial fields are considered to be homogeneous if one-dimensional probability distributions of all flow quantities are independent of spatial location while their multi-dimensional distributions depend only on the spatial separation between observations. These conditions are known as stationarity and homogeneity in the strict sense. It is impossible, however, to rigorously test these hypotheses in the experiments with turbulent flows and therefore relaxed notions of weak stationarity and homogeneity are used instead. These notions, also known as stationarity and homogeneity in a wide sense, require validity of two conditions only: mean values should be independent of time or spatial location while the correlation functions should depend on time and spatial separations (lags) only. There is a variety of statistical tests, specifically designed for checking stationarity and homogeneity conditions (in a wide sense) which are available to experimentalists (see Section 6.12 in Volume I). In data treatment, stationarity and homogeneity properties need to be complemented with the ergodicity condition when ensemble averages may well be approximated with time and/or spatial averages from long enough data records. If stationarity and homogeneity of turbulent flows can be tested (see Section 6.12.1 in Volume I), the ergodicity assumption has to be taken for granted as rigorous tests of this condition for three-dimensional turbulent flows are not yet available (*e.g.*, Tsinober, 2009). There are of course situations when the focus is on non-stationary flows for which the ergodic hypothesis is inapplicable and therefore an experimentalist needs to employ ensemble averaging or methodologies designed for non-stationary processes (*e.g.*, wavelet analysis).

Another frequently used hypothesis used in turbulence experiments is known as Taylor's frozen turbulence hypothesis. This hypothesis emerged as a by-product of Taylor's study of a relationship between velocity correlation function and spectrum (Taylor, 1938). To test this relationship, Taylor needed to transform available spatial correlation function into time correlation function (so it could be used with the frequency spectrum). He proposed that the temporal change in velocity measured at a fixed point in the flow is identical to spatial change along the line which crosses this point and is oriented along the mean flow. Taylor (1938) hypothesized that the longitudinal eddy transfer at a fixed point occurs with the local mean velocity without appreciable deformation of eddy structure and, therefore, this suggestion is known as the 'frozen' turbulence hypothesis. Taylor's frozen turbulence hypothesis appeared to be in high demand among experimentalists since it relates statistical characteristics of turbulence in the time (frequency) domain (typically measured in the experiments) to those in the spatial (wavenumber) domain (typically used in theoretical considerations) and *vice-versa*.

In analytical form, Taylor's hypothesis states that:

$$\frac{\partial \zeta}{\partial t} = -U_E \frac{\partial \zeta}{\partial x} \quad \text{or} \quad \overline{\left(\frac{\partial \zeta}{\partial t}\right)^2} = -U_E^2 \overline{\left(\frac{\partial \zeta}{\partial x}\right)^2} \tag{2.3.58}$$

$$R(\tau) = R(r = U_E \tau) \tag{2.3.59}$$

$$D(\tau) = D(r = U_E \tau) \tag{2.3.60}$$

$$U_E S(\omega) = S(k = \omega/U_E) \tag{2.3.61}$$

where the eddy convection velocity U_E is assumed to be equal to the local mean streamwise velocity $\bar{u}$; ζ is flow quantity of interest (*e.g.*, velocity components, pressure or substance concentration); $R(\tau)$ and $R(r)$ are temporal and spatial correlation functions; $D(\tau)$ and $D(r)$ are temporal and spatial structure functions; $S(\omega)$ and $S(k)$ are frequency and wavenumber spectra, respectively; τ and r are temporal and spatial lags. Although the validity of Equations (2.3.58) to (2.3.61) has been extensively studied for various quantities and various flows, the active research of Taylor's assumption continues as its underpinning mechanisms may provide new insights in turbulence mechanics (*e.g.*, Dennis & Nickels, 2008; del Alamo & Jimenez, 2009). For example, for open channel flows it was found that Taylor's hypothesis is fairly valid away from a smooth or rough bed ($z > 0.10H$-$0.20H$). However, in the near-bed region the convection velocity appeared to be much higher than the local mean velocity, approaching the bulk flow velocity (Nikora & Goring, 2000; Cameron & Nikora, 2008). This effect still awaits an in-depth experimental study.

2.4 OPEN-CHANNEL FLOWS

In Section 2.2.1, Volume I, *open-channel flows* (OCFs) were defined as a class of free-surface flows driven by gravity, bounded by channel boundaries, and having an upper interface (free surface) exposed to air. Many books describe OCFs in detail (*e.g.*, Chow,

1959; Henderson, 1966; French, 1985; Yen, 1993; Chaudhry, 1993; Jain, 2001; Julien, 2002; Chanson, 2004; Akan, 2006; Sturm, 2011; Moglen, 2015). This section gives a brief introduction to OCF as it pertains to experimental hydraulics and fluid mechanics. Additionally, this section leads to more complex flow situations such as those associated with mobile channel boundaries, flow through vegetation, aerated flows, and ice-laden flows.

2.4.1 Classification

In addition to the boundary-layer classifications described in section 2.2.1, Volume I, OCFs comprise a wide variety of channel and flow conditions, as this section outlines. Open channels can be either *natural channels* (*e.g.*, rivers, streams, etc.) or *artificial channels* (*e.g.*, manmade canals, drainage channels, non-pressurized culverts, laboratory flumes). Open channels are called *prismatic channels* when their cross-section and bed slope are constant, and *non-prismatic channels* when their cross section and/or slope change in the main flow direction. From a morphologic point of view, an open channel is *rigid* (strictly or practically speaking) if erosion or deposition processes do not alter its geometry (cross section, slope, alignment), or *deformable* if its geometry changes due to erosion or deposition. Most natural channels are non-prismatic, deformable-bed channels, and most prismatic channels, such as irrigations canals, are manmade.

Open channel flows are also classified in terms of flow behavior. A *Steady flow* is a flow whose mean depth and velocity do not vary with time. A flow with temporal variations in depth and/or velocity is a *time-varying* or *unsteady flow*. A *uniform flow* is a flow that has constant velocity, depth, slope and cross section over a given length (or reach) of channel; *i.e.*, the downslope gravitational driving force and the flow resistance force are in balance. When any of these flow quantities varies in space the flow is *non-uniform*. Uniform flows are very rare in nature and can only be observed in manmade prismatic channels or under well-controlled conditions in laboratory flumes.

Non-uniform flows can be further classified as *gradually varied* (GVF) and *rapidly varied flow* (RVF). In GVF, velocity and water depth vary gradually, and the free surface is stable (*e.g.*, flow in canals and most river reaches). Another characteristic of GVF, in addition to the steady flow conditions, is that the streamlines of these flows are practically parallel and thus, a hydrostatic pressure distribution can be assumed in each section of the flow. On the other hand, in RVF both flow velocity and water depth vary significantly -usually- over a relatively short reach distance (*e.g.*, flow through hydraulic jumps, sluice gates, and around obstacles) and high flow streamlines curvature prevails over the flow section rendering the hydrostatic pressure assumption invalid. These changes in streamlines curvature may cause the presence of sudden water profiles discontinuities (*e.g.*, hydraulic jump) or separation/recirculation zones and eddies in which the flow patterns are complex and the velocity distributions are confined by the separation zones rather than the by the rigid bed or side-walls solid boundaries (Chow, 1959). A schematic of the aforementioned OCF classification can be found in Figure 2.4.1.

Most OCFs are primarily governed by the effects of gravity relative to the inertial forces of the flow (Chow, 1959); viscous forces can be important, but usually to a lesser extent in hydraulic engineering. An OCF may be either *laminar* or *turbulent*, based on

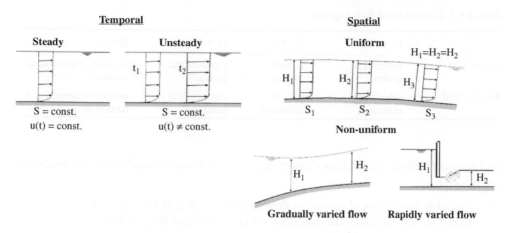

Figure 2.4.1 Schematic of different temporal and spatial open channel flow cases.

the *Reynolds number*, a dimensionless ratio of inertial to viscous forces. In OCF the *Reynolds number* is defined as $Re_{R_h} = UR_h/\nu$ where U is the mean or bulk, cross-sectional averaged velocity of the flow, R_h is the hydraulic radius of the flow (cross-sectional area of flow/length of wetted perimeter of flow), and ν is kinematic viscosity. OCF is often assumed to be fully laminar if $Re_{R_h} < 500$ and fully turbulent if $Re_{R_h} > 2000$. For $500 < Re_{R_h} < 2000$ the flow is in the *transitional state*, displaying intermittent characteristics of laminar and turbulent flow behavior.

The effect of gravity upon the state of a free-surface flow is represented by the *Froude Number, $Fr = U/\sqrt{gD}$*. The *hydraulic depth D* is the cross-sectional area of the water normal to the direction of the flow in the channel divided by the width of the free surface (Chow, 1959). For rectangular channels D is flow depth. When $Fr = 1$, the flow is in a *critical state*. When $Fr < 1$, the flow is *subcritical*, and gravitational forces dominate inertia forces; the flow is deeper and has a slower velocity than when $Fr = 1$, and are often described as tranquil or streaming. *Subcritical flow* occurs in most manmade canals and rivers (except in steep terrain). When $Fr > 1$, the flow is known as *supercritical*. In this state the inertial forces dominate; so the flow is shallower and faster than when $Fr = 1$, and is often described as rapid, shooting, or torrential. *Supercritical flows* are commonly observed during floods in mountain streams as well as in steep urban storm-water runoff channels, chutes, and spillways.

The combined effect of viscosity and gravity leads to four OCF *regimes*, namely,

1) subcritical-laminar flow, when $Fr < 1$ and $Re_{R_h} < 500$
2) supercritical-laminar flow, when $Fr > 1$ and $Re_{R_h} < 500$
3) subcritical-turbulent flow, when $Fr < 1$ and $Re_{R_h} > 2000$
4) supercritical-turbulent flow, when $Fr > 1$ and $Re_{R_h} > 2000$

Additionally, there are two *flow regimes* associated with the occasional instability of supercritical flows. This flow instability results from conditions (see Sturm, 2011) producing roll waves, which completely change flow behavior. Roll waves occur when the $Fr \geq 1.5 - 2$, depending on the geometry of the channel, the bottom roughness

Table 2.4.1 Open channel flow regimes

	$Fr < 1$	$Fr = 1$	$Fr > 1$	$Fr > 2$
$Re_{R_b} < 500$	subcritical laminar	critical laminar	supercritical laminar	unstable-supercritical laminar
$Re_{R_b} > 2000$	subcritical turbulent	critical turbulent	supercritical turbulent	unstable-supercritical turbulent

and the unsteadiness of the flow (Yen *et al.*, 1977). Two additional flow regimes can be introduced,

5)   unstable-supercritical-turbulent flow, when $Fr > 2$ and $Re_{R_b} > 2000$
6)   unstable-supercritical-laminar flow, when $Fr > 2$ and $Re_{R_b} < 500$.

In Table 2.4.1 the aforementioned flow regimes are summarized in tabulated format. Unstable free-surface flows are very common in the mining industry but usually not desired in hydraulic engineering since they can lead to overbank flow followed by flooding.

The relationship between flow depth and velocity that Figure 2.4.2 shows for the different types of steady and uniform flow was first proposed by Robertson and Rouse (1941) and restated by Chow (1959). The diagram was extended by Yen *et al.* (1977) to

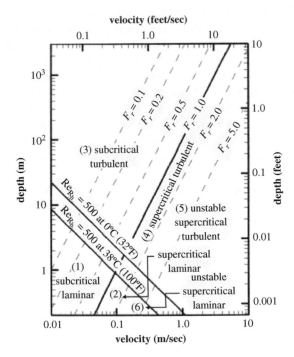

Figure 2.4.2 Diagram for depth-velocity variation with OCF regimes for wide channels (adapted from Yen et al., 1977).

include roll waves in the case of overland flow over steep slopes. Julien & Hartley (1986) have also observed roll waves in laminar sheet flows in laboratory experiments. Regimes (1) and (2) in the diagram are not commonly encountered in open-channel hydraulics, which mainly deals with turbulent flow conditions. Regimes (1) and (2) occur in thin layers, such as sheet flows typical of overland flows, such as over road surfaces.

OCF can be treated as *three dimensional* (3D), *two dimensional* (2D) or *one dimensional* (1D), depending on flow characteristics and the flow information sought. In *three dimensional* OCF, the velocity components in all directions are important; *e.g.*, local transport of pollutants by secondary currents. When the effects of one velocity components are negligible, the OCF can be considered as *two dimensional*. This is the case for shallow flows in which the vertical velocity is much smaller than the velocity components in the horizontal plane, allowing a *depth-averaged* approach. Alternatively, there are flows for which the velocities and concentration gradients are primarily in the streamwise and the vertical direction. For them, the *horizontally averaged* treatment of the flow may be proper. This is the case when lateral fluxes and buoyancy related phenomena are driving the flow (*e.g.*, turbidity currents, heated water discharge, etc.). An OCF can be treated as *one dimensional* when it has a strong dominant direction and curvature effects can be ignored.

2.4.2 Boundary layer concepts applied to open-channel flows

2.4.2.1 General concepts

OCFs are a subset of the wall-bounded shear flows described in Section 2.2.1, Volume I. There is a close relationship between the distributions of flow velocity and, therefore, shear stress in a channel and the resistance a channel boundary exerts in opposing flow. The region in which velocity and shear stress vary with elevation above the channel boundary is known as the boundary layer. As velocity and shear stress usually vary with flow depth for OCF, boundary layer concepts are important when analyzing OCF distributions of velocity and shear stress. The ensuing discussion considers relatively wide channels, and thus focused on the boundary layer developed only from the bottom or bed of a channel.

Consider the case of steady, turbulent, uniform, wide ($B/H \gg 1$) OCF of Figure 2.4.3 where H is the water depth, B is the channel width, S is the bed slope and k_s is the equivalent roughness height. For natural or artificial channels with a flat sediment bed, k_s values can be calculated as proportional to a representative sediment diameter, D_x, as $\Delta = a_s D_x$, where a_s is a constant that ranges from 1 to 6.6 but in general a value of ~2.0–2.5 is usually adopted. Appendix 2.A.1 suggests values of the coefficient a_s for channels and rivers for various characteristic diameters D_x (Yen, 2002; Garcia, 2008).

In addition to the skin friction caused by the bed surface roughness, bed forms associated with movable bed (see Section 2.5.1.6 in Volume I) can influence the bed roughness. The hydraulic roughness in presence of bedforms is a very dynamic parameter which depends on the flow characteristics (mean flow velocity, depth etc.) and the sediment properties (sediment diameter, density etc.). Brownlie (1981), van Rijn (1982, 1984c) and later Fedele and Garcia (2001) are some of the researchers

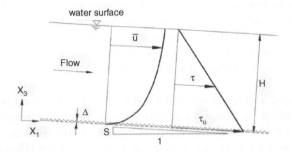

Figure 2.4.3 Definition diagram steady, uniform open-channel flow in a very wide channel.

attempting to associate bedform-induced drag with a friction parameter in order to combine it with the "skin" surface friction and estimate the total flow resistance.

2.4.2.2 Bed shear stress

In uniform OCF, the bed shear stress counteracts the streamwise component of gravitational force driving the flow. Using the momentum equation, the average bed shear stress (τ_o) for steady, uniform, wide OCF is:

$$\tau_o = \rho g H S \tag{2.4.1}$$

where ρ is the fluid density and g is the acceleration of gravity. Also, for uniform flow the shear stress varies linearly with distance above the bed (see Figure 2.4.3): *i.e.*,

$$\tau = \tau_o\left(1 - \frac{x_3}{H}\right) \tag{2.4.2}$$

Shear velocity u_* is defined as follows:

$$u_* = \sqrt{\frac{\tau_o}{\rho}} \tag{2.4.3}$$

This parameter is commonly used as a surrogate for boundary shear stress and is used to scale velocity distributions in boundary-layer flows. Different ways to estimate the value of the shear velocity in open channel flows can be found in Lopez and Garcia (1999).

2.4.2.3 Velocity distribution

The velocity distribution normal to a channel bottom can be predicted using the "law of the wall" which for the flow above the viscous sublayer is expressed as the "logarithmic law" (Schlichting, 1979; Nezu & Rodi, 1986):

$$\frac{\bar{u}}{u_*} = \frac{1}{\kappa}\ln\left(\frac{x_3}{z_0}\right) \tag{2.4.4}$$

where $\bar{u}$ is the time-averaged local velocity at the distance x_3 above the bed, z_0 is the hydrodynamic bed roughness length that represents the distance from the bed where the velocity $\bar{u}$ becomes zero (it is proportional to the roughness height Δ) and κ is the empirical von Kármán's constant which is approximately 0.41 (Nezu & Rodi, 1986). Although the "law of the wall" is valid for a region close to the bed ($x_3/H < 0.2$, Nezu & Nakagawa, 1993), it is commonly used to approximate the velocity profile across the entire water column in an OCF.

OCFs can be divided into flows over smooth beds and flows over rough beds. For channels over smooth beds, the influence of water viscosity dominates that of turbulence in a thin, near-bed layer called the inner region. This very thin layer has important implications for turbulence generation, and is called the *viscous sublayer*, whose thickness, δ_v, is usually estimated as $\delta_v = 11.6v/u_*$ (e.g., Rouse, 1939).

For smooth boundaries, $\delta_v >> \Delta$, roughness plays no part in determining the velocity distribution of flow, because the roughness elements are enclosed by the viscous sublayer. For rough boundaries, $\delta_v << \Delta$ and the viscous sublayer plays no part in determining the velocity distribution of flow, because the sublayer is completely disrupted by the roughness elements. For values of δ_v and Δ within these limits, a transitional range exists for which both viscous and roughness effects must be taken into account.

2.4.2.3.1 Velocity distributions in channels with smooth beds

Within the viscous sublayer of the velocity profile near a smooth bed ($x_3 < \delta_v$), the velocity has a linear profile (O'Connor, 1995) referred to as the *inner law*:

$$\frac{\bar{u}}{u_*} = \frac{u_* x_3}{v} \tag{2.4.5}$$

The dimensionless velocity shown on the left side of the equation is sometimes denoted as u^+. The dimensionless distance shown on the right side of the equation is sometimes denoted as x_3^+ and is called the *inner* or *wall length scale* (usually it is found in textbooks as y^+ using y as the vertical axis). This dimensionless wall distance x_3^+, expressed in "wall units", is usually used for the definition of the range x_3^+ for which a velocity distribution expression is valid. For example, the aforementioned *inner law* is usually considered to be valid inside the range from $0 < x_3^+ < 5.0$.

In the *outer region* (for $\Delta > \delta_v$), also known as logarithmic region, the following velocity distribution applies for a dynamically smooth boundary layer:

$$\frac{\bar{u}}{u_*} = \frac{1}{\kappa}\ln\left(\frac{9u_* x_3}{v}\right) \tag{2.4.6}$$

Comparing Equations (2.4.4) and (2.4.6), indicates $z_0 = v/9u_*$ for dynamically smooth flows. Typically, Equation (2.4.6) is considered to be valid for $x_3^+ > 30.0$. Finally, there is a *transitional* or *buffer layer* ($5.0 < x_3^+ < 30.0$) in which the velocity profile deviates from the Equations (2.4.5) and (2.4.6).

Figure 2.4.4, shows Equations (2.4.6) and (2.4.5) plotted along with LDV data obtained in a laboratory for the case of a smooth OCF (Möller, 2014).

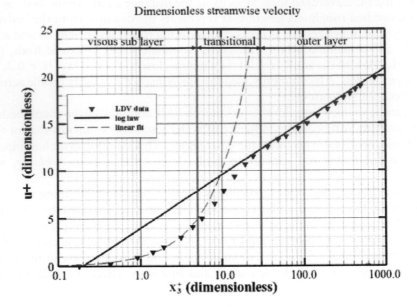

Figure 2.4.4 Experimental data obtained with a LDV in an open-channel laboratory flow (Möller, 2014).

2.4.2.3.2 Velocity distributions in channels with rough beds

Most river and channel flow are dynamically rough or transitional. If the bed roughness height is much larger than the viscous sublayer thickness ($\Delta/\delta_v > 1$), the bed is rough and the logarithmic law is given by:

$$\frac{\bar{u}}{u_*} = \frac{1}{\kappa}\ln\left(\frac{30x_3}{\Delta}\right) \tag{2.4.7}$$

From Equations (2.4.7) and (2.4.4), $z_0 = \Delta/30$ and elevation which extends to the average bed surface, since the inner law is no longer valid.

2.4.2.3.3 Velocity defect law and log-wake law

A drawback for the use of the logarithmic law for flow over a dynamically rough bed is that it requires "a priori" knowledge of the roughness height. An alternative formula is the *velocity-defect law* (*e.g.*, Schlichting, 1979):

$$\frac{\bar{u}_{\max} - \bar{u}}{u_*} = -\frac{1}{\kappa}\ln\left(\frac{x_3}{H}\right) \tag{2.4.8}$$

The velocity-defect law assumes that the maximum velocity $\bar{u}_{\max}$ occurs at the free surface and can be used for any bottom roughness. Although the logarithmic law is generally used to calculate velocities for the entire water column, it is valid only over a much smaller range ($x_3/H < 0.2$, Nezu & Nakagawa, 1993). Coleman and Alonso (1983) and Sarma *et al.* (1983) have shown that the logarithmic law is not valid close to the water surface due to wake effects associated with disturbances of the water surface (Coleman, 1981; Lyn, 1991). For hydraulically smooth beds the logarithmic law can be rewritten as (Nezu & Nakagawa, 1993):

$$\frac{\overline{u}}{u_*} = \frac{1}{\kappa}\ln\left(\frac{u_* x_3}{v}\right) + w\left(\frac{x_3}{H}\right) \tag{2.4.9a}$$

where $w\left(\frac{x_3}{H}\right)$, a wake correction for turbulent boundary layer flows (Coles, 1956), is estimated as

$$w\left(\frac{x_3}{H}\right) = \frac{2W_0}{\kappa}\sin^2\left(\frac{\pi x_3}{2H}\right) \tag{2.4.9b}$$

where W_0 is the Coles wake strength parameter. Through trigonometric substitution, Equation (2.4.8) also can be written in log-wake form (Coleman, 1981; Coleman & Alonso, 1983)

$$\frac{\overline{u}_{\max} - \overline{u}}{u_*} = -\frac{1}{\kappa}\ln\left(\frac{x_3}{H}\right) + \frac{2W_0}{\kappa}\cos^2\left(\frac{\pi x_3}{2H}\right) \tag{2.4.10}$$

A procedure to estimate Coles wake parameter, originally proposed by Coleman (1981), can be found in Julien (2002). Nezu and Rodi (1986), in experiments on flat-bed, smooth-bed, turbulent flows, found W_0 to vary from 0 to 0.253, with a mean value of $W_0 \approx 0.2$. This result was confirmed independently by Lyn (1991). Coleman (1981) and Parker and Coleman (1986) demonstrated for the case of sediment-laden flows over flat-beds, W_0 increases with increasing sediment concentration, ranging from 0.191 to 0.861. Lyn (1991) found that for flow over artificial bed forms, W_0 ranged from –0.05 to 0.1, and suggested that negative values of W_0 are the result of strong, favorable pressure gradients. Lyn also found good results in replicating measured velocity profiles over bed forms with the log-wake law.

Neglecting wake effects, the logarithmic laws for dynamically smooth ($\Delta/\delta_v << 1$) and dynamically rough ($\Delta/\delta_v >> 1$) beds have been combined to form a composite equation that covers both ranges and the transition between them (Yalin, 1992):

$$\frac{\overline{u}}{u_*} = \frac{1}{\kappa}\ln\left(\frac{x_3}{\Delta}\right) + B_s \tag{2.4.11}$$

where the *Roughness function* B_s is dependent of specific Reynolds number $\mathrm{Re}_* = u_*\Delta/v$:

$$B_s = 8.5 + [2.5ln(\mathrm{Re}_*) - 3]e^{-0.212[ln(\mathrm{Re}_*)]^{2.42}} \tag{2.4.12}$$

Equation (2.4.12) is obtained from experiments and is shown in Figure 2.4.5 (Garcia, 2008 with permission from ASCE).

An alternative to Equation (2.4.12) has been given by Swamee (1993),

$$\frac{\overline{u}}{u_*} = \left\{\left(\frac{v}{u_* x_3}\right)^{10/3} + \left[\frac{1}{\kappa}ln\left(1 + \frac{9(u_* x_3/v)}{1 + 0.3(u_*\Delta/v)}\right)\right]^{-10/3}\right\}^{-0.30} \tag{2.4.13}$$

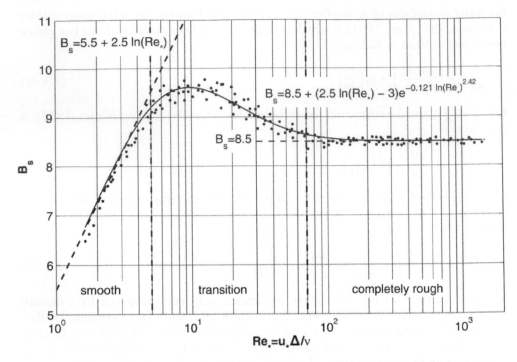

Figure 2.4.5 Roughness function B_s plotted against the roughness Reynolds number (adapted from Garcia, 2008 with permission from ASCE).

Most of the knowledge about flow velocity distribution in turbulent, free-surface flows stems from laboratory studies (*e.g.*, Nezu & Rodi, 1986; Song *et al.*, 1994; Best *et al.*, 1996; Lemmin & Rolland, 1997; Muste & Patel, 1997; Graf & Cellino, 2002). In the last few years, however, new acoustic technology for flow measurement has made possible the observation of velocity profiles in streams and rivers as well (*e.g.*, Kostaschuk *et al.*, 2004). With the help of observations made in the Missouri river, Holmes and Garcia (2008) has found that the velocity-defect law, Equation (2.4.13), works well for field conditions and the Coles wake parameter takes values ranging from -0.035 to 0.36. In all cases, dune-like bed forms were present, suggesting that such features are indeed responsible for the deviations from the logarithmic velocity distribution, observed away from the bottom.

2.4.2.4 Mean velocity and flow resistance coefficients

Velocity profiles are useful for analyzing OCF flow, however for many hydraulic applications the depth-averaged velocity, U, is more practical. As first shown by Keulegan (1938), the logarithmic law for the case of a hydraulically rough bed can be substituted into the definition of depth-averaged velocity ($U = \frac{1}{H}\int\limits_{0}^{H} \overline{u}\,dz$), yielding the following:

$$\frac{U}{u_*} = \frac{1}{\kappa} ln\left(11\frac{H}{\Delta}\right) \tag{2.4.14}$$

This equation is known as *Keulegan's resistance law for dynamically rough beds.*

Another relation for the depth-averaged velocity can be found by integrating a power law fit of the velocity profile across the entire depth. The integration result in a Manning-Strickler relation can be expressed as (Keulegan, 1938; Garcia, 1999):

$$\frac{U}{u_*} = 8.1\left(\frac{H}{\Delta}\right)^{1/6} \tag{2.4.15}$$

Like in most boundary-layer flows, the local bed shear stress on OCF is proportional to the square of the depth-averaged velocity:

$$\tau_o = \rho C_f U^2 \tag{2.4.16}$$

where C_f is called a friction coefficient. Comparing (2.4.14) with (2.4.16):

$$C_f = \left[\frac{1}{\kappa} ln\left(11\frac{H}{\Delta}\right)\right]^{-2} \tag{2.4.17}$$

Note that while Equation (2.4.16) can be used to determine the local bed shear stress based on local properties of the flow (*e.g.*, local depth-averaged velocity, water depth, and roughness height). Equation (2.4.17) assumes uniform flow. Its use to calculate C_f implies the use of reach-averaged shear stress. Both equations assume negligible channel bank effects. C_f can also be found using the power-law fit given by Equation (2.4.15):

$$C_f = \left[8.1\left(\frac{H}{\Delta}\right)^{1/6}\right]^{-2} \tag{2.4.18}$$

2.4.2.5 Uniform open channel flow equations

Commonly used OCF equations relate C_f with other resistance coefficients. Combining Equations (2.4.1) and (2.4.16) lead to a form of Chezy's equation (Chow, 1959):

$$U = C_z H^{1/2} S^{1/2} \tag{2.4.19}$$

where C_z is a dimensionless Chezy coefficient:

$$C_z = \left(\frac{g}{C_f}\right)^{1/2} \tag{2.4.20}$$

Equations (2.4.17) and (2.4.20) show the Chezy coefficient is not constant, but varies with the logarithm of H/Δ, that represents a relative roughness. In some OCFs, the variation of the relative roughness can be significant. Combining Equations (2.4.18) and (2.4.19) lead to a Manning's relationship for wide rectangular channels:

$$U = \frac{1}{n} H^{2/3} S^{1/2} \tag{2.4.21}$$

where n is the Manning's roughness coefficient (Strickler, 1923; Brownlie, 1983):

$$n = \frac{\Delta^{1/6}}{8.1g^{1/2}} \tag{2.4.22}$$

The 1/6 exponent in Equation (2.4.22) implies that Manning's n is not very sensitive to changes in roughness height. On the other hand, Δ values can change dramatically with small changes in Manning's n (Yen, 1973). Thus, given a Manning's n value, Equation (2.4.22) can show if the corresponding estimate of Δ is reasonable (*e.g.*, if Δ is found to be larger than the water depth, it is likely that the value of n is unrealistic).

For steady, uniform flows in prismatic channels the hydraulic radius has to be used with a correction coefficient, K_n, accounting for dimensions being used (Yen, 1991):

$$U = \frac{K_n}{n} R_h^{2/3} S^{1/2} \tag{2.4.23}$$

When the hydraulic radius and the velocity magnitude are expressed in metric units (*m* and *m/s*); and, K_n is 1.0, and when the hydraulic radius and the velocity magnitude are expressed in English units (*ft* and *ft/s*), K_n is 1.49. In practice, Manning's values are found in tables based on qualitative stream descriptions (*e.g.*, bank and bed material, sinuosity, channel debris, etc.). Detailed tables for Manning's n values can be found, for example, in Chow (1959) Yen (1991) and USGS (1989).

Another well-known equation applied to uniform OCF is Darcy-Weisbach, written as:

$$U = \sqrt{\frac{8}{f}} \sqrt{gR_h S} = \sqrt{\frac{8}{f}} u_* \tag{2.4.24}$$

where f is a friction factor (or resistance coefficient) that is a function of Reynolds number and relative roughness (*e.g.*, Brownlie, 1981).

Keulegan (1938), proposed the following equation for hydraulically rough, turbulent OCF:

$$f = \left[2.21 + 2.031 log \left(\frac{R_h}{\Delta} \right) \right]^{-2} \tag{2.4.25}$$

Alternative relationships have been proposed by Strickler (1923) and Christensen (1993):

$$f = \alpha \sqrt[3]{\frac{\Delta}{H}} \tag{2.4.26}$$

where $\alpha = 0.178$ (Strickler, 1923) or 0.1173 (Christensen, 1993).

Brownlie (1981) has demonstrated how the Nikuradse friction factor diagram for pipe flow presented in Figure 2.4.6 can be used to analytically remove side wall effects from bed shear stress measurements when conducting laboratory flume experiments.

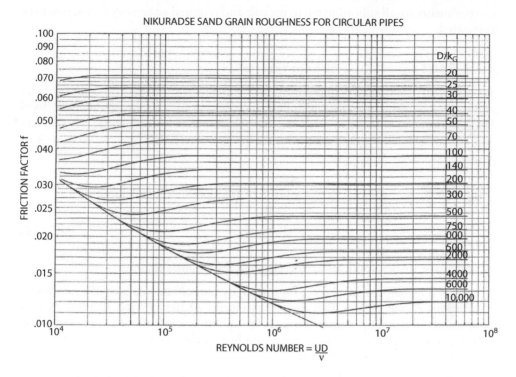

Figure 2.4.6 Revised Nikuradse friction factor diagram for pipes of diameter D or open-channel flows with hydraulic radius $R_h=D/4$ (after Brownlie, 1981 with permission from ASCE).

Similar side wall correction procedures have been proposed by Einstein (1942), Vanoni and Brooks (1957), Williams (1970) and others. An excellent account about resistance in open channel flows can be found in Yen (1991).

2.4.3 Relevant 1D, 2D, and 3D equations for open-channel flows

Figure 2.4.1 shows that most natural OCF are turbulent in nature. Given the incomplete knowledge on the turbulence-related processes in the most of the environmental fluid mechanics and hydraulic engineering problems, practical problems in these areas cannot be directly solved (*e.g.*, Marchi & Rubatta 1981, Rodi *et al.*, 2013). Most often, the OCF analytical studies and numerical simulations make recourse to the Reynolds-averaged Navier-Stokes (RANS) equations introduced in Section 2.3.1, Volume I. The subsequent discussion on RANS equations is relevant for this handbook because many of the terms in these equations are determined and/or verified experimentally. Moreover, there is an increased number of investigations of practical problems that use numerical simulations in combination with experiments for providing the missing information in the analytical posing of the problem. Providing relationships for these terms over a range of flow conditions is an on-going activity in many laboratories. Therefore it is useful to understand the physical meaning of the terms, their role and the accuracy required for their estimation. Subsequent decisions on the selection of the facilities and instrumentation and the conduct and processing of the data acquired by the experiments are all subordinated to the role and required accuracy of the terms to be investigated.

2.4.3.1 RANS equations with heat transfer

Semi-empirical techniques based on closure models are among the most popular methods of modeling the Navier-Stokes equations using the mean values of local velocity and pressure $(\overline{u}_i, \overline{p})$ $(i = 1, 2, 3)$. For the case in which there are temperature variations in the flow, the gravitational force in Equation (2.3.3) is extended to include buoyancy effects:

$$\frac{\partial \overline{u}_i}{\partial t} + \overline{u}_j \frac{\partial \overline{u}_i}{\partial x_j} = -\frac{1}{\rho_0} \frac{\partial \overline{p}}{\partial x_i} + \frac{\partial}{\partial x_i} \left(v \frac{\partial \overline{u}_i}{\partial x_i} \right) + \frac{1}{\rho_0} \frac{\partial}{\partial x_l} \left(-\rho_0 \overline{u_i' u_l'} \right) + g_i \frac{\rho}{\rho_0} \qquad (2.4.27)$$

$$\frac{\partial \overline{u}_i}{\partial x_i} = 0 \qquad (2.4.28)$$

Most often the RANS application to OCF analyses assumes that:

1   the flow is incompressible except for the "buoyancy effect" associated with modest variations of the density that induce variations of the gravity. In other words, variations in the fluid density (ρ) have a negligible effect on all terms except on the term in which the variations are multiplied by the gravitational acceleration;
2   Coriolis acceleration is neglected.

It has to be stressed here that the gravitational term is only important for flows in which there are temperature variations. This buoyancy term is simplified to a gravity term if the fluid flow is isothermal. Assumption (1) is commonly attributed to Boussinesq (1903) and couples the hydrodynamic problem to the thermodynamic problem, written from the First Principle of Thermodynamics and a (linearized) "equation of state" for the fluid. For the cases in which small variations of temperature are present, the equation of state can be expresses as a function of a coefficient of isobaric expansion (as pressure variations do not have significant impact) α_r in the form

$$\rho = \rho_o [1 + \alpha_r (T - T_r)] \qquad (2.4.29)$$

where ρ_o and T_r are reference density and reference temperature respectively.

Fluids are thermally conductive materials. Therefore, additional equations are needed for process description. The "heat equation" neglecting the effect of the dissipated mechanical power and fluid compressibility can be written as:

$$\frac{\partial \overline{T}}{\partial t} + \overline{u}_j \frac{\partial \overline{T}}{\partial x_j} = \frac{\partial}{\partial x_i} \left(k_T \frac{\partial \overline{T}}{\partial x_i} - \overline{u_i' t'} \right) + \frac{s_T}{\rho c_P} \qquad (2.4.30)$$

where $\overline{T}$ is a mean value and t' is the fluctuating component of temperature, k_T is the (molecular) "thermal diffusivity" of the liquid, s_T is the specific thermic power associated to a distributed heat source (if present) and c_P is the specific heat.

Equations analogous to Equation (2.4.30) can be written for mass balances applied to the transport of pollutants or dissolved substances (e.g., sea salt) in water. This is done by replacing $\overline{T}$ with the volumetric concentration of the substance $\overline{C}$, k_T with the molecular diffusivity of the substance k_C, t' with c' and s_T with s_C that is the time derivative of the concentration input or output by distributed sources or sinks. In natural channels the boundaries can consist of mobile material that continuously interacts with the flow. The mechanism is very complex requiring formulation of additional mass-balance equations for the material transported by the flow, at local and global scales, that is, with reference to the entire flow depth.

Finding solutions for Equations (2.4.27) to (2.4.30) requires determining the correlation terms $\overline{u_i' u_j'}$ and $\overline{t' u_j'}$ (and $\overline{c' u_j'}$ when considering material fluxes) with $(i, j = 1, 2, 3)$. The available turbulence descriptions cannot provide analytical expressions for these correlation terms. Moreover, these equations contain 3rd-order correlations which must be determined from equations containing 4th order correlations and so on. This dependency makes direct solution of the RANS equations unfeasible (or at least very challenging when neglecting the effects of higher order correlations). Alternatively, semi-empirical turbulence models (a.k.a "closure" relationships)have been developed for Equations (2.4.27) to (2.4.30) to provide an approximate description of the unknown correlations up to a certain order as a function of lower order correlations and/or mean flow characteristics. Such models are expressed by algebraic or differential relations that, together with Equations (2.4.27) to (2.4.30), allow the system of equations to be numerically solved.

2.4.3.2 Closure relationships for turbulence models

Research over the last 50 years has resulted in the successful development of "turbulence models" that, although not universal, provide reasonable solutions to many real world applications (*e.g.*, Rodi, 1984; Lopez & Garcia, 1997; Lyn, 2008). From a historic point of view, the origin of these methods dates back to the late 1800's when Reynolds published his results on turbulent flows. Among the first attempts of a mathematical description of turbulent stresses (commonly referred to as Reynolds stresses), lies the concept of turbulent viscosity, introduced by Boussinesq (1877). This concept is based on the hypothesis that, analogous to laminar flow, viscous stresses, the turbulent stress tensor is proportional to the mean velocity gradients.

One commonly used closure approach expresses the Reynolds stresses as a series expansion of velocity gradients up to the first order:

$$-\overline{u_i' u_j'} = v_{ij}^T \left(\frac{\partial \overline{u_i}}{\partial x_j} + \frac{\partial \overline{u_j}}{\partial x_i} \right) - \frac{2}{3} k \delta_{ij} \qquad (2.4.31)$$

where v_{ij}^T is an "eddy viscosity" tensor and k is the specific turbulent kinetic energy. Equation (2.4.31) was first introduced by Boussinesq (1877) who expressed v_{ij}^T as v_T with no dependence on i or j (*i.e.*, isotropic macroturbulence). He postulated an analogy between the mechanism of generation and the consequent structure of the viscous stresses. Note that the term in Equation (2.4.31) containing the turbulent kinetic energy is necessary to satisfy the condition:

$$k = \frac{1}{2} \overline{u_{ii}'}^2 \qquad (2.4.32)$$

Neither Reynolds nor Boussinesq attempted to solve the averaged Navier-Stokes equations and it wasn't until the early 1900's, when Prandtl (1904) introduced the boundary layer concept, that the essence of viscous flow physics began to be understood. Since then, different turbulence closure schemes have been proposed to model Reynolds stresses. They can be classified as follows:

(a)   Algebraic models (Zero-equation models)
(b)   One-equation models
(c)   Two-equation model
(d)   Higher order models

2.4.3.2.1 Algebraic models

As mentioned before, the concept of turbulent viscosity was conceived in analogy to the molecular motion theories that derive in the well-known Stokes viscosity law for laminar flows. According to that hypothesis, turbulent eddies are seen as fluid entities that collide and exchange momentum (as molecules do). Therefore, just as molecular viscosity is proportional to mean velocity and the molecules' mixing path, turbulent viscosity is considered proportional to the characteristic velocity fluctuations and a typical length associated to the turbulence (which Prandtl called mixing length). Since its formulation, many objections have been made to this analogy (some of which will be mentioned below) but in a practical sense, the results obtained in a vast majority of studies are sufficiently good to justify its application in applied problems. From a dimensional perspective, the turbulent viscosity, v_T, is proportional to the product between a characteristic velocity scale of the flow, $\hat{V}$, and a characteristic length scale of the macro-vorticity component of the flow, $\hat{L}$, (Rodi, 1984):

$$v_t \propto \hat{V}\hat{L} \tag{2.4.33}$$

Turbulence models have evolved from models in which v_T is directly expressed as a function of the mean flow properties (i.e., the velocity gradient) and assume that turbulence is dissipated where it is generated, to more recent models that account for the transport of turbulence. One or more differential equations for the $\hat{V}$ and $\hat{L}$ scales (or equivalent parameters) are used for such accounting.

Similarly, in direct analogy to the turbulent transport of momentum, the turbulent transport of a passive tracer ϕ such as heat t or solute concentration c (turbulent transport of scalar properties) is assumed to be proportional to the gradients of the specie being transported:

$$-\overline{u_i'\phi'} = \Gamma \frac{\partial \phi}{\partial x_i} \tag{2.4.34}$$

where Γ represents the turbulent diffusivity (thermal or mass diffusivity). It is not a property of the fluid; it depends on the turbulent properties of the particular flow. The Reynolds analogy between heat and mass transport with momentum suggests that Γ is closely related to v_T:

$$\Gamma = \frac{v_T}{\sigma_t} \tag{2.4.35}$$

where σ_t is the Prandtl number (for heat transport) or the Schmidt number (for mass transport). The modeling issues regarding v_T can therefore be extended to the modeling of transport of a passive tracer.

The main difference between the models with different number of equations lies in the use of transport equations to characterize the variables $\hat{V}$ and $\hat{L}$, needed to determine the turbulent viscosity. Algebraic (zero-equation) models do not use transport equations to determine the velocity and characteristic lengths. Instead, the turbulent viscosity is directly specified empirically, or estimated by trial and error or by empirical formulations, or by relating it to the mean velocities. These models may in turn be classified into:

1. Constant turbulent viscosity
2. Mixing length models
3. Prandtl's model for unconfined flows

The mixing length model is perhaps the most well-known and it is actually the first ever proposed model to describe the distribution of the turbulent viscosity (Prandtl, 1925). Considering the case of boundary layers with only one significant Reynolds stress $(\overline{u'_1 u'_3})$, Prandtl postulated that $\hat{V}$ is equal to the product of a certain mixing length times the mean velocity gradient (Schlichting, 1979):

$$\hat{V} = l_m \left| \frac{\partial \overline{u}_1}{\partial x_3} \right| \tag{2.4.36}$$

Assuming that $\hat{L} = l_m$, the turbulent viscosity is then given by:

$$v_T = l_m^2 \left| \frac{\partial \overline{u}_1}{\partial x_3} \right| \tag{2.4.37}$$

This is the so-called Prandtl's mixing length hypothesis, which has been used satisfactorily in a large number of applications. Specifying l_m for every particular case is the problem with this formulation. For example, in unconfined flows (plumes, jets, etc.) l_m may be assumed as constant and proportional to the local thickness δ. The proportionality factor or empirical constant in this type of model depends on the particular type of flow. The mixing length hypothesis as presented above applies only to two-dimensional boundary layer flows (including asymmetric flows). For other flows, the hypotheses can be written in general form as:

$$v_T = l_m^2 \left[\left(\frac{\partial \overline{u}_i}{\partial x_j} + \frac{\partial \overline{u}_j}{\partial x_i} \right) \frac{\partial \overline{u}_i}{\partial x_j} \right]^{1/2} \tag{2.4.38}$$

In practice, however, very limited used has been found for general expressions, mainly due to the difficulties involved in the specification of l_m.

In the case of heat and/or mass transfer problems, the mixing length hypothesis requires determining the Prandtl and/or Schmidt number which usually adopt values close to 0.5 in planar jets and mixing layers, 0.7 in circular jets and 0.9 in wall flows.

2.4.3.2.2 One-equation models

As mentioned before, models that solve the transport differential equations have been developed to overcome the limitations of the algebraic models. One-equation models provide a direct connection between the velocity fluctuation scale (characteristic velocity $\hat{V}$) and the mean velocity gradients, by solving transport differential equations. The most common models are based on the turbulent viscosity concept and they will be presented here. The natural way to characterize the velocity fluctuations' scale is by making it proportional to the square root of the mean turbulent kinetic energy by unit mass ($k=1/2\overline{u_i' u_i'}$). Given that the largest amount of turbulent kinetic energy is found in the largest scales (small wave numbers), $\sqrt{k}$ is a length scale for the large eddies that scale with $\hat{L}$. Therefore, the turbulent viscosity results in:

$$\nu_T = c'_\mu \sqrt{k} \hat{L} \tag{2.4.39}$$

where c'_μ is an empirical constant. This expression is known as the Kolmogorov-Prandtl formula since both authors proposed it independently. In the same way, both authors suggested that k could be determined by solving a transport differential equation for the turbulent kinetic energy which can be derived exactly from the Navier-Stokes equations,

$$\frac{\partial k}{\partial t} + \overline{u}_j \frac{\partial k}{\partial x_j} = -\left(\frac{1}{\rho} \overline{u_i' \frac{\partial p'}{\partial x_i}} + \overline{u_i' u_j'} \frac{\partial \overline{u}_i}{\partial x_j} + \overline{u_j' \frac{\partial k}{\partial x_j}}\right) + \nu \overline{u_i' \frac{\partial^2 u_i'}{\partial x_j \partial x_i}} \tag{2.4.40}$$

However, the exact form of this equation is of little practical interest due to the appearance of new unknowns in the production and dissipation terms. To close the problem, other models involving these terms must be assumed. Using the gradient hypothesis again, the diffusive flux of k is usually assumed to be proportional to its gradient, this is:

$$-\overline{u_i' \left(\frac{u_j' u_j'}{2} + \frac{p'}{\rho}\right)} = \frac{\nu_T}{\sigma_k} \frac{\partial k}{\partial x_i} \tag{2.4.41}$$

where σ_k is an empirical diffusion coefficient. The dissipation term ε is modeled as:

$$\varepsilon = c_\varepsilon \frac{k^{3/2}}{\hat{L}} \tag{2.4.42}$$

where c_ε represents another empirical constant. Note that even though dissipation occurs at the smaller turbulent scales, the rate of dissipation is governed by the larger scales. Using the above hypotheses and the turbulent viscosity concept to model the Reynolds stresses, the transport energy for k becomes:

$$\frac{\partial k}{\partial t} + \overline{u}_i \frac{\partial k}{\partial x_i} = \frac{\partial}{\partial x_i}\left(\frac{\nu_T}{\sigma_k} \frac{\partial k}{\partial x_i}\right) + \nu_T \left(\frac{\partial \overline{u}_i}{\partial x_j} + \frac{\partial \overline{u}_i}{\partial x_i}\right) \frac{\partial \overline{u}_i}{\partial x_i} + \beta g_i \frac{\nu_T}{\sigma_k} \frac{\partial \overline{\phi}}{\partial x_i} - c_\varepsilon \frac{k^{3/2}}{L} \tag{2.4.43}$$

It can be seen in the previous equation that the buoyancy term $\beta g_i \frac{\nu_t}{\sigma_t} \frac{\partial \overline{\phi}}{\partial x_i}$, which corresponds to the diffusive flux $\overline{u'\phi'}$ assuming that it is proportional to the gradient of $\overline{\phi}$, has been included. The variable ϕ is responsible for producing the density difference (e.g., temperature, salinity, suspended sediment concentration) and β is a transformation constant that relates ϕ with variations in density.

It must be noted that the expression above is valid only for large Reynolds numbers and cannot be used in the viscous sublayer close to the wall. For low Reynolds numbers a molecular diffusivity term must be included in the expression, with constants whose values are a function of the Reynolds number itself defined as $R_{et} = \frac{\sqrt{k}\hat{L}}{v}$. As a particular case, it is interesting to apply the equation above to conditions in which k is time invariant and therefore the convective and diffusive terms of the expression are negligible. Under such circumstances, the turbulence is in a local equilibrium, where, in the absence of buoyant terms, for boundary layers, a balance between production and dissipation of turbulent energy is obtained:

$$v_T \left(\frac{\partial \overline{u_1}}{\partial x_3}\right)^2 = c_\varepsilon \frac{k^{3/2}}{L} \tag{2.4.44}$$

Substituting the above equality in the Kolmogorov-Prandtl expression yields:

$$v_T = \left(\frac{c_\mu'}{c_D}\right)^2 \hat{L}^2 \left|\frac{\partial \overline{u_1}}{\partial x_3}\right| \tag{2.4.45}$$

which is none other than the mixing length expression introduced before where

$$l_m = \left(\frac{c_\mu'^3}{c_\varepsilon}\right)^{1/4} \hat{L} \tag{2.4.46}$$

One of the mixing length hypothesis limitations becomes apparent from the above: the mixing length hypothesis is only applicable for turbulence in local equilibrium. It is evident from the expression above that, to complete the modeling effort, the characteristic length $\hat{L}$ must be specified. Its determination is precisely what distinguishes the different one-equation models. Unfortunately, $\hat{L}$ is as difficult to determine as is l_m. More details may be found in Rodi (1984).

2.4.3.2.3 Two-equation turbulence models

These type of models try to determine the characteristic length of the large turbulent scales L, by solving an additional transport equation. The general idea is based on assuming that the size of the turbulent eddies depends on their initial dimensions while being advected by the flow, the turbulent dissipation (which by destroying the small eddies increases the effective size of the larger scales) and the processes related to the energy cascade. Therefore, the balance between these processes might be modeled with a differential equation for L. This results in two transport equations, giving name to these type of models. Most available models do not solve transport equations for $\hat{L}$ specifically. Rather, a variable $Z = k^n \hat{L}^m$ is specified which serves the purpose given that k is known due to the solution of the kinetic energy transport equation. Different k–Z two-equation models are available in the literature. The most popular ones are:

k-e model: Here $n = 3/2$ and $m = -1$, with Z being proportional to the turbulent energy dissipation ε. This is the most popular model, and therefore it has been applied to the majority of applied problems.

k-ω model: In this model $n = 1/2$ and $m = -1$. Z represents what Kolmogorov originally referred to as ω, understanding by this the rate of energy dissipation per unit volume per unit time. Physically ω represents certain mean characteristic frequency of the large turbulence scales.

$k\text{-}k_L$ model: In this model $n = 1$ and $m = 1$.

$k\text{-}\tau$ model: In this model $n = -1/2$ and $m = -1$, where $\tau = k/\varepsilon$.

For all models above the overall results are similar, thus a common transport relation exists for Z which may be expressed as:

$$\frac{\partial Z}{\partial t} + \bar{u}_i \frac{\partial Z}{\partial x_i} = \frac{\partial}{\partial x_i}\left(\frac{\sqrt{k}\hat{L}}{\sigma_Z}\frac{\partial Z}{\partial x_i}\right) + c_{Z1}\frac{z}{k}P - c_{Z2}Z\frac{\sqrt{k}}{\hat{L}} + S \tag{2.4.47}$$

where σ_Z, c_{Z1} and c_{Z2} are empirical constants, P is the turbulent kinetic energy production and S_z represents a secondary source term that differs according to the election of the Z from one model to the other.

Due to its popularity, the k–ε model is presented below in more detail.

2.4.3.2.3.1 k – ε TURBULENCE MODEL

Simply put, this model consists in solving two transport equations, one for the turbulent kinetic energy (k) and one for the dissipation (ε). After solving both equations, local and instantaneous values for k and ε for the whole flow field are obtained. With these values, it is then possible to estimate the instantaneous turbulent viscosity (ν_T) for the entire domain. Finally, the Reynolds stresses can be determined with the values of ν_T, closing the turbulence problem (same number of equations and unknowns) and leaving only the Navier-Stokes equations to be solved to determine the entire flow field. The previous steps are meant to be explanatory since as it will be shown below, in order to obtain k and ε, ν_T is required and therefore, it is necessary to solve all equations simultaneously or iteratively. The semi-empirical transport equations for k and ϵ are respectively:

$$\frac{\partial k}{\partial t} + \bar{u}_i \frac{\partial k}{\partial x_i} = \frac{\partial}{\partial x_i}\left(\frac{\nu_T}{\sigma_k}\frac{\partial k}{\partial x_i}\right) + \nu_T\left(\frac{\partial \bar{u}_i}{\partial x_j} + \frac{\partial \bar{u}_i}{\partial x_i}\right)\frac{\partial \bar{u}_i}{\partial x_i} + \beta g_i\frac{\nu_T}{\sigma_k}\frac{\partial \bar{\phi}}{\partial x_i} - \varepsilon \tag{2.4.48}$$

$$\frac{\partial \varepsilon}{\partial t} + \bar{u}_i \frac{\partial \varepsilon}{\partial x_i} = \frac{\partial}{\partial x_i}\left(\frac{\nu_T}{\sigma_\varepsilon}\frac{\partial \varepsilon}{\partial x_i}\right) + c_{1\varepsilon}\frac{\varepsilon}{k}(P+G)(1+c_{3\varepsilon}R_f) + c_{2\varepsilon}\frac{\varepsilon^2}{k} \tag{2.4.49}$$

Where

$$P = \nu_T\left(\frac{\partial \bar{u}_i}{\partial x_j} + \frac{\partial \bar{u}_i}{\partial x_i}\right)\frac{\partial \bar{u}_i}{\partial x_i} \tag{2.4.50}$$

$$G = \beta g_i\frac{\nu_T}{\sigma_k}\frac{\partial \bar{\phi}}{\partial x_i} \tag{2.4.51}$$

$$\nu_T = c_\mu\frac{k^2}{\varepsilon} \tag{2.4.52}$$

while σ_ε, $c_{1\varepsilon}$, $c_{2\varepsilon}$ and $c_{3\varepsilon}$ are empirical constants and R_f is a Richardson number which in the case of horizontal shear layers takes the form $R_f = G/((P+G))$ (Rodi, 1984).

2.4.3.2.3.2 VALUES OF CONSTANTS IN THE k-ε MODEL

In order to solve the transport equations presented above, all constants must be determined first. Their value has been obtained by applying the model to simple cases where due to the simplifications inherent to the problem, only a few of the constants actually appear in the equations allowing a calibration. A few examples are presented below.

Obtaining c_μ:
For the case of shear layers in local equilibrium, $P = \varepsilon$ thus it can be shown with the equations that:

$$c_\mu = \left(\frac{\overline{u_1' u_3'}}{k}\right)^2 \tag{2.4.53}$$

Measurements in such flows give values of $\frac{\overline{u_1' u_3'}}{k} = 0.30$ from where

$$c_\mu = 0.09 \tag{2.4.54}$$

Obtaining $c_{2\varepsilon}$:
In the particular case of grid turbulence, the diffusion and production terms cancel out leaving only $c_{2\varepsilon}$ in the transport equations. Therefore:

$$u_1 \frac{\partial k}{\partial x_1} = \varepsilon \tag{2.4.55}$$

and

$$\overline{u}_1 \frac{\partial \varepsilon}{\partial x_1} = -\frac{\varepsilon^2}{k} \tag{2.4.56}$$

Which has a general solution of the form:

$$k = const. x_1^{-m} \tag{2.4.57}$$

$$m = \frac{1}{c_{2\varepsilon} - 1} \tag{2.4.58}$$

After comparing with experimental results, values of $c_{2\varepsilon}$ between 1.80 and 2.00 are obtained.

Obtaining $c_{1\varepsilon}$:
Close to the wall, in boundary layer flows $\left(\frac{\partial}{\partial x_1} = 0\right)$, the velocity can be approximated with the semi-logarithmic profile; on the other hand, the production is equal to the dissipation. Therefore, for steady flows, the dissipation equation reduces to:

$$0 = \frac{\partial}{\partial x_3}\left(\frac{v_t}{\sigma_\varepsilon}\frac{\partial \varepsilon}{\partial x_3}\right) + c_{1\varepsilon}\frac{\varepsilon^2}{k} - c_{2\varepsilon}\frac{\varepsilon^2}{k} \tag{2.4.59}$$

And from the logarithmic velocity distribution hypothesis the following result is obtained

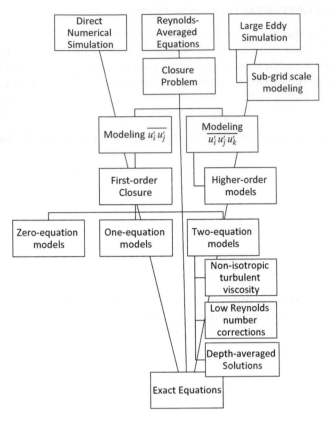

Figure 2.4.7 The diagram above serves as a summary of the material presented in this section. It provides a general overview of the problem related to the numerical modeling of turbulence (Garcia, 2000b).

$$\varepsilon = \frac{u_*^3}{kx_3} \tag{2.4.60}$$

$$k = \frac{u_*^2}{\sqrt{c_\mu}} \tag{2.4.61}$$

Substituting the last two expressions in the transport equation, it can be shown that:

$$c_{1\varepsilon} = c_{2\varepsilon} - \frac{k^2}{\sigma_\varepsilon} \frac{1}{\sqrt{c_\mu}} \tag{2.4.62}$$

And after evaluating the different parameters $c_{1\varepsilon} = 1.44$.

2.4.3.2.4 Higher-order models

One way of avoiding the modeling effort involved in estimating the Reynolds stresses (and avoiding the use of the Boussinesq hypothesis) consists in deriving an exact equation for these stresses (algebraic details may be found in Hinze, 1975). Obviously, this does not close the turbulence problem but it changes the unknowns to higher order moments such as $\overline{u_i' u_j' u_k'}$, which also need to be modeled giving origin to the name of such models. In specific applications where the momentum gradient fluxes oppose the mean velocity gradients (*e.g.*, Raupach & Thom, 1981 for more details), these models have proven to effectively describe the flow field.

2.4.3.3 Determination of the boundary conditions

Another aspect of the governing equations where experiments are needed is to provide boundary conditions and the associated closure relationships for the problem addressed analytically. The definition of the wall boundary conditions for 3D problems is often crucial for the "effectiveness" of the particular turbulence model that is adopted. The theoretical condition at a wetted solid boundary (non-moving boundary) is that of "no slip". This condition should be expressed as

$$\overline{u_1} = \overline{u_2} = \overline{u_3} = 0 \tag{2.4.63}$$

However, the turbulent nature of the flow makes the application of Equations (2.4.63) problematic. In fact, the extreme difficulties encountered when describing flow close to roughness elements, and the rapidity with which velocity varies in the "equilibrium" layer next to the bed, indicate that, for unidirectional flows, the velocity should be set to zero at an empirically-defined distance from the boundary. The ideal location of the boundary condition depends on the shear Reynolds number or the relative roughness, or both, depending on the turbulence regime (smooth, rough, or in transition). This approach ignores the actual flow structure (strongly unsteady and unstable) close to the roughness elements but can be effective for engineering purposes (Lyn, 2008).

For strongly three-dimensional configurations, Equations (2.4.63) are replaced by conditions for the velocity parallel to the bed at a specific normal distance from the bed. A logarithmic velocity profile is imposed, based on the assumption that even complex turbulence flows are governed by the same universal characteristics that result in the logarithmic distributions observed in steady, uniform flows. In the case of the $k - \varepsilon$ turbulence model, wall functions are commonly used to estimate the boundary conditions for k and ε near a boundary (Lopez & Garcia, 1997).

The condition of zero velocity normal to the boundary must also be imposed to satisfy mass conservation laws. The formulation of the boundary condition at the free surface $h(x_1, x_2, t)$ is relatively easier. The kinematic condition valid for the instantaneous flow is

$$\left(\frac{dF_s}{dt}\right)_{F=0} = \left\{\frac{d}{dt}[x_3 - h(x_1, x_2, t)]\right\}_{F=0} = 0 \tag{2.4.64}$$

where $F_s = x_3 - h$ is a function that has a value of zero at the free surface. Applying the chain rule to 2.4.64 and averaging the resulting equation yields:

$$\frac{\partial h}{\partial t} + \overline{u_1}\frac{\partial h}{\partial x_1} + \overline{u_2}\frac{\partial h}{\partial x_2} - \overline{u_3} = 0 \tag{2.4.65}$$

In addition to Equation (2.4.65), the condition of continuity pertains for flow stresses at the free surface

$$\underline{t} \cdot \underline{n} = \underline{n} \cdot \underline{\underline{T}} \cdot \underline{n} = 0 \tag{2.4.66}$$

where $\underline{n}$ is the tensor normal to the free surface, defined in the form

$$\underline{n} = \frac{\nabla F}{|\nabla F|} \tag{2.4.67}$$

$\underline{\underline{T}}$ is the tensor of the turbulent stresses whose components are generically expressed by Equation (2.4.31). Note that the continuity of the normal stresses is expressed by Equation (2.4.66) in absence of significant variations of the atmospheric pressure at the free surface; moreover, the free surface must not be characterized by curvatures that can induce significant effects due to surface tension.

If $\underline{\tau}_1$ and $\underline{\tau}_2$ indicate two vectors, normal to each other, on the plane tangent to the free surface, then

$$\underline{n} \cdot \underline{\underline{T}} \cdot \underline{\tau}_1 = \underline{n} \cdot \underline{\underline{T}} \cdot \underline{\tau}_2 = 0 \tag{2.4.68}$$

in the absence of any significant wind effect on the free surface.

Appropriate boundary conditions must also be set for temperature (or solute concentration). They are problem-specific but they can be generally expressed in terms of assigned distribution of temperature or specific heat flux (concentration or specific mass flux) at the initial time and at the boundary. Often, in hydraulic applications, the distribution of temperature (or concentration) is assigned at control sections, and the thermal (or mass) flux is practically zero along the remaining portion of the boundary.

2.4.3.4 One-dimensional shallow-water flow equations: the St. Venant equations

One-dimensional models can be used to effectively describe the behavior of many OCF. Such models treat OCF as a "current" in a "dominant direction" of flow, and are applicable for channels much longer than they are wide or deep. The length of a modeled channel is divided into segments called reaches. At the end of each reach, "cross sections" of the flow (*i.e.*, planes that are orthogonal to the longitudinal direction of the flow) are defined. For each cross section, cross-sectional averages of boundary conditions and dynamic quantities (flow area, roughness, velocity, momentum, energy, etc.) are considered. These quantities provide a synthetic description of the flow that is sufficient for many practical purposes. They are a function of the longitudinal coordinate, s, and temporal coordinate, t, of flow.

To apply a one-dimensional model to a particular flow, two conditions are required. First, the longitudinal axis of the flow must be characterized by small curvature, limiting the effects of secondary flows generated by the disequilibrium between the transversal pressure gradient and centrifugal force, and justifying the assumption that the flow has a "dominant direction." Second, spatial and temporal variations of the cross-section shape must be sufficiently gradual, so that the assumption of quasi-one-dimensional flow is not violated. Under such conditions, selection of a "dominant direction" of flow, and the longitudinal axis, is relatively straightforward, and the degree of arbitrariness of its selection has a negligible impact on the solution.

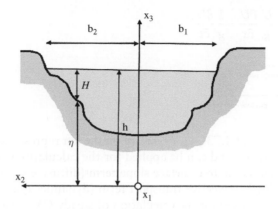

Figure 2.4.8 Definition diagram for the OCF cross section.

Figure 2.4.8 defines the cross section of a typical one-dimensional flow, and indicates the depth of the cross section, H, and the bed elevation, η. Integrating the continuity equation in the transverse direction, gives

$$\int_{-b_1}^{b_2} \frac{\partial H}{\partial t} dx_2 + \int_{-b_1}^{b_2} \frac{\partial (HU_1)}{\partial x_1} dx_2 + \int_{-b_1}^{b_2} \frac{\partial (HU_2)}{\partial x_2} dx_2 = 0 \tag{2.4.69}$$

which reduces to the following continuity equation for 1D open-channel flows.

$$\frac{\partial A}{\partial t} + \frac{\partial Q}{\partial x_1} = 0 \tag{2.4.70}$$

where A is the cross section area of the flow which is a function of the longitudinal position x ($A(x,t)$) and Q is the flow rate (also $Q(x,t)$).

Once again, assuming no lateral inflows and no other external force other than gravity, the momentum equation can be written as (*e.g.*, Sturm, 2011):

$$\frac{\partial Q}{\partial t} + \frac{\partial (QU + gAH)}{\partial x} = gA(S - S_f) \tag{2.4.71}$$

where S is the bed slope, S_f is the friction slope ($S_f = \tau_0/(\rho g R_h)$, where R_h is the hydraulic radius).

For a wide rectangular channel, Equation (2.4.71) can be written as:

$$\frac{\partial U}{\partial t} + g \frac{\partial}{\partial x}\left(\frac{U^2}{2g} + H\right) = g(S - S_f) \tag{2.4.72}$$

The pair of conservation equations, Equations (2.4.70) and (2.4.71) is known in OCF literature as the "Saint Venant Equations" in honor of the French mathematician who first introduced them (De Saint Venant, 1871). Equation (2.4.72) is also called *dynamic wave equation* and is widely applied in the literature for one dimensional analysis of OCF. This momentum equation can be written as:

$$\underbrace{\underbrace{\underbrace{\underbrace{S_f = S}_{\substack{\textit{Steady uniform} \\ \textit{flow}}} - \frac{\partial H}{\partial x} - \frac{U}{g}\frac{\partial U}{\partial x}}_{\substack{\textit{Steady nonuniform / gradually varied} \\ \textit{flow}}} - \frac{1}{g}\frac{\partial U}{\partial t}}_{\substack{\textit{Unsteady nonuniform / gradually varied} \\ \textit{flow}}}}$$

(2.4.73)

As is shown in Equation (2.4.73), the first term on the left represents the channel slope (kinematic wave equation) and can be applied for the calculations of steady, uniform OCF. The second term is the free surface slope term (diffusive wave equation) which combined with the third convective transport term (dynamics quasi-steady term) and the channel slope can be used for the calculation of steady GVF. The forth term is the local acceleration term which together with the third one are inertial terms of the momentum equations that complete the dynamic wave equations. The full version of Equation (2.4.73) can be applied for unsteady GVF and is rewritten below (Yen, 1973):

$$\underbrace{\frac{1}{g}\frac{\partial U}{\partial t} + \underbrace{\frac{U}{g}\frac{\partial U}{\partial x} + \underbrace{\frac{\partial H}{\partial x} = \underbrace{S}_{\substack{\textit{kinematic} \\ \textit{wave}}} - S_f}_{\substack{\textit{diffusive} \\ \textit{wave}}}}_{\substack{\textit{dynamic} \\ \textit{quasi-steady} \\ \textit{wave}}}}_{\substack{\textit{dynamic} \\ \textit{wave}}}$$

(2.4.74)

2.4.4 Secondary flows in open-channel flows

2.4.4.1 Secondary flows classifications

Secondary flows play an important role in altering the patterns of streamwise velocities, boundary shear stresses, turbulence, mixing, reaeration and sediment transport in OCF. Secondary flows fall into two categories. The first category, based on the mechanics-based classification first proposed by Prandtl (1926), is usually called Prandtl's secondary currents of the first kind. They are due to the mean streamwise vorticity being enhanced through vortex stretching. This kind of flow is common in river bends and meandering rivers in which the centrifugal forces on the flow. Centrifugal forces may produce super-elevation, in which the water surface at the outer bank is elevated with an accompanying lowering at the inner bank. A characteristic example of the first kind secondary flows as well as the super-elevation caused by the centrifugal force can be seen in Figure 2.4.9. Research has also shown that even the fairly simple case of a uniform, straight, channel flow can present cross sectional non uniformities in the principal velocity as a result of secondary currents (Prandtl, 1952; Perkins, 1970; Bradshaw, 1987; Tominaga *et al.*, 1989; Nezu & Nakagawa, 1993; Tamburrino & Gulliver, 1999; Rodriguez & Garcia, 2008; Nikora & Roy 2012). These secondary currents originate at the channel boundaries due to turbulence anisotropy and are referred to as Prandtl's secondary currents of the second kind (Figure 2.4.10). Prandtl's secondary currents of the second kind are also known as "corner flows" based on the analysis of non-circular duct flows (Schlichting,

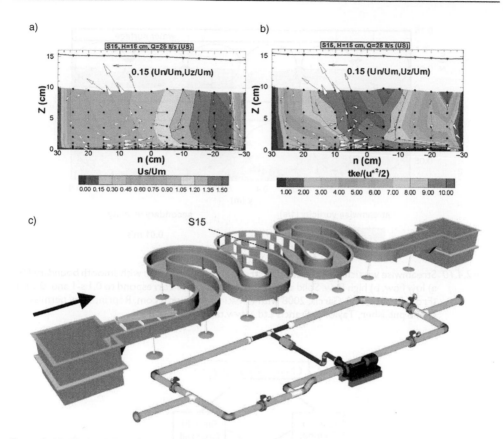

Figure 2.4.9 Illustration of first kind secondary flows (adapted from Abad & Garcia, 2009): a) measurements of normalized velocity components observed in the Kinoshita meandering flume. Cross section S15 is located at the apex of the middle bend of three upstream skewed meanders. Velocity is shown in terms of streamwise (s), transverse (n), and vertical (z) components; these velocity components are normalized by the mean cross-sectional velocity in the streamwise (s) direction, U_m. Colored contours show normalized streamwise velocity (U_s/U_m), while the vectors show the normalized transversal (U_n/U_m) and normalized vertical (U_z/U_m) velocity components. The scatter points (triangles) at the top of each plot define the measured water surface elevation. b) Measurements of normalized turbulent kinetic energy (tke/u*2/2) are plotted as colored contours. Vectors show the normalized transversal (U_n/U_m) and normalized vertical (U_z/U_m) velocity components. c) Water recirculation system of the Kinoshita flume with flow direction and bisecting line at S15. Shown at the top is the open-channel used for the experiment and at the bottom is the recirculation system.

1979). These secondary flows have velocities on the order of 5 % of the mean streamwise velocity and their size scales with the flow depth (Nezu & Nakagawa, 1993). They originate at the channel boundaries where the wall and bottom roughness can affect the relative strength of the vortices (Naot, 1984). Thus, while turbulence in general acts as a dissipative mechanism for secondary currents of the first kind, it can generate secondary currents of the second kind.

Bradshaw (1987) introduced the first topological classification of secondary flows as follows:

1 Non-helical cross-flows
2 Helical flows.

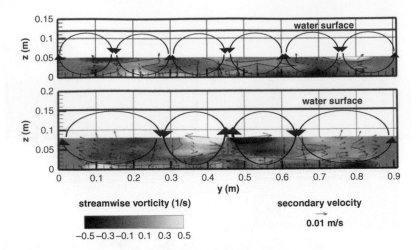

Figure 2.4.10 Streamwise vorticity distribution is a straight laboratory flume with smooth boundary for: a) low flow, b) high flow. Solid and dashed contour lines correspond to 0.1 s-1 and -0.1 s-1 (From Rodriguez & Garcia, 2008 with www.informaworld.com. Reprinted by permission of the publisher, Taylor & Francis Ltd, www.tandfonline.com).

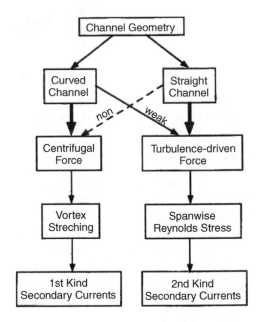

Figure 2.4.11 Classification of secondary currents (Nezu, 2005 with permission from ASCE).

Nezu (2005) proposed the secondary flow classification illustrated in Figure 2.4.11. More recently, Nikora and Roy (2012) combined Bradshaw's and Prandtl (1926) classifications to conclude that it is possible to distinguish at least the following four kinds of secondary flows:

1 Prandtl's first kind of cross flow (non-helical);
2 Prandtl's second kind of cross-flow (non-helical);
3 Prandtl's first kind of helical flow; and,
4 Prandtl's second kind of helical flow.

2.4.4.2 Origin of secondary flows in channels: vorticity conservation

The investigation of the details of the secondary flows are most often made using the RANS equations for vorticity provided in Section 2.3.1, Volume I. A more complex form of Equation (2.3.5) accounting for additional relevant processes but eliminating the pressure term can be written for the streamwise component of the time-averaged vorticity, ω_1, through cross differentiation (Nezu, 1994):

$$\underbrace{\frac{\partial \overline{\omega_1}}{\partial t} + \overline{u_1}\frac{\partial \overline{\omega_1}}{\partial x_1} + \overline{u_2}\frac{\partial \overline{\omega_1}}{\partial x_2} + \overline{u_3}\frac{\partial \overline{\omega_1}}{\partial x_3}}_{(1)} = \underbrace{\nu\nabla^2\overline{\omega_1}}_{(2)} + \underbrace{\left(\frac{\partial \overline{u_1}}{\partial x_1}\overline{\omega_1} + \frac{\partial \overline{u_1}}{\partial x_2}\overline{\omega_2} + \frac{\partial \overline{u_1}}{\partial x_3}\overline{\omega_3}\right)}_{(3)} +$$

$$+ \underbrace{\frac{\partial^2}{\partial x_1 \partial x_2}\left(\overline{u_1'^2} - \overline{u_3'^2}\right)}_{(4)} + \underbrace{\frac{\partial^2}{\partial x_2^2}(-\overline{u_2'u_3'}) - \frac{\partial^2}{\partial x_3^2}(-\overline{u_2'u_3'})}_{(5)} + \underbrace{\frac{\partial}{\partial x_1}\left[\frac{\partial}{\partial x_2}\left(-\overline{u_1'u_3'}\right) - \frac{\partial}{\partial x_3}\left(-\overline{u_1'u_2'}\right)\right]}_{(6)}$$

$$(2.4.75)$$

where $\overline{\omega_1} \equiv \dfrac{\partial \overline{u_3}}{\partial x_2} - \dfrac{\partial \overline{u_2}}{\partial x_3}$, $\overline{\omega_2} \equiv \dfrac{\partial \overline{u_1}}{\partial x_3} - \dfrac{\partial \overline{u_3}}{\partial x_1}$ and $\overline{\omega_3} \equiv \dfrac{\partial \overline{u_2}}{\partial x_1} - \dfrac{\partial \overline{u_1}}{\partial x_2}$.

Terms (1) and (2) represent the advection and viscous diffusion of ω_1, term (3) represents possible amplification of ω_1 by vortex stretching and terms (4), (5) and (6) are relevant to the Reynolds stresses and the turbulence of the flow. The classification of Figure 2.4.11 was done on the basis of term (3), which represents the centrifugal force along the curvilinear axes. For the case of uniform flow in straight channels, term (3) equals zero as $\partial \overline{u_1}/\partial x_1 = 0$ and $(\partial \overline{u_1}/\partial x_2)\overline{\omega_2} = -(\partial \overline{u_1}/\partial x_3)\overline{\omega_3}$; however secondary currents can be still generated by terms (4), (5) and (6), which take non-zero values only in case of nonhomogeneous and anisotropic turbulence.

In OCF, turbulence anisotropy is present at all flow boundaries (bottom, side-walls and free-surface). Side-wall effects are more important for the case of narrow channels. The distinction between wide and narrow channels is arbitrary and depends on the width to depth ratio (b/h). Thus, channels with $b/h < 5$ are considered to be narrow while $b/h > 10$ corresponds to wide channels. Narrow channels have stronger secondary currents which cause the maximum velocities to occur below the free surface, a phenomenon called *velocity dip* (Chow, 1959; Gibson, 1909).

Naot and Rodi (1982) showed that the secondary currents close to walls for the case of rectangular open channels are different from those of a square duct. In a square duct, the secondary currents spans out from the center toward the corner forming a pair of vortices at each corner that convey the flow back into the main stream while in OCF the secondary currents form a surface vortex paired with a relatively weaker bottom vortex (Figure 2.4.12). The roughness of the wall and the bed may affect the relative strength of the vortices. Wide channels may have well-defined secondary currents in the region

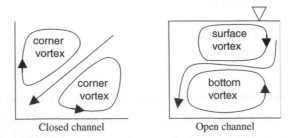

close to the walls. However, the strength of secondary currents is usually diminished close to the channel centerline due to the absence of spanwise variations in bed shear stress (Nezu & Rodi, 1985). Nevertheless, the bed shear stress may still show periodic variations due to the presence of flow instabilities, flow separation and interactions with a mobile-bed that can generate stable secondary circulation patterns, even in wide channels (McLelland *et al.*, 1999).

2.4.4.3 Analytical models for secondary currents

As mentioned above, the 3D flow patterns occurring in straight channels are due to secondary currents. Einstein and Li (1958) proposed to study different mechanisms producing OCF secondary circulation using the streamwise vorticity $\overline{\omega_1} \equiv \frac{\partial u_3}{\partial x_2} - \frac{\partial u_2}{\partial x_3}$ as a starting point. For straight uniform flow in an infinitely wide channel $\overline{\omega_1}$ is purely induced by turbulence, hence the following relation holds (Ikeda, 1981):

$$\frac{\partial^2}{\partial x_2 x_3}\left(u_2'^2 - u_3'^2\right) = \left(\frac{\partial^2}{\partial x_2^2} - \frac{\partial^2}{\partial x_3^2}\right)\overline{u_2' u_3'} \tag{2.4.76}$$

where the production of $\overline{\omega_1}$ by turbulence anisotropy (left hand side) is balanced by the suppression of $\overline{\omega_1}$ by the Reynolds stresses $\overline{u_2' u_3'}$. Using an eddy viscosity model the higher order terms $\overline{vw}$ can be approximated as:

$$-\overline{u_2' u_3'} = v_{T23}\left(\frac{\partial \overline{u_3}}{\partial x_2} + \frac{\partial \overline{u_2}}{\partial x_3}\right) \tag{2.4.77}$$

with v_{T23} being the *y-z* component of the eddy viscosity tensor. For the vorticity generation term, Ikeda (1981) used the following equation assuming a linear distribution of the Reynolds stresses $\overline{u_1' u_3'}$ (see Equation (2.4.2)):

$$\frac{u_2'^2 - u_3'^2}{u_*^2} \cong -\frac{\overline{u_1' u_3'}}{u_*^2} = 1 - \frac{x_3}{h} \tag{2.4.78}$$

Ikeda (1981) solved theoretically the set of Equations (2.4.76) and (2.4.78) using a depth-averaged eddy viscosity $v_{T23} = \kappa u_* h/6$ and a sinusoidal spanwise distribution of the bottom shear stress (which enters the equations through its surrogate the shear velocity, u_*), as follows,

$$\frac{\tau}{\langle \tau \rangle} = 1 + \delta_{\sin} \cos \left(\frac{k_b x_2}{h} \right) \tag{2.4.79}$$

where δ is the amplitude of the sinusoidal distribution and k_b is the wave number. For the dominant cells ($k_b = \pi$), Ikeda obtained a pattern of cellular secondary circulation given by:

$$\frac{\overline{u_2}}{\langle u_* \rangle} = \frac{6\delta}{\kappa \pi^2} \sin \frac{\pi x_2}{h} \left[\frac{2}{\pi} \cos \frac{\pi x_3}{h} - \left(2 \frac{x_3}{h} - 1 \right) \sin \frac{\pi x_3}{h} \right] \tag{2.4.80a}$$

$$\frac{\overline{u_3}}{\langle u_* \rangle} = \frac{6\delta}{\kappa \pi^2} \cos \frac{\pi x_2}{h} \left[\left(2 \frac{x_3}{h} - 1 \right) \cos \frac{\pi x_3}{h} + 1 \right] \tag{2.4.80b}$$

Equations (2.4.80a) and (2.4.80b) describe circular counter rotating cells of diameter h. Although the resulting cells correspond to an idealized case, they reproduce key elements observed in boundary layers, closed conduits and open channel flows. Of particular importance is the correlation between bottom shear stresses and the vertical velocity component given by a shear-stress increase in regions of downwelling and a decrease in regions of upwelling, which is in agreement with the ideas of Perkins (1970) and Townsend (1976) for boundary layers and experimental evidence in closed flows (Nezu & Nakagawa, 1984) and open-channel flows (Nezu & Rodi, 1985). Note that a number of simplifications have been made in order to arrive at the solution and perhaps the most drastic is the feedback effect of secondary circulation on the wall shear stress (Nezu & Nakagawa, 1993), which could provide a mechanism for the self-sustainability of secondary currents. That feedback is through the Reynolds stresses $\overline{u_1' u_3'}$, which are influenced by turbulence anisotropy as shown by Equation (2.4.78). In fact, $\overline{u_1' u_3'}$ deviates from Equation (2.4.77) in the presence of secondary currents. The deviation is such that points in the upflow regions have higher values of $\overline{u_1' u_3'}$ than those predicted by Equation (2.4.77), whereas points in the downflow regions have lower values (Nezu & Nakagawa, 1993). More sophisticated models exist for the description of turbulence-induced secondary circulation, but they rely on semi-empirical equations for the transverse Reynolds stresses (e.g., Tominaga et al., 1989; Jin et al., 2004).

Both laboratory and field measurements of in the presence of secondary flows are extremely challenging (Julien, 2002; Rodriguez & Garcia, 2008; Abad & Garcia, 2009). However, the experimental techniques presented later in this handbook should benefit from the material presented above. In particular, the interpretation of flow observations made with acoustic technologies should account for the potential presence of secondary flows and other 3D flow patterns (Muller et al., 2007).

2.5 COMPLEX FLOWS

2.5.1 Mobile boundary channels and sediment transport

2.5.1.1 Introduction to sediment transport and related phenomena

The experiments with consideration of sediment transport are among the most complex that a hydraulic experimentalist has to face. The analytical considerations provided in this section assembles the relevant and essential information that one has to use in order to conduct this type of complex experiments. In what follows, the main focus is on the special case when OCFs are transporting sediment throughout the water column resulting in bed morphology variations due to the strong and complex interactions between flow and sediment. The field of "sediment transport" might just as well be called "transport of granular particles by fluids." As such, it embodies a type of two-phase flow, where one phase is fluid and the other phase is solid. The prototype for the field condition is the natural river. Here, the fluid phase is river water, and the solid phase is sediment grains, *e.g.*, quartz sand.

The most common modes of sediment transport in rivers are those of bedload and suspended load. In the former case, the particles roll, slide, or saltate over each other, never deviating too far above the bed. In the latter case, the fluid turbulence comes into play carrying the particles well up into the water column. In both cases, the driving force for sediment transport is the action of gravity on the fluid phase; this force is transmitted to the particles via drag. Whether the mode of transport is saltation or suspension, the volume concentration of solids anywhere in the water column tends to be rather dilute in the case of rivers. As a result, it is generally possible to treat the two phases separately. The solid phase can vary greatly in size, ranging from clay particles to silt, sand, gravel, cobbles, and boulders. Rock types can include quartz, feldspar, limestone, granite, basalt, and other less common types such as magnetite. The fluid phase can, in principle, be almost anything that constitutes a fluid. In the geophysical sense, however, the two fluids of major importance are water and air.

Another mode of sediment transport, covered in less detail in this section, is called "washload". The floodplains of most sand-bed rivers often contain copious amounts of silt and clay finer than about 50–60 μm. This material is often called "washload" because it moves through the river system without being deposited in the bed in significant quantities. Increased washload does not increase deposition on the bed, and decreased washload does not increase erosion, because it is transported at well below power stream capacity. This is not meant to imply the washload does not interact with the river system. Washload in the water column exchanges with the banks and the floodplain rather than the bed. Greatly increased washload, for example, can lead to thickened floodplain deposits with a consequent increase in bankfull channel depth. The washload is controlled by land surface erosion (rainfall, vegetation, land use) and not by channel bed erosion. However, cohesive stream banks can contribute to the washload during bank full flow events. Washload often plays an important role in the sustainability of deltaic systems such as the Mississippi River delta, since it's the main source of sediment supply.

The study of the movement of grains under the influence of fluid drag and gravity constitutes a fascinating field in its own right. The subject becomes even more interesting when one considers the link between sediment transport and morphology. In the laboratory, the phenomenon can be studied in the context of a variety of facilities, such as flumes and wave tanks, specified by the experimentalist (see also Section 4.3 in Volume I). In field conditions, however, the fluid-sediment mixture shapes continuously the river morphology. This degree of freedom opens up a variety of intriguing possibilities.

Depending on the existence of lack thereof of a viscous sublayer and the relative importance of bedload versus suspended load, a variety of rhythmic structures can form on a river or channel bed. These include ripples, dunes, antidunes, and alternate bars. The first three of these can have a profound effect on the resistance to flow offered by the river bed. Thus, they act to control river depth. River banks themselves can also be considered to be a self-formed morphological feature, thus specifying the entire container. The container itself can deform in plan. Alternate bars cause rivers to erode their banks in a rhythmic pattern, thus allowing for the onset of meandering. Fully-developed river meandering implies an intricate balance between sediment erosion and deposition. If a stream is sufficiently wide, it will braid rather than meander, dividing into several intertwining channels.

After considering some of the fundamental sediment and fluid properties, an introduction is made of important phenomena related to sedimentation mechanics, such as the initiation of sediment motion, bed forms and some more special topics related to sediment stratification effects, non-equilibrium sediment suspension, sediment mixtures and unsteady flows. The material presented should provide useful background and guidance for conducting both laboratory and field measurements in channel, streams and rivers.

2.5.1.2 *Sediment and fluid properties relevant for sediment transport*

It helps to consider the sediment transport process as a general two phase problem (fluid phase – solid phase). In this two-phase flow, sediment particles are subject to gravity, friction and fluid forces (drag and lift). In general, the first two forces hold the grains in place while fluid forces try to move the particles. The motion of each sediment grain is affected by different parameters which can be categorized as (Garcia, 1999):

1) parameters related to the randomness of the flow and the grain placement
2) parameters of fluid and sediment that affect the force balance on the grain

Flow turbulence and its "random" spatial and temporal fluctuations introduce a stochastic characteristic on sediment transport. Similarly, the placement of a single particle (*e.g.*, whether it is exposed to the flow or it is hidden by other particles) changes the probability of grain movement. Although these random processes play important role sediment transport they can be neglected for many practical computations in which the flow effect is expressed through the flow shear stress.

The solid phase of the problem embodied in sediment transport can be any granular substance. In terms of engineering application, however, the granular substance in question typically consists of fragments ultimately derived from rocks—hence the name "sediment" transport. The properties of these rock-derived fragments, taken singly or in groups of many particles, all play a role in determining the transportability of the grains under fluid action. The important properties of groups of particles include porosity and size distribution. The most common rock type one is likely to encounter in the river or coastal environment is quartz. Quartz is a highly resistant rock and can travel long distances or remain in place for long periods of time without losing its integrity. Another highly resistant rock type that is often found together with quartz is feldspar. Other common rock types include limestone, basalt, granite, and more esoteric types such as magnetite. Limestone is not a resistant rock; it tends to abrade to silt rather easily. Silt-sized limestone particles are susceptible to solution unless the water is sufficiently buffered. As a result, limestone is not typically found to be a major component of sediments at locations distant from its source. On the other hand, it can

Table 2.5.1 Specific gravity of rock types and artificial material

Rock type or material	Specific gravity ρ_s/ρ
Quartz	2.60 ~ 2.70
Limestone	2.60 ~ 2.80
Basalt	2.70 ~ 2.90
Magnetite	3.20 ~ 3.50
Bakelite	1.30 ~ 1.45
Coal	1.30 ~ 1.50
Ground walnut shells	1.30 ~ 1.40
PVC	1.14–1.25

often be the dominant rock type in mountain environments.Sediments in the fluvial or coastal environment in the size range of silt, or coarser, are generally produced by mechanical means, including fracture or abrasion. The clay minerals, on the other hand, are produced by chemical action. As a result, they are fundamentally different from other sediments in many ways. Their ability to absorb water means that the porosity of clay deposits can vary greatly over time. Clays also display cohesiveness, which renders them more resistant to erosion.

Sediment specific gravity is defined as the ratio between the sediment density ρ_s and the density of water ρ. Some typical specific gravities for various natural and artificial sediments are listed in Table 2.5.1.

In the laboratory, it is often of value to employ light-weight "model" sediment. In order to see the utility of this, it is useful to consider a movable-bed scale model of an actual river (ASCE, 2000). Consider a reach of the Minnesota River, Minnesota, with a bankfull width of 90 m, a bankfull depth of 4 m, a streamwise slope of 0.0002, and a median sediment size D_{50} of 0.5 mm. The reach is scaled down by a factor of 100 so as to fit into a typical laboratory model basin, resulting in a bankfull width of 90 cm and a bankfull depth of 4 cm. In an undistorted model, slope remains constant at 0.0002. If the sediment employed in the model were to be the same as in the field, it would most likely not move at all in the scale model. Carrying the analogy to its logical conclusion, it would be as if the sediment in the field Minnesota River had a median size of 0.5 mm x 100 = 0.5 m, i.e., boulders. It should be clear that, in this case, the field sediment cannot be employed directly in the model. The obvious alternative is to scale down sediment size by the same factor as all other lengths, i.e., by a factor of 100. This would yield a size of 5 μ, which is so close to the clay range that it can be expected to display some kind of pseudo-cohesiveness. In addition, viscous effects are expected to be greatly exaggerated due to the small size. The net result is model sediment that is much less mobile than it ought to be and, in addition, behaves in ways radically different from the prototype sediment.

There are several ways out of this dilemma (see also Section 3.5 in Volume I). One of them involves using artificial sediment with a low specific gravity. Let ρ denote the density of water, and ρ_s denote the specific gravity of the material in question. The weight W of a particle of volume V_p is given by

$$W = \rho_s g V_P \qquad (2.5.1)$$

where g denotes the acceleration of gravity. Quartz, for example, is natural sediment with a specific gravity ρ_s/ρ near 2.65. If a grain of the same volume were modeled in the laboratory using crushed coal with a specific gravity of 1.3, it would follow from Equation (2.5.1) that the coal grain would be only 1.3/2.65 or 0.49 times the weight of the quartz grain. Rephrasing, the coal grain is 2.04 times lighter than the quartz grain, and thus, in some sense, twice as mobile.

In fact, the benefit of using lightweight material is much larger than this, because the effective weight determining the mobility of a grain is the submerged weight W_s, i.e., actual weight minus the buoyant force associated with the hydrostatic pressure distribution about the particle. That is,

$$W_S = (\rho_S - \rho)gV_P = \rho R g V_P \qquad (2.5.2)$$

where

$$R = \left(\frac{\rho_S}{\rho} - 1\right) \qquad (2.5.3)$$

denotes the "submerged specific gravity" of the sediment. Comparing coal and quartz again in terms of submerged weight, it is seen that,

$$\frac{(W_S)_{coal}}{(W_S)_{quartz}} = \frac{(R)_{coal}}{(R)_{quartz}} = \frac{0.30}{1.65} = 0.18$$

It follows that under water, the coal grain is 1/0.18 = 5.5 times lighter than a quartz grain of the same size. Lightweight model sediments are thus a very effective way of increasing mobility in laboratory experiments (Zwamborn, 1981; ASCE, 2000, p.105).

Sediment Size: the notation D is commonly used to denote sediment size, the typical units of which are millimeters (mm) for sand and coarser material or microns (μ) for clay and silt. Another standard way of classifying grain sizes is the sedimentological Φ scale, according to which

$$D = 2^{-\Phi} \qquad (2.5.4a)$$

Taking the logarithm of both sides, it is seen that

$$\Phi = -\log_2(D) = -\frac{\ln(D)}{\ln(2)} \qquad (2.5.4b)$$

Note that the size $\Phi = 0$ corresponds to $D = 1$ mm. The utility of the Φ scale becomes apparent when analyzing size distributions. The minus sign has been inserted in Equation (2.5.4b) simply as a matter of convenience to sedimentologists who are more accustomed to working with material finer than 1 mm rather than coarser material. The reader should always recall that larger Φ implies finer material. The Φ scale provides a very simple way of classifying grain sizes into the following size ranges in descending order: boulders, cobbles, gravel, sand, silt, and clay. This is illustrated in Table 2.5.2.

It should be noted that the definition of clay according to size ($D < 2\mu$) does not always correspond to the definition of clay according to mineral. That is, some clay mineral particles can be coarser than this limit, and some silt-sized particles produced

Table 2.5.2 Sediment grade scale (adapted from Vanoni, 1975)

Class Name	Size Range				Approximate Sieve Mesh Openings per Inch	
	Millimeters	Φ	Microns	Inches	Tyler	U.S. standard
Very large boulders	4096~2048			160~80		
Large boulders	2048~1024			80 ~ 40		
Medium boulders	1024~512			40~20		
Small boulders	512~256	-9~-8		20~10		
Large cobbles	256~128	-8~-7		10~5		
Small cobbles	128~64	-7~-6		5~2.5		
Very coarse gravel	64~32	-6~-5		2.5~1.3		
Coarse gravel	32~16	-5~-4		1.3~0.6	2~1/2	5
Medium gravel	16~8	-4~-3		0.6~0.3	5	10
Fine gravel	8~4	-3~-2		0.3~0.16	9	18
Very fine gravel	4~2	-2~-1		0.16~0.08	16	35
Very coarse sand	2.000~1.000	-1~0	2000~1000		32	60
Coarse sand	1.000~0.500	0~1	1000~500		60	120
Medium sand	0.500~0.250	1~2	500~250		115	230
Fine sand	0.250~0.125	2~3	250~125		250	
Very fine sand	0.125~0.062	3~4	125~62			
Coarse silt	0.062~0.031	4~5	62~31			
Medium silt	0.031~0.016	5~6	31~16			
Fine silt	0.016~0.008	6~7	16~8			
Very fine silt	0.008~0.004	7~8	8~4			
Coarse clay	0.004~0.002	8~9	4~2			
Medium clay	0.002~0.001		2~1			
Fine clay	0.001~0.0005		1~0.5			
Very fine clay	0.0005~0.00024		0.5~0.24			

by grinding can be finer than this. In general, however, the effect of viscosity makes it quite difficult to grind up particles in water to sizes finer than 2μ. In practical terms, there are several ways to determine grain size. The most popular way for grains ranging from $\Phi = 4$ to $\Phi = -4$ (0.0625 to 16 mm) is with the use of sieves. Each sieve has a square mesh, the gap size of which corresponds to the diameter of the largest sphere that would fit through. The grain size D so measured thus exactly corresponds to diameter only in the case of a sphere. In general, the sieve size D corresponds to the smallest sieve gap size through which a given grain can be fitted. For grains in the silt and clay sizes, many methods (hydrometer, sedigraph, etc.) are based on the concept of equivalent fall diameter. That is, the terminal fall velocity v_s of a grain in water at a standard temperature, is measured. The equivalent fall diameter D is the diameter of the sphere having exactly the same fall velocity under the same conditions. Sediment fall velocity is discussed in more detail below.

Sediment Porosity: The porosity λ_p quantifies the fraction of a given volume of sediment which is composed of void space. That is,

$$\lambda_p = \frac{\text{volume of voids}}{\text{volume of total space}} \tag{2.5.5}$$

If a given mass of sediment of known density is deposited, the volume of the deposit must be computed assuming that at least part of it will consist of voids. In the case of well-sorted sand, the porosity can often take values between 0.3 and 0.4. Gravels tend to be more poorly-sorted. In this case, finer particles can occupy the spaces between coarser particles so reducing the void ratio to as low as 0.2. So-called open-work gravels are essentially devoid of sand and finer material in their interstices; these may have porosities similar to sand. Freshly deposited clays are notorious for having high porosities. As time passes, the clay deposit tends to consolidate under its own weight so that porosity slowly decreases.

Fall Velocity: a fundamental property of sediment particles is their fall or settling velocity. The fall velocity of sediment grains in water is determined by their diameter and density, and by the viscosity of the water. Falling under the action of gravity, a particle will reach a constant, terminal velocity, once the drag equals the submerged weight of the particle. The relation for terminal fall velocity for a spherical particle in quiescent fluid, v_s, can be presented as

$$R_{f v_s} = \left[\frac{4}{3} \frac{1}{C_D(R_p)} \right]^{1/2} \tag{2.5.6a}$$

Where

$$R_{f v_s} = \frac{v_s}{\sqrt{gRD}}; R_p = \frac{v_s D}{v} \tag{2.5.6bc}$$

and the functional relation $C_D = f(R_p)$ denotes the drag coefficient for spheres (Rouse, 1938). Here g is the acceleration of gravity, $R = (\rho_s - \rho)/\rho$ is the submerged specific gravity of the sediment, and v is the kinematic viscosity of water. This relation is not very useful because it is not explicit in v_s; one must compute fall velocity by trial and error. One can use the equation for the drag coefficient C_D given below

$$C_D = \frac{24}{R_p}\left(1 + 0.152R_p^{1/2} + 0.0151R_p\right)$$

(2.5.7a)

And the following Reynolds number defined with sediment particle properties,

$$R_{ep} = \frac{\sqrt{gRD}D}{v}$$

(2.5.7b)

to obtain an explicit relation for fall velocity in the form of R_{fv_s} versus R_{ep}. Such diagram is presented in Figure 2.5.1, where the ranges for silt, sand, and gravel are plotted for a kinematic viscosity $v = 0.01$ cm^2/s (clear water at 20 degrees Centigrade) and a submerged specific gravity $R = 1.65$ (quartz). Notice that for fine silts, R_p is smaller than one and the drag coefficient given by Equation (2.5.4a) reduces to

$$C_D = \frac{24}{R_p}$$

(2.5.7c)

Substitution of (2.5.7c) into (2.5.3a) yields the well-known Stokes law for settling velocity of very fine particles,

$$v_s = \frac{gRD^2}{18v}$$

(2.5.8)

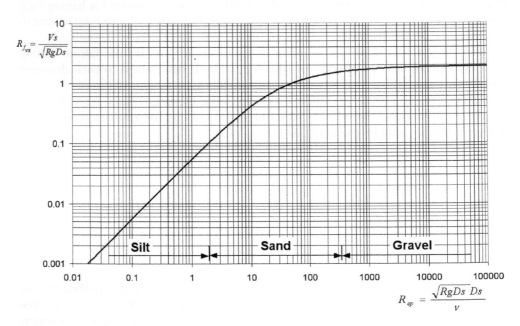

Figure 2.5.1 Diagram of R_{fv_s} versus R_{ep} calculated from the drag coefficient for spheres (Garcia, 2008 with permission from ASCE).

A good summary of relations for terminal fall velocity for the case of non-spherical (natural) particles can be found in Dietrich (1982). A simple fit to fall velocity data of natural sand particles has been proposed by Soulsby (1997), as

$$v_s = \frac{v}{D} \left[(10.36^2 + 1.049 \, D_*^3)^{1/2} - 10.36 \right] \qquad (2.5.9a)$$

where

$$D_* = \left[\frac{gR}{v^2} \right]^{1/3} \qquad (2.5.9b)$$

Equations very similar to (2.5.9a, 2.5.9b) have been proposed independently by Zanke (1977) and Van Rijn (1984b). A useful fit to estimate the kinematic viscosity of clear water both in the laboratory and in the field is provided by Raudkivi (1998), as

$$v = \frac{1.79 \; 10^{-6}}{1 + 0.03368 \, T + 0.00021 \, T^2} \; (m^2/s) \qquad (2.5.10)$$

Where T = temperature of the water in degrees centigrade ($^{\circ}C$).

Other useful relations to estimate sediment fall velocity can be found in Ahrens (2000, 2003), Chang and Liou (2001), and Cheng (1997). Jimenez and Madsen (2003) have used the work of Dietrich (1982) to advance an empirical formula to estimate settling velocity of natural particles which takes into account both the shape and the roundness of the particles.

At high concentrations the flows around adjacent settling grains interact resulting in a larger drag than for the same grain in isolation. This phenomenon is known as hindered settling and results in the hindered settling velocity, v_{sC}, for high sediment concentrations to be smaller that the fall velocity (v_s) at low sediment concentrations (less than 0.05 by volume). Applying reasoning similar to the one that led to Equation (2.5.9a), Soulsby (1997) proposed the following relation for the hindered fall velocity v_{sC} of grains in a dense suspension having concentration a volumetric sediment concentration C:

$$v_{sC} = \frac{v}{D} \left[(10.36^2 + 1.049 \, (1 - C)^{4.7} D_*^3)^{1/2} - 10.36 \right] \qquad (2.5.11)$$

which is valid for all values of D_* and C. When C = 0, Equation (2.5.11) reduces to Equation (2.5.9a).

2.5.1.3 Initiation of motion, bedload rates, and bedforms

When a granular bed is subjected to a turbulent flow, it is found that virtually no motion of the grains is observed at some flows, but that the bed is noticeably mobilized at other flows (Cheng & Chiew, 1998; Papanicolau et al., 2002, Nino et al., 2003). Literature reviews on incipient motion can be found in Miller et al. (1977), Lavelle and Mofjeld (1987) and Buffington and Montgomery (1997). Factors that affect the mobility of grains subjected to a turbulent flow include grain placement, fluid forces such as mean drag and lift, and turbulence intensity.

In the presence of turbulent flow, random fluctuations typically prevent the clear definition of a critical, or threshold condition for motion: the probability of grain movement is never precisely zero (Paintal, 1971; Graf & Paziz, 1977; Lopez & Garcia, 2001). However, it is possible to define a condition below which movement can be neglected for many practical purposes. For the case of a non-cohesive sediment grain lying in a bed of similar grains, there is a critical condition during which the fluid flow forces become just than the forces that hold the grain in place. This condition is usually called incipient motion threshold larger. Such force balance on single sediment particles has been studied by Ikeda (1982) based on the analysis by Iwagaki (1956) &Coleman (1967) and by Wiberg and Smith (1987), among others. These studies became the basis for the derivation of the so called *mechanistic or deterministic relationships for the threshold of motion* of sediment Using an Ikeda-Coleman-Iwagaki model (Garcia, 2008) calculate the critical dimensionless Shields stress that corresponds to the critical flow velocity needed for sediment particles motion:

$$\tau_c^* = \frac{4}{3} \frac{(\mu\cos\theta - \sin\theta)}{c_D + \mu c_L} \frac{1}{F^2(u_*D/v)} \qquad (2.5.12)$$

where θ is the downstream slope angle, μ is the coefficient of Coulomb friction which is defined as the ration between the tangential resistive forces and the downward normal force and it is a material property usually calculated as the tangent of the angle of repose φ (φ is near 30° for sand and it is gradually increases to 40° for gravels), C_D and C_L are the drag and lift coefficients and F is a function of $\frac{u_*D}{v}$ that represents the ratio between the effective fluid velocity that a particle experience and the friction velocity and it can be calculated using a continuous formula such as Equation (2.4.13) proposed by Swamee (1993).

The most relevant fact about the mechanistic approach to the problem of initiation of motion presented above, relates to the possibility of obtaining an explicit formulation of the relation explored by Shields with the help of dimensional analysis and experiments. Furthermore, Equation (2.5.12) makes it possible to visualize the sources of uncertainty (*i.e.*, angle of repose, drag and lift coefficients, particle location, etc.) and helps to understand why it is so difficult to characterize the threshold condition with a deterministic model (Neill & Yalin, 1969; Bettess, 1984; Lavelle & Mojfeld, 1987; Komar, 1996; Papanicolau *et al.*, 1999; Shvidchenko & Pender, 2000; Dancey *et al.*, 2002; Nino *et al.*, 2003). Interested readers can read details and the derivation of Equation (2.5.12) in the work by Ikeda (1982), Iwagaki (1956), Coleman (1967), Wiberg and Smith (1987) or in the review by Garcia (2008). The calculated critical stress from Equation (2.5.12) corresponds to a critical shear stress τ_c required for the initiation of sediment motion. Such critical bed shear stress required to initiate the motion of a given particle from rest, is commonly known as the critical Shields stress.

2.5.1.3.1 Modified Shields diagram

Shields (1936) conducted a set of pioneering experiments to elucidate the conditions for which sediment grains would be at the verge of moving. Shields' observations have become legendary in the field of sediment transport (Kennedy, 1995; Buffington, 1999, Garcia, 2000a). Shields deduced from dimensional analysis and fluid mechanical considerations that τ^* should be a function of shear Reynolds number $u_* D/v$, as implied by Equation (2.5.12). The Shields diagram is expressed by dimensionless combinations of critical shear stress τ_c, sediment and water specific weights γ_s and γ, respectively, sediment size D, critical shear velocity $u_{*c} = \sqrt{\tau_c/\rho}$ and kinematic viscosity of water v. These quantities can be expressed in any consistent set of units. Shields dimensionless parameters are related by a simple expression

$$\tau_c^* = \frac{\tau_c}{\rho g R D} = f\left(\frac{u_{*c} D}{v}\right) \tag{2.5.13}$$

The classic Shields diagram can be found in Vanoni (1975). Shields values of τ_c^* are used commonly to denote conditions under which bed sediments are stable but on the verge of being entrained. The curve in the Shields diagram was first introduced by Rouse (1939) while the auxiliary scale was proposed by Vanoni (1964) to facilitate the determination of the critical shear stress τ_c once the submerged specific gravity, the particle diameter D and the kinematic viscosity of water v are specified. Not all workers agree with the results given by the Shields curve. For example, some researchers, e.g., Neill (1968), give $\tau_c^* = 0.03$ for the dimensionless critical shear stress for values of $\text{Re}_* = u_* D/v$ in excess of 500 instead of 0.06.

The value of τ_c^* to be used in design depends on the particular case at hand. If the situation is such that grains that are moved can be replaced by others moving from upstream, some motion can be tolerated, and the values from the Shields curve may be used. On the other hand, if grains removed cannot be replaced as on a stream bank, the Shields value of τ_c^* are too large and should be reduced. From the work of Gessler (1970) and Neill and Yalin (1969) it is well known that Shields original values for initiation of motion of coarse material are too high and should be divided by a factor of 2 for engineering purposes.

The Shields diagram is sometimes not directly applicable because in order to find t_c, one must know the shear velocity $u_* = \sqrt{\tau_c/\rho}$ beforehand. The relation can be cast in explicit form by plotting τ_c^* versus R_{ep}, noting the internal relation

$$\frac{u_* D}{v} = \frac{u_*}{\sqrt{Rg D}} \frac{\sqrt{Rg D}\ D}{v} = (\tau^*)^{1/2} R_{ep} \tag{2.5.14}$$

where $R = \frac{\rho_s - \rho}{\rho}$ is the submerged specific gravity of the sediment.

Brownlie (1981) used the Shields data transformed by Vanoni (1964) with the help of Equation (2.5.14) to prepare the Modified Shields diagram shown in Figure 2.5.2. A useful fit to the Shields data plotted in the form suggested by Vanoni is given by Brownlie (1981, p.161):

$$\tau_c^* = 0.22\, R_{ep}^{-0.6} + 0.06 \exp\left(-17.77\, R_{ep}^{-0.6}\right) \tag{2.5.15a}$$

With this relation, the value of τ_c^* can be readily computed when the properties of the water and the sediment are given. As already mentioned, to be on the safe side the values given by Equation (2.5.15a) should be divided by 2 for engineering purposes, resulting in the following expression

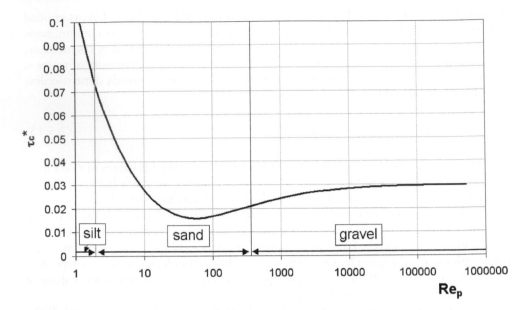

Figure 2.5.2 Modified Shields diagram (after Parker, 2005).

$$\tau_c^* = \frac{1}{2}\left[0.22R_{ep}^{-0.6} + 0.06\ exp\left(-17.77R_{ep}^{-0.6}\right)\right] \hspace{2cm} (2.5.15b)$$

This equation is plotted in the modified Shields diagram shown in Figure (2.5.2). With this relation, the value of t_c^* can be readily computed when the properties of the water and the sediment are given.

For fine-grained sediments, Shields observations and Equations (2.5.15a, 2.5.15b) do not provide realistic results. Mantz (1977) conducted a series of experiment and observed that for fine-grained sediments the critical shear stresses can be estimated with the following relationship,

$$\tau_c^* = 0.135R_{ep}^{-0.261} \hspace{2cm} (2.5.15c)$$

which is valid for the range $0.056 < R_{ep} < 3.16$. Equations (2.5.15b) and (2.5.15c) merge for a value of $R_{ep} = 4.22$.

In a study of temperature effects on initiation of motion, Taylor and Vanoni (1972) reported that small but finite amounts of fine-grained sediment were transported in flows with values of t_c^* well below those given by the Shields curve. They found that as the size of sediment grains decreases, the dimensionless critical shear stress increases more slowly than one would infer by extrapolating the Shields curve. Similar observations were made by Mantz (1977, 1980) but the most conclusive evidence for such behavior was provided by Yalin and Karahan (1979) through carefully conducted experiments. Yalin and Karahan (1979) compiled a substantial amount of data and conducted their own set of experiments with sand sizes ranging from 0.10 *mm* to 2.86 *mm*, for both laminar and turbulent flow conditions. They used glycerine in some of the experiments to increase the thickness of the viscous sublayer, thus making it possible to observe initiation of motion under laminar and turbulent flow conditions. They were able to elucidate the nature of transport inception conditions for a wide range of grain Reynolds number $Re_* = u_* D/v$. For $Re_* > 70$, hydraulically rough conditions, t_c^* takes a value of about 0.045. For values of $Re_* < 10$, the relation between t_c^* and Re_* depends on the flow regime, *i.e.*, whether the flow is laminar or turbulent.

Most of the work on initiation of motion has been done for uniform size sediment. One exception is the model advanced by Wiberg and Smith (1987). They derived an expression for the critical shear stress of noncohesive sediment using a balance of forces on individual particles very similar to the one shown above. For a given grain size and density, the resulting equation depends on the near-bed drag force, lift force to drag force ratio, and particle angle of repose. They were able to reproduce the observations of Shields for uniform size sediments as well as initiation of motion in the case of sediment mixtures. They found that for mixed grain sizes the initiation of motion also depends on the relative protrusion of the grains into the flows and the particle angle of repose.

2.5.1.3.2 Shields-Vanoni-Parker (SVP) river regime diagram

Alluvial rivers that are free to scour and fill during floods can broadly be divided into two types: sand bed streams and gravel bed streams. Sand bed streams typically have values of median bed sediment size between 0.1 *mm* and 1 *mm*. The sediment tends to be relatively well sorted, with values of geometric standard deviation of the bed sediment size varying from 1.1 to 1.5. Gravel bed streams typically have values of median size of the bed sediment exposed on the surface of 15 *mm* to 200 *mm* or larger; the substrate is usually finer by a factor of 1.5 to 3. The geometric standard deviation of the substrate sediment size is typically quite large, with values in excess 3 being quite common. Although gravel and coarser material constitute the dominant sizes, there is usually a substantial amount of sand stored in the interstices of the gravel substrate.

Two dimensionless parameters provide an effective delineator of rivers into the above two types (Garcia, 1999, 2000a). The first of these is the dimensionless Shields stress for steady, uniform, bank-full flow conditions, defined as:

$$\tau^* = \frac{\tau_o}{\rho g R D} = \frac{HS}{RD} \qquad (2.5.16)$$

where τ_o is the bed shear stress, g is the gravitational acceleration, ρ is the water density, $R = (\rho_s/\rho - 1)$ is the submerged specific gravity of the sediment, D is the sediment size (mean diameter), H is the flow depth and S is the stream slope which for steady, uniform flow is the same as the energy gradient. The second of these is a particle Reynolds number R_{ep} defined as:

$$R_{ep} = \frac{\sqrt{gRD}D}{v} \qquad (2.5.17)$$

where v is the kinematic viscosity of water. This second parameter can be considered as a dimensionless surrogate for grain size. Figure 2.5.3 shows a plot of the values of the Shields stress evaluated at bankfull flow versus particle Reynolds number for six sets of field data: a) gravel bed rivers in Wales, UK (Wales); b) gravel bed rivers in Alberta,

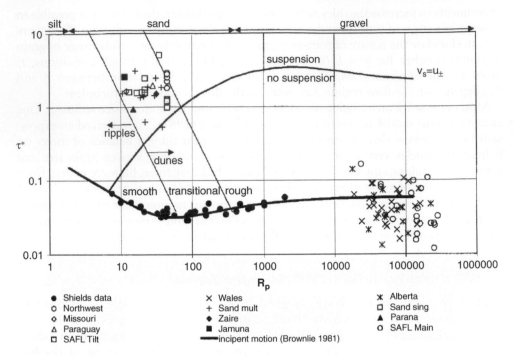

Figure 2.5.3 Shields-Vanoni-Parker (SVP) river sedimentation regimediagram (Garcia, 2000 with permission from ASCE).

Canada (Canada); c) gravel bed rivers in the Pacific Northwest, USA (Pacific); d) single-thread sand streams (Sand sing); e) multiple-thread sand streams (Sand mult); f) large sand-bed rivers (Parana, Missouri, etc.); and g) large-scale laboratory experiments conducted at St Anthony Falls Laboratory (SAFL), University of Minnesota (Parker *et al.*, 1998). There are three curves in the diagram that make it possible to know, for different values of τ^* and R_{ep}, if a given bed sediment will go into motion, and if this is the case whether or not the prevailing mode of transport will be in suspension or as bedload. The diagram also can be used to predict what kind of bedforms can be expected. For example, ripples will develop in the presence of a viscous sublayer and fine-grained sediment. If the viscous sublayer is disrupted by coarse sediment particles, then dunes will be the most common type of bedform. The Shields regime diagram also shows a clear distinction between the conditions observed in sand-bed rivers and gravel-bed rivers at bankfull stage. As one would expect, in gravel-bed rivers sediment is predominantly transported as bedload, while in sand-bed streams sediment is transported mainly in suspension. The regime diagram is valid for steady, uniform, turbulent flow conditions, where the bed shear stress can be estimated with $\tau_o = \rho g H S$ (Equation (2.4.1)). The ranges for silt, sand, and gravel are also included.

2.5.1.3.3 Movable bed laboratory models

The SVP regime diagram also shows a very clear distinction between the conditions observed in sand-bed and gravel-bed rivers at bank-full stage. If one wanted to model in the laboratory sediment transport in rivers, the experimental conditions would be quite different depending on the river type in question (Garcia, 2000adelinea). In order to satisfy similarity in a small-scale, river model, it would be necessary to satisfy the following identities (*e.g.*, Zwamborn, 1981),

$$\tau^*|_{\text{mod } el} = \tau^*|_{prototype} \tag{2.5.18a}$$

$$R_{ep}|_{\text{mod } el} = R_{ep}|_{prototype} \tag{2.5.18b}$$

for bankfull flow conditions. In most movable-bed model, Froude similarity is enforced and Equation (2.5.18a) is used to achieve sediment transport similarity. However, sediment transport conditions and the associated bed morphology in a model seldom precisely reproduce prototype conditions since the second condition given by Equation (2.5.18b) is rarely satisfied. Thus, in order to reduce scale effects it is common practice the use light-weight material in an attempt to reproduce prototype conditions in small-scale models. However, this does not imply that the bedforms and general morphology observed in the model will be the same as those in prototype. The SVP diagram provides a tool to quickly determine potential scale effects in movable-bed model studies, by simply plotting the values of $(\tau^*; R_{ep})$ for model and prototype conditions in Figure 2.5.4. As discussed in Henderson (1966, p.504), and pointed out earlier by Bogardi (1959), the condition given by Equation (2.5.18b) can be relaxed for suffi-ciently large values of R_{ep} (*i.e.*, hydraulically rough flow) in both model and prototype. It is clear from Figure 2.5.4, that this would be possible only for the case of gravel bed streams.

2.5.1.3.4 Bedload transport

Bed load particles, roll, slide and saltate at the vicinity of the bed. Most of the models for the estimation of bed load transport under uniform equilibrium conditions can be categorized in those who follow the approach proposed by Einstein (1950) and those based on the work by Bagnold (1956). Bagnold (1956) defines the bedload transport as that in which the successive contacts of the particles with the bed are strictly limited by the effect of gravity, while the suspended load transport is defined as that in which the excess weight of the particles is supported by random successions of upward impulses induced by turbulence eddies. Einstein (1950), however, elucidates a somewhat differ-ent view of the phenomenon. Einstein defines bedload transport as the transport of sediment particles in a thin layer about two particle diameters thick just above the bed by, sliding, rolling, and making jumps with a longitudinal distance of a few particle diameters. The bedload layer is considered to be a layer in which mixing due to turbulence is so small that it cannot directly influence the sediment particles, and therefore suspension of particles is impossible in the bedload layer. Further, Einstein assumes that the average distance traveled by any bedload particle (as a series of successive movements) is a constant distance of about 100 particle diameters, indepen-dent of the flow condition, transport rate, and bed composition. In Einstein's view, saltating particles belong to the suspension mode of transport, because the jump heights and lengths of saltating particles are larger than a few grain diameters. On the other hand, Bagnold (1956) regards saltation as the main mechanism responsible for bedload transport.

The Einstein-based models require the determination of a *pick-up function* based on which an entrainment rate of particles into bedload can be defined as a function of bed shear and other parameters and an average length of bedload particles travel from entrainment to deposition. A characteristic of those models is that there is no concept of a threshold/critical value of the bed shear that must be exceeded to have sediment transport. Examples of formulations based on the Einstenean approach can be found in the works by Nakagawa and Tsujimoto (1980), van Rijn (1984a), and Tsujimoto (1991).

On the other hand, Bagnold-based models typically use a dynamic condition according to which the fluid bed shear stress drops to the critical value for the initiation of sediment motion while an area concentration of bedload particles can be calculated from the bed shear stress value. The above, means that the moving bedload sediment particles extract momentum from the fluid in such a way that the fluid stresses at the bottom remains at the critical initiation of motion shear stress value. This dynamic condition is usually found into the literature as *Bagnold's hypothesis*. While criticized in the past regarding its applicability in certain conditions, *e.g.*, in the work by Fernandez-Luque and van Beek (1976), Nino and Garcia (1998), McEwan *et al.* (1999), Seminara *et al.* (2002) and Parker *et al.* (2003), Bagnold's hypothesis has been used as a basis for the development and derivation of new models and calculations of sediment transport, *e.g.*, in the work by Owen (1964),Wiberg and Smith (1989), Sekine and Kikkawa (1992),Sekine and Parker (1992), Niño and Garcia (1994, 1998) and other publications.

In general, bedload transport can be defined as:

$$q_b = u_b c_b \delta_b \tag{2.5.19}$$

where q_b is the volumetric bed load transport rate (m^2/s), u_b is the particle velocity (m/s), c_b is the volumetric sediment concentration and δ_b is the thickness of the bedload layer (m). Bagnoldean bedload transport models use this definition of the bedload transport rate (Ashida & Michiue, 1972; Engelund & Fredsoe, 1976; Van Rijn, 1984a; Wiberg & Smith, 1987; Sekine & Kikkawa, 1992; Nino & Garcia, 1994, 1998; Lee & Hsu, 1994; Lee *et al.*, 2000). A large number of bedload relations have been proposed in the literature for the estimation of bedload transport typically as a function of dimensionless Shield stress and some other sediment characteristics. Typically, these formulas are introduced in the following form:

$$q^* = q^* \left(\tau^*, \ R_{ep}, \ R \right) \tag{2.5.20a}$$

here, q^* is a dimensionless bedload transport rate known as the Einstein bedload number and is defined as,

$$q^* = \frac{q_b}{D\sqrt{gRD}} \tag{2.5.20b}$$

One of the most well-known relations for the calculation of bed load was developed by Meyer-Peter and Muller (1948):

$$q^* = 8\left(\tau^* - \tau_c^*\right)^{3/2} \tag{2.5.21}$$

where $\tau_c^* = 0.047$. The formula by Meyer-Peter and Muller (1948) is empirical and has been verified with data for uniform coarse sand and gravel. Wong and Parker (2006) reanalyzed the data of Meyer-Peter & Muller, proposed two alternative forms that fit better with their data:

$$q^* = 4.93(\tau^* - 0.047)^{1.6} \tag{2.5.22a}$$

$$q^* = 3.97(\tau^* - 0.0495)^{3/2} \tag{2.5.22b}$$

After Meyer-Peter & Muller's work, many formulas have been proposed into the literature based on mathematical derivation and/or experimental observations. A summary of some of the most commonly applied bedload equations proposed into the literature can be found in Table 2.5.3. It is clear that the empirical nature of these formulas and the limitations and assumptions related to their derivation make them to perform with different levels of accuracy depending on the range of conditions. For example in Figure 2.5.4, values of q^* as a function of t^* for some of the formulas in

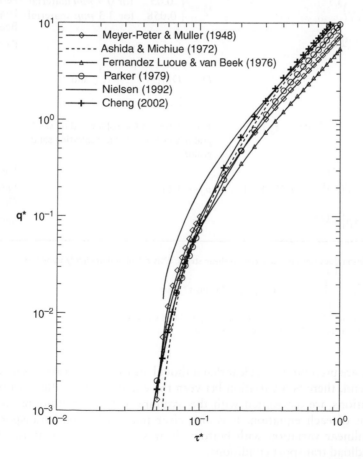

Figure 2.5.4 Plot of several bed load functions found in the literature (Garcia, 2008 with permission from ASCE).

Table 2.5.3 Commonly used bedload formulas

Equation	Parameters	Reference
$q^* = 8(\tau^* - \tau_c^*)^{3/2}$	$\tau_c^* = 0.047$	Meyer-Peter & Muller (1948)
$q^* = q^*(\tau^*)$	$\frac{43.5q^*}{1+43.5q^*} = 1 - \frac{1}{\sqrt{\pi}} \int\limits_{-(0.413/\tau^*)-2}^{(0.413/\tau^*)-2} e^{-t^2}\,dt$	Einstein (1950)[1]
$q^* = 0.635s(\tau^*)^{1/2}\left[1 - \frac{\ln(1+a_2s)}{a_2s}\right]$	$a_2 = 2.45(R+1)^{0.4}(\tau_c^*)^{1/2}$; $s = \frac{\tau^* - \tau_c^*}{\tau_c^*}$	Yalin (1963)
$q^* = 12(\tau^* - \tau_c^*)^{3/2}$	τ_c^* was determined from the Shields diagram	Wilson (1966)
$q^* = 6.56\,10^{18}\,\tau^{*16}$	–	Paintal (1971)
$q^* = 18.74(\tau^* - \tau_c^*)\left[(\tau^*)^{1/2} - 0.7(\tau_c^*)^{1/2}\right]$	$\tau_c^* = 0.05$	Engelund and Fredsoe (1976)
$q^* = 5.7(\tau^* - \tau_c^*)^{3/2}$	$\tau_c^* = \begin{cases} 0.05 & \text{for 0.9 } mm \text{ material} \\ 0.058 & \text{for 3.3 } mm \text{ material} \end{cases}$	Fernandez Luque and van Beek (1976)
$q^* = 11.2\frac{(\tau^* - 0.03)^{4.5}}{\tau^{*3/2}}$	–	Parker (1979)[2]
$q^* = 0.053\frac{T^{2.1}}{D_*^{0.3}}$	$D_* = D_{50}\left(\frac{gR}{\nu^2}\right)^{1/3} = R_{ep}^{2/3}$ and $T = \frac{\tau_s^* - \tau_c^*}{\tau_c^*}$	Van Rijn (1984a)[3]
$q^* = F_M(\tau^{*1/2} - 0.7\tau_c^{*1/2})(\tau^* - \tau_c^*)$	$F_M = 8/\tan\varphi$ for rolling/sliding sand grains and $F_M = 9.5$ for saltating sand grains	Madsen (1991)
$q^* = 12\tau^{*1/2}(\tau^* - \tau_c^*)$		Nielsen (1992)
$q^* = \frac{12}{\mu_d}(\tau^* - \tau_c^*)[(\tau^*)^{1/2} - 0.7(\tau_c^*)^{1/2}]$	$\mu_d = 0.23$	Niño and Garcia (1998):
$q^* = 13\tau^{*3/2}\exp\left[-\frac{0.05}{\tau^{*3/2}}\right]$	–	Cheng (2002)

1 Einstein's relation does not include a critical shear stress. q* is calculated implicitly using the relation $1 - \frac{1}{\sqrt{\pi}} \int\limits_{-(0.413/\tau^*)-2}^{(0.413/\tau^*)-2} e^{-t^2}\,dt = \frac{43.5q^*}{1+43.5q^*}$ for any τ* value

2 simplified fit to the relation of Einstein (1950) for gravel-bed streams

3 τ_s^* is the bed shear stress due to skin or grain friction

Table 2.5.3, are presented. It is clear that although most all the equations tend to follow the same trend, there is a variation between the equations for different ranges of t^*. These variations are associated with the assumptions and conditions used for the development of each equation. It is also clear that all bedload transport relations show a nonlinear variation with bottom shear stress and this is most obvious for incipient bedload transport conditions.

The bedload transport rate can also be defined as the product of the number of moving particles per unit area, the particle volume and the particle velocity, as follows

$$q_b = N_b V_p u_b \tag{2.5.23}$$

in which N_b is the number of particles per unit bed area (m^{-2}), V_p is the particle volume (m^3), and u_b is the particle velocity (m/s). If the particle velocity is defined as the ratio of the saltation or step length λ_{sal} and the saltation or movement period T_{sal} (*i.e.* $u_b = \lambda/T$), then

$$q_b = N_b V_p \lambda_{sal}/T_{sal} = E_p \lambda_{sal} = D_p \lambda_{sal} \tag{2.5.24}$$

Here E_p and D_p are the eroded and deposited volume of particles per unit bed area per unit time (m/s), respectively. The idea of a "pick-up rate" and a step length was first proposed by Einstein (1942, 1950) and constitutes the basis of Einsteinian bedload transport models (Nakagawa & Tsujimoto, 1980; Tsujimoto, 1992). Pick-up rate functions and their applications can be found in Van Rijn (1984a) and Van Rijn (1986).

Very recently, Cheng and Emadzadeh (2015) conducted a study showing that the sediment pickup rate is better correlated with a densimetric Froude number rather than the Shields number. With laboratory data collected in open channel flows, an empirical formula for estimating the pickup rate was proposed. Because it does not involve the critical condition for the incipient sediment motion, the formula can be applied to sediment entrainment of different stages including weak sediment motion. In spite of the fact that the experiments were carried with a limited range of sediment particle diameters, ranging from 0.23 to 0.86 mm, the results obtained are also supported by several series of data reported in the literature.

2.5.1.3.5 Two-dimensional (2D) transport of bedload

The relations presented above for bedload transport are all one-dimensional (1D) in their analytical approach. That is, they provide the magnitude of a bedload transport vector that is oriented in the direction of the boundary shear stress. That is, if the s is a boundary-embedded centerline streamwise component and n is a boundary-embedded transverse component, the bedload rate is:

$$\vec{q} = (q_s, q_n) = (q, 0) \tag{2.5.25}$$

where $\vec{q}$ denotes the two-dimensional (2D) vector of bedload transport rate and q denotes the magnitude of that vector, which is computed using one of the relations presented above.

As a matter of fact, bedload transport is fundamentally 2D in nature even if the reach is straight, as seen in Figure 2.5.5a. For a river bend (see Figure 2.5.5b), the bend generates secondary flow in addition to the downstream primary flow, hence the boundary shear stress vector $\vec{\tau}_o$ is not parallel to the s direction, but is skewed somewhat inward. This drives a component of bedload transport in the $-n$ direction, *i.e.* inward. The bed itself has slopes downward from inside to outside in the transverse direction with a magnitude $|\partial \eta_b/\partial n|$. As a result, gravity pulls bedload particles down the slope, driving a component of bedload transport in the $+n$ direction. Depending on the magnitude of the forces involved, the bedload vector $\vec{q}$ may have a positive or negative component in the transverse direction.

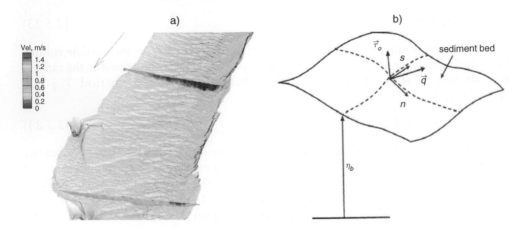

Figure 2.5.5 Illustration of 2D bedload transport: a) straight channel reach (adapted from Baranya et al., 2016); b) river bend (adapted from Parker, 2005).

These competing transverse effects play an important role in determining the morphology of rivers in meander bends (Engelund, 1974; Falcon & Kennedy, 1983; Ikeda & Nishimura, 1986; Bridge, 1992). Secondary flow tends to drive erosion at the outside of a bend and deposition on the inside (Johanesson & Parker, 1989). This creates a transverse component to the bed slope, which in turn acts to drive sediment down the slope from inside to outside. An equilibrium condition can be obtained by which secondary forces and gravity forces balance. A second problem for which 2D bedload relations are needed is the quantification of the erosion of a river bank composed of non-cohesive sediment. Banks in non-cohesive material form side slopes; as bedload is moved downstream by the flow, gravity pulls it down the side slope, accomplishing bank erosion.

Most 2D bedload relations were developed based on a linearized formulation for small transverse slope and streamwise bed slopes and they take the general form,

$$\vec{q} = |\vec{q}| \left[\frac{\vec{\tau}_o}{|\vec{\tau}_o|} - \beta_q \left(\frac{\tau^*_{mag}}{\tau^*_c} \right)^{-n_q} \vec{\nabla} \eta_b \right] \qquad (2.5.26a)$$

where the absolute values denote the magnitude of the bedload vector, β_q and n_q are constants and

$$\vec{\tau}_b = (\tau_{os}, \tau_{on}) \,, \tau^*_{mag} = \frac{|\vec{\tau}_b|}{\rho R g D} \,, \vec{\nabla} \eta_b = \left(\frac{\partial \eta_b}{\partial s}, \frac{\partial \eta_b}{\partial n} \right) \qquad (2.5.26b, c, d)$$

where $\vec{\tau}_b = (\tau_{os}, \tau_{on})$ is the 2D vector of bed shear stress and its streamwise and spamwise components, τ^*_{mag} is the Shields stress and $\vec{\nabla} \eta_b = \left(\frac{\partial \eta_b}{\partial s}, \frac{\partial \eta_b}{\partial n} \right)$ is the bed slope components in the streamwise and spamwise direction. The derivations of these formulations are based on the following constraints,

$$\frac{\tau_{bn}}{\tau_{bs}} << 1 \,, \frac{\partial \eta_b}{\partial n} << 1, \quad \frac{\partial \eta_b}{\partial s} << 1 \qquad (2.5.27)$$

If in turn the streamwise slope $\partial\eta/\partial s$ is so small that direct streamwise gravitational forces on bedload can be neglected, Equation (2.5.26a) can be further cast in the approximate form

$$|\vec{q}| = q_s, \quad q_n = q_s \left[\frac{\tau_{bn}}{\tau_{bs}} - \beta_q \left(\frac{\tau_s^*}{\tau_c^*} \right)^{-n_q} \right] \frac{\partial\eta_b}{\partial n}, \quad \tau_s^* = \frac{\tau_{bs}}{\rho RgD} \qquad (2.5.28a, b, c)$$

For example, Engelund (1974) proposed the following values of β_q and n_q,

$$\beta_q = \frac{1}{\mu_d}, \quad n_q = 0 \qquad (2.5.29)$$

where μ_d denotes a dimensionless coefficient of dynamic Coulomb friction for particles in bedload transport. Many other values for β_q and n_q have been proposed by a number of researchers and can be found in (Garcia, 2008, p.75).

2.5.1.3.6 Conservation of sediment mass: the Exner equation

To estimate morphological changes associated with sediment transport, an equation for bed sediment mass balance is needed. As shown in Figure 2.5.5, a datum of constant elevation is located well below the bed level, and the elevation of the bed with respect to such datum is given by η_b. The change in the bed level with time t due to bedload transport, sediment entrainment into suspension, and sediment deposition onto the bed, can be predicted with the help of the Exner (1925):

$$(1 - \lambda_p) \frac{\partial\eta_b}{\partial t} = - \frac{\partial q_{bs}}{\partial s} - \frac{\partial q_{bn}}{\partial n} + v_s (\bar{c}_b - E_s) \qquad (2.5.30)$$

where λ_p is the bed porosity, η_b is the bed elevation with respect to an arbitrary reference elevation, q_{bs} and q_{bn}, are the bedload transport in the longitudinal and transverse direction respectively and the term $v_s(\bar{c}_b - E_s)$ is associated with the net suspended sediment particles that are entrained into suspension. To solve the Exner equation, it is necessary to have relations to compute bedload transport (i.e., q_{bs} and q_{bn}), near-bed suspended sediment concentration $\bar{c}_b$ and sediment entrainment into suspension E_s (Garcia & Parker, 1991; Garcia, 2001). The net suspended sediment flux equal the amount of sediment that deposit to the bed which can be calculated as the product of fall velocity v_s and the near-bed suspended sediment concentration $\bar{c}_b$ minus the sediment entrainment into suspension E_s. Equation (2.5.30) can be obtained from the mass balance equation of sediments for a unit of sediment bed (Garcia, 2008) The basic form of Equation (2.5.30), without the suspended sediment component, was first proposed by Exner (1925).

2.5.1.3.7 Bedforms

The flow interacts strongly with sediment transport. As the flow increases, this interaction often results to the formation of *bedforms* which are features that develop along a moveable bed. The most common bedforms are *ripples, dunes* and *antidunes*. Figure

2.5.6 provides a schematic of the most common bedforms characterized by the flow condition and bed sediment diameter.

These bedforms have an important influence on sediment transport and on flow resistance. Best (2005) highlights recent year's great progress in our knowledge and understanding regarding bedform characteristics due to the increased ability to monitor flow and dune morphology in the laboratory and the field as well as the development of increased sophistication numerical modeling to capture not only the characteristics of the mean flow field but realistically simulate the origins and motions of coherent flow structures above bedforms. These advances are summarized in the work by Reynolds (1976), Engelund and Fredsøe (1982), Ikeda and Parker (1989), McLean (1990), Southard (1991), Kennedy and Odgaard (1991), Seminara (1995), Best (1996), Seminara and Blondeaux (2001), Yalin and da Silva (2001), ASCE (2002), Bridge (2003), Best (2005), and Parker and Garcia (2006) and Garcia (2008). Since the effect of the bedforms both on sediment and flow is so important, it is impossible to cover all the material needed in detail. Instead, a brief introduction on the major bedform types and some of their characteristics are presented below. Interested readers are guided to consult the references of this section for more details.

The origin of bed forms has been studied, among others, by Yalin (1992), Nelson and Smith (1989), McLean (1990), Southard (1991), Bennett and Best (1996), Coleman and Melville (1996) Nikora and Hicks (1997) Gyr and Kinzelbach (2004). Some of these works (*e.g.*, Coleman & Melville, 1996; Coleman & Fenton 2000; Coleman *et al.*, 2003; Zhou & Mendoza, 2005) focus mainly on the conditions that cause the development of wavelets that may lead to the development of ripples (Coleman & Eling, 2000). After their formation, bedforms interact strongly with the flow and their stability and transformations have been studied by Leeder (1983), Bennett and Best (1996), Robert and Ulhman (2001), Schindler and Robert (2004). Their use on the estimation of bed load is discussed by other researchers, *e.g.*, the work by Engel and Lau (1980), Mohrig and Smith (1996), Vionnet *et al.* (1998) and Zhou and Mendoza (2005).

Although the mechanics of bedform development (*e.g.*, Coleman & Melville, 1994) and growth (*e.g.*, Nikora & Hicks, 1997) under steady flow conditions have been explored in the past there are still knowledge gaps related to the application of these finding in common engineering practice as well as the modification required in order to apply in the field scale findings that have been extracted in laboratory or theoretically. Detailed field studies (*e.g.*, Kostachuk & Villard, 1996; Holmes, 2003; Kostachuk *et al.*, 2004; Parsons *et al.*, 2005) of dune's characteristics, flow field and sediment transport demonstrate the complexity of the field scale bedforms which combined with the difficulties and practical constraints of the field measurements make their analysis very challenging (van den Berg & van Gelder, 1993).

Ripples are characteristic of very low transport rates and rivers with sediment size less than 0.6 mm. Typically, ripples are small scale wavelike bedforms that migrate downstream. They show a characteristic asymmetry having a pronounced slip face and their existence is usually associated with the presence of a viscous sublayer. In general, the existence of the viscous sublayer is not related to the flow Reynolds number (laminar or turbulent) but whether the bed can be considered hydrodynamically smooth or rough. Engelund and Hansen (1967) have suggested the following condition for ripple formation: $D \leq \delta_v$, where $\delta_v = 11.6v/u*$ denotes the thickness of the viscous sublayer. This threshold value can be rearranged in order to be expressed in terms of a

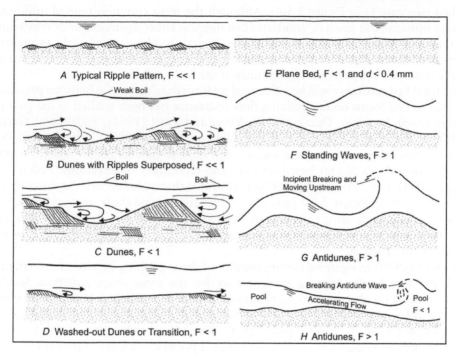

Figure 2.5.6 Schematic of different beforms. Note: F is the Froude number and d is the sediment diameter (after Simons & Richardson, 1961).

dimensionless Shields stress as $\tau^* = \left(11.6/\text{Re}_p\right)^2$. This threshold value is plotted in the Shields-Vanoni-Parker (SVP) river regime diagram of Figure 2.5.4. Ripples are, in general, more 3D than the rest of the bedforms presented here (dunes and antidunes) and they have typical wavelengths in the order of tens of centimeters and wave heights on the order of some centimeters (Parker, 2005). Due to their relatively small scale ripples do not interact with the water surface. Ripples can coexist with dunes (case B in Figure 2.5.6).

Dunes are common bedforms in sand-bed rivers although they can be found in gravel-bed rivers too (Parker, 2005). Similarly to the ripples, dunes are characteristic of subcritical flows ($Fr \ll 1$) but are typically larger than ripples. They are wavelike features that migrate downsteam and their shape sometimes become approximately triangular causing flow separation at the progressing edge side. Dunes' wavelength size can range up to hundreds of meters depending on the flow depth ($0.25 < \frac{2\pi H}{\lambda} < 4.0$, where λ is the wavelength and H is the water depth) and their wave height can range up to 5 *m* or more for the case of large rivers ($\frac{\Delta_w}{H} \leq \frac{1}{6}$, where Δ_w is the wave height). Dunes interact weakly with the water free surface, such that the flow accelerates over their crests, where water surface elevation is slightly reduced. This out of phase between bed and water free surface is demonstrated in case B and C of Figure 2.5.6.

Antidunes exist in rivers with high (but not necessarily supercritical) Froude number. Similarly to dunes they can be found in sand-bed and gravel-bed rivers but they distinguished from dunes by the fact that water free surface is nearly in phase with the bed (see case G in Figure 2.5.6). Although the most common type of antidunes migrate upstream, in general, antidunes can migrate either upstream or downstream. Another difference compared to the dunes is that antidunes are nearly symmetric wavelike features. Thus, the observation of a sequence of symmetrical surface waves usually indicated the presence of antidunes at the bed.

The total bed resistance will be influenced by form drag due to the beforms presence. The role of bed forms on determining flow resistance has been studied in the past by various researchers e.g., Ogink (1988), Yoon and Patel (1996), Fedele and Garcia (2001), Julien et al. (2002) and Wilbers and ten Brinke (2003). Van Rijn (1982, 1984c, 1996) proposed a formula for an equivalent roughness height that takes into consideration the bedforms based on field and laboratory observations. Fedele and Garcia (2001) used a boundary-layer based formula that includes both the skin friction and the form drag. Figure 2.5.7 provides a schematic of total shear stress as a function of mean velocity following the work by Raudkivi (1990). From this plot is becomes clear that the form drag (associated with the pressure) can exceed a lot the effect by the surface "skin" friction.

Einstein (1950) recognizing the importance of distinguishing between form friction and form drag he developed a methodology for the *shear stress partition*. Similar methodologies for the shear stress partitioning were proposed by Nelson and Smith (1989) and Fredsøe and Deigaard (1992) based on the dimensionless bed stress due to skin friction and form drag proposed by Fredsøe (1982). The methods by Einstein, Nelson-Smith and Engelund-Fredsøe require an a priori knowledge of the total boundary shear stress which is in general not known. Thus, other empirical techniques have been developed for the estimation of shear stress. Einstein and Barbarossa (1952) developed a method applicable for the case of dune resistance in sand-beds. Engelund

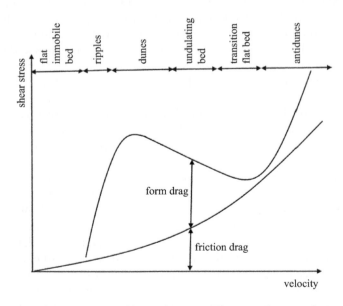

Figure 2.5.7 Variation of shear stress or friction factor as a function of mean velocity over a fine-sand bed.

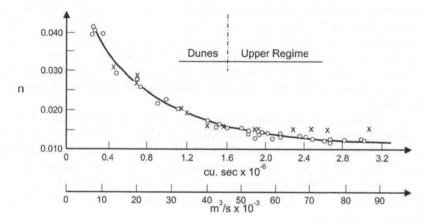

Figure 2.5.8 Variation of Manning's n with flow discharge for the Padma River, Bangladesh. (o) observations of Stevens *et al.* (1976), and (x) computations of Chollet & Cunge, 1980; reprinted with permission from Elsevier.

and Hansen (1967) also developed a method applicable for sand-bed streams. Wright and Parker (2004) modified the method by Engelund & Hansen to perform better in field conditions. Other methods were proposed by Brownlie (1981) and Cruickshank and Maza (1973), Karim and Kennedy (1981) and others.

The presence or absence of bed forms on the bed of a river can lead to curious effects on river stage, as implied in the friction factor diagram for a movable shown in Figure 2.5.8. According to the Manning-type relation for a non-erodible bed presented above, the following should hold:

$$U = \frac{1}{n}H^{2/3}S^{1/2} \qquad\qquad (2.5.31)$$

Here, the channel is assumed to be wide enough to allow the hydraulic radius to be replaced with the depth H. According to this equation, if the energy slope remains relatively constant, depth should increase monotonically with increasing velocity. This would indeed be the case for a rigid bed. In a sand-bed stream, however, resistance decreases as U increases over a wide range of conditions, therefore at equilibrium we can assume that

$$\tau_o = \rho\, C_f\, U^2 = \rho g H S \qquad\qquad (2.5.32)$$

This decrease in resistance implies that depth does not increase as rapidly in U in a movable-bed stream as it would for a rigid-bed open channel. In fact, as transition to upper regime is approached, the bed forms can be wiped out quite suddenly, resulting in a dramatic decrease in resistance. The result can be an actual decrease in depth as velocity increases. This phenomenon was documented for the case of the Padma River, Bangladesh, by Stevens *et al*, (1976) and predicted numerically by (Chollet & Cunge, 1980), as shown in Figure 2.5.8. This plot shows how Manning's n decreases as the flow discharge increases and the dunes are first elongated and finally washed out. Notice that the numerical model predictions agree very well with the observations for the lower regime conditions but overestimate the value of Manning's n in the upper regime where most of the flow resistance should be mainly due to grain friction. Even for one-dimensional computations, this remains a challenging problem for river engineers. For field measurements, the presence of bedforms makes measurements a very challenging undertaking (*e.g.*, Kostachuk & Villard, 1996; Holmes & Garcia, 2008). A method to estimate the equivalent roughness of bedforms is presented in Appendix 2.A.2.

While the mechanics of bed-form development (Coleman & Melville, 1994) and rates of bed-form growth for steady flows (Nikora & Hicks, 1997) have been clarified, practical implications and models accessible to the engineer remain to be elaborated. For example, how and at what rates bed forms change for increasing and decreasing flows remains to be quantified. The problem of transitions to a dune bed from a rippled or plane bed and from a dune bed to an upper-regime plane bed or antidune bed is also of much practical interest.

From an experimental/measurement perspective, and probably from a theoretical modeling perspective, the transient problem in which dune characteristics change over time poses *additional* severe difficulties beyond those of the equilibrium case. For sediment transport engineering, a minimum contribution desired of a theory (model, understanding) for bed-form development would be a reliable means of determining which equilibrium would be established, *i.e.*, delineating stability boundaries. Attempts have been made to base such boundaries on theoretical stability models, as proposed by Kennedy (1969), but engineering approaches (*e.g.*, van Rijn, 1982) have been primarily based on dimensional analysis and empiricism.

2.5.1.4 Suspended sediment transport

Suspended sediment differs from bedload sediment in that it may be diffused throughout the vertical column of fluid via turbulence. Here the local instantaneous volume concentration of suspended sediment at point xi and time t is denoted as $c(xi,t)$. The Eulerian particle velocity field is denoted as $vi(xj,t)$. For fine particles the following approximation holds:

$$v_i = u_i - v_s \delta_{i3} \tag{2.5.33}$$

Where ui is the Eulerian flow velocity field, vs is the sediment fall velocity in quiescent water, and δ_{i_3} is the Kronecker delta. The above velocities are instantaneous values. As long as the suspended sediment under consideration is sufficiently coarse so as not to undergo Brownian motion (*i.e.*, silt or coarser), molecular effects can be neglected. Suspended particles are transported solely by convective fluxes. Using a transport equation for the suspended sediment (convected by the velocity field in Equation (2.5.33)) and after applying Reynolds averaging:

$$\frac{\partial \bar{c}}{\partial t} + \overline{u_1}\frac{\partial \bar{c}}{\partial x_1} + \overline{u_2}\frac{\partial \bar{c}}{\partial x_2} + (\overline{u_3} + v_s)\frac{\partial \bar{c}}{\partial x_3} = -\frac{\partial \overline{u_1'c'}}{\partial x_1} - \frac{\partial \overline{u_2'c'}}{\partial x_2} - \frac{\partial \overline{u_3'c'}}{\partial x_3} \tag{2.5.34}$$

The Reynolds flux $\overline{u_i'c'}$ in the above relation is clearly diffusive in nature. The simplest closure assumption one could make for these terms is

$$\overline{u_i'c'} = -D_d \frac{\partial \overline{c}}{\partial x_i} \tag{2.5.35}$$

In Equation (2.5.35), the kinematic eddy diffusivity Dd is assumed to be a scalar quantity. For the case of non-isotropic turbulence, Equation (2.5.35) must be generalized to the form

$$\overline{u_i'c'} = -D_{d_{ij}} \frac{\partial \overline{c}}{\partial x_j} \tag{2.5.36}$$

Here $D_{d_{ij}}$ is a tensor quantity. It is often assumed to represent a diagonal matrix, such that $D_{d_{ij}} = 0$ if $i \neq j$, and $D_{d_{11}} \neq D_{d_{22}} \neq D_{d_{33}}$.

The component of the Reynolds flux of suspended sediment near the bed that is directed upward normal to the bed may be termed the rate of *erosion*, or more accurately, *entrainment* of bed sediment into suspension per unit bed area per unit time. The entrainment rate E is thus given by

$$E_s = \overline{u_i'c'}|_{near\ bed} \cdot n_i^b \tag{2.5.37}$$

where n_i^b is the normal to the bed vector. The terminology "near bed" is employed to avoid possible singular behavior at the bed (located at $x_3 = 0$).

The net upward normal flux $\overline{F_{si}}|_{near\ bed} \cdot n_i^b$ of suspended sediment at (or rather just above) the bed is given by

$$\overline{F_{si}}|_{near\ bed} \cdot n_i^b = v_s \left(E_s - \overline{c_b} k \cdot n^b \right) \tag{2.5.38}$$

where

$$E_s \equiv \frac{E}{v_s} \tag{2.5.39}$$

denotes a dimensionless rate of entrainment of bed sediment into suspension and k is the parallel to the gravity vector. The required bed boundary condition, then, is a specification of E. Typically a relation of the following form is assumed:

$$E_s = E(\tau_{bs}, other\ parameters) \tag{2.5.40}$$

where τ_{bs} denotes the boundary shear stress due to skin friction.

In the special case of steady, uniform flow over a flat (averaged over bed forms) bed, with z defined to be (nearly) vertical, Equation (2.5.38) reduces to

$$\overline{F_{sz}}|_{near\ bed} = v_s(E_s - \overline{c_b}) \tag{2.5.41}$$

It is furthermore assumed that an equilibrium steady, uniform suspension has been achieved, it follows that there should be neither net deposition on ($\overline{F_{sz}} < 0$) or erosion from ($\overline{F_{sz}} > 0$) the bed. That is, $\overline{F_{sz}} = 0$, yielding the result

$$E_s = \overline{c_b} \tag{2.5.42}$$

This relation simply states that the entrainment rate equals the deposition rate, so that there is no net normal flux of suspended sediment at the bed (Garcia & Parker, 1991).

A number of relations are available in the literature for estimating the entrainment rate of sediment into suspension E_s (and thus the reference concentration $\overline{c_b}$ for the equilibrium case). Table 2.5.4 summarizes most of the available relations (Garcia, 2001). The relations were checked against a carefully selected set of data pertaining to equilibrium suspensions of uniform sand. In such case, it is possible to measure $\overline{c_b}$ directly at some near-bed elevation z = b, and to equate the result to E_s according to Equation (2.5.42).

2.5.1.4.1 Rousean profile

Let's now consider normal flow in a wide, rectangular open channel. The bed is assumed to be erodible and has no curvature when averaged over bedforms. The x_3 coordinate is quasi-vertical, implying low channel slope S. The suspension is likewise assumed to be in equilibrium. That is, $\overline{c}$ is a function of x_3 alone. Thus, under the assumption that the flow and suspension are uniform in the streamwise and transverse direction and steady in time, so that Equation (2.5.34) can be written as:

$$v_s \frac{\partial \overline{c}}{\partial x_3} = -\frac{\partial \overline{u_3' c'}}{\partial x_3} \tag{2.5.43}$$

Equation (2.5.43) can be written as:

$$\overline{u_3' c'} - v_s \overline{c} = 0 \tag{2.5.44}$$

An eddy diffusion assumption is required for the modeling of the Reynold stress term. The simple approach taken here is that of Rouse (1937). It involves the use of the *Prandtl analogy*. The argument is as follows. Fluid mass, heat, momentum, etc. should all diffuse at the same kinematic rate due to turbulence and thus have the same kinematic eddy diffusivity, because each is a property of the fluid particles, and it is the fluid particles that are being transported by Reynolds fluxes. The velocity profile is approximated as logarithmic throughout the depth.

The vertical kinematic eddy viscosity takes the form:

$$D_d = \kappa u_* x_3 \left(1 - \frac{x_3}{H}\right) \tag{2.5.45}$$

Equation (2.5.36) can be combined with Equation (2.5.44) and then integrated from the nominal bed level to distance x_3 above the bed. The resulting form can be cast as

$$\overline{c} = \overline{c_b} \left[\frac{(H - x_3)/x_3}{(H - b)/b}\right]^{Z_R} \tag{2.5.46}$$

where Z_R denotes the **Rouse Number**, a dimensionless number given by

$$Z_R = \frac{v_s}{\kappa u_*} \tag{2.5.47}$$

Sediment concentration profiles of suspended sediment observed in the Amazon River, Brazil, are plotted in Rousean form in Figure 2.5.9 (Vanoni, 1980). Despite the wide range of sediment sizes present in the water column, it is clear that Equation (2.5.46)

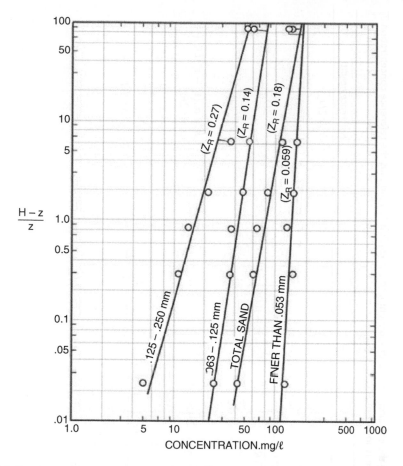

Figure 2.5.9 Distribution of suspended sediment plotted in Rousean Coordinates for the Amazon River at Manacapuru, where flow depth is 44m (note that here z=x₃). Each line shows the particle size range and the corresponding value of the Rouse number (Vanoni, 1980).

provides a good predictor for sediment concentration profiles under equilibrium conditions. Notice that the Rouse developed the concepts leading to Equation (2.5.46) in a glass jar agitated with an oscillating grid. Yet the equation is found to hold well with data obtained at a depth of 44m in the Amazon River, which is remarkable.

2.5.1.4.2 Rousean profile in stratified flows

River flows carrying suspended sediment are self-stratifying. As illustrated in Figure 2.5.10, the Rousean profile predicts a concentration of suspended sediment that decreases with increasing elevation above the bed. It follows that the density of the water-sediment mixture also decreases with increasing elevation above the bed. This stable stratification inhibits turbulent mixing of both flow momentum and sediment mass in the vertical. The result is a modification of the vertical distributions of both streamwise momentum and suspended sediment concentration. More specifically, stratification effects lead to a streamwise velocity profile that increases more rapidly

Table 2.5.4 Existing Relations in the literature for estimating the entrainment rate of sediment into suspension E_s (and thus the reference concentration $\overline{c}_b$ for the equilibrium case) (Garcia, 2008 with permission from ASCE).

Reference	Formula	Parameters	Reference height
Einstein (1950)	$\overline{c}_b = \dfrac{q_*}{23.2\left(\tau_s^*\right)^{0.5}}$	—	$b = 2D_s$
Engelund and Fredsøe (1976, 1982)	$\overline{c}_b = \dfrac{0.65}{\left(1 + \lambda_b^{-1}\right)^3}$	$\lambda_b = \left[\dfrac{\tau_s^* - 0.06 - \frac{\beta p \pi}{6}}{0.027(R+1)\tau_s^*}\right]^{0.5}$; $p = \left[1 + \left(\dfrac{\frac{\beta p \pi}{6}}{\tau_s^* - 0.06}\right)^{-0.25}\right]$; $\beta = 1.0$	$b = 2D_s$
Smith and McLean (1977)	$\overline{c}_b = \dfrac{0.65 \gamma_o T}{1 + \gamma_o T}$	$T = \dfrac{\tau_s^* - \tau_c^*}{\tau_c^*}$; $\gamma_o = 2.4 \times 10^{-3}$	$b = a_o(\tau_s^* - \tau_c^*)D_s + k_s$ $a_o = 26.3$
Itakura and Kishi (1980)	$\overline{c}_b = k_1\left(k_2 \dfrac{u_*}{v_s}\dfrac{\Omega}{\tau^*} - 1\right)$	$\Omega = \dfrac{\tau^*}{k_3}\left[k_4 + \dfrac{\exp\left(-A_o^2\right)}{\displaystyle\int_{A_o}^{\infty}\exp\left(-\zeta^2\right)d\zeta}\right] - 1$; $A_o = \dfrac{k_3}{\tau^*} - k_4$; $k_1 = 0.008$; $k_2 = 0.14$; $k_3 = 0.143$; $k_4 = 2.0$; $v_s = $ fall velocity	$b = 0.05H$

Van Rijn (1982)

$$\overline{c_b} = 0.015 \frac{D_s}{b} \frac{T^{1.5}}{D_*^{0.3}}$$

$$D_* = D_s \left(\frac{gR}{v^2}\right); \quad \text{where } \Delta_b \text{ is the mean dune height}$$

$$b = \frac{\Delta_b}{2} \text{ if; } \Delta_b \text{ is known}$$
$$\text{else } b = k_s \cdot b_{min} = 0.01H$$

Celik and Rodi (1984)

$$\overline{c_b} = \frac{k_o C_m}{I}$$

$$C_m = 0.034 \left[1 - \left(\frac{k_s}{H}\right)^{0.06}\right] \frac{u_*^2}{gRH} \frac{U_m}{v_s}; \quad I = \int_{0.05}^{1} \left(\frac{1-\eta}{\eta} \cdot \frac{\eta_b}{1-\eta_b}\right)^{v_s/0.4u_*} d\eta;$$

$$b = z/H; h_b = 0.05; k_o = 1.13$$

$$b = 0.05H$$

Akiyama and Fukushima (1986)

$$E_s = 0; \quad Z < Z_u$$
$$E_s = 3 \times 10^{-12} Z^{10}\left(1 - \frac{Z_c}{Z}\right);$$
$$Z_c < Z$$
$$E_s = 0.3; Z > Z_m$$

$$Z = \frac{u_*}{v_s} R_p^{0.5}; Z_c = 5; Z_m = 13.2$$

$$b = 0.05H$$

Garcia and Parker (1991)

$$E_s = \frac{A Z_u^5}{1 + \frac{A}{0.3} Z_u^5}$$

$$Z_u = \frac{u_{*s}}{v_s} R_p^{n}; u_{*s} = \frac{g^{0.5}}{C'} U_m; C' = 18 \cdot \log\left(\frac{12R_b}{3D_s}\right); n = 0.6; A = 1.3 \times 10^{-7}$$

$$b = 0.05H$$

Zyserman and Fredsøe (1994)

$$\overline{c_b} = \frac{0.331(\tau_s^* - 0.045)^{1.75}}{1 + \frac{0.331}{0.46}(\tau_s^* - 0.045)^{1.75}}$$

$$\tau_s^* = \frac{(u_{*s})^2}{RgD_s}$$

$$b = 2D_s$$

in the vertical than the logarithmic profile, and a suspended sediment profile that decreases more rapidly in the vertical than the Rousean formulation. Smith and McLean (1977), Gelfenbaum and Smith (1986), and McLean (1992) offer quantitative formulations of stratification effects based on simple algebraic closures. The kinematic eddy diffusivity D_d given in Equation (2.5.45) is now denoted as D_{do}, where the subscript "o" denotes the absence of stratification effects. The value D_d in the presence of stratification effects is given as

$$D_d = D_{do}F_{strat}\left(RI_g\right), D_{do} = \kappa u_* x_3\left(1 - \frac{x_3}{H}\right), RI_g = \frac{-Rg\frac{d\bar{c}}{dx_3}}{\left(\frac{d\bar{u}_1}{dx_3}\right)^2} \qquad (2.5.48)$$

Here RI_g denotes a gradient Richardson number and F_{strat} is a function that decreases with increasing gradient Richardson number, so capturing the effect of damping of the turbulence due to flow stratification. Using Equations (2.5.48) the following equation for the time averaged suspended load profile and velocity profile that takes into consideration the sediment stratification can be derived (Smith & McLean, 1977; Wright & Parker, 2004)

$$\bar{c} = E_s \exp\int_b^{x_3}\left[-\frac{v_s}{\kappa u_* z\left(1 - \frac{x_3}{H}\right)F_{strat}\left(RI_g\right)}dx_3\right] \qquad (2.5.49a)$$

and

$$\bar{u} = \frac{u_*}{k}\left[\ln\left(\frac{b}{z_o}\right) + \int_b^{x_3}\frac{1}{x_3 F_{strat}\left(RI_g\right)}dx_3\right] \qquad (2.5.49b)$$

The above two equations do not constitute an explicit solution for the concentration and velocity profiles, because RI_g is a function of the concentration gradient $d\bar{c}/dx_3$ in accordance with Equation (2.5.48). When $RI_g \to 0$, however, Equation (2.5.49a) converges to the Rousean solution described by Equation (2.5.46) and Equation (2.5.49b) converges to the logarithmic profile. These two unstratified profiles can be used as base forms for an iterative solution for Equations (2.5.49a) and (2.5.49b). When conducting field measurements of suspended sediment concentration with acoustic methods, sediment stratification affects the observations. The theory presented above can be used to interpret acoustic measurements of velocity profiles and suspended sediment which might not follow the logarithmic law and Rousean distribution, respectively.

2.5.1.4.3 Non-equilibrium suspensions

The relations for E_s shown in Table 2.5.3 can be combined with Equations (2.5.46) and (2.5.47) to provide a complete treatment for the case of equilibrium suspensions. However there are cases of flows for which the equilibrium assumption for the bed is not valid. Two examples of non-equilibrium suspension, referred to as the *pick-up* and *pull-down* problems, are illustrated in Figure 2.5.10.

In the *pick-up* problem, sediment-free water flows from an non erodible bed to an erodible bed (Figure 2.5.10 top). Sediment is gradually entrained, such that far

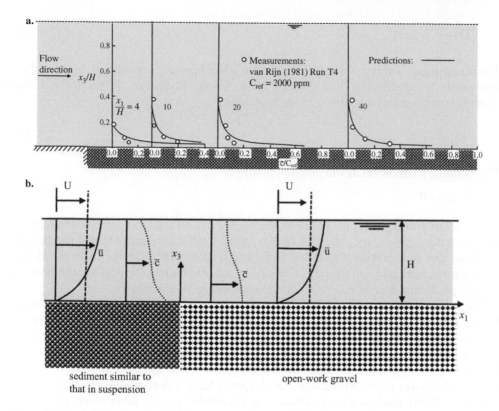

Figure 2.5.10 a): Downstream development of concentration profiles over loose sand bed- net-entrainment case 6 (reproduced from Celik & Rodi, 1988 with permission from ASCE) b): Equilibrium sediment-laden flow ranging from Sediment-Covered Bed to Open-Work Gravel Bed (reproduced from Huang & Garcia, 2000 with permission from ASCE).

downstream an equilibrium suspension is attained. That is, where $\overline{c}_e(x_3)$ denotes the equilibrium profile,

$$\lim_{x_1 \to \infty} \overline{c}(x_1, x_3) = \overline{c}_e(x_3) \qquad (2.5.50)$$

In order to simplify the problem, it is assumed that the composite bed roughness of the non-erodible portion is identical to that of the erodible portion. The suspension is assumed to be developing in the s-direction (streamwise direction) but to be constant in time and uniform in the transverse direction. The origin of the s-coordinate is taken at the junction between the non-erodible and erodible portions. In fact, local scour should occur just downstream of the junction. The bed elevation, however, can be taken to be constant over time scales that are long compared to the characteristic time of the flow but short compared to that required for noticeable bed erosion. With the flow at equilibrium the sediment mass conservation can be written as

$$\overline{u}_1 \frac{\partial \overline{c}}{\partial x_1} - v_s \frac{\partial \overline{c}}{\partial x_3} = \frac{\partial}{\partial x_3} \left(D_d \frac{\partial \overline{c}}{\partial x_3} \right) + \frac{\partial}{\partial x_1} \left(D_d \frac{\partial \overline{c}}{\partial x_1} \right) \qquad (2.5.51)$$

The boundary condition at the water surface is:

$$\left[D_d\frac{\partial \bar{c}}{\partial x_3} + v_s\bar{c}\right]\Big|_{x_3=H} = 0 \tag{2.5.52}$$

The Equations in Table 2.5.3 cannot be used as a bed boundary condition, because they apply only to equilibrium flow (Parker, 1978; Garcia & Parker, 1991). Instead, the appropriate boundary condition at the bottom is:

$$-D_d\frac{\partial \bar{c}}{\partial x_3}\Big|_{x_3=b} = v_s E \tag{2.5.53}$$

The boundary condition at the junction is simply

$$\bar{c}\big|_{x_3=0} = 0 \tag{2.5.54}$$

The pick-up non-equilibrium problem (see Figure 2.5.10 top) has attracted the attention of many authors since the 1930's (e.g., Mei, 1969). Two more recent treatments are those due to Van Rijn (1982) and Celik and Rodi (1988). Van Rijn uses the above formulation, with a closure for D_d afforded by a k-ε turbulence model for the flow. A sample prediction from Celik & Rodi is shown in Figure 2.5.10a .

The ather classical non-equilibrium case is the **put-down** problem illustrated in Figure 2.5.10b. In this problem, a steady uniform suspension flows from a sediment-covered bed to an open-gravel bed. As the sediment deposits on the gravel pores it cannot be resuspended by the flow turbulence and eventually the water column will be free of suspended sediment. Observations have shown that accumulation of fine sediment in the pores of spawning open-work gravel have a detrimental effect on biota. Einstein (1968) studied this problem experimentally in the laboratory and more recently, Huang and Garcia (2000) have found an analytical solution that provides estimates of vertical sediment concentration profiles which compare reasonably well with laboratory observations by Wang and Ribberink (1986).

2.5.1.4.4 Transport of sediment mixtures in suspension

Let the grain size range of interest be divided into N subranges, each with mean size ϕ_j on the phi scale and geometric mean diameter $D_j = 2^{-\phi_j}$, where $j = 1\ldots N$. Furthermore, let $\bar{c}_j$ denote the volume concentration of sediment in the j-th subrange. It follows that

$$\bar{c} = \sum_{j=1}^{N} \bar{c}_j \tag{2.5.55}$$

As long as the suspension remains dilute, the equation for mass conservation generalizes easily; for example, Equation (2.5.34) generalize to

$$\frac{\partial \bar{c}_j}{\partial t} + (\bar{u}_i - v_s k_i)\frac{\partial \bar{c}_i}{\partial x_i} = \frac{\partial}{\partial x_i}\left(D_d\frac{\partial \bar{c}_i}{\partial x_i}\right) \tag{2.5.56}$$

The boundary condition at the water surface similarly generalizes in a straight forward manner: the flux of suspended sediment normal to the water surface should vanish for each grain size range.

The boundary condition at the bed can be formulated by generalizing the relations for sediment entrainment rate normal to the bed embodied in Equation (2.5.37); where Ej denotes the volume entrainment rate for the j-th subrange,

$$E_j = \overline{u_i' c_i'}|_{near\ bed} n_i^b = -D_d \frac{\partial \overline{c_j}}{\partial z_n}|_{near\ bed} \tag{2.5.57}$$

In the above relation, z_n is specifically defined to be a coordinate defined upward normal from the bed (averaged over bedforms). Garcia and Parker (1991) have provided a generalized treatment for E_j in the case of mixtures, for which the $E_s = \frac{A Z_{uj}^5}{1 + \frac{A}{0.3} Z_{uj}^5}$ is used for each parameter Z_{uj} that can be specified as

$$Z_{uj} = \lambda_m \frac{u_{*s}}{v_{sj}} Re_{pj}^{0.6} \left(\frac{D_j}{D_{50}}\right)^{0.2} \tag{2.5.58}$$

In the above relations, v_{sj} denotes the fall velocity of grain size D_j in quiescent water, D_{50} denotes the median size of the surface material in the bed,

$$Re_{pj} = \frac{\sqrt{RgD_jD_i}}{v} \tag{2.5.59}$$

and the parameter λ_m is given by $\lambda_m = 1 - 0.288\sigma_\phi$. Here, σ_ϕ denotes the arithmetic standard deviation of the bed surface material on the phi scale. No summation in j is implied in the above equations. Comparisons of values estimated with the above sediment mixture entrainment formulation against observations made under equilibrium conditions in the Niobrara River (USA) indicate good agreement (Garcia & Parker, 1991).

2.5.1.4.5 Sediment resuspension by unsteady flows

Past sedimentation research has focused primarily on steady flows. However, a number of important problems occur in unsteady flows. Examples include sediment transport by floods, boat wakes, and flow surges. As shown above, there are a number of entrainment relations available for predicting the upward flux of sediment from a mobile bed, but as flow unsteadiness increases, the reliability of such relations becomes doubtful. Motivated by the need to assess the environmental impact of increasing navigation in the Upper Mississippi River Basin, a set of experiments was conducted in a custom made rectangular duct using an acoustic velocity profiler (Garcia et al., 1999). The profiler also provided suspended sediment concentrations for the coarse and fine sand in suspension. Figure 2.5.11 visualizes time lags observed between the bed shear stress and the upward flux (entrainment) of sand from the bed (Admiraal & Garcia, 2000). The phase lags were larger for tests with fine sand than for tests with coarse sand (Admiral et al., 2000). The finer sediment seems to be more protected from the flow turbulence by the presence of a viscous sublayer, and does not respond to flow changes as readily as the coarser material that is more exposed to the flow turbulence. The Garcia-Parker relation in Table 2.5.3 was found to predict well sediment entrainment rates for weakly

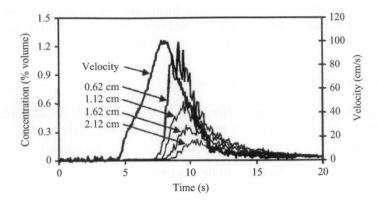

Figure 2.5.11 Time series of flow velocity and sediment concentration at different elevation above the bottom in unsteady flow condition. Notice the lag between peak velocity and maximum concentration at different heights (from Admiraal & Garcia, 2000, with permission of Springer).

unsteady flows. However for rapidly accelerating flows, the Garcia-Parker equation had to be corrected to account for time lags.

2.5.2 Flows in vegetated channels

The role of aquatic vegetation is crucial in preserving the balance of an ecosystem by improving water quality due to nutrient uptake, oxygen production and hosting crucial fauna species. The additional resistance caused by vegetation, and the corresponding decrease in flow capacity, are the focus of studies of vegetation in OCFs. Also, vegetation is known to reduce the water turbidity by trapping the near canopy suspended load as well as the bed load by reducing the flow speed that keeps the sediments in suspension and causes the high bed shear stress required for bed load transport. Because of the important effects of aquatic vegetation on OCF hydraulics, water quality and sediment transport, a better understanding of vegetation interaction with the flow is needed. Consequently, vegetated OCF studies have increased in number during recent years. A brief introduction of some of the previously applied approaches ensues. An extensive presentation of the available literature can be found in Nepf (2012), which reflects contemporary efforts to understand flow and transport in vegetated channels.

Hydraulic engineers usually try to include the effects of vegetation in OCF calculations by increasing the flow resistance coefficient (*e.g.*, the Manning-Strickler roughness coefficient). Among the first to study OCF through vegetation were Palmer (1945) and Ree and Palmer (1949). They used Manning's coefficient and they suggested that the n-UR_h relationship depends only on the physical properties of the grass and is independent of channel geometry and flow conditions. Other studies,*e.g.*, by Petryk and Bosmajian (1975), proposed a quantitative procedure for estimating the Manning's-Strickler coefficient that considers flow depth and the vegetation characteristics for cases in which flow depth is less than maximum plant height.

Vegetation canopies exhibit a high range of dimensions and scales related to the canopy structure as well as to the geometry characteristics of the individual plants (stems, blades etc.), as illustrated in Figure 2.5.12.a. These different dimensions

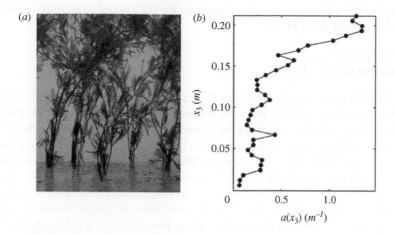

(a)

(b)

Figure 2.5.12 (a) Photograph of Eurasian water milfoil (*Myriophyllum spicatum*) in a laboratory flume, and (b) vertical profile of frontal area density $a(z)$, for the milfoil canopy at a stem density of $n = 50 \ m^{-2}$, where z is the elevation above the bed (King et al., 2012, reproduced with permission).

introduce length scales which affect the canopy flows. Important scales in canopy flows include water depth H, plant height h_P, range of stem diameters d_P, spacing between plants ΔS, and frontal area per canopy volume a $(a = d_p/\Delta S^2)$. The latter scale is usually called "leaf area index" in the terrestrial canopy literature (Nepf, 2012). King *et al.* (2012) introduced another, less obvious, length scale; the inverse of the plant frontal area per unit volume a^{-1} (for the general case $a \equiv \frac{1}{V}\sum_{i=1}^{N} A_i$; where V is the volume which includes both fluid and plants, A_i is the frontal area of plant i perpendicular to the mean flow and N is the number of plants). A vertical profile of the frontal area density of Eurasian water milfoil (*Myriophyllum spicatum*) is shown in Figure 2.5.12b. Another dimensionless measure of the canopy density is the frontal area per bed area, also called roughness density (Wooding *et al.*, 1973), λ_f $(\lambda_f = \int_{x_3=0}^{H} adx_3)$.

Another quantity related to canopy density is the solid volume fraction occupied by plants, ϕ_p (*e.g.*, for the case of plants that can be approximated as circular cylinder, $\phi_P \approx (\pi/4)ad)$.

Vegetation canopies may be emergent or fully submerged. King *et al.* (2012) organized flow through vegetation canopies into three categories:

1. flow through emergent vegetation, for which $H/h_P \leq 1$ (here, H = water depth and h_p = height of the plants);
2. flow through deeply submerged vegetation $(H/h_P > 2)$ and,
3. flow through intermediate submerged vegetation $(1 < H/h_p < 2)$.

For the case of submerged vegetation, the flow within the canopy is driven by the turbulence stress at the top of the canopy, the pressure gradient and gravitational acceleration (channel slope). Nepf and Vivoni (2000) showed that the relative importance of the aforementioned mechanisms for the case of submerged canopies can be defined using the ratio H/h_P:

$$\frac{\text{turbulent stress}}{\text{pressure gradient}} \propto \frac{H}{h_p} - 1 \qquad (2.5.60)$$

Another classification may be based on the canopy density. For emergent vegetation, the point where the transition between sparse and dense canopies takes place is not immediately clear. Within an emergent aquatic canopy the characteristic turbulence length scale is defined using the stem diameter and the spacing between the plants (Tanino & Nepf, 2008). The mean flow is defined using the distribution of the frontal area of the plant canopy (Nepf, 2012). For submerged vegetation canopies the transition between sparse and dense cases occurs at a roughness density of $\lambda_f \sim 0.1$ ($\lambda_f = ah_P$ for vertically uniform$a$). When $ah_p \ll 0.1$ (sparse canopy case) the contribution of the canopy drag is small compared to the bed drag. For this case, a relatively well-defined turbulent boundary layer profile is developed. The effect of vegetation can be included in an effective roughness of the bed (Figure 2.5.12 a). However, when the density of the canopy increases, the drag of the canopy becomes important and a drag discontinuity forms at the top of the canopy, leading to the development of a shear layer and producing canopy-scale turbulence. This mixing layer reaches a fixed scale and has fixed penetration into the canopy δ_e (Figures 2.5.13. b and c) at the distance from the canopy leading edge where the shear production of canopy-scale vortices equals the kinetic energy dissipation due to the canopy drag. Nepf $et\ al.$ (2007) used laboratory experiments to show that for dense canopies this length scale can be predicted as:

$$\delta_e = \frac{0.23 \pm 0.6}{C_D a} \qquad (2.5.61)$$

where C_D is the drag coefficient. For the range of $0.1 < C_D ah_p < 0.23$ the shear layer vortices penetrate to the bed ($\delta_e = h_p$) causing a highly turbulent condition within the canopy (Figure. 2.5.13 b). For higher canopy densities the shear layer vortices do not penetrate to the bed (Figure 2.5.13 c).

Kowen and Unny (1973) did a series of pioneering experiments using plastic strips of different thickness to represent vegetation, and outlined several flow regimes: a)

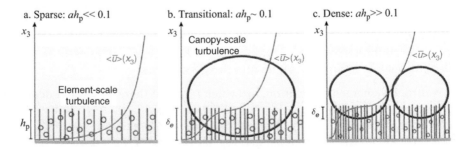

Figure 2.5.13 Vertical velocity profile and dominant turbulence scales for (a) sparse canopy ($ah_P \ll 0.1$), (b) transitional canopy ($ah_P \approx 0.1$) and (c) dense canopy ($ah_P \gg 0.1$) (adapted from Nepf, 2012. Reproduced with permission of Annual Review of Fluid Mechanics, Volume 44 by Annual Reviews, www.annualreviews.org).

erect, when the plastic strips are erect and stationary; b) waving, when the strips oscillate with a waving motion; and c) prone, when the strips are bent over. The regimes concur with the classification by Gourlay (1970) for Kikuyu grass. However, as the hydraulic behavior and turbulence structure for erect and waving regimes are similar, only two hydraulic regimes may be considered; erect and prone. The frictional coefficients are similar for the erect and waving regimes compared to the prone vegetation, for which the friction coefficient is considerably less (by a factor of five). Reduced turbulent diffusion at the top of the canopy is a further consideration for prone vegetation, as bent vegetation partially blocks vertical movement of the flow. Kowen and Unny (1973) were among the first investigators to quantify the effects of the mechanical characteristics of the plants, doing so by introducing the MEI parameter, which is the product of the relative density M and the flexural rigidity EI (the product of the plant modulus of elasticity and the moment of inertia of the plant stem).

Early laboratory studies of sediment transport through vegetated OCF included those by Tollner (1974) and Tollner et al. (1982). They reported good sediment capacity predictions using parameters similar to those proposed by Graf (1971). The predictive parameters were based on vegetation spacing instead of channel width. However, for their experiments they used a relatively narrow and short flume (2.10m × 0.13m) for which equilibrium conditions (at least for suspended sediments) may be questionable. Li and Shen (1973) used a method similar to the technique proposed by Petryk (1969) and estimated the sediment transport capacity in OCF numerically. The method was based on superposition of the wakes generated behind isolated plants. For their calculations they assumed a local drag coefficient of 1.2 *a priori*, and their results show a mean coefficient close to 1.1.

Lopez and Garcia (1997) studied flow and sediment transport in vegetated OCF experimentally. They used an approach similar the approach of Raupach and Shaw (1982), originally developed for atmospheric boundary layer flow though plant canopies. They showed that both temporal and spatial averaging of the measured data are required to correctly describe the flow in vegetated channels using a horizontally-averaged one-dimensional model. For a 1D description of the vertical profiles of flow quantities within the canopy, Raupach and Shaw (1982) proposed two schemes for the horizontal averaging of the conservation equations. Lopez and Garcia (1997) checked both schemes in their analysis concluding in similar results. Similar methodologies were later used by Lopez and Garcia (1999, 2001), Nikora et al. (2001, 2007a, 2007b) for the analysis of OCF over a rough bed and by Nepf (2012) for vegetated flows.

In the last decade, there has been an increasing need to understand sediment processes in wetlands, floodplains and rivers with vegetated banks, by which suspended solids and chemical contaminants (pesticides, heavy metals, etc.) are deposited and retained within a natural or man-made waterway. When properly formulated, mathematical models can be quite useful to assess the role played by vegetation in reducing both the capacity of water flows to transport sediment in suspension as well as the ability of turbulence to entrain sediment into suspension. A two-equation turbulence model based on the k-ε closure scheme was developed to simulate the flow and turbulence characteristics of open-channel flows through non emergent vegetation (Lopez & Garcia, 1997). Once the performance of the model was verified, the flow structure of vegetated open channels was studied with the help of numerical experiments. Simulated rigid and flexible plants were used to validate the model. Dimensional analysis allowed identification of the dimensionless parameters that govern suspended sediment transport processes in the presence of vegetation, and thus helped in the design of numerical experiments to investigate the role of different flow properties, sediment characteristics, and vegetation parameters upon the transport capacity. King *et al.* (2012) used a modified version of the aforementioned k-e turbulence closure which takes into consideration two parts of kinetic energy; Shear Kinetic Energy (*SKE*) to another Wake Kinetic Energy (*WKE*) and include a stem diameter scale dissipation mechanism for the divert *SKE* to *WKE*. Their results compare well with the results of previous work by Lopez and Garcia (2001) and a terrestrial model developed by Katul *et al.* (2004).

The two equation turbulence model was found to accurately represent the mean flow and turbulence structure of open channels through simulated vegetation, thus providing the necessary information to estimate suspended sediment transport processes (Lopez & Garcia, 1998). A reduction of the averaged streamwise momentum transfer toward the bed (*i.e.*, shear stress) induced by the vegetation was identified as the main reason for lower suspended sediment transport capacities in vegetated waterways compared with those observed in non-vegetated channels under similar flow conditions (Lopez & Garcia, 1998). Simulated profiles of kinematic eddy viscosity were used to solve the sediment diffusion equation, yielding distributions of relative sediment concentration slightly in excess of the ones predicted by the Rousean formula. Also, a power law was found to provide a very good collapse of all the numerically generated data for suspended sediment transport rates in vegetated channels.

2.5.3 Aerated flows

Aerated flows occur when bubbles become distributed in the flow of a water body, resulting in a type of multiphase flow field. Flow aeration can be natural (*e.g.*, breaking waves, white water, plunging falls) or man-made (*e.g.*, bubble plume injection in estuaries or reservoirs, aeration at dams or weirs). Previously, aspects of multiphase flow were introduced in Section 2.5.1, Volume I, for the special case of sediment transport. In either case, the dispersed phase (whether bubbles or sediment particles) introduces an agitation, sometimes referred to as particle turbulence or pseudo turbulence, to the background flow, which can alter its mean and turbulent flow statistics and transport properties (Riboux, *et al.*, 2010). This agitation results from both the turbulent wakes following the dispersed phase as well as from the effects of fluid advection around each bubble or particle as they slip through the fluid.

Physically, bubbles or particles exert a drag force on the water, generally balanced by their buoyancy. Likewise, the water body alters the pathways of bubbles or particles by their transport with the local currents. The effect of particle drag can be considered through the drag coefficient, the particle size, or the particle slip velocity. Each of these parameters introduces different units to the analysis, but since they are related to one another, they should not be considered as independent parameters. For analysis of bubble plumes, for instance, the slip velocity has traditionally been used to parameterize the effect of the bubbles (Socolofsky & Adams, 2005). When slip velocity or particle size is used, the associated dimensions (L/T or L) can normally be combined with the other parameters governing the fluid flow to form a characteristic length scale of particle motion (see, e.g., Bombardelli, *et al.*, 2007 for analysis of bubble plumes). Hence, multiphase flows always have at least two length scales: one associated with the flow domain and one associated with the characteristic range of influence of the particle motion. This fact complicates solutions to multiphase flow problems by removing the option of applying self-similarity.

The effect of local currents is normally addressed by the Stokes number, which is the ratio of the relaxation time of the particle motion to the time scale of the local water acceleration (Balachandar & Eaton, 2010). In a physical sense, this is a non-dimensional measure of a particle's inertia. Due to the large mass difference between air and sediment, the Stokes numbers of bubbles and sediment particles of equal size are much different: air bubbles are more easily deflected and transported by ambient motion than sediment particles when all other parameters are held constant. This has important impact on the forces that must be considered in the governing equations of particle motion and ultimately on the spreading of particles.

2.5.4 Ice-laden flows

Flow situations involving ice are direct extensions of modeling free-surface flows, with the added complexities of the flow boundaries imposed by ice covers, and the effects that ice-piece drift and accumulation may exert. Substantial further complexities soon arise when thermal and strength processes need to be simulated. Special laboratory facilities may be needed for replicating some of these processes.

Common situations of laboratory investigation and hydraulic modeling involve free-surface flows with ice as a floating solid boundary retarding water flow, or ice occurs as solid pieces conveyed and accumulated by water flow. Such flows primarily concern patterns and profiles of water flow, and possibly how they interactively affect patterns and profiles of ice movement and accumulation. The usual forces associated with water flow have to be considered in models concerning ice effects on water flow: *i.e.*, water momentum, gravity, the viscosity and surface-tension properties of ice, and boundary resistance. When investigating or simulating ice movement and accumulation, additional forces are associated with ice momentum, hydrodynamic forces exerted against ice pieces, friction between ice pieces as well as between ice and other solid boundaries. These forces are normally handled in terms of the same parameters and similitude relationships used for simulating free-surface flows in channels of fixed or mobile (sediment) boundaries.

The strength and deformation behavior of ice and accumulations of ice pieces, such as form ice covers or ice jams, are determined by geometric and material factors. Depending on the combination of these factors, the strength and deformation behavior can be relatively simple or complicated to formulate and replicate in the laboratory. At best they can be reproduced or simulated within a fairly narrow range of length scales or facility extents, and sometime require special materials and additives, and procedures, for modifying the strength and material properties of ice. Thermal factors, such as freezing, freeze-bonding and the frictional behavior of ice pieces, can be difficult to simulate, because they involve processes at the molecular scale (notably the phase change of water) that cannot be further reduced in scale. Useful references for additional information about ice formation, ice transport, and ice strength behavior and their effects on water flow are Ashton (1986) and Beltaos (1995; 2013).

2.A APPENDIX

2.A.1 Ratio of Nikuradse equivalent roughness size and sediment size for rivers (Garcia, 2008 with permission from ASCE)

Investigator	Measure of Sediment Size, D_x	$a_s = \Delta/D_x$
Ackers and White (1973)	D_{35}	1.23
Aguirre-Pe andFuentes (1990)	D_{84}	1 - 6
Strickler (1923)	D_{50}	3.3
Katul et al. (2004)	D_{84}	3.5
Keulegan (1938)	D_{50}	1
Meyer-Peter and Muller (1948)	D_{50}	1
Thompson and Campbell (1979)	D_{50}	2.0
Hammond et al. (1984)	D_{50}	6.6
Einstein and Barbarossa (1952)	D_{65}	1
Irmay (1949)	D_{65}	1.5
Engelund and Hansen (1967)	D_{65}	2.0
Lane and Carlson (1953)	D_{75}	3.2
Gladki (1979)	D_{80}	2.5
Leopold et al. (1964)	D_{84}	3.9
Limerinos (1970)	D_{84}	2.8
Mahmood (1971)	D_{84}	5.1
Hey (1979); Bray (1979)	D_{84}	3.5
Ikeda (1981)	D_{84}	1.5
Colosimo et al. (1986)	D_{84}	3–6
Whiting and Dietrich (1990)	D_{84}	2.95
Simons and Richardson (1966)	D_{85}	1
Kamphuis (1974)	D_{90}	2.0
Van Rijn (1982)	D_{90}	3.0

2.A.2 Equivalent roughness of bed forms

As the flow intensity increases, bed forms such as ripples and dunes can develop (*e.g.*, Raudkivi, 1997). In this situation, the bed roughness also will be influenced by form drag due to the presence of bed forms. The fundamental problem is that the bed form characteristics and, hence, the bed roughness depend on the main flow characteristics (*e.g.*, mean velocity, depth) and sediment characteristics (*e.g.*, grain size, density). Thus, the hydraulic roughness in the presence of bed forms is a dynamic parameter that depends strongly on flow conditions as well as on the bed sediment properties. The equivalent roughness of alluvial beds in the presence of ripples and dunes was addressed with the Nikuradse hydraulic roughness approach by Brownlie (1981) and van Rijn (1982, 1984c). In van Rijn's approach, the height due to grain-induced roughness is added to an estimate of the equivalent roughness height produced by ripples and dunes obtained from field and laboratory observation, to obtain a measure of the total (grain plus form resistance) effective roughness, resulting in the following

$$k_s = a_s D_{90} + \gamma_{sf} 1.1 \Delta \left(1 - e^{-25\Delta/\lambda}\right) \qquad (2.A.2.1)$$

where $a_s = 3$; D_{90} = grain size for which 90% of the bed material is finer; γ_{sf}= dune shape factor = 1; Δ and λ = bed form height and length, respectively; and Δ/λ = bedform steepness.

The effective roughness height is then used to estimate the Chezy friction coefficient,

$$C_Z = (\frac{g}{C_f})^{1/2} = 18 \log (\frac{12\,R_{bb}}{k_s}) \qquad (2.A.2.2)$$

In this equation, R_{bb} = hydraulic radius of the river bed (i.e., substracting streambank effects on flow resistance) according to Vanoni and Brooks (1957) (see Vanoni, 2006, p.91). Notice that the Chezy coefficient is not dimensionless. Application of Equation (2.A.2.1) to field conditions has resulted in considerable overestimation of the hydraulic roughness (van Rijn, 1996). Further analysis showed that the lee-side slopes of natural sand dunes in rivers were less steep than those of dunes in the laboratory and a shape factor γ_{sf}= 0.7 was recommended for application to natural river dunes.

A different approach based on boundary-layer theory and measured velocity profiles was proposed by Fedele and Garcia (2001). When spatially-averaged velocity profiles of flow (Nikora et al., 2001) over dunes are available, this method can be used to estimate a spatially-averaged composite roughness k_c due to the combined effect of both grain friction and form drag due to bed forms in large sand-bed rivers. Boundary layer studies have shown that an alternative approach for describing the vertical flow velocity distribution in flows where the geometry and size of the roughness elements is such that skin friction and form drag are present, is given by the following Equation

$$\frac{u}{u_*} = \frac{1}{\kappa} ln\left(\frac{x_3 u_*}{v}\right) + A - \frac{\Delta u}{u_*}\left[\frac{u_* k_c}{v}\right] \qquad (2.A.2.3)$$

In Equation (2.5.1A.3), $\kappa = 0.41$ and $A = 5.5$ are universal constants previously introduced, and $\Delta u/u*$ is a **roughness function** which is equal to zero for smooth walls (square brackets indicate functional relationship). When plotting u/u_* versus $ln(u_* x_3/v)$, this equation represents a family of parallel lines, each being displaced downwards from the smooth-wall velocity profile by an amount $\Delta u/u*$ (Schlichting, 1979).

The roughness function for alluvial streams with dunes is shown in Figure 2.A.2.1. It shows $\Delta u/u_*$ as a function of the parameter $k_c u_*/v$ for laboratory and field streams with fully-developed dunes (Fedele & Garcia, 2001). It is observed that for values of the roughness Reynolds number $k_c u_* T/v$ larger than 100–200, most of the data collapse along a straight line, along the fully-rough hydraulic regime, which is well represented by the following fit,

$$\frac{\Delta u}{u_*} = 2.43 ln\left(\frac{u_* k_c}{v}\right) - 3.24 \qquad (2.A.2.4)$$

An application of the alluvial roughness function is its potential use to assess the effect of temperature changes on flow structure and bed morphology, a phenomenon commonly observed in the Missouri River, USA. It is observed in Figure 2.A.2.1 that even

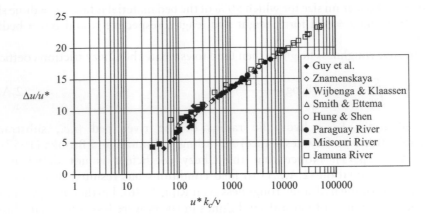

Figure 2.A.2.1 Roughness function for alluvial streams with dunes (republished with permission of Copyright Clearance Center, Inc., from Fedele & Garcia, 2001).

though the flows are under fully-rough hydraulic conditions, temperature variations will affect the viscosity of the water and this in turn will cause variations in the roughness Reynolds number and the flow structure. If spatially average profiles are available (Holmes & Garcia, 2008), the roughness function can provide an estimate of the equivalent roughness induced by the presence of bedforms.

NOTATION

Symbols and indexes (for averaging etc.)

$\theta = \bar{\theta} + \theta'$ = Reynolds decomposition of a variable θ

$\theta = \langle \bar{\theta} \rangle + \widetilde{\bar{\theta}} + \theta'$ = modified Reynolds decomposition of a variable θ

$\bar{\theta} = \langle \bar{\theta} \rangle + \widetilde{\bar{\theta}}$ = decomposition of a time-averaged variable $\bar{\theta}$ into its double-averaged value $\langle \bar{\theta} \rangle$ and fluctuation $\widetilde{\bar{\theta}}$

Latin letters

A = cross section area of the flow
B = channel width
b = body width (or diameter)
b_1 and b_2 = the distance of a channel centerline from the channel side walls
$B_p(\tau)$ = time correlation function
B_s = roughness function
$B_{ij}(\mathbf{x}, \mathbf{x}', t, t')$ = three-dimensional correlation function of a vector field
c_b = volumetric sediment concentration near the bed
$\bar{c}_e(x_3)$ = equilibrium sediment volume concentration profile
c_P = the specific heat
$\bar{c}_j$ = volume concentration of sediment in the *j-th* subrange

C = substance concentration or suspended sediment concentration
C_D = body drag coefficient
C_f = friction coefficient
C_L = body lift coefficient
$Co_{ij}(k)$ = coherence function
C_z = Chezy coefficient
$c_\varepsilon, c_{1\varepsilon}\ c_{2\varepsilon}\ c_{3\varepsilon}$ = empirical constant for the estimation of the TKE dissipation rate
c'_μ = empirical constant for the estimation to "eddy viscosity"
d = zero-plane displacement (defines the origin of the vertical coordinate for the log velocity profile)
d_P = stem diameter of a plant
D = sediment grain size
D_d = vertical kinematic eddy viscosity
D_p = eroded volume of particles per unit area per unit time
D_x = a characteristic diameter of a grain ($e.g.$, D_{50})
$D_{ij}^{p}(\mathbf{r}, \tau)$ = p-order structure function
$D_* = \left[\frac{gR}{v^2}\right]^{1/3}$ = parameter used for the estimation of sediment particle fall velocity
E = rate of sediment entrainment
E_p = eroded volume of particles per unit area per unit time
E_s = dimensionless sediment entrainment rate
$E(k)$ = energy spectrum function
f = friction factor (or resistance coefficient)
f_s = vortex shedding frequency
$F = \overline{u'^4_i}/\overline{u'^2_i}^2$ = flatness
$Fr = U/\sqrt{gD}$ = Froude number
$F_s = x_3 - h$ = a function that has a value of zero at the free surface
F_{strat} = function for the correction of kinematic eddy diffusivity under stratified conditions
g = magnitude of the gravity acceleration
$g_i = i^{th}$ component of the gravity acceleration
H = flow depth
h = water surface elevation
h_p = plant height
$k = \frac{1}{2}\overline{u'^2_{ii}}$ = turbulent kinetic energy
$k = 2\pi/\lambda$ = wavenumber
k_C = molecular diffusivity of a substrate C
k_T = thermal diffusivity
$Ku = F - 3$ = kurtosis
K_n = correction coefficient of Manning's equation
l_m = mixing length
l_w = half-width of the wake region
L = boundary layer length
L_{ii} = integral turbulence scale
$\mathcal{L}$ = external turbulence scale
n = Manning's coefficient

n_i = unit vector normal to the interface surface and directed into the fluid

N_b = particles per unit bed area

p = instantaneous pressure

$\bar{p}$ = time-averaged pressure (sometimes presented *as* P)

p' = fluctuating pressure ($p' = p - \bar{p}$)

P = turbulent kinetic energy production

q_b = volumetric bed load transport rate

q_{bs} and q_{bn} = bedload transport in the longitudinal and transverse direction respectively

$\vec{q} = (q_s, q_n)$ = 2D vector of bedload transport rate

q^* = dimensionless bedload transport rate (Einstein number)

Q = flow rate

$\mathbf{r} = \mathbf{x} - \mathbf{x}'$ = distance between two points in the flow

R = submerged specific gravity

R_f = Richardson number

R_h = hydraulic radius of the flow

$R_p = \frac{v_s D}{v}$ = particle Reynolds number based on the falling velocity

$R_{ep} = \frac{\sqrt{gRDD}}{v}$ = Reynolds number defined with sediment particle properties

$Re_{R_h} = UR_h/v$ = Reynolds number based on R_h

$Re_* = u_* \Delta/v$ = roughness Reynolds number

$Re_x = U_m x/v$ = boundary layer Reynolds number based on x

$Re_\delta = U_m \delta/v$ = boundary layer Reynolds number based on δ

$Re_L = U_m L/v$ = boundary layer Reynolds number based on L

$Re_W = \bar{u}_e b/v$ = wake Reynolds number

$R_{ij} = \overline{u'_i u'_j}/(\sigma_i \sigma_j)$ = velocity correlation coefficient

RI_g = gradient Richardson number

u_b = particle velocity

u_i = instantaneous velocity component with i=1 for longitudinal (u), i=2 for transverse (v), and i=3 for vertical (w) directions

$\bar{u}_i$ = time-averaged velocity component with i=1 for longitudinal ($\bar{u}$), i=2 for transverse ($\bar{v}$), and i=3 for vertical ($\bar{w}$) directions (sometimes presented as U_i)

u'_i = velocity fluctuation ($u'_i = u - \bar{u}$)

$\bar{u}_c$ = flow velocity at the wake centerline

$\bar{u}_e$ = velocity of the external (or approach) flow

$u_* = (\tau_o/\rho)^{0.5}$ = shear velocity

$u^+ = \bar{u}/u_*$ =dimensionless velocity expressed in wall units

u_{*c} = critical shear velocity

$\bar{u}_{max}$ = time averaged maximum velocity at the free surface (velocity-defect law)

U = mean or bulk velocity

U_E = eddy convection velocity

U_m = external (or maximum) velocity

S = bed slope

S_f = friction slope

$S(k)$ = wavenumber spectrum

$S(\omega)$ = frequency spectrum

S_{ij} = strain, $S_{ij} = 0.5(\partial \bar{u}_i/\partial x_j + \partial \bar{u}_j/\partial x_i)$

S_{int} = total area of the fluid-boundary interface within the averaging domain

$Sk = \overline{u_i'^3} / \overline{u_i'^2}^{3/2}$ = skewness

St = Strouhal number

s_T = heat equation sink/source term

s_C = sick/source term of the dissolved substrate transport equation

t' = fluctuating of temperature

T_o = total averaging period

T_f = period of time (not necessarily continuous) when fluid passes through a fixed point x_i

T_r = reference temperature

T_{sal} = particle saltation or movement period

$\overline{T}$ = mean value of temperature

$\underline{T}$ = tensor of the turbulent stresses

$\overline{TKE} = 0.5(\overline{u_1'^2} + \overline{u_2'^2} + \overline{u_3'^2})$ = total turbulent kinetic energy (sometimes presented as k)

v_s = particles falling velocity

v_{sC} = hindered particles falling velocity

V_m = the part of the total averaging domain of volume V_o that has been 'visited' by fluid

V_p = volume of a particle

$\hat{V}$ = characteristic velocity scale of the flow

$w\left(\frac{x_3}{H}\right) = \frac{2W_0}{\kappa} sin^2\left(\frac{\pi x_3}{2H}\right)$ = wake correction for turbulent boundary layer flows

W = weight of a particle

Ws = submerged weight of a particle

W_0 = Coles wake strength parameter

x_1, x_2, x_3 = longitudinal x, transverse y, and vertical z coordinates, respectively

$x_3{}^+ = u_* x_3 / v$ = inner or wall length

$Z = k^n \hat{L}^m$ = turbulent quantity Z for a k-Z turbulent closure (e.g., k-e, k-ω etc.)

$Z_R = \frac{v_s}{\kappa u_*}$ = Rouse number

zn = coordinate defined upward normal from the bed (averaged over bedforms).

Z_{uj} = Garcia-Parker (1991) entrainment coefficient for sediment mixtures

z_0 = bed roughness length that represents the distance from the bed where the velocity becomes zero

Greek letters

$a = d/\Delta S^2$ = frontal area per canopy volume

a_s = constant for the estimation of equivalent roughness height for open channel flows

α_r = coefficient of isobaric expansion

β = transformation constant that relates ϕ with variations in density

$\gamma(x_i, t)$ = a 'clipping' or 'distribution' function equal to 1 in the fluid and 0 otherwise

Γ = turbulent diffusivity (thermal or mass diffusivity)

δ = boundary layer thickness

δ_v = thickness of the viscous sublayer, $\delta_v \sim 5(v/u_*)$

δ_b = thickness of the bedload layer

δ_e = mixing length penetration into a canopy

δ_{sin} = amplitude of a sinusoidal distribution

δ_{ij} = Kronecker's delta

Δ = roughness height

Δ_w = wave height

ΔS = spacing between plants

$\Delta u(r, \tau)$ = velocity increment

ε = energy dissipation

$\bar{\varepsilon}$ = mean energy dissipation, $\bar{\varepsilon} = 0.5 v \sum_{i,j} \overline{(\partial u'_i/\partial x_j + \partial u'_j/\partial x_i)^2}$

η = dissipative scale, $\eta = (v^3/\bar{\varepsilon})^{1/4}$

η_b = bed elevation with respect to an arbitrary reference elevation

θ = downstream slope angle

κ = von Kármán constant

λ = wavelength (or a spatial scale)

λ_f = canopy density or roughness density

λ_p = bed or sediment porosity

λ_{sal} = sediment saltation or step length

μ = coefficient of Coulomb friction

v = kinematic fluid viscosity

v_T = "eddy viscosity"

v_η = Kolmogorov's velocity scale, $v_\eta = (v\bar{\varepsilon})^{1/4}$

v^T_{ij} = "eddy viscosity" tensor

ρ = fluid density

ρ_o = reference density

ρ_s = sediment density

$-\rho\overline{u'_i u'_j}$ = Reynolds (turbulent) stress tensor

σ_i = standard deviation

σ_t = Prandtl number (for heat transport) or the Schmidt number (for mass transport)

σ_k = empirical diffusion coefficient

σ_z = empirical constant of transport equation of quantity Z

c_μ = empirical constant for the computation of "eddy viscosity"

$\tau = t - t'$ = time lag

τ_η = Kolmogorov's time-scale, $\tau_\eta = (v/\bar{\varepsilon})^{1/2}$

τ_o = wall (or bed) shear stress

τ_{bs} = boundary shear stress due to skin friction

τ^* = dimensionless Shields stress

τ_c^* = critical dimensionless Shields stress

τ_c = critical shear stress

v_i = velocity vector of the fluid at the fluid-boundary interface

ϕ = passive tracer concentration

ϕ_p = solid volume occupied by plants

φ = angle of repose

$\Phi = -\log_2(D) = -\frac{\ln(D)}{\ln(2)}$ = sedimentological scale

$\Phi_{ij}(k_1, k_2, k_3)$ = three-dimensional spectral tensor of a vector field

$\phi_T(x_i, t)$ = local time porosity defined as the ratio T_f/T_o

$\phi_{Vm} = V_m/V_o$ = space porosity

$\phi_{VT} = \phi_{Vm}\langle\phi_T\rangle$ = combined space-time porosity

χ_m = molecular diffusion coefficient
ω = angular frequency
ω_k = vorticity vector, $\omega_k = (\partial u_i/\partial x_j - \partial u_j/\partial x_i)$
$\omega_i\omega_i/2$ = enstrophy

REFERENCES

Abad, J.D. & Garcia, M.H. (2009) Experiments in a high-amplitude Kinoshita meandering channel: 1. Implications of bend orientation on mean and turbulent flow structure. *Water Resour Res.*, 45, W02401, 1–19.

Admiraal, D., Garcia, M.H., & Rodriguez, J.F. (2000) Entrainment response of bed sediment to time-varying flows. *Water Resour Res.*, 36(1), 335–348.

Admiraal, D. & Garcia, M.H. (2000) Laboratory measurements of suspended sediment concentration using an acoustic concentration profiler (ACP). *Exp Fluids.*, 28, 116–227.

Adrian, R.J. (2007) Hairpin vortex organization in wall turbulence. *Phys Fluids.*, 19, 041301.

Adrian, R.J. & Marusic, I. (2012) Coherent structures in flow over hydraulic engineering surfaces. *J Hydraul Res.*, 50(5), 451–464.

Adrian, R.J., Meinhart, C.D., & Tomkins, C.D. (2000) Vortex organization in the outer region of the turbulent boundary layer. *J Fluid Mech.*, 422, 1–54.

Ahrens, J.P. (2000) The fall-velocity equation. *J Waterw Port C-ASCE.*, 126(2), 99–102.

Ahrens, J.P. (2003) Simple equations to calculate fall velocity and sediment parameter. *J Waterw Port C-ASCE.*, 126(3), 146–150.

Akan, O. (2006) *Open Channel Hydraulics*. Butterworth-Heinemann. Oxford (UK).

Akiyama, J. & Fukushima, Y. (1986) Entrainment of noncohesive bed sediment into suspension. In: S.Y. Wang, Shen, H.W., & Ding, L.Z. (eds.) *Proceedings of the Third International Symposium on River Sedimentation*, The University of Mississippi, Mississippi, 804–813.

ASCE Task Committee (2000) *Hydraulic Modeling: Concepts and practice. Manuals and Reports on Engineering Practice No 97*, ASCE, Reston, Virginia.

ASCE Task Force (2002) Flow and transport over dunes. *J Hydraul Eng-ASCE.*, 128(8), 726–728.

Ashida, K. & Michiue, M. (1972) Study on hydraulic resistance and bed-load transport rate in alluvial streams. *T Jpn Soc Aeronaut S.*, 206, 59–69.

Bagnold, R.A. (1956) The flow of cohesionless grains in fluids. *Philos T Roy Soc A.*, 249, 315–319.

Balachandar, S. & Eaton, J.K. (2010) Turbulent Dispersed Multiphase Flow. *Annu Rev Fluid Mech.*, 42, 111–133.

Baranya, S., Muste, M., Abraham, D., & Pratt, T.C. (2016) Acoustic mapping velocimetry for in-situ bedload transport estimation, *River Flow Conference*, July 12–15, 2016, St Louis, MO, USA.

Batchelor, G.K. (1970) *The Theory of Homogeneous Turbulence*. Cambridge University Press.

Bennett, S.J. & Best, J.L. (1995) Mean flow and turbulence structure over fixed, two dimensional dunes: implications for sediment transport and dune stability. *Sedimentology*, 42, 491–514.

Bennett, S.J. & Best, J.L. (1996) Mean flow and turbulence structure over fixed ripples and the ripple-dune transition. In: P.J. Ashworth, S.J. Bennett, J.L. Best, & S.J. McLelland, (eds.) *Coherent Flow Structures in Open Channels*. John Wiley and Sons, Chichester, U.K., pp. 281–304.

Bernard, P.S. & Wallace, J.M. (2002) *Turbulent Flow: Analysis, Measurement, and Prediction*. John Wiley and Sons, New York.

Best, J.L. (1996) The fluid dynamics of small-scale alluvial bedforms. In P.A. Carling & M.R. Dawson, (eds.) *Advances in Fluvial Dynamics and Stratigraphy*. John Wiley and Sons, Chichester, U.K., pp. 67–125.

Best, J. (2005) The fluid dynamics of river dunes: a review and some future research directions. *J Geophys Res.*, 110, F04S02, doi:10.1029/2004JF000218.

Best, J., Bennett, S., Bridge, J., & Leeder, M. (1997) Turbulence modulation and particle velocities over flat sand beds at low transport rates. *J Hydraul Eng.*, 10.1061/(ASCE)0733–9429(1997)123:12(1118), 1118–1129.

Bettess, R. (1984) Initiation of sediment transport in gravel streams. In: *Proceedings Institute of Civil Engineers*, 77(2), 79–88.

Bigillon, F., Nino, Y., & Garcia, M.H. (2006) Measurements of turbulence characteristics in an open-channel flow over a transitionally-rough bed using particle image velocimetry, *Exp Fluids.*, 41(6), DOI 10.1007/s00348-006-0201-2, 857–867.

Bogardi, J. (1959) Hydraulic similarity of river models with movable bed. *Acta Tech Budapest.*, 24(3/4), 417–446.

Bombardelli, F.A., Buscaglia, G.C., Rehmann, C.R., Rincon, L.E., & Garcia, M.H. (2007) Modeling and scaling of aeration bubble plumes: A two-phase flow analysis. *J Hydraul Res.*, 45(5), 617–630.

Boussinesq, J. (1877) *Essay sur la Theorie des Eaux Courantes*. 23 Paris, Mem. Presentes par Divers Savants l'Acadameie Sci., pp. 46–50

Boussinesq, J. (1903) *Théorie analytique de la chaleur*, 2, Paris, Gauthier-Villars. p. 172.

Bradshaw, P. (1987) Turbulent secondary flows. *Annu Rev Fluid Mech.*, 19, 351–368.

Bridge, J.S. (1992) A revised model for water flow, sediment transport, bed topography, and grain size sorting in natural river bends. *Water Resour Res.*, 28, 999–1013.

Bridge, J.S. (2003) *Rivers and floodplains: Forms, processes, and sedimentary record*. Blackwell, Oxford, United Kingdom.

Brown, G.L. & Roshko, A. (1974) On density effects and large structure in turbulent mixing layers. *J Fluid Mech.*, 64, 775–816.

Brown, G.L. & Roshko, A. (2012) Turbulent shear layers and wakes. *J Turbul.*, 13(51), 1–32.

Brownlie, W.R. (1981) Re-examination of Nikuradse roughness data. *J Hydr Eng Div-ASCE.*, 107(1), 115–119.

Brownlie, W.R. (1983) Flow depth in sand-bed channels. *J Hydraul Eng-ASCE.*, 109(7), 959–990.

Buffington, J.M. (1999) The legend of A. F. Shields. *J Hydraul Eng-ASCE.*, 125(4), 376–387.

Buffington, J.M. & Montgomery, D.R. (1997) A systematic analysis of eight decades of incipient motion studies, with special reference to gravel-bedded rivers. *Water Resour Res.*, 33, 1993–2029.

Cameron, S. & Nikora V. (2008) Eddy convection velocity for smooth- and rough-bed open-channel flows: Particle Image Velocimetry study. In: *Proceedings of the 4th International Conference on Fluvial Hydraulics "River Flow 2008"*, Turkey, 3–5 September, 2008, Volume 1, 143–150.

Celik, I. & Rodi, W. (1984) A deposition-entrainment model for suspended sediment transport. Report SFB 210/T/6, Germany, University of Karlsruhe.

Celik, I. & Rodi, W. (1988) Modeling suspended sediment transport in nonequilibrium situations. *J Hydr Eng.*, 114, 1157.

Chanson, H. (2004) *Hydraulics of Open-Channel Flow*. Elsevier, Oxford.

Chaudhry, M.H. (1993). *Open Channel Flow*. Prentice-Hall, Englewood Cliffs (NJ)

Cheng, N.S. (1997) A simplified settling velocity formula for sediment particle. *J Hydraul Eng-ASCE.*, 123(2), 149–152.

Cheng, N.S. (2002) Exponential formula for bedload transport. *J Hydraul Eng-ASCE.*, 128 (102), 942–946.

Cheng, N.S. & Chiew, Y. (1998) Pickup probability for sediment entrainment. *J Hydraul Eng-ASCE.*, 124(2), 232–235.

Cheng, N. & Emadzadeh, A. (2015) Estimate of sediment pickup rate with the densimetric froude number. *J Hydraul Eng.*, 10.1061/(ASCE)HY.1943-7900.0001105,06015024.

Chollet, J.P. & Cunge, J.A. (1980). Simulation of unsteady flow in alluvial streams. *Appl Math Model.*, 4(8). 234–244.

Chow, V.T. (1959). *Open-Channel Hydraulics*. The Blackburn Press. New York (NY).

Christensen, B.A. (1993) Discussion of "Dimensionally Homogeneous Manning's Formula" by Ben Chie Yen. *J Hydraul Eng-ASCE.*, September, 1992, 118(9), 12/1993; 119(12DOI: 10.1061/(ASCE)0733–9429(1993)119:12(1442).

Coleman, N.L. (1967) A theoretical and experimental study of drag and lift forces acting on a sphere resting on a hypothetical stream bed. In:*Proceedings of the 12th Congress, International Association for Hydraulic Research*, Fort Collins Colorado, 185–192.

Coleman, N.L. (1981) Velocity profiles with suspended sediment. *J Hydraul Res.*, IAHR, 19, 211–229.

Coleman, N.L. & Alonso, C.V. (1983) Two-dimensional channel flows over rough surfaces. *J Hydraul Eng-ASCE.*, 109, 175–188.

Coleman, S.E. & Eling, B. (2000) Sand wavelets in laminar open-channel flows. *J Hydraul Res.*, IAHR, 38, 331–338.

Coleman, S.E. & Fenton, J.D. (2000) Potential flow instability theory and alluvial stream bed forms. *J Fluid Mech.*, 418, 101–117.

Coleman, S.E., Fedele, J.J., & Garcia, M.H. (2003) Closed conduit bed-form initiation and development. *J Hydraul Eng-ASCE.*, 129(12), 956–965.

Coleman, S.E. & Melville, B.W. (1996) Initiation of bed forms on a flat bed. *J Hydraul Eng-ASCE.*, 122(6), 301–310.

Coles, D.F. (1956) The law of the wake in turbulent boundary layer. *J Fluid Mech.*, 1, 191–226.

Cruickshank, C. & Maza, J.A. (1973) *Flow resistance in sand bed channels*. International Symposium on River Mechanics, Thailand, IAHR, Bangkok, A30, pp. 1–9.

Dancey, C.L., Diplas, P., Papanicolau, A., & Bala, M. (2002) Probability of individual grain movement and threshold condition. *J Hydraul Eng-ASCE.*, 128(12), 1069–1075.

Davidson, P.A. (2004) *Turbulence*. Oxford, Oxford University Press.

De Saint Venant (1871) *Comptes Rendus de l'Acad.* Des Sciences, 3, 4.

del Alamo, J.C. & Jimenez, J. (2009) Estimation of turbulent convection velocities and corrections to Taylor's approximation. *J Fluid Mech.*, 640, 5–26.

Dennis, D.J.C. & Nickels, T.B. (2008) On the limitations of Taylor's hypothesis in constructing long structures in a turbulent boundary layer. *J Fluid Mech.*, 614, 197–206.

Dietrich, W.E. (1982). Settling velocities of natural particles. *Water Resour Res.*, 18(6), 1615–1626.

Dwyer, H.A. (1996) Aspects of time dependent and three dimensional boundary layer flow. In: *Fluid dynamics and fluid machinery. Volume 1. Fundamentals of Fluid Dynamics*, New York, John Wiley, pp. 389–396.

Einstein, H.A. (1942) Formulas for the transportation of bed load. *T Am Soc Civ Eng.*, 117, 561–597.

Einstein, H.A. (1950) *The Bedload Function for Bedload Transportation in Open Channel Flows*. Technical Bulletin No. 1026, U.S.D.A., Soil Conservation Service, 1–71.

Einstein, H.A. (1968) Deposition of suspended particles in a gravel bed. *J Hydr Div.*, ASCE, 94 (5), 1197–1205.

Einstein, H.A. & Barbarossa, N.L. (1952) River channel roughness. *T Am Soc Civ Eng.*, 117, 1121–1146.

Einstein, H.A. & Li, H. (1958) Secondary currents in straight channels. *Eos T Am Geophys Un.*, 39, 1085–1088.

Engel, P. & Lau, Y.L. (1980) Computation of bed load using bathymetric data. *J Hydr Eng Div-ASCE.*, 106(3), 639–480.

Engelund, F. (1974) Flow and bed topography in channel bends. *J Hydr Eng Div-ASCE.*, 100, 1631–1648.

Engelund, F. & Hansen, E. (1967) *A Monograph on Sediment Transport in Alluvial Streams.* Copenhagen, Denmark, Teknisk Vorlag.

Engelund, F. & Fredsøe, J. (1976) A sediment transport model for straight alluvial channels. *Nordic Hydrology*, 7, 293–306.

Engelund, F. & Fredøse, J. (1982) Sediment ripples and dunes, *Annual Reviews in Fluid MechanicsAnnu Rev Fluid Mech.*, 14, 13–37.

Ettema, R. (2008) Ice effects on sediment transport in rivers. *Sedimentation Engineering*: In Garcia, M.H, (ed.) *Processes, Measurements, Modeling, and Practice.*, ASCE Manuals and Reports on Engineering Practice No. 110, pp. 613–648.

Exner, F.M. (1925) *Uber die Wechselwirkung zwischen Wasser und Geschiebe in Flussen.* Sitzber. Akad. Wiss Wien, Part IIa, Bd. 134 (in German).

Falcon, M.A. & Kennedy, J.F. (1983) Flow in alluvial-river curves. *J Fluid Mech.*, 133, 1–16.

Fedele, J.J. & Garcia, M.H. (2001) Hydraulic roughness in alluvial streams: a boundary layer approach. In G. Seminara & P. Blondeaux, (eds.) *Riverine, Coastal, and Estuarine Morphodynamics*, Springer-Verlag, Berlin, pp. 37–60.

Fernandez Luque, R., & van Beek, R. (1976) Erosion and transport of bed sediment. *J Hydraul Res.*, IAHR, 14(2), 127–144.

Finnigan, J.J. (2000) Turbulence in plant canopies. *Annu Rev Fluid Mech.*, 32, 519–571.

Fisher. H.B., List, E.J., Koh, R.C.Y., Imberger, J., & Brooks, N.H. (1979) *Mixing in Inland and Coastal Waters.* New York, Academic Press.

French, R.H. (1985) Open-Channel Hydraulics. McGraw-Hill Book Co, New York, NY.

Fredsøe, J. (1982) Shape and dimensions of stationary dunes in rivers. *J Hydr Eng Div-ASCE.*, 108(8), 932–947.

Fredsøe, J. & R. Deigaard (1992) *Mechanics of Coastal Sediment Transport.* World Scientific, Hackensack (NJ).

Frisch, U. (1995) *Turbulence. The Legacy of A.N. Kolmogorov.* Cambridge University Press, Cambridge (UK).

Garcia, M.H. (1999) Sedimentation and Erosion Hydraulics. In: Larry Mays, (ed.) *Hydraulic Design Handbook*, Chapter 6, McGraw-Hill, Inc.

Garcia, M.H. (2000a) Discussion of "The Legend of A. F. Shields," *J Hydraul Eng-ASCE.*, 126 (9), 718–720.

Garcia, M.H. (2000b) *Hidrodinamica Ambiental. Universidad Nacional del Litoral*, Centro de Publicaciones, Santa Fe, Argentina (in Spanish).

Garcia, M.H. (2001) Modeling Sediment Entrainment into Suspension, Transport, and Deposition in Rivers, Chapter In: Paul Bates and Malcolm Anderson (Eds.), *Model Validation in Hydrologic Science*, United Kingdom, Wiley and Sons.

Garcia, M.H. (2008) Sedimentation Engineering (Manual 110) Processes, Measurements, Modeling, and Practice. *ASCE Manuals and Reports on Engineering Practice.* ASCE.

Garcia, M.H., Admiraal, D.M., & Rodriguez, J.F. (1999) Laboratory experiments on navigation-induced bed shear stresses and sediment resuspension. *Int J Sediment Res.*, 14(2), 303–317.

Garcia, M.H. & Parker, G. (1991) Entrainment of Bed Sediment into Suspension. *J Hydraul Eng-ASCE.*, 117(4), 414–435.

Gelfenbaum, G. & Smith, J.D. (1986) Experimental evaluation of a generalized suspended-sediment transport theory. In: R.J. Knight, & J.R. McLean, (eds.) *Sedimentology of Shelf Sands and Sandstones*, Canadian Society of Petroleum Geologists, Memoir II, pp. 133–144.

Gessler, J. (1970) Self-stabilizing tendencies of sediment mixtures with large range of grain sizes. *J Waterway Div-ASCE*, 96(2), 235–249.

Gibson, A.H., (1909) On the depression of the filament of maximum velocity in a stream flowing through an open channel. In: *Proceedings of Royal Society of London*, Series A, 82, 149–159.

Gimenez-Curto, L.A., & Corniero Lera, M.A. (1996) Oscillating turbulent flow over very rough surfaces. *J Geophys Res.*, 101(C9), 20,745–20,758.

Goto, S. (2008) A physical mechanism of the energy cascade in homogeneous isotropic turbulence. *J Fluid Mech.*, 605, 355–366.

Gourlay, M.R. (1970) Discussion on: "Flow retardance in vegetated channels." In: N. Kowen & Unny, T.E. *Journal of the Irrig., And Drainage Div.*, Proceedings Paper No. 7498, 96, 351.

Graf., W.H. (1971) Hydraulics of sediment transport. McGraw-Hill, New York.

Graf, W.H. & Cellino M. (2002) Suspension flows in open channels; experimental study, *J Hydraul Res.*, 40(4), 435–447, DOI: 10.1080/00221680209499886.

Graf, W.H. & Pazis, G.C. (1977) Les phenomenes de deposition et d'erosion dans un canal alluvionnaire. *J Hydraul Res.*, IAHR, 15, 151–165.

Grinvald, D.I. & Nikora, V.I. (1988) *River Turbulence* (in Russian). Hydrometeoizdat, Leningrad, Russia.

Gyr, A. & Kinzelbach, W. (2004) Bed forms in turbulent channel flow. *Appl Mech Rev.*, 57, 77–93.

Henderson, F.M. (1966), Open Channel Fow. *MacMillan Series in Civil Engineering*, Prentice Hall, New York (NY).

Hinze, J.O. (1975) *Turbulence*. McGrawHill, New York (NY).

Holmes, Jr., R.R. (2003) *Vertical Velocity Distributions in Sand-Bed Alluvial Rivers*. Ph.D. Dissertation, Department of Civil and Environmental Engineering, University of Illinois at Urbana-Champaign, Urbana, Illinois, p. 325.

Holmes R.R. Jr. & Garcia, M.H. (2008) Flow over bedforms in a large sand-bed river: A field investigation, *J Hydraul Res.*, 46(3), 322–333

Huang, X. & Garcia, M.H. (2000) Pollution of gravel spawning grounds by deposition of suspended sediment. *J Environ Eng-ASCE.*, 126(10), 963–967.

Ikeda, S. (1981) Self-formed straight channels in sandy beds. *J Hydr Eng Div-ASCE.*, 107(Hy4), April, 389–406.

Ikeda, S. (1982) Incipient motion of sand particles on side slopes. *J Hydr Eng Div-ASCE.*, 108(1), 95–114.

Ikeda, S. & Nishimura, T. (1986) Flow and bed profile in meandering sand-silt rivers. *J Hydr Eng Div-ASCE.*, 112, 562–579.

Ikeda, S. & Parker, G. (1989) *River Meandering*. Water Resources Monograph 12, American Geophysical Union, Washington D.C.

Itakura, T. & Kishi, T. (1980) Open channel flow with suspended sediments. *J Hydr Eng Div-ASCE.*, 106(8), 1325–1343.

Iwagaki, Y. (1956) Fundamental study on critical tractive force. *Proceedings of Japan Society of Civil Engineering*, 41, 1–21.

Jain, S.C., (2001) *Open-Channel Flow*. New York, John Wiley & Sons.

Janna, W.S. (2009) *Introduction to Fluid Mechanics*. CRC Press, Boston.

Jin, Y., Zarrati, A., & Zheng, Y. (2004) Boundary Shear Distribution in Straight Ducts and Open Channels. *J Hydraul Eng.*, 10.1061/(ASCE)0733-9429(2004)130:9(924), 924–928.

Jiménez, J. (2004) Turbulent flows over rough walls. *Annu Rev Fluid Mech.*, 36, 173–196.

Jiménez, J.A. & Madsen, O.S. (2003) A simple formula to estimate settling velocity of natural sediments. *J Waterw Port C-ASCE.*, 129(2), 70–78.

Johanesson, H. & Parker, G. (1989a) Secondary flow in mildly sinuous channel. *J Hydraul Eng-ASCE.*, 115, 289–308.

Johanesson, H. & Parker, G. (1989b) Linear theory of river meanders. In S. Ikeda & G. Parker, (eds.) *River Meandering*, AGU Water Resources Monograph 12, 181–213.

Johanesson, H. & Parker, G. (1989c) Velocity redistribution in meandering rivers. *J Hydraul Eng-ASCE.*, 115, 1010–1039.

Julien, P.Y. & Hartley, D.M. (1986) Formation of roll waves in laminar sheet flow. *J Hydraul Res.*, IAHR, 24(1), 5–17.

Julien, P.Y. (2002) *River Mechanics*, Cambridge University Press, Cambridge, U.K.

Julien, P.Y., Klaassen, G.J., ten Brinke, W.B.M., & Wilbers, A.W.E. (2002) Case study: Bed resistance of Rhine River during 1988 flood. *J Hydraul Eng-ASCE.*, 128, 1042–1050.

Karim, F. & Kennedy, J.F. (1981) *Computer-based predictors for sediment discharge and friction factor of alluvial streams*. Report 242, Iowa Institute of Hydraulic Research, University of Iowa, Iowa City, Iowa.

Katul, G.G., Mahrt, L., Poggi, D., & Sanz, C. (2004) One- and two-equation models for canopy turbulence. *Boundary Laer Met.* 113, 81–109.

Kennedy, J.F. (1969) The formation of sediment ripples, dunes and antidunes. *Annu Rev Fluid Mech.*, 1, 147–168.

Kennedy, J.F. (1995) The Albert Shields story. *J Hydraul Eng-ASCE.*, 121, 766–772.

Kennedy, J.F. & Odgaard, A.J. (1991) *Informal monograph on riverine sand dunes*. Contract Report CERC-91-2, Coastal Engineering Research Center, Vicksburg, Mississippi, US Army Waterways Experiment Station.

Keulegan, G.H. (1938) Laws of turbulent flow in open channels. *Journal J Res Nat Bur Stand*, Research Paper 1151, 21, Washington, D.C, pp. 707–741.

Kim, J. (2012) Progress in pipe and channel flow turbulence, 1961–2011. *J Turbul.*, 13(45), 1–19.

King, A.T., Tinoco, R.O., & Cowen, E., (2012) A k-e turbulence model based on the scales of vertical shear and stem wakes valid for emergent and submerged vegetated flows. *J Fluid Mech.*, 701, 1–39.

Kline, S.J., Reynolds, W.C., Schraub, F.A., & Runstadler, P.W. (1967) The structure of turbulent boundary layers. *J Fluid Mech.*, 30, 741–773.

Kolmogorov, A.N. (1941) The local structure of turbulence in incompressible viscous fluid for very large Reynolds numbers. *Dokl Akad Nauk SSSR.*, 30, 301–305.

Kolmogorov, A.N. (1962) A refinement of previous hypotheses concerning the local structure of turbulence in a viscous incompressible fluid at high Reynolds number. *J Fluid Mech.*, 13, 82–85.

Komar, P.D. (1996) Entrainment of sediments from deposits of mixed grain sizes and densities. In: P.A. Carling, & M.R. Dawson, (eds.) *Advances in Fluvial Dynamics and Stratigraphy*, Chichester, Wiley, pp. 127–181.

Kostaschuk, R.A. & Villard, P. (1996) Turbulent sand suspension events: Fraser River, Canada. In: P.J. Ashworth, S.J. Bennett, J.L. Best, & S.J. McLelland, (eds.) *Coherent Flow Structures in Open Channels*, Chichester, Wiley & Sons, pp. 305–319.

Kostaschuk, R., Villard, P., & Best, J.L. (2004) Measuring velocity and shear stress over dunes with acoustic Doppler profiler. *J Hydraul Eng-ASCE.*, 130, 932–936.

Kowen, N. & Unny, T.E. (1973) Flexible roughness in open channels, *J Hydr Div.*, 99, 713.

Kundu, P.K., Cohen, I.M., & Dowling, D.R. (2012) *Fluid Mechanics*. Elsevier, Oxford.

Lavelle, J.W., & Mofjeld, H.O. (1987) Do critical stresses for incipient motion and erosion really exit? *J Hydraul Eng-ASCE.*, 113(3), 370–385.

Lee, H-Y. & Hsu, I-S. (1994) Investigation of saltating particle motion. *J Hydraul Eng-ASCE.*, 120, 831–845.

Lee, H-Y., Chen, Y-H., You, J-Y., & Lin, Y-T. (2000) Investigations of continuous bed load saltating process. *J Hydraul Eng-ASCE.*, 126, 691–700.

Leeder, M.R. (1983) On the interactions between turbulent flow, sediment transport and bed-form mechanics in channelized flows. In: J.D. Collinson and J. Lewin, (eds.) *Modern and*

Ancient Fluvial Systems, International Association of Sedimentologists, Special Publication, 6, pp. 5–18.

Li, R. & Shen, H. (1973) Effect of tall vegetation on flow and sediment. *J Hydr Div.*, 99, 793.

Lopez, F. & Garcia, M.H. (1997) Open-channel flow through simulated vegetation: turbulence modeling and sediment transport. *Wetlands Research Progr, Technical Report WRP-CP-10*, Washington DC, US Army Corps of Engineers.

Lopez, F. & Garcia, M.H. (1998) Open-channel flow through simulated vegetation: Suspended sediment transport modelling. *Water Resour Res.*, 34(9), 2341–2352.

Lopez, F. and Garcia, M. (1999) Wall similarity in turbulent open-channel flow. *J Eng Mech.*, 10.1061/(ASCE)0733-9399(1999)125:7(789), 789–796.

Lopez, F. & Garcia, M.H. (2001) Mean flow and turbulence structure of open-channel flow through emergent vegetation. *J Hydraul Eng-ASCE.*, 127(5), 392–402.

Lumley, J.L. (1989) Whither Turbulence? Turbulence at Crossroads. Springer.

Lyn, D.A. (1991) Resistance in flat-bed sediment-laden flows. *J Hydraul Eng-ASCE.*, 117, 94–114.

Lyn, D.A. (2008) Turbulence Models for Sediment Transport Engineering. In: Garcia M.H., (ed.) *Sedimentation Engineering (Manual 110) Processes, Measurements, Modeling, and Practice. ASCE Manuals and Reports on Engineering Practice.* ASCE.

Madsen, O.S. (1991) Mechanics of cohensionless sediment transport in coastal waters. In: N.C. Kraus, K.J. Gingerich, and D.L. Kriebel, (eds.) *Coastal Sediments 91*, ASCE, New York pp. 15–27.

Mantz, P.A. (1977) Incipient transport of fine grains and flakes by fluids-extended Shields diagram. *J Hydraul Eng-ASCE.*, 103(6), 601–616.

Mantz, P.A. (1980) Semi-empirical correlations for fine and coarse cohesionless sediment transport. *P I Civil Eng.*, 75(2), 1–33.

Marchi, E. & A. Rubatta, (1981) *Meccanica dei Fluidi: principi e applicazioni idrauliche*, Torino, Italy (in Italian).

Marion, A., Nikora, V., Puijalon, S., Bouma, T., Koll, K., Ballio, F., Tait, S., Zaramella, M., Sukhodolov, A., O'hare, M., Wharton, G., Aberle, J., Tregnaghi, M., Davies, P., Nepf, H., Parker, G., & Statzner, B. (2014). Aquatic interfaces: a hydrodynamic and ecological perspective. *J Hydraul Res.*, 52(6), 744–758.

Marusic, I., & Adrian, R.J. (2013) Eddies and scales of wall turbulence. In: P.A. Davidson, Y. Kaneda & K.R. Sreenivasan, (ed.) *Ten Chapters in Turbulence*, Cambridge UK, Cambridge University Press.

Marusic, I., McKeon, B.J., Monkewitz, P.A., Nagib, H.M., Smits, A.J., & Sreenivasan, K.R. (2010a). Wall-bounded turbulent flows: Recent advances and key issues. *Phys Fluids*, 22, 065103.

Marusic, I., Mathis, R., & Hutchins, N. (2010b) Predictive model for wall-bounded turbulent flow. *Sci.*, 329, 193–196.

McEwan, I.K., Jefcoate, B.J., & Willetts, B.B. (1999) The grain-fluid interaction as a self-stabilizing mechanics in fluvial bed load transport. *Sedimentology*, 46, 407–416.

McLean, S.R. (1990) The stability of ripples and dunes. *Earth Sci.*, 29, 131–144.

McLean, S.R. (1992) On the calculation of suspended load for non-cohesive sediments. *J Geophys Res.*, 97, 5759–5770.

McLean, S.R., Nelson, J.M., & Gary, L. 2007. Suspended sediment in the presence of dunes. In: C. Marjolein Dohmen-Janssen & Suzanne J.M.H. Hulscher (eds.) *River, Coastal, and Estuarine Morphodynamics*, Leiden, Taylor & Francis Group, pp. 611–618.

McLelland, S.J., Ashworth, P.J., Best J.L., & Livesey, J.R. (1999) Turbulence and secondary flow over sediment stripes in weakly bimodal bed material. *J Hydraul Eng.* ASCE, 125(5), 463–473.

Mei, C.C. (1969) Nonuniform diffusion of suspended sediment, *J Hydr Eng Div-ASCE.*, 95(1), January/February 1969, 581–584

Meyer-Peter, E. & Muller, R. (1948) Formulas for bedload transport. In: *Proceedings of the 2nd Congress, IAHR*, Stockholm, 39–64.

Miller, M.C., McCave, I.N., & Komar, P.D. (1977) Threshold of sediment motion in unidirectional currents. *Sedimentology*, 24, 507–528.

Mohrig, D. & Smith, J.D. (1996) Predicting the migration rates of subaqueous dunes. *Water Resour Res.*, 10, 3207–3217.

Möller, N. (2014) *Effects of boundary roughness on turbulence in uniform open channel flow*. MS Thesis. Ven Te Chow Hydrosystems Lab, Urbana, IL, University of Illinois at Urbana-Champaign.

Moglen, G.E. (2015), *Fundamentals of Open Channel Flow*, CRC Press, Boca Raton (FL).

Monin, A.S. & Yaglom, A.M. (1971) *Statistical Fluid Mechanics: Mechanics of Turbulence*. Volume. 1, Boston, MA, MIT Press.

Monin, A.S. & Yaglom, A.M. (1975) *Statistical Fluid Mechanics: Mechanics of Turbulence*. Volume. 2, Boston, MA, MIT Press.

Muller, D., Abad, J.D., Garcia, C., Gatner, J.W. & Garcia, M.H. (2007) Errors in acoustic doppler velocity measurements caused by flow disturbance. *J Hydraul Eng-ASCE.*, 133(12), 1411–1420.

Muste, M. & Patel, V. (1997) Velocity profiles for particles and liquid in ppen-channel flow with suspended sediment. *J Hydraul Eng.*, 10.1061/(ASCE)0733-9429(1997)123:9(742), 742–751.

Naot, D. (1984) Response of channel flow to roughness heterogeneity. *J Hydr Div.*, ASCE, 110 (11), 1568–1587.

Naot, D. & Rodi, W. (1982) Calculation of secondary currents in channel flow. *J Hydr Eng Div-ASCE.*, 108(8), 948–968.

Nakagawa, H. & Tsujimoto, T. (1980) Sand bed instability due to bed load motion. *J Hydr Eng Div-ASCE.*, 106(12), 2029–2051.

Neill, C.R. (1968) Note on initial movement of coarse uniform material. *J Hydraul Res.*, IAHR, 6 (2), 157–184.

Neill, C.R. & Yalin. M.S. (1969) Qualitative definition of beginning of bed movement. *J Hydr Eng Div-ASCE.*, 95(1), 585–587.

Nelson, J.M. & Smith, J.D. (1989) Flow in meandering channels with natural topography. In: S. Ikeda & G. Parker, (eds.) *River Meandering*, Water Resources Monograph No. 12, American Geophysical Union, Washington, D.C., 69–102.

Nepf, H.M. (2012) Flow and transport in regions with aquatic vegetation. *Annu Rev Fluid Mech.*, 44(1), 123–142.

Nepf, H. (2012) Hydrodynamics of vegetated channels. *J Hydraul Res.*, 50(3):262–279.

Nepf, H.M. & Vivoni, E.R. (2000) Flow structure in depth-limited, vegetated flow. *J Geophys Res.*, 105 (C12), 28547–28557.

Nepf, H., Ghisalberti, M., White, B., & Murphy, E. (2007) Retention time and dispersion associated with submerged aquatic canopies. *Water Resour Res.*, 43, W04422.

Nezu, I (1994) Compound open-channel turbulence and its role in river environment. *Keynote Address of 9th APD-IAHR Congress* Delft, The Netherlands, pp. 1–24.

Nezu, I. (2005) Open-channel flow turbulence and its research prospect in the 21st century. *J Hydraul Eng-ASCE.*, 131, 229–246

Nezu, I. & Nakagawa, H. (1984) Cellular secondary currents in straight conduit. *J Hydraul Eng.*, 10.1061/(ASCE)0733-9429(1984)110:2(173), 173–193.

Nezu, I. & Rodi, W. (1985) Experimental study on secondary currents in open channel flow. In: *Proceedings of 21st IAHR Congress*, Melbourne, 2, 19–23.

Nezu, I. & Rodi, W. (1986) Open-channel flow measurements with a laser Doppler anemometer. *J Hydraul Eng-ASCE.*, 112, 335–355.

Nezu, I. & Nakagawa, H. (1993) *Turbulence in Open-Channel Flows*. A.A. Balkema, Rotterdam, Netherlands.

Nielsen, P. (1992) *Coastal Bottom Boundary Layers and Sediment Transport*. River Edge, N.J, World Scientific.

Nikora, V.I. & Goring, D.G. (2000) Eddy convection velocity and Taylor's hypothesis of 'frozen' turbulence in a rough-bed open-channel flow. *J Hydrosci Hydraul Eng, JSCE*, 18(2), 75–91.

Nikora, V.I., Goring, D.G., McEwan, I., & Griffiths, G. (2001) Spatially-averaged open-channel flow over a rough bed. *J Hydraul Eng-ASCE.*, 127(2), 123–133.

Nikora, V.I. & Hicks, D.M. (1997) Scaling relationships for sand wave development in unidirectional flow. *J Hydraul Eng-ASCE.*, 123(12), 1152–1156.

Nikora, V.I., McEwan, I.K., McLean, S.R., Coleman, S.E., Pokrajac, D., & Walters, R. (2007a) Double-averaging concept for rough-bed open-channel and overland flows: Theoretical background. *J Hydraul Eng-ASCE.*, 133(8), 873–883.

Nikora, V., McLean, S., Coleman, S., Pokrajac, D., McEwan, I., Campbell, L., Aberle, J., Clunie, D., & Koll, K. (2007b) Double-averaging concept for rough-bed open-channel and overland flows: Applications. *J Hydraul Eng-ASCE.*, 133(8), 884–895.

Nikora, V. & Rowinski, P. (eds.) (2008) Rough-bed flows in geophysical, environmental, and engineering systems: Double-Averaging approach and its applications. Special Issue. *Acta Geophysica*, 56(3), 529–934.

Nikora, V. & Roy, A.G. (2012) A.G. Secondary flows in rivers: Theoretical framework, recent advances, and current challenges. In: Church M., Biron P.M. & A.G. Roy (eds.), *Gravel Bed Rivers: Processes, Tools, Environments*. London, Wiley and Sons, pp. 3–22.

Nikora, V., Ballio, F., Coleman, S.E., & Pokrajac, D. (2013) Spatially-averaged flows over mobile rough beds: definitions, averaging theorems, and conservation equations. *J Hydraul Eng-ASCE.*, 139(8), 803–811.

Niño, Y., Garcia, M.H., & Ayala, L. (1994) Gravel saltation I: Experiments. *Water Resour Res., AGU*, 30(6), 1907–1914.

Niño, Y. & M.H. Garcia. (1994) Gravel saltation II: Modeling. *Water Resour Res.*, AGU, 30(6), 1915–1924.

Niño, Y. & Garcia, M.H. (1998) Using lagrangian particle saltation observations for bedload sediment transport modeling. *Hydrol Process.*, 12, 1197–1218.

Niño, Y. & Garcia, M.H. (1998) Experiments on saltation of fine sand. *J Hydraul Eng-ASCE.*, 124(10), 1014–1025.

Niño, Y., Lopez, F., & Garcia, M. (2003) Threshold for particle entrainment into suspension. *Sedimentology*, 50, 247–263.

O'Connor, D.J. (1995) Inner region of smooth pipes and open channels. *J Hydraul Eng-ASCE.*, 121, 555–560.

Ogink, H.J.M. (1988) *Hydraulic Roughness of Bedforms*. Delft, The Netherlands, Delft Hydraulics Report M2017.

Owen, P.R. (1964) Saltation of uniform grains in air. *J Fluid Mech.*, 20(2), 225–242.

Paintal, A.S. (1971) Concept of critical shear stress in loose boundary open channels. *J Hydraul Res., IAHR*, 9, 91–113.

Palmer, V.J. (1945) A method for designing vegetated waterways. *Agric Eng.*, 26, 516.

Papanicolaou, A.N., Diplas, P., Balakrishnan, M., & Dancey, C.L. (1999) Computer vision technique for tracking bed load movement. *J Comput Civil Eng, ASCE*, 13(2), 71–79.

Papanicolaou, A., Diplas, P., Evaggelopoulos, N., & Fotopoulos, S. (2002) Stochastic incipient motion criterion for spheres under various packing conditions. *J Hydraul Eng-ASCE.*, 128(4), 369–380.

Parker, G. (1978) Self-formed straight rivers with equilibrium banks and mobile bed, Part 2. The gravel river. *J Fluid Mech.*, 89(1), 127–146.

Parker, G. (1979) Hydraulic geometry of active gravel rivers, *J Hydraul Eng-ASCE.*, 105(9), 1185–1201.

Parker, G. (2005) *1D Morphodynamics of Rivers and Turbidity Currents.* http://cee.uiuc.edu/people/parkerg/morphodynamics_e-book.htm>.

Parker, G. (2008) Transport of Gravel and Sediment Mixtures, Chapter 3 of: *Sedimentation Engineering, ASCE Manual No. 110, Sedimentation Engineering: Processes, Measurements, Modeling, and Practice*, ISBN 10 # 0784408149, ISBN 13 # 978078440814, May, 2008, 1050, pp. 165–252.

Parker, G. & N.L. Coleman (1986) Simple model of sediment-laden flows, *J Hydraul Eng.*, 112 (5), 356–375.

Parker, G., Solari, L., & Seminara, G. (2003) Bedload at low Shields stress on arbitrarily sloping beds: Alternative entrainment formulation. *Water Resour Res.*, 39(7), 1183, doi:10.1029/2001WR001253.

Parker, G., Toro-Escobar, C., Voigt, Jr., R. *et al.* (December 1998) *Countermeasures to Protect Bridge Piers from Scour*, by Project Report No. 433, St. Anthony Falls Laboratory, University of Minnesota.

Parker, G. & Garcia, M.H. (eds.) (2006) River, coastal and estuarine morphodynamics: RCEM 2005 (Vols. I & II), Taylor & Francis/Balkema, Leiden, The Netherlands.

Parsons, D.R., Best, J.L., Hardy, R.J., Kostaschuk, R., Lane, S.N., & Orfeo, O. (2005) Morphology and flow fields of three-dimensional dunes, Rio Paraná, Argentina: Results from simultaneous multibeam echo sounding and acoustic Doppler current profiling. *J Geophys Res.*, 110(F4), F04S04, DOI 10.1029/2004JF000231.

Perkins, H.J. (1970) The formation of streamwise vorticity in turbulence flow. *J. Fluid Mech.*, 44, part 4, 721–740.

Petryk, S. (1969) *Drag on cylinders in open channel flow.* Ph.D. diss, Colorado State University, Fort Collins, CO.

Petryk, S. & Bosmajian, G. (1975) Analysis of flow through vegetation. *Journal of Hydr. Engrg.* 101(7), 871–884.

Pope, S.B. (2005) *Turbulent Flows*. Cambridge University Press.

Prandtl, L. (1904) *Uber Flussigkeitsbewegung bei sehr kleiner Reibung*. In: *Proceedings of the 3rd International Math Congress*, Heidelberg.

Prandtl, L. (1926) *Über ein neues formelsystem für ausgebuldete turbulenz*. Nachrichten von der Akad. Der Wissenschaft in Göttingen, Math-Phys. K1.6.

Prandtl, L. (1952) *Essentials of Fluid Dynamics*, London, Blackie and Son Ltd.

Raudkivi, A.J. (1990) *Loose Boundary Hydraulics*, 3rd edn., New York, Pergamon Press.

Raudkivi, A.J. (1997) Ripples on stream bed. *J Hydraul Eng-ASCE.*, 16, 58–64.

Raudkivi, A.J. (1998) *Loose Boundary Hydraulics*, 3rd edn., Balkema, The Netherlands.

Raupach, M.R. & Shaw, R.H. (1982) Averaging procedures for flow within vegetation canopies. *Bound-Lay Meteorol.*, 22, 79–90.

Raupach, M.R. & Thom, A.S. (1981) Turbulence in and above canopies. *Ann Rev Fluid Mech.*, 13, 97–129.

Reynolds, O. (1895) On the dynamical theory of turbulent incompressible viscous fluids and the determination of the criterion, *Phil Trans R Soc.*, London A 186, 123–161.

Reynolds, A.J. (1976) A decade's investigation of the stability of erodible stream beds. *Nord Hydrol*, 7, 161–180.

Ree, W.O. & Palmer, V.J. (1949) Flow of water in channels protected by vegetative linings. *USDA Tech Bul* No. 967, February, p. 115.

Riboux, G., Risso, F., & Legendre, D. (2010) Experimental characterization of the agitation generated by bubbles rising at high Reynolds number. *J Fluid Mech.*, 643, 509–539.

Richardson, L.F. (1922) *Weather Prediction by Numerical Process*. Cambridge University Press, New York (NY).

Robert, A. & Uhlman, W. (2001) An experimental study on the ripple-dune transition. *Earth Surf Proc Land.*, 26, 615–629.

Robertson, J.M. & Rouse, H. (1941) On the four regimes of open-channel flow. *Civil Eng.*, 11(3) 169–171.

Robinson, S.K. (1991) Coherent motion in the turbulent boundary layer. *Annu Rev Fluid Mech.*, 23, 601–639.

Rodi, W. (1984) *Turbulence Models and Their Application in Hydraulics.* IAHR Monographs, CRC Press.

Rodi, W., Constantinescu, G., & Stoesser, T. (2013) *Large-Eddy Simulation in Hydraulics.* IAHR Monographs. CRC Press, Taylor & Francis Group.

Rodriguez, J.F. & Garcia, M.H. (2008) Laboratory measurements of 3-D flow patterns and turbulence in straight open channel with rough bed. *J Hydraul Res.*, 46, 454–465.

Rouse, H. (1937) Modern conceptions of the mechanics of turbulence. *T Am Soc Civ Eng.*, 102, Paper No. 1965, 463–543.

Rouse, H. (1938) Experiments on the mechanics of sediment suspension. In: *Proceedings of the Fifth International Congress for Applied Mechanics.* New York, John Wiley & Sons, 55, 550–554.

Rouse, H. (1939) *An analysis of sediment transport in the light of fluid turbulence.* Technical Report. Sedimentation Division SCS-TP-25. Presented before the Waterways Division at the Annual Meeting of the American Society of Civil Engineers New York City, January 19, 1939.

Rumelin, Th. (1913) *Wie Bewegt Sich Fliebendes Wasser.* Verlag von v. Zahn & J aensch, Dresden.

Rutherford, J.C. (1994) *River Mixing.* New York, Wiley.

Sarma, K.V.N., Lakshminarayana, P., & Rao, N.S.L. (1983) Velocity distribution in smooth rectangular channels. *J Hydraul Eng-ASCE.*, 109, 270–289.

Schindler, R.J. & Robert, A. (2004) Suspended sediment concentration and the ripple-dune transition. *Hydrol Process.*, 18, 3215–3227.

Schlichting, H. (1979) *Boundary Layer Theory*, 7th edn, New York, McGraw-Hill.

Schlichting, H. & Gersten, K. (2003) *Boundary Layer Theory.* 8th Revised and Enlarged Edition. Berlin, Springer.

Schetz, J.A. & Fuhs, A.E. (1996) *Wakes. In: Fluid dynamics and fluid machinery. Volume 1. Fundamentals of Fluid Dynamics*, John Wiley, pp. 448–455.

Sekine, M. & Kikkawa, H. (1992) Mechanics of saltating grains. II. *J Hydraul Eng-ASCE.*, 118 (4), 536–558.

Sekine, M. & Parker, G. (1992) Bed-load transport on transverse slope. I. *J Hydraul Eng-ASCE.*, 118(4), 513–535.

Seminara, G. (1995) Effect of grain sorting on the formation of bedforms. *Appl Mech Rev.*, 48, 549–563.

Seminara, G. & Blondeaux, P. (eds.) (2001) *River, Coastal and Estuarine Morphodynamics*, Springer-Verlag, Berlin.

Seminara, G., Solari, L., & Parker, G. (2002) Bed load at low Shields stress on arbitrarily sloping beds: Failure of the Bagnold hypothesis. *Water Resour Res.*, 38(11), 1249. doi:10.1029/2001WR000681

Shames, I.H. (2003) *Mechanics of Fluids.* New York, McGrow-Hill.

Shields, A., (1936) *Anwendung der Aechichkeits-Mechanic und der Turbuleng Forschung auf dir Geschiebewegung Mitt Preussische*, Versuchsanstalt für Wasserbau and Schiffbau, Berlin, Germany (translated to English by W.P. ott and J.C. van Uchelen, California Institute of Technology, Pasadena, California).

Shvidchenko, A.B. & Pender, P. (2000) Flume study of the effect of relative depth on the incipient motion of coarse uniform sediments. *Water Resour Res.*, 36, 619–628.

Simons, D.B. & Richardson, E.V. (1961) Forms of bed roughness in alluvial channels. *J Hydr Eng Div-ASCE.*, 87(3), 87–105.

Smith, J.D. & McLean S.R. (1977a). Spatially averaged flow over a wavy surface. *J Geophys Res.*, 83, 1735–1746.

Smith, J.D., & McLean, S.R. (1977b). Boundary layer adjustments to bottom topography and suspended sediment. *Bottom Turbulence:* In: J.C.J. Nihoul (ed.) *Proceedings of the 8th International Liege Colloquium on Ocean Hydrodynamics*, Elsevier Scientific Publishing Company, Liege, Belgium, 112, 123–151.

Smits, A.J., McKeon, B.J., & Marusic, I. (2011) High Reynolds number wall turbulence. *Annu Rev Fluid Mech.*, 43, 353–375.

Socolofsky, S.A. & Adams, E.E. (2005) Role of slip velocity in the behavior of stratified multi-phase plumes. *J. Hydraul. Eng.-ASCE*, 131(4), 273–282.

Song, T., Graf, W.H., & Lemmin U. (1994) Uniform flow in open channels with movable gravel bed, *J Hydraul Res.*, 32(6), 861–876, DOI: 10.1080/00221689409498695

Soulsby, R.L. (1997) Dynamics of marine sands, Thomas Telford, London, p. 249.

Southard, J.B. (1991) Experimental determination of bed-form stability. *Annual Review of Earth Sciences*, 19, 423–455.

Stevens, M.A., Simons, D.B., & Lewis, G.L. (1976) Safety factors for riprap protection. *J Hydr Eng Div-ASCE.*, 102(5), 637–655.

Stone, H.A. (2010) Interfaces: in fluid mechanics and across disciplines. *J Fluid Mech.*, 645, 1–25.

Strickler, A. (1923). Beiträge zur Frage der Geschwindigkeitsformel und der Rauhigkeitszahlen für Steüme, Kanäle und geschlossene Leitungen. Mitteilungen des Eidgenossischen Amtes fur Wasserwirtschaft [Contributions to the question of a velocity formula and roughness data for streams, channels and closed pipelines]. *Mitt Eidgeno assischen Amtes Wasserwirtschaft* 16, Bern (Switzerland), W. M. Keck Lab of Hydr and Water Res Translation, T-10. Cal. In. of Tech., 1981, Pasadena (CA). Available from: http://resolver.caltech.edu/CaltechAUTHORS: 20120202–142837599

Sturm, T.W. (2011) *Open channel hydraulics.* McGraw-Hill Education.

Swamee, P.K. (1993) Generalized inner region velocity distribution equation. *J Hydraul Eng-ASCE.*, 119(5), 651–656.

Tamburrino, A. & Gulliver, J.S. (1999) Large flow structures in a turbulent open channel flow, *J Hydraul Res.*, 37(3), 363–380, DOI: 10.1080/00221686.1999.9628253

Tanino, Y. & Nepf, H.M. (2008) Lateral dispersion in random cylinder arrays at high Reynolds number. *J Fluid Mech.*, 600, 339–371.

Taylor, G.I. (1915) Eddy motion in the atmosphere. In: *Proceedings of Royal Society of London Series A*, 215, 1–26.

Taylor, G.I. (1938) The spectrum of turbulence. In: *Proceedings of Royal Society of London Series A*, 164(919), 476–490.

Taylor, G.I. (1954) The dispersion of matter of turbulent flow through a pipe. *Proceedings of Royal Society of London Series A*, I, 446–468.

Taylor, B.D. & Vanoni, V.A. (1972) Temperature effects in flat-bed flows. *J Hydrl Div.*, ASCE, 98(8), 1427–1445.

Tollner, E.W. (1974) *Modeling the sediment filtration capacity of simulated rigid vegetation.* MS thesis, Lexington, KY, University of Kentucky.

Tollner, E.W., Barfield, B.J., & Hayes, J.C. (1982) Sedimentology of erect vegetal filters. *J Hydr Div.*, 108(12), 1518.

Tominaga, A., Nezu, I., Ezaki, K., & Nakagawa, H. (1989) Three-dimensional turbulent structure in straight open channel flows, *J Hydraul Res.*, 27(1), 149–173, DOI: 10.1080/00221688909499249

Townsend, A.A. (1976) *The Structure of Turbulent Shear Flow*. Cambridge, MA, Cambridge University Press.

Tsinober, A. (2009) *An Informal Conceptual Introduction to Turbulence*. Springer.

Tsujimoto. T. (1991) *Mechanics of Sediment Transport of Graded Materials and Fluvial Sorting*. Report, Project 01550401, Kanazawa University, Japan, p. 126.

Van den Berg, J.H. & Van Gelder, A. (1993) *A new bedform stability diagram, with emphasis on the transition of ripples to plane bed in flows over fine sand and silt*. In: M. Marzo & C. Puidefabregas, (eds.) Alluvial sedimentation, International Association of Sedimentologists, Special Publication 17, 11–21.

Van Rijn, L.C. (1982) Equivalent roughness of alluvial bed. *J Hydr Eng Div-ASCE.*, 108(10), 1215–1218.

Van Rijn, L.C. (1984a) Sediment transport, part I: bed load transport. *J Hydraul Eng-ASCE.*, 110(10), 1431–1456.

Van Rijn, L.C. (1984b) Sediment transport, part II: suspended load transport. *J Hydraul Eng-ASCE.*, 110(11), 1613–1641.

Van Rijn, L.C. (1984c) Sediment transport, part III: Bed forms and alluvial roughness. *J Hydraul Eng-ASCE.*, 110(12), 1733–1754.

Van Rijn, L.C. (1986) Application of sediment pick-up functions. *J Hydraul Eng-ASCE.*, 112(9).

Van Rijn, L.C. (1996) *Combining laboratory, field, and mathematical modeling research for bed forms, hydraulic roughness, and sediment transport during floods*. In: T. Nakato & R. Ettema, (eds.) *Issues and Directions in Hydraulics*, Balkema, Rotterdam, pp. 55–73.

Vanoni, V.A. (1964) *Measurements of critical shear stress for entraining fine sediments in a boundary layer*. Report KH-R- 7, W. M. Keck Laboratoty of Hydraulics and Water Resources, California Institute of Technology, Pasadena, California.

Vanoni, V.A., (ed.) (1975) *Sedimentation Engineering, ASCE Manual and Reports on Engineering Practice 54*, ASCE/EWRI, Reston, Virginia.

Vanoni, V.A. (1980) *Sediment studies in the Brazilian Amazon River basin*. United Nations Development Program, *Report KHP-168*, W. M. Keck Laboratoty of Hydraulics and Water Resources, California Institute of Technology, Pasadena, California.

Vanoni V.A., (ed.) (2006) *Sedimentation Engineering, ASCE Manual and Reports on Engineering -Practice, 54*, New York.

Vanoni, V. A. & Brooks, N.H. (1957) *Laboratory Studies of the Roughness and Suspended Load of Alluvial Streams. Sedimentation Laboratory*, Pasadena, California, USA., California Institute of Technology

Vionnet, C.A., Marti, C., Amsler, M.L., & Rodriguez, L. (1998) *The use of relative celerities of bedforms to compute sediment transport in the Paraná River* Modelling Soil Erosion, Sediment Transport and Closely Related Hydrological Processes, International Association of Hydrological Sciences Special Publication, 249, pp. 399–406.

USGS (1989), *Guide for Selecting Manning's Roughness Coefficients for Natural Channels and Flood Plains*. Water Supply Paper 2339, US Geologcal Survey, Washington, D.C.

Wallace, J.M. (2013) Highlights from 50 years of turbulent boundary research. *J Turbul.*, 13 (53), 1–70.

Wang, Z.B. & Ribberink, J.S. (1986) The validity of a depth-integrated model for suspended sediment transport. *J Hydr Res.*, Delft, The Netherlands, 24(1), 53–67.

Wiberg, P.L. & Smith, J.D. (1987) Calculations of the critical shear stress for motion of uniform and heterogeneous sediments. *Water Resour Res.*, 23(8), 1471–1480.

Wiberg, P.L. & Smith, J.D. (1989) Model for calculating bedload transport of sediment. *J Hydraul Eng-ASCE.*, 115(1), 101–123.

Wilbers, A.W.E. & ten Brinke, W.B.M. (2003) The response of subaqueous dunes to floods in sand and gravel bed reaches of the Dutch Rhine. *Sedimentology*, doi: 10.1046/j.1365–3091.2003.000585.x.

Williams III, J.C. (1996) Separated flows. *In: Fluid Dynamics and Fluid Machinery.* 1. *Fundamentals of Fluid dynamics,* John Wiley, pp. 397–421.

Williams, G.P. (1970) *Flume width and water depth effects in sediment transport experiments.* U. S. Geological Survey, Professional Paper 562-H

Wilson, K.C. (1966) Bedload transport at high shear stresses. *J Hydr Eng Div-ASCE.,* 92(HY6), 49–59.

Wilson, N.R. & Shaw, R.H. (1977) A higher order closure model for canopy flow. *J Appl Meteorol.,* 16, 1197–1205.

Wong, M. & Parker, G. (2006) Re-analysis and correction of bedload relation of Meyer-Peter and Muller using their own database. *J Hydraul Eng-ASCE.,* 132(11), 1159–1168.

Wooding, R.A., Bradley, E.F., & Marshall, J.K. (1973) Drag due to regular arrays of roughness elements of varying geometry. *Bound-Lay Meteorol.,* 5(3), 285–308.

Wright, S. & Parker, G. (2004) Density stratification effects in sand-bed rivers. *J Hydraul Eng-ASCE.,* 130(8), 783–795.

Wright, S. & Parker, P. (2004) Flow resistance and suspended load in sand-bed rivers: simplified stratification model density. *J Hydraul Eng-ASCE.,* 130(8), 796–805.

Yalin, M.S. (1963) An expression for bedload transportation. *J Hydr Eng Div-ASCE.,* 89(HY3), 221–250.

Yalin, M.S. (1992) *River mechanics,* Pergamon Press, New York.

Yalin, M.S. & da Silva, A.M. (2001) *Fluvial Processes,* IAHR Monograph, International Association of Hydraulic Engineering and Research, Delft, The Netherlands.

Yalin, M.S. & Karahan, E. (1979) Inception of sediment transport. *J Hydraul Eng-ASCE.,* 105 (HY11), 1433–1443.

Yen, B.C. (1973) Open-channel flow equations revisited. *J Eng Mech Div., Am Soc Civ Eng.,* 99 (EM5), 979–1009.

Yen, B.C. (1991) *Channel Flow Resistance: centennial of Manning's Formula,* Highlands Ranch, Colorado, Water Resources Publications. 453.

Yen, B.C. (1993) Closure to discussion by B.A. Christensen of "Dimensionally homogeneous Manning's formula," *J Hydraul Eng-ASCE.,* 119(12), 1442–1443.

Yen, B.C. (2002) Open channel flow resistance. *J Hydraul Eng-ASCE.,* 128, 20–39.

Yen, B.C., Chow V.C., & Akan A.O. (1977) Stormwater runoff on urban areas of steep slope. EPA report, Environmental Protection Technology Series, EPA-600/2-77-168.

Yoon, J.Y. & Patel, V.C. (1996) Numerical model of turbulent flow over sand dune. *J Hydraul Eng-ASCE.,* 122, 10–18.

Zanke, U. (1977) Berechnung der Sinkgeschwindigkeiten von Sedimenten, *Mitteilungen des Franzius-Institutes,* 46, 231–245

Zdravkovich, M.M. (1997) Flow around Circular Cylinders. Volume. 1, Oxford University Press, Oxford.

Zdravkovich, M.M. (2003) *Flow around Circular Cylinders.* Volume. 2, Oxford University Press, Oxford.

Zhou, D. & Mendoza, C. (2005) Growth model for sand wavelets. *J Hydraul Eng-ASCE.,* 131 (10), 866–876.

Zwamborn, J.A. (1981), Umfolzi road bridge hydraulic model investigation. *J Hydr Eng Div-ASCE,* ASCE, 107(11), 1317–1333.

Zyserman, J. & Fredsøe, J. (1994) Data analysis of bed concentration of suspended sediment. *J Hydraul Eng-ASCE.,* 120, 1021–1042.

Chapter 3

Similitude

3.1 INTRODUCTION

Consideration of similitude is centrally important in experimental hydraulics, because similitude provides the physics-based foundation necessary to design an experiment, and for interpreting the data and observations it produces. Moreover, similitude usefully promotes early thinking about the most important processes an experiment must replicate, and the appropriate or practicable scales (and the experiment layout) needed for a successful experiment. These considerations help to define the role of an experiment, possibly in conjunction with other investigative approaches such as analytical formulation, numerical modeling, and field work. An essential aspect of similitude considerations is that they enable processes occurring at different scales, and possibly involving different materials, to be described using meaningful variables assembled as key non-dimensional parameters. Figure 3.1.1, for example, illustrates the similitude between a laboratory model of a braided channel and its prototype. The model, correctly calibrated, replicates the main processes at play in the prototype even though the model did not include all of the prototype's details, such as sediment size variation and vegetation presence.

This chapter discusses similitude and experiment planning, focusing on core similitude aspects common to hydraulic experiments, and their implications for experiment planning. Dimensional analysis is a key aspect of similitude, as it helps in identifying non-dimensional parameters pertinent to interpreting general trends an experiment may reflect. Because civil engineering hydraulics involves water movement and flow, similitude of single-phase (liquid) water flow is at the core of hydraulic experiments. Similitude considerations for more complex, multi-phase hydraulic processes build upon similitude of liquid water flow. Whether an experiment focuses on the fundamentals of a physical process or, by means of a hydraulic model, on the design of specific hydraulic structures, similitude is the path indicating generality and scalability of results.

Ideally, geometric, kinematic (time and velocity) and dynamic (force) similarity should be maintained between processes occurring in an experiment setup, a hydraulic model, and in the life-size (or prototype) flow. Practical constraints usually make this ideal difficult to achieve – the fundamental difficulty arises because of the difficulties encountered in reducing the magnitude of three length scales (flow depth, some aspect of a hydraulic structure or boundary, and material likely transported by water; *e.g.*, sediment particles) and the material properties of water and other materials of interest

(a)   (b)

Figure 3.1.1 Similitude in hydraulic modeling of braided channel morphology: (a) hydraulic model built at a length scale of 80; and (b) prototype site. (Source: Colorado State University)

(*e.g.*, surface-tension behavior of water, or the elasticity of a boundary element). Explicit satisfaction of more than one dynamic-similitude criterion associated with water flow can require the model fluid to differ from water. Satisfaction of two dynamic-similitude criteria associated with particle transport, for example, would require the use of a model particle of density differing from that of the prototype particle. Different model fluids and particles having the needed physical properties, however, may be difficult to obtain, especially in sufficient quantity. Strict satisfaction of three similitude criteria requires use of a model scale of 1:1, or very nearly so.

Consequently, it is necessary to identify the processes of primary importance and determine the dominant forces at play. Then, it is necessary to scale the model and model-material properties so as to maintain, as closely as practicable, the same ratios between the primary forces in the model as in the prototype. In some cases, experimental or analytical corrections can be implemented to correct for a model's inability to scale the less important forces by the same ratio. For a hydraulic model or experimental setup to be quantitatively successful, it is important that the experimenter understands the essential processes to be replicated, and adequately appreciate the limitations of the experiment setup and measurement equipment used; an experienced experimenter may in various ways be able to work around an experiment's shortcomings.

To aid readers, this chapter includes two tables that concisely summarize similitude considerations and indicate similitude limits for avoiding unwanted scale effects. Table 3.3.1 summarizes the main similitude considerations associated with hydraulic experiments and hydraulic models; single-phase flows, multi-phase flows (mixing, stratified flows, density plumes), sediment transport, wave dynamics, and hydro-elastic interaction. Typically, the more complex flow processes (or set of processes) require more complex similitude considerations, and must address similitude shortcomings. Table 3.6.1 is intended to help experimenters assess modeling limits for various hydraulic processes under Froude number similitude. Citing experience found in hydraulic engineering literature, Table 3.6.1 suggests approximate limits or guidelines that may help experimenters avoid significant scale effects. It is important to consider the prototype feature mentioned, particularly for these cases where the limitations are not directly expressed in terms of similitude parameters. Scale effects typically are unwanted deviations in flow behavior, and occur because of similitude short-comings. The two tables are elaborated further during this chapter.

The reader is encouraged to read additional books or papers elaborating aspects of similitude and potential scale effects; *e.g.*, Kobus (1980, 1984), USBR (1981), Martins (1989), Shen (1990), Hughes (1993), ASCE (2000) and Heller (2011). These sources recount modeling experiences beyond this chapter.

3.2 BASICS

For many experiments it is useful to begin with a dimensional analysis of the variables involved. This step helps to ensure that similitude conditions are appropriately considered, and may reduce the required number of experiments.

The important principle of the dimensional homogeneity of equations describing a process underpins similitude. When relevant flow equations are normalized, or made non-dimensional, the relative importance of each variable in a given process is elucidated, making it easier to identify and discard negligible variables. The units of a dimensionally homogeneous equation are consistent, causing the values of coefficients in the equation to be independent of the system of units used. A useful outcome of dimensional homogeneity is that it unifies flow-related processes into a relatively compact system of equations or relationships.

Complete similitude requires satisfaction of the following conditions:

1. Geometric similitude, whereby the ratios of all homologous (geometrically equivalent) length dimensions are equal. Geometric similitude involves only similarity in form;
2. Kinematic similitude, whereby at geometrically homologous points in model and prototype, velocities and accelerations are in a constant ratio; and,
3. Dynamic similitude whereby, in addition to kinematic similitude, the force polygons are similar at geometrically equivalent points for model and prototype.

Figure 3.2.1 illustrates these conditions.

If dynamic similitude is satisfied, kinematic similitude automatically follows. In the following discussion the subscripts r, m, and p denote ratio, experiment (or hydraulic model) value, and prototype value, respectively. Here r = prototype/model values; *e.g.*, if $r = 10$, the prototype value is 10 times larger than the model value. The discussion pertains to situations of geometric similitude, for which all length scales are equal. The added complexity of geometric distortion (usually vertical distortion) is discussed in Section 3.6, Volume I.

The primary parameter for geometric similitude is the length ratio

$$L_r = \frac{L_p}{L_m} \tag{3.2.1}$$

which must be constant for all parts of the model and prototype. As a consequence of geometric similitude, the area, A, and volume, $\forall$, ratios follow as $A_r = L_r^2$ and $\forall_r = L_r^3$.

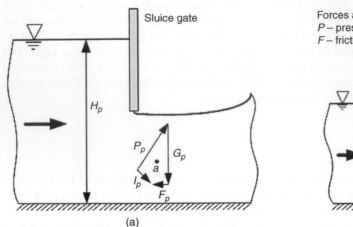
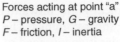
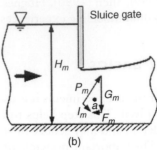

Figure 3.2.1   An illustration of the three basic similarity conditions for a prototype flow under a sluice gate (a), and a one-tenth scale model of the flow (b). The flows are geometrically, kinematically and dynamically similar, and thus the force polygons should be similar. Here, the subscripts *p* and *m* relate to prototype and model, respectively (adapted from White, 1986).

For kinematic similitude the velocity ratio, V_r, and the acceleration ratio, a_r, must be constant at all homologous points of the model and the prototype. The commensurate ratios are

$$V_r = \frac{L_r}{T_r} \tag{3.2.2a}$$

and

$$a_r = \frac{V_r}{T_r} = \frac{L_r}{T_r^2} \tag{3.2.2b}$$

with the time ratio being

$$T_r = \frac{T_p}{T_m} \tag{3.2.2c}$$

Dynamic similitude involves the force ratio, F_r. Forces arise from multiple physical phenomena: notably, gravity, inertia, viscosity, vorticity generation, pressure, and surface tension. The inertial force is always important when flows accelerate or decelerate due to changes in flow area or turbulence.

Newton's second law relates the inertial force to the mass, M, and acceleration. Expressed in ratio form,

$$F_r = M_r a_r. \tag{3.2.3}$$

The mass ratio can be written in terms of a density (ρ) ratio and the length ratio,

$$M_r = \rho_r \forall_r = \rho_r L_r^{3}.$$ (3.2.4)

Thus Newton's second law can be expressed as

$$F_r = \rho_r L_r^{3} \frac{V_r}{T_r}.$$ (3.2.5)

From Equation (3.2.2a), Equation (3.2.5) reduces to

$$F_r = \rho_r L_r^{2} V_r^{2}.$$ (3.2.6)

The inertial force ratio expressed in Equation (3.2.6) is relevant to any hydraulic experiment.

The basic nature of these similitude expressions (and scale ratios) enable the similitude principles to provide scaling laws with which the data obtained from laboratory experiments can be extrapolated to many full-life situations.

A necessary early step in experiment design is to determine parameters defining equivalence of dynamic similitude for process(es) of focal interest for the experiment. Three approaches are often used for taking this step:

1. Scale analysis of the equations used to describe a flow or fluid-transport process;
2. Direct comparison of the ratios of pertinent variables, such as forces or flux rates; and,
3. Dimensional analysis, whereby pertinent variables are grouped without direct regard for their mathematical relationship.

Approach 1 is the ideal and most rigorous approach, but requires that the equations be known, which often is not the case. Approach 2 involves understanding the critical processes in order to arrive at the main non-dimensional parameters needed for characterizing a process. Approach 3, dimensional analysis, based on the theory of dimensions, is a convenient and practical way to assemble dimensionless parameters from a listing of pertinent variables. The ensuing section of this chapter outlines Approach 1, and Appendix 3.A elaborates Approach 3. Approach 2 essentially relies on the judgment of the modeler to identify the main (or useful) parameters, and may be viewed as an informal or partial version of Approach 3.

3.3 DYNAMIC SIMILITUDE FROM FLOW EQUATIONS

The requirements for dynamic similitude, together with physical insight into the relative magnitudes of various non-dimensional parameters, can be deduced by normalizing the basic equations for water flow or transport of some substance or property. A limitation of this approach is that the equations governing many processes (*e.g.*, turbulence, bedload transport of sediment) are inadequately known.

The ensuing outline discusses normalizing the equations describing the single-phase flow of liquid water. For conservation of mass or flow continuity, written in vector form,

$$\nabla \cdot \vec{U} = 0 \qquad (3.3.1)$$

For conservation of momentum,

$$\rho \frac{d\vec{U}}{dt} = \nabla(-p + \rho gy) + \mu \nabla^2 \vec{U} \qquad (3.3.2)$$

Here, x, y, and z being the horizontal, vertical, and transverse directions, respectively, and μ is dynamic viscosity. Their corresponding velocity components, and the components of velocity vector, $\vec{U}$, are u, v, and w.

The solutions to these equations depend on flow boundary conditions. Typical boundary conditions for turbulent flow without surface wind are

1. At a fixed solid surface,

$$\vec{U} = 0 \qquad (3.3.3)$$

2. At a free surface

$$y = y_o; \qquad v = \frac{dy_o}{dt} \qquad (3.3.4)$$

and

$$p = p_{atmospheric} - \sigma\left(\frac{1}{R_x} + \frac{1}{R_y}\right) \qquad (3.3.5)$$

In which y_o = flow depth; and R_x and R_y = the radii of curvature of the water surface in the x and y directions, respectively. For a planar water surface, $p = p_{atmospheric}$.

Eqs (3.3.1) through (3.3.5) contain three basic dimensions, mass (or force), length, and time. These dimensions can be restated as dimensionless ratios:

$$\vec{U}^* = \vec{U}/U, \quad x^* = x/L, \quad y^* = y/L, \quad z^* = z/L, \quad t^* = (U/L)t, \quad p^* = p/p_o,$$

$$R_x^* = R_x/L, \ R_y^* = R_y/L$$

in which U = a reference velocity, L = a reference length, and p_o = a reference pressure (usually atmospheric pressure). Substitution of these ratios into the equation of motion and their boundary conditions yields

$$\nabla^* \cdot \vec{U}^* = 0 \qquad (3.3.6)$$

For conservation of momentum,

$$\frac{d\vec{U}^*}{dt^*} = \frac{gL}{U^2} - \left(\frac{p_o}{\rho U^2}\right)\nabla^* p^* + \frac{\mu}{\rho UL}\nabla^{*2}\vec{U}^* \qquad (3.3.7)$$

Now, the boundary conditions can be restated as follows:

1. At a fixed solid surface,

$$\vec{U}^* = 0 \qquad\qquad (3.3.8)$$

2. At a free surface,

$$v^* = \frac{dy_o{}^*}{dt^*} \qquad\qquad (3.3.9)$$

and

$$p^* = p^*_{atmospheric} + \frac{\sigma}{\rho U^2 L}\left(\frac{1}{R_x^*} + \frac{1}{R_y^*}\right) \qquad\qquad (3.3.10)$$

Eqs (3.3.7) through (3.3.10) state that dynamic similitude requires keeping the following dimensionless groups constant:

$$\frac{p_o}{\rho U^2} = Eu \qquad\qquad (3.3.11)$$

$$\frac{\mu}{\rho U L} = \frac{1}{Re} \qquad\qquad (3.3.12)$$

$$\frac{gL}{U^2} = \frac{1}{Fr^2} \qquad\qquad (3.3.13)$$

and

$$\frac{\sigma}{\rho U^2 L} = \frac{1}{We} \qquad\qquad (3.3.14)$$

This similitude consideration leads to four important, non-dimensional parameters:

1. Euler number, $Eu = \frac{p_o}{\rho U^2}$
2. Reynolds number, $Re = \frac{UL}{v}$
3. Froude number, $Fr = \frac{U}{\sqrt{gL}}$
4. Weber number, $We = \frac{\rho U^2 L}{\sigma}$

Similitude of the Euler number, Eu, is readily satisfied for incompressible flows involving two length scales; e.g., flow depth, and width of some hydraulic structures. Values of Eu usually are preserved once Reynolds number, Re, or Froude number, Fr, similitude is prescribed. The Reynolds number is an important as a similitude criterion for water flow with or without a free surface, but, generally speaking, the Froude number is the dominant similitude parameter for water flows with a free surface (e.g., open channel flow, water waves). The Froude number and Reynolds number each define unique relationships between the scale ratios L_r, T_r, and U_r. They cannot be simultaneously satisfied without manipulating fluid properties, which at best is a difficult proposition. For certain ranges of Re values, and many flow situations (*e.g.*, fully turbulent flow in closed conduits and open channels), explicit satisfaction of Reynolds number similitude can be relaxed, because the effects of water viscosity do not significantly alter flow behavior (e.g., Heller, 2017).

When experiments focus on three-dimensional flow fields involving three independent lengths – approach flow depth, width of a hydraulic structure in the flow, and diameter of material (*e.g.*, sediment) transported in the flow – similitude of Eu, may not be preserved when prescribing values of Fr and Re (of flow around or through a hydraulic structure). In these situations, it is especially difficult to simulate pressure gradients, flow field vorticity, and flow forces responsible for transporting material (*e.g.*, Ettema *et al.*, 2006). A case in point is flow around a cylinder, in which similitude of Fr alone does not result in similitude of the wake vortices generated by flow around the cylinder.

Additional parameters expressing dynamic similitude arise from the non-dimensionalization of the conservation-of-momentum equation when additional forces related to other processes are important; *e.g.*, fluid compressibility and hydroelastic vibration. For compressible flow, the Euler number, Eu, is a special form of the Mach number, Ma, which has the same importance in compressible flow as the Froude number has in free surface flow; *i.e.*,

$$Eu = \frac{p_o}{\rho_o U^2} = \frac{1}{JM_a^2} \tag{3.3.15}$$

Here, Mach number

$$Ma = \frac{U}{c_S} \tag{3.3.16}$$

and specific-heat ratio

$$J = \frac{c_p}{c_v} \tag{3.3.17}$$

In Eqs (3.3.15) through (3.3.17), c_S is the speed of sound in water, ρ_o is the reference density, and c_p and c_v are the specific heats of water at constant pressure and volume, respectively. Compressibility effects become important when the Mach number is greater than about 0.3, which rarely occurs in hydraulic engineering, because c_s is 1,482 m/s for freshwater at 20°C.

An important, practical variant of the Euler number is the cavitation index

$$Ca = \frac{p_o - p_v}{0.5\rho U^2} \tag{3.3.18}$$

in which p_o and p_v = a reference pressure and the vapor pressure, respectively. This parameter is important if anywhere in a flow the pressure is reduced sufficiently to cause vapor formation (cavitation).

If the flow is oscillating, in the wake of a body or structure, an additional parameter is the Strouhal number,

$$St = \frac{\omega L}{U} \tag{3.3.19}$$

Here ω is the characteristic frequency of oscillation. This kinematic parameter, though, usually is not the prescribing similitude criterion, because it usually arises as a dependent variable that depends on the Reynolds number for flow around the body. Its replication is consequent to satisfaction of a dynamic-similitude criterion such as Froude number similitude, though being mindful of Re similitude. As an example, for flows with $10^2 < Re < 10^7$, the frequency of vortex shedding produced by flow around a cylinder of diameter, d, is characterized non-dimensionally as

$$St = \frac{\omega}{2\pi} \frac{d}{U} = 0.21 \tag{3.3.20}$$

Non-dimensionalization of the equation of conservation of thermal energy and its boundary conditions leads to additional parameters. At least four of them are important for describing flow and heat-transfer processes: Prandtl number, Eckert number, Grashof number, and wall-temperature ratio. Table 3.3.1 lists and defines these and other parameters ordinarily associated with experimental hydraulics.

Table 3.3.1 Annotated summary of similitude considerations for experimental hydraulics. The Notation List at the beginning of this book defines the symbols used in this table.

Hydraulics Processes and Similitude Considerations

Single-Phase Flow

Processes

Single-phase flow and movement of water occurs in all engineering hydraulics. Therefore, the processes and similitude considerations listed here underpin the other processes listed in this table. The main single-phase processes of interest are:

1. Flow field structure and distribution at various hydraulic structures or boundary forms in free-surface flow;
2. Flow resistance along conduits of various form; and,
3. Flow field structure and distribution at various hydraulic structures or boundary forms in closed-conduit flow.

Similitude considerations

Inertia force relative to gravitational force (Froude number), $Fr = U/(gL)^{0.5}$ (See Equation (3.3.13))
Inertia force relative to viscous force (Reynolds Number), $Re = UL/\nu$ (See Equation (3.3.12))
Inertia force relative to pressure force (Euler Number), $Eu = (p - p_o)/(\rho U^2)$ ((See Equation (3.3.11))

Loose-Boundary Flow and Sediment Transport

Processes

Modeling of flow over a loose boundary, such as the bed of an alluvial channel, usually aims at simulating and illuminating any or combinations of the following four groups of processes:

1. Flow over a loose planar bed;
2. Flow with bedforms;
3. Rates of sediment transport (bedload and suspended load); and,
4. Local patterns of flow and sediment movement in the vicinity of hydraulic structures.

Table 3.3.1 (continued)

Hydraulics Processes and Similitude Considerations

Similitude considerations

Flow resistance for a planar loose bed, f

Flow resistance for a bed with bedforms, $\sqrt{\left(\dfrac{\theta}{\theta_c}\right)} = \dfrac{u_*}{u_{*c}} = \dfrac{Y}{\sqrt{X(\Delta\rho/\rho)d\theta_c}}$

With incipient sediment motion assessed using the Shields number, $\theta = \dfrac{\rho u_*^2}{g\Delta\rho d}$ (See Equation (3.5.1))

Bed sediment transport, $Q_S = \dfrac{Y^3}{X^{1/2}(\Delta\rho/\rho)}$

Suspended transport of bed sediment, $\Phi = \left(\dfrac{q_s}{\rho_s\sqrt{(\Delta\rho/\rho)gd^3}}\right) = function\left(\dfrac{u_*}{w}\right)$ (See Equation (3.5.2))

Mixed Flows of Water and Air

Processes

Hydraulic engineering often involves mixed flows of gas (notably air) and water, either by design or as an inadvertent consequence of free-surface flow. Such flows require careful consideration of fluid properties (water and air).

1. Air entrainment in free-surface flows (e.g., spillways of various form);
2. Bubble plumes and aerators; and,
3. Cavitation.

Similitude considerations

Gravity force relative to surface-tension force (Eötvos number), $E\ddot{o} = \gamma L^2/\sigma$
Viscous force relative to gravity force (Stokes number), $St = \mu U/\gamma L^2$
Inertia force relative to gravity force (Froude number), $F = U/(\gamma L/\rho)^{0.5}$
Inertia force relative to viscous force (Reynolds Number), $Re = UL/v$ (See Equation (3.3.12))
Inertia force relative to surface tension force (Weber Number), $We = U^2/(\sigma/\rho L)$ (See Equation (3.3.14))
Inertia force relative to pressure force (Euler Number), $Eu = (p - p_o)/(\rho U^2)$ (See Equation (3.3.11))

Environmental Flows

Processes

Three types of processes are of particular importance:

1. Mixing and density stratification;
2. Spreading of immiscible fluids; and,
3. Water-quality and flow requirements of aquatic life.

Similitude considerations

Inertia force relative to density force (densimetric Froude number), $Fr_d = \dfrac{U}{\sqrt{g(\Delta\rho/\rho_o)L}}$

Inertia force relative to viscous force (Reynolds Number), $Re = UL/v$ (See Equation (3.3.12))
Ratio of molecular momentum diffusivity to molecular thermal diffusivity (Prandtl number), $Pr = v/\alpha$

Ratio of molecular momentum diffusivity to molecular mass diffusivity (Schmidt number), $Sc = \dfrac{v}{D_f}$

Ratio of inertia force to Coriolis force (Rossby number), $Ro = \dfrac{U}{\Omega L_f}$

Table 3.3.1 (continued)

Hydraulics Processes and Similitude Considerations

Coastal and Estuarine Flows

Processes
The main processes are:

1. Waves;
2. Oscillatory flows;

Similitude considerations
Because waves and tidally induced circulations in coastal regions, harbors, and estuaries, are dominated by gravity and inertia forces, the primary consideration is Froude number similitude;

$$(Fr)_r = \left(\frac{U}{\sqrt{gY}}\right)_r = 1$$

Hydroelastic Vibrations

Processes
Experiments on hydroelastic vibration usually investigate the susceptibility of a structure (or structural element) to flow induced excitation. Of primary importance is simulating the dynamic response, or vibration behavior, of the structure under investigation.

Similitude considerations
Inertia force relative to viscous force (Reynolds Number), $Re = UL/\nu$ (See Equation (3.3.12))
Water inertia force relative to solid elastic force (Cauchy number), which varies in accordance with structure type:

(a) For structures of continuous elasticity, $C_a = \frac{V}{\sqrt{E/\rho}}$

(b) For structures as discrete oscillators $C_a = \frac{V}{\sqrt{K/\rho L}}$

Ice

Processes
Most hydraulic experiments regarding ice are a straightforward extension of Single-phase flow experiments, with the additional considerations of:

1. Ice-piece drift and accumulation;
2. Complications arising from thermal (e.g., ice growth) and strength processes (e.g., ice bonding); and,
3. The breaking of ice accumulations and ice sheets.

Similitude considerations
Ice-piece inertia force to buoyancy force (densimetric Froude number), $Fr_D = \dfrac{U}{\sqrt{\left(\frac{\rho-\rho_i}{\rho}\right)gL}}$

To ensure ice in an experiment deforms in the same manner as ice at full scale, it is customary to consider ratio of ice strength, σ, and ice elastic modulus, E.

Note:
u_* = shear velocity; u_{*c} = critical value of u_* associated with incipient motion of bed particles; γ = specific weight; σ = coefficient of surface tension (water); ρ_s = sediment particle density; w = particle fall velocity; α = thermal diffusion coefficient; D_f = molecular diffusion coefficient; Ω = local angular velocity of the earth; L_f = specific length; E = elastic modulus; V = mean velocity of flow; K = structural spring stiffness; ρ_i = density of ice

3.4 WATER FLOW

The preceding section shows how the equations of motion of liquid water can be non-dimensionalized and how additional parameters sometimes need to be taken into account. Considerations of dynamic similitude for water flow form a useful similitude base for all experimental hydraulics, because civil engineering hydraulics involves water flow, movement, or displacement. This section discusses water flow processes and leads to Table 3.3.1, which gives an annotated summary of similitude considerations for various hydraulic experiments, including more complicated flow processes such as sediment transport, buoyancy-modified flow, and hydroelastic vibration.

3.4.1 Flow processes

Water flow experiments typically concern the following processes:

1. Flow field structure and distribution of free-surface flows;
2. Flow resistance and profiles of free-surface flows; and,
3. Structure and distribution of flow around immersed objects or in closed conduits. Often of interest are hydrodynamic forces exerted against immersed bodies, energy losses associated with flow through conduits, and the diverse influences of flow patterns on habitat, mixing or transport processes (without quantitatively replicating these processes).

The forces associated with single-phase flows generally are attributable to fluid inertia, gravity, the physical properties of the fluid, and boundary drag or friction. Fluid inertia is an important force in almost all situations involving fluid movement. Gravity is of prime importance for free-surface flows in which simulation of the water-surface profile is a modeling goal. Forces associated with the material properties of water (notably: viscosity, density, surface tension, and vapor pressure) increase in importance for flow situations where flow behavior is influenced by fluid properties; *e.g.*, when drag or shear forces are important, fluid density and viscosity must be considered. Cavitation is of concern when local pressures approach the vapor pressure for the fluid. Surface tension is important when modeling flows prone to air-entraining vortices. Failure to simulate the forces associated with fluid properties (especially viscosity and surface tension) may cause a small model to exhibit substantially different flow behavior than that which occurs at full scale.

The principal similitude criteria used to determine the kinematic and dynamic scales for experiments are selected in accordance with the forces dominating each flow process. Flow processes 1 through 3 require similitude of the motivating forces (inertia and gravity) and of the influences of water viscosity and surface tension. Experiments regarding flow resistance and flow (and flow energy) profiles, however, require stricter attention to similitude of channel or conduit resistance to flow. This requirement can be relaxed when experiments focus on local flow behavior because the flow lengths usually are sufficiently short that differences in water-surface elevation due to flow resistance are insignificant.

Wave motion may also be classed as a single-phase flow, and indeed experiments on wave motion largely use the similitude criteria prescribed for single-phase flows, along with similitude criteria stemming from boundary conditions. Experiments on buoyancy-modified flows and flow of immiscible fluids build on similitude of single-phase flow, but face additional complications posed by variable fluid properties, notably fluid density and immiscibility.

3.4.2 Dynamic similitude

This section shows how useful non-dimensional parameters can be developed by direct comparison of the forces relevant to a water flow process.

When pressure and inertia are considered, the ratio of those forces can be expressed as a pressure coefficient (a form of Euler number, Eu),

$$\frac{\Delta p L^2}{0.5 \rho L^2 U^2} = \frac{\Delta p}{0.5 \rho U^2} = C_P. \tag{3.4.1}$$

Common variants of C_P are drag coefficient, $C_D = F_D/(0.5\rho U^2 A)$, and the lift coefficient, $C_L = F_L/(0.5\rho U^2 A)$; in which F_D and F_L are drag and lift forces, respectively. The appropriate representation of area, A, depends on the specific drag or lift problem and definition of the coefficient. Typically, projected area or surface area is used in the definition of C_P. When equal pressure coefficients are maintained in model and prototype,

$$\Delta p_r = \rho_r U_r^2, \tag{3.4.2}$$

which relates pressure to density and kinematic scale ratios.

Examining the ratio of inertial forces to gravity forces leads to

$$\frac{\rho L^2 U^2}{\gamma L^3} = \frac{U^2}{gL} \equiv Fr^2. \tag{3.4.3}$$

Froude number similitude sets the main similitude criterion for free-surface flows; $i.e.$,

$$Fr_r = 1. \tag{3.4.4}$$

If the Froude numbers, Fr, are equal in the model and prototype;

$$U_r = \frac{L_r}{T_r} = \sqrt{g_r L_r}, \tag{3.4.5}$$

and, if the ratio of gravitational acceleration, g_r, is assumed equal to unity,

$$T_r - \sqrt{L_r}, \tag{3.4.6}$$

and

$$U_r = \sqrt{L_r}, \tag{3.4.7}$$

thus establishing a unique relation between the kinematic and geometric variables.

These are the most common similitude (and scale) expressions to be found in hydraulic laboratory practice. However, when the ratio of inertial forces to viscous forces must be considered, the following non-dimensional parameter results:

$$\frac{\rho L^2 U^2}{\mu UL} = \frac{\rho LU}{\mu} = Re. \tag{3.4.8}$$

If the Reynolds numbers, Re, are equal in model and prototype, a different relationship between variables results. The scales for velocity and time are

$$U_r = \frac{v_r}{L_r},$$

$$(3.4.9)$$

and

$$T_r = \frac{L_r}{v_r},$$

$$(3.4.10)$$

where v_r is the ratio of kinematic viscosity of prototype and model fluids.

Other non-dimensional force parameters arise by considering the ratios of other forces. For example, the Weber number parameter results from the ratio of inertial to surface tension forces; *i.e.,*

$$\frac{\rho U^2 L^2}{\sigma L} = \frac{\rho U^2 L}{\sigma} = We.$$

$$(3.4.11)$$

In this case, the velocity and time scales are

$$U_r = \sqrt{\frac{\sigma_r}{\rho_r L_r}},$$

$$(3.4.12)$$

and

$$T_r = \sqrt{\frac{\rho_r L_r^3}{\sigma_r}}.$$

$$(3.4.13)$$

Additional non-dimensional force parameters and consequent similitude ratios can be developed in accordance with the type of phenomenon investigated.

It is soon evident when considering similitude of Eu, Fr, Re, and We that multiple and conflicting scale ratios arise for kinematic and dynamic variables. As strict similitude of all these parameters cannot be achieved, it is important to know which forces are of paramount importance and to set similitude and scales in accordance with the appropriate force ratio.

Experiments often are used to determine water-surface profiles in channels with complicated roughness patterns or bathymetry; *e.g.,* flow in vegetated floodplains, or along geometrically complicated channels. The Froude-number criterion, Equation (3.4.4), prescribes similitude for the dominant forces, which are attributable to fluid inertia and gravity.

But this criterion by itself may be insufficient for prescribing similitude of flow resistance. Flow resistance can be described by relationships such as the Darcy-Weisbach, Manning, Manning-Strickler, and Chézy equations, which inter-relate.

The Darcy-Weisbach equation states

$$U = \sqrt{\frac{8gR_h S_f}{f}},$$

$$(3.4.14)$$

in which S_f = slope of the energy gradient of the flow, and the dimensionless resistance coefficient, f, can be written in functional form as

$$f = \varphi(k/4R_h,\ Re,\ channel\ shape). \tag{3.4.15}$$

Here, R_h = hydraulic radius; and, k = surface roughness height, often related to a characteristic bed-particle diameter, d (Henderson, 1966; Reynolds, 1976). The Moody diagram shows this functional relationship for pipes, and indicates zones of laminar flow and fully turbulent or hydraulically rough flow conditions. The zone of fully rough flow is delineated by the line

$$Ref^{1/2}\left(\frac{k}{4R_h}\right) \geq 200 \tag{3.4.16}$$

which can be used to estimate the minimum model-scale Reynolds number needed to ensure fully turbulent flow.

In combination with Equation (3.4.14) the Manning-Strickler equation for flow resistance produced by fully rough flow over a planar bed of particle roughness gives

$$U \propto \frac{R_h^{2/3}S_f^{1/2}}{d^{1/6}}, \tag{3.4.17}$$

can be used to express f;

$$f \propto \left(\frac{d}{R_h}\right)^{1/3}. \tag{3.4.18}$$

From Equation (3.4.18), f varies as

$$f_r = \left(\frac{d_r}{R_{hr}}\right)^{1/3}. \tag{3.4.19}$$

Ideally the scale of the roughness must be the same as the scale of the flow geometry if the friction factors in the model and the prototype are to be the same; for most flow situations, it is important that $f_r \approx 1$. It is possible in some situations, though, to attain $f_r \approx 1$ by appropriate combination of reduced model roughness and Reynolds number; as is evident from the Moody diagram.

Besides Equation (3.4.19), power-law approximations of the form $f \propto (k/R_h)^\beta$, or $(k/Y)^\beta$, have been proposed for relating friction factor or boundary shear stress to relative roughness; here, Y = flow depth. Different values of the exponent, β, may occur in accordance with differences in the ranges of k/R_h (or k/Y) and Reynolds-number values considered, the closeness of a power-law approximation to a log-law expression for velocity distribution in a turbulent boundary layer, and with the accuracy of the data to which the power law was fitted (e.g., see Chen, 1992).

For channels of complex geometry and whose beds and sides are not planar, flow resistance is attributable to form drag as well as surface friction. In modeling these channels, resistance similitude should still be based on Equation (3.4.19), which can be restated in terms of the alternate, though dimensional, Manning and Chézy resistance coefficients, *n* and *C*, respectively. Considerable trial and error adjustment may be needed to calibrate experiments concerning flow in complex channels, especially when there are pronounced transverse as well as streamwise gradients in water level. Such calibration may typically seek to match water levels with flow rates, or velocity distribution with flow rate.

3.5 MULTI-PHASE FLOW AND TRANSPORT PROCESSES

Experimental hydraulics becomes more complicated when the topic of investigation extends from single-phase flow of liquid water to multi-phase (liquid, solid, gas) flow processes, and includes flows involving transport processes, and flows with moving boundaries. Further similitude considerations are needed to characterize the additional processes to be replicated in an experiment.

3.5.1 Other processes

Hydraulic experiments and hydraulic models fairly commonly focus on the following further processes:

1. Flow over loose-boundaries, such as alluvial sediment;
2. Sediment transport;
3. Mixed flows of water and air;
4. Environmental flows, notably various forms of mixing and density-driven flows;
5. Coastal and estuarine processes;
6. Hydro-elastic vibrations; and,
7. Ice.

Table 3.3.1 lists the main processes and considerations associated with these processes, and serves as a concise summary of the relevant similitude concerns. It indicates how similitude considerations form a hierarchy of priorities aligned with the additional process of experimental interest. For most experimental hydraulics, similitude of single-phase flow is the basis from which additional similitude consideration extend. In some experiments, a higher similitude consideration may be selected to supersede a lower consideration. On occasion, an experiment or hydraulic model may pose a question as to the greater importance of one process versus another, and thus require either a decision as to greater importance or the conduct of additional tests to determine the effect of demoting one process, and similitude criterion, relative to another. For example, experiments or models for investigating bed sediment movement may place less weight on Froude number similitude than on similitude of bed sediment mobility.

Space herein precludes an extended discussion of the processes and the similitude considerations they invoke. However, the ensuing sub-section briefly elaborates similitude associated with sediment transport.

3.5.2 Dynamic similitude for sediment transport

Flow over a loose boundary and bed sediment transport are frequent topics of laboratory experiments, and need additional similitude considerations related to sediment movement in water. Dimensional analysis of the variables influencing the movement of uniform bed sediment along a channel leads to the parameters (*e.g.*, ASCE, 2000)

$$\frac{u_*d}{v} \quad \text{and} \quad \frac{\rho u_*^2}{g\Delta\rho d}$$

where u_* is a shear velocity, d is a particle diameter, and $\Delta\rho$ is the density difference between a sediment particle, ρ_s, and water, ρ; the specific gravity for alluvial sediment is typically $\rho_s/\rho = 2.65$.

The parameter, $\frac{u_*d}{v}$, known conventionally as the particle Reynolds number, Re_*, (*e.g.*, ASCE, 1975, 2000), relates the particle size to the thickness of the viscous length scale developed by a flow. The particle Reynolds number plays an important role in flow entrainment of particles. The parameter,

$$\theta = \frac{\rho u_*^2}{g\Delta\rho d}, \tag{3.5.1}$$

often termed the Shields number, expresses the ratio of the average or nominal shear force acting on the bed surface, $\tau_o = \rho u_*^2$, to the submerged weight of a bed-particle of average size (d). It also is termed particle mobility number, flow intensity, particle Froude number, Fr_*, or densimetric Froude number. It is useful for characterizing the condition of incipient motion of particles on a bed, and for describing the intensity of bed particle movement.

When the mass rate of sediment transport is of primary concern, a useful parameter is

$$\Phi = \frac{g_s}{\rho_s\sqrt{g(\Delta\rho/\rho)d^3}} = \frac{q_s}{\sqrt{g(\Delta\rho/\rho)d^3}} \tag{3.5.2}$$

in which ρ_s is the density of sediment, Φ expresses g_S and q_S, the mass and volumetric rates, respectively, of sediment transport per unit width of channel in terms of the water and sediment properties (*e.g.*, ASCE, 1975). Hydraulic modeling may seek to achieve $\Phi_r \approx 1$, but this may be difficult to achieve, and modeling may settle for just having a loose bed sufficiently mobile for the purpose of an experiment; *e.g.*, an experiment whose purpose is to design a water diversion that excludes sediment.

The particle fall velocity, w, is sometimes used in establishing scales for sediment transport; notably, when suspended motion of sediment is important. Now, the relationship for Φ_r becomes

$$\Phi_r = \left(\frac{q_S}{\sqrt{g(\Delta\rho/\rho)d^3}}\right)_r = 1 = \left(\frac{u_*}{w}\right)_r, \tag{3.5.3}$$

The modeling of some processes remains reliant on single-phase flow similitude, but includes additional water movement processes that require variations of this similitude. Coastal and estuarine processes, for instance, involve the additional complications of waves and oscillatory flow. The complexities of flow-induced structural vibration include a flexible boundary, and flow in hydro-machines (principally, turbines and pumps) involve moving flow boundaries. In Table 3.3.1, the term "environmental flow" signifies flows involving water quality and mixing processes. It especially includes buoyancy-modified flows, the mechanics of jets and plumes, turbulent mixing phenomena, and heat transport.

3.6 ADDRESSING SIMILITUDE SHORTCOMINGS

It is a simple truism that hydraulic experiments should be designed to replicate the flow and transport process(es) of focal interest. However, experiment design and conduct can soon face complications, as experimenters often run into various practical limits that constrain achievable similitude. Such limits may produce scale effects, and are attributable to material limitations and laboratory limitations.

Material limitations relate to –

1. water properties;
2. properties of the material to be transported by water flow (*e.g.*, sediment particles), or moving in water (*e.g.*, an immiscible fluid like oil); and,
3. strength-related properties of material in experiments involving deformation and possible failure of a structure (*e.g.*, hydroelastic flexure of structures, and ice-breaking).

Laboratory considerations relate to –

1. space limitations in a laboratory or a flume;
2. instrument limitations; and,
3. time limitations.

Lack of similitude due to variables not being scaled in accordance with similitude criteria, or to inadequacies in space and time, risks diminishing or compromising the reliability of results produced from an experiment. Figure 3.6.1 illustrates a scale-effect in the modeling of an overflow spillway of Gebidem Dam. Scale effects in this example are responsible for different air entrainment under similar flow conditions. There often are ways to lessen such risks of reducing experiment or modeling accuracy, as the next subsections explain.

3.6.1 Avoidance

Experimenters should be aware of the suggested guidelines as to similitude limits associated with laboratory experiments on flow processes. Table 3.6.1 lists suggested limits, determined from experience, intended to avoid significant scale effects for various hydraulic engineering phenomena under Froude number similitude. The limits are expressed in non-dimensional and dimensional terms. Further research is needed to confirm and explain many of the limits.

It is important to consider the prototype relation in the fourth column particularly for cases where the limitations are not directly expressed with force ratios Re, We, C_a or Eu. Scale effects due to Re can be avoided if water is replaced with air, but air models are unable to reproduce density, free surface and cavitation effects (*e.g.*, Westrich, 1980).

Further, to avoid compressibility effects, air velocity should be limited to 50 m/s according to Westrich (1980), or 60 to 90 m/s according to Rouse *et al.* (1959). Studies

Figure 3.6.1 Overflow spillway of Gebidem Dam, Valais, Switzerland: (a) hydraulic model built at a length scale of 30; and, (b) prototype operating in 1967. The air entrainment of the free jet differs considerably between model and prototype due to non-identical values of We (from Heller, 2011, with permission from Taylor and Francis Ltd., www.tandfonline.com).

of sediment transport may avoid significant scale-effects with the replacement of both fluid and sediment; an example mentioned in Kobus (1980) used coal dust as the sediment in glycerin as the fluid. Other authors change the fluid properties, such as Ghetti and D'Alpaos (1977) and Miller (1972), who added a surfactant to the water, or Stagonas *et al.* (2011) who used a mixture of 90% distilled water and 10% isopropyl alcohol solution, to avoid or lower surface-tension effects.

3.6.2 Hybrid approaches

When practical constraints limit or diminish information produced from a laboratory experiment, consideration should be given to augmenting the experiment with information obtained by other means, such as field work, numerical modeling, or additional laboratory experiments configured to obtain the needed information.

Many processes in hydraulic engineering require a combination of approaches, or a hybrid approach, to obtain the needed information. For instance, this consideration may arise when investigating processes occurring over large areas as well as in important local areas of concern; or over differing periods of time; or, involving complexities (*e.g.*, biological) not readily handled in a laboratory flume. Developments in numerical modeling and field instrumentation make hybrid approaches increasingly attractive as part of an investigative strategy in experimental hydraulics.

A further approach sometimes used involves "scale-series" testing. Lack of similitude may cause so-called scale effects. To determine how scale effects influence model results, it is possible to use a series of scale models, where by models of the same

Table 3.6.1 Suggested guidelines for avoiding significant Reynolds and Weber number scale effects in various hydraulic flow processes under Froude number similitude (from Heller, 2011, with permission from Taylor and Francis Ltd., www.tandfonline.com). Note that the suggested limits include non-dimensional and dimensional bounds.

Investigation topic	Flow process	Guideline	Related prototype feature	Reference
Air-entraining free vortex at horizontal intake	Flow conditions	$Q_i/(z_i\nu) > 30{,}000$ and $\rho Q_i^2 z_i/(A_i^2\sigma) > 10{,}000$	Scale series: largest discharge 0.009 m³/s, largest submergence depth 0.8 m, constant intake diameter = 0.0762 m	Anwar et al. (1978)
Air-entraining free vortex at horizontal intake	Air entrainment rate	$Vd/\nu \geq 600{,}000$ and $V(\rho d/\sigma)^{1/2} \geq 120$	Scale series: largest $d = 0.484$ m and $0.4 \leq V/(gd)^{1/2} \leq 1.4$	Möller et al. (2012)
Broad-crested weir	Discharge coefficient	Overfall height ≥0.07m	Weir length 0.15 to 5 m, weir width 4.88 m	Hager (1994)
Chute aerator	Local bottom air concentration	$Vh/\nu \geq 220{,}000$ and $(\rho V^2 h/\sigma)^{0.5} \geq 140$	Assessed in relation to trend line of physical model test data for $(\rho V^2 h/\sigma)^{0.5} \geq 140$	Pfister and Hager (2010)
Dambreak wave	Sudden failure, dam in smooth rectangular channel	Still water depth ≥ 0.30m	n.a.	Lauber and Hager (1998)
Dam with ski jump	Lateral overfall weir, spillway capacity, ski jump, flow conditions in stilling basin	Scale 1:30	Upper Siebertalsperre, dam height 90 m, $Q_d = 140$ m³/s	Kobus (1980)
Dike breaching	Hydraulics over dike consisting of uniform and non-cohesive material	Unit discharge ≥ 20 l/s and $d_g \geq 1$ mm	Scale series: largest scale with dike height = width = 0.40 m and $d_g = 8$ mm	Schmocker and Hager (2009)
High-speed two-phase flows	Air concentration	$Vh/\nu > 200{,}000$ to $300{,}000$ and $(\rho V^2 h/\sigma)^{0.5} > 140$ for $5 \leq V/(gh)^{1/2} \leq 15$	Based on literature review of physical model tests	Pfister and Chanson (2013)
Hydraulic jump	Sequent depths ratio h_2/h_1	$V_1 h_1/\nu > 100{,}000$ for $Fr_1 < 10$ and $h_1/b < 0.1$	Scale series: maximum $h_1 = 0.063$ m, maximum $Fr_1 = 42.7$	Hager and Bremen (1989)
Hydraulic jump	Void fraction and bubble count rate distributions, bubble chord time	$\rho V_1 h_1/\mu > 100{,}000$	Scale series: maximum $h_1 = 0.024$ m, maximum $b = 0.50$ m, maximum $Fr_1 = 8.5$	Chanson (2009)

Application	Purpose	Condition	Parameters	Reference
Impulse wave	Generation by subaerial landslide	$g^{1/2}h^{3/2}/\nu \geq 300,000$ and $\rho g h^2/\sigma \geq 5,000$ resulting in $h \geq 0.200$ m	Scale series: maximum $h = 0.60$ m, $1.7 \leq V_s/(gh)^{1/2} \leq 4.3$	Heller et al. (2008)
Mountain river	Bed morphology	Scale 1:10 to 1:20	Steinibach, $d_m = 0.20$ m, $d_{90} = 0.52$ m, $d_{max} = 0.90$ m, slope 3-13%	Weichert (2006)
Rubble mound breakwater	Stability	Limiting scale as a function of prototype wave height in Fig. 1 of Oumeraci (1984)	Prototype wave height up to 13 m	Oumeraci (1984)
Scour	Bridge pier and abutment scour depth development prediction with Equation (1) of Oliveto and Hager (2005)	$0.60 <$ threshold Froude number < 1.20, width pier/$b \geq 0.05$, width abutment/$b \geq 0.05$	Model tests: width pier = 0.022 – 0.500 m, width abutment = 0.05 – 0.20 m, 0.45 $\leq t \leq 21.0$ days, $b = 1.0$ m, $0.03 \leq h \leq 0.18$ m, $d_{50} \geq 0.80$ mm, $1.07 \leq V/(g'd_{50})^{1/2} \leq 4.26$ with $g' = [(\rho_s - \rho)/\rho]g$	Oliveto and Hager (2005)
Scour	Effect of large-scale turbulence on equilibrium scour depth at cylinders	Cylinder diameter ≥ 0.400 m for scale effect $\leq 5\%$	Model tests: cylinder diameter = 0.064 – 0.406 m, average velocity = 0.46 m/s, $v^*/v_c^* = 0.80$, $d_m = 1.05$ mm, $h = 1.000$ m	Ettema et al. (2006)
Ski jump	Jet throw distance	Approach flow depth ≥ 0.04 m	Scale series: largest water depth 0.07 m	Heller et al. (2005)
Skimming flow on stepped spillway	Turbulence level, entrained bubble sizes and interfacial areas	$\rho(gh_c^3)^{1/2}/\mu_s > 500,000$ and step height > 0.02 m	Scale series: maximum step height 0.143 m, spillway slopes 3.4 to 50°	Chanson (2009)
Solitary wave run-up	Run-up height	$[g(h + H)]^{1/2}(h + H)/\nu > 70,000$ and $[g(h + H)]^{1/2}\{\sigma/[\rho(h + H)]\}^{1/2} > 30$ resulting in $h \geq 0.08$ m	Scale series: maximum $h = 0.200$ m, slope 1:5, maximum $H/h = 0.7$	Fuchs and Hager (2012)
Spillway	Amount of air entrainment from aerator	$V/[\sigma/(\rho h)]^{1/2} > 110$	Measurements in models and a small prototype and consideration of further prototype data	Rutschmann (1988)
Stepped spillway	Flow velocity profile air-water mixture	Scale $\geq 1:15$	Step height 0.6 m, maximum specific discharge 20 m²/s	Boes (2000)
Surf zone beach profile	Volume of transported sand	Scale $\geq 1:7.5$, $d_{50} = 0.13$ mm, significant wave height 0.20 m, peak wave period 2.0 s	$d_{50} = 0.335$ mm, significant wave height 1.5 m, peak wave period 6.0 s	Ranieri (2007)

Table 3.6.1 (continued)

Investigation topic	Flow process	Guideline	Related prototype feature	Reference
Vertical plunging circular jet	Void fraction and bubble count rate distributions, bubble size	$\rho V_j^2 d_j/\sigma > 1,000$	Scale series: largest jet diameter 0.025 m, jet Froude number up to 10	Chanson (2009)
Wave overtopping at coastal structures	Overtopping velocity	$2(R - R_c)^2/(vT) > 1,000$ and $V_R^2 h_R \rho/\sigma > 10$	Theoretically deduced	Schüttrumpf and Oumeraci (2005)
Wave run-up	Overland flow depth and front velocity	$V_f h_{max}/v > 6,300$ and $V_f[\sigma/(\rho h_{max})]^{1/2} > 10$	Scale series: maximum h = 0.200 m, shore height = 1.25h, slope 1:5, maximum H/h = 0.7	Fuchs and Hager (2012)
Wave run-up	Run-up velocity	$2(R - R_c)^2/(vT) > 1,000$ and $V_R^2 h_R \rho/\sigma > 10$	Theoretically deduced	Schüttrumpf and Oumeraci (2005)
Water wave	Force on slope during wave breaking	Wave height $\geq$ 0.50 m	Scale series: maximum wave height 1.25 m	Skladnev and Popov (1969)
Water wave	Theoretical effect of surface tension	$T > 0.35$ s, $h > 0.02$ m	Wave with wave length where surface tension effects contribute less than 1%	Hughes (1993)

Note:
A_i = cross sectional area intake; b = channel width; d = pipe diameter; d_g = grain diameter; d_j = jet diameter; d_m = mean grain diameter; d_{max} = maximum grain diameter; d_{50} = grain diameter at 50 weight percent; d_{90} = grain diameter at 90 weight percent; Fr_1 = Froude number at sequent depth 1; g = gravitational acceleration; g' = relative gravitational acceleration; h = water depth; h_c = critical flow depth; h_{max} = maximum flow depth; h_R = run-up water depth; h_1 = water depth at sequent 1; h_2 = water depth at sequent depth 2; H = wave height; n.a. = not available; Q_d = design discharge; Q_i = discharge intake; R = unlimited run-up height; R_c = run-up height to crest of structure; t = time; T = wave period; v^* = shear velocity; v_c^* = critical shear velocity; V = velocity; V_f = flow front velocity; V_j = jet velocity; V_R = run-up velocity at still water level; V_1 = flow velocity at sequence 1; V_s = slide impact velocity; ρ = water density; ρ_s = slide density; μ = dynamic viscosity; v = kinematic viscosity; σ = surface tension between air and water; z_i = submergence depth intake.

prototype are built at different scales. "Scale-series" testing can be useful if inadequate prototype information is available. Prototype results then may be inferred from the model studies by extrapolation.

3.6.3 Similitude compromise

At times, some compromise in similitude must be considered so that experiments can be conducted within practical limits imposed by local facilities. Practical limits commonly relate to available space, flow capacity, and instrumentation at hand. Such compromise concentrates on replicating the main physical processes of concern, and relies on careful interpretation of the results of an experiment in order to discern the information sought.

Most experimental hydraulics is carried out using water to simulate water flow. Occasionally, for reasons mainly of practicality, and to avoid *Re*-related scale effects, air flow is used to simulate water flow and vice-versa. Provided the similitude considerations are satisfied and cavitation is not a concern, water flow around a fully immersed body is adequately investigated using airflow in a wind tunnel. However, the buoyancy-driven flow associated with the release of a steam plume from a cooling tower may be more conveniently modeled using a water tank with heated water replicating heated air and steam. When using air to simulate water, or vice-versa, care is needed to ensure that undesirable additional processes associated with material properties (*e.g.*, viscosity or heat capacities) do not introduce inadvertent effects.

3.6.4 Geometric distortion

To achieve a desired flow behavior, such as ensuring flow velocities are of sufficient magnitude and distribution, the geometry of an experiment may be adjusted or distorted in several ways. Although distortion usually involves exaggeration of vertical scale, at times the slope, velocity, or roughness may be altered to achieve a desired flow behavior in an experiment setup or model. Slope distortion entails adjusting the slope of the model so that it differs from the prototype slope, though retaining the length scales. Velocity distortion may entail directing flow, by means of altered flow distribution or magnitude in a portion of the model. There is no rule that forbids such distortions, but experimenters should be wary of inadvertent effects that may cause the model to behave differently than the prototype.

Geometric distortion, alternatively called vertical distortion, is a common departure from complete similitude, and thus deserves discussion here. In many hydraulic applications, the vertical dimensions are much smaller than the horizontal dimensions; *e.g.*, rivers are much wider than they are deep. Consequently to produce a viable laboratory model, it may be necessary to use a vertical scale ratio, Y_r, smaller than the horizontal scale ratio, X_r. The experiment layout, or hydraulic model, is then vertically distorted by an amount $G = X_r/Y_r$; *e.g.*, if $Y_r = 10$, and $X_r = 40$, then $G = 40/10 = 4$.

For experiments examining processes occurring in relatively large, wide flow fields, it sometimes is necessary to resort to geometric distortion so as to attain adequate dynamic similitude, or to facilitate sufficiently accurate measurement of flow properties. Geometric distortion is especially common in hydraulic modeling situations where a three-dimensional flow field can be treated as being approximately two-dimensional. Vertical distortion and slope distortion are particularly common for hydraulic models of loose-bed hydraulics, as a fairly common tactic used in modeling flow resistance in loose-bed channels. Experiments on markedly three-dimensional flows (*e.g.*, jets, local scour, cavitation, mixing) should be undistorted, unless an experiment is intended to be preliminary and qualitative; even then, care is needed in interpreting results.

Suggested limits for vertical distortion depend on acceptable distortions related to specific experiment purposes, like sediment transport, ice transport and jamming, dispersion and mixing of effluents, and waves. On the whole, experiments encompassing large areas of flow, of which extensive areas are shallow, may require larger distortion. For instance, hydraulic models of estuaries typically have distortions in the range $G = 5$ to 10, and sometimes more.

The main advantages of vertical distortion are:

1. Reduced expenses incurred with constructing and running an experiment, which, if built undistorted, would occupy a larger area;
2. Increased values of Re (instead of a scale ratio of $Xr^{1.5}$, the ratio is reduced to $(X_r/G)^{1.5}$; and,
3. Greater depths enabling increased accuracies of flow-velocity and depth measurements in the model.

Vertically distorted experiment configurations or models are distorted to achieve a specific purpose, such as replication of streamwise flow profiles. The main drawbacks of vertical distortion are that three-dimensional and two-dimensional (across the depth of flow) flow patterns and pressure distributions are distorted in response to the altered aspect (width-to-depth) ratio of the flow. Therefore, vertical distortion must be used cautiously in situations where flow fields are markedly three dimensional. It is commonly used when channel resistance is the primary concern, but is rarely used to investigate the flow fields in the vicinity of hydraulic structures such as, for example, water intakes, spillways, bridge piers and abutments.

Vertical distortion may produce the following effects in an experiment or hydraulic model:

1. Exaggeration of secondary currents;
2. Distortion of eddies;
3. Increase in the vorticity and shedding frequency of wake eddies;
4. Because of items 1 and 2, differing head-discharge relationships may occur in channels and through hydraulic structures. Exaggerated secondary flows in contractions, expansions, bends, etc., generally require proportionately more head (energy) to pass the modeled flow;
5. Occurrence of flow separation on inclined boundaries, whose slope is increased, where separation would not occur at full scale;
6. In consequence to, and interacting with, items 1 through 3, lateral distributions of flow in the model may differ from lateral flow distributions at full scale;
7. The ratios between vertical and horizontal forces at full scale will not be preserved in the experiment or model; and,
8. For hydraulic models, there may be an unfavorable psychological effect on the sponsor or client of the hydraulic model if its geometry appears out of proportion.

The Darcy-Weisbach resistance equation, Equation (3.4.14), alters in response to vertical distortion:

$$f_r = (S_f)_r \left(\frac{(R_h)_r}{U_r^2} \right) = \frac{Y_r}{X_r} = \frac{1}{G} \qquad (3.6.1)$$

Eqs. (3.4.19) and (3.6.1) indicate that the scale of the roughness elements, d_r, varies with the third power of vertical distortion, G. The important conclusion here is that the greater the distortion of vertical and horizontal scales, the greater is the required exaggeration of the model roughness. Because it usually is not practical to select the roughness elements in exact accordance with Equation (3.6.1), considerable trial-and-error adjustment of boundary conditions may be needed to calibrate the model. Modeling of flow in channels with large roughness elements, such as exist in rapids, can be especially challenging, and flow calibration is of paramount importance.

3.A APPENDIX: DIMENSIONAL ANALYSIS

Dimensional analysis utilizes the principle of dimensional homogeneity – *If an equation truly expresses a proper relationship between variables in a physical process, it will be dimensionally homogeneous; i.e., each additive term in the equation will have the same units.*

This principle undergirds the Π theorem, which states that a dimensionally homogeneous linear equation is reducible to a functional relationship among a set of dimensionless parameters. The theorem leads to the proposition that a process involving n variables, with m basic dimensions reduces to $n-m$ dimensionless parameters, or Π numbers. In mathematical terms, Π means "product of variables." The basic dimensions in fluid flow (*i.e.*, the m terms) are mass, M, length, L, and time, T; or, force, length, and time, F, L, T. The two systems of dimensions relate via Newton's Second Law, $F = ML/T^2$ or $M = FT^2/L$. Buckingham (1921), Langhaar (1951), and Sedov (1959), for example, give useful proofs of the theorem. Buckingham (1921) made the theorem popular, such that it now is often called the Buckingham Π theorem; he formalized what earlier was known as Raleigh's method of dimensional analysis.

There are several ways to identify meaningful non-dimensional parameters. The most direct method is by direct inspection of variables, but the experimenter must have a good grasp of the physical processes in question to do this. The formal procedure uses the Π theorem, which reduces n dimensional variables into $n - m$ dimensionless Π parameters, as illustrated below.

For a process involving a set of n variables, $a_1, a_2, \ldots a_n$, in which variable a_1 depends on only the independent variables $a_2, \ldots a_n$, a general functional relationship can be written as

$$f_1(a_1, a_2, \ldots, a_n) = 0 \qquad (3.A.1a)$$

or

$$a_1 = f_2(a_2, \ldots, a_n) \qquad (3.A.1b)$$

The n dimensional variables can be combined as a non-dimensional product of the form

$$\Pi = a_1{}^{x_1} a_2{}^{x_2} a_3{}^{x_3} ... a_n{}^{x_n} \tag{3.A.2}$$

in which the exponents, x_i, are pure numbers and the a_i variables are written in terms of their fundamental dimensions L, T, and M. The net exponent of each of the m fundamental dimensions involved must be zero as Π is by definition dimensionless.

In accordance with the Π theorem, a process may be described using the following functional relationship of $n–m$ Π parameters:

$$\Phi\left(\Pi_1, \Pi_2, \Pi_3, \cdots, \Pi_{n-m}\right) = 0 \tag{3.A.3a}$$

or

$$\Pi_1 = \Phi_1\left(\Pi_2, \Pi_3, \Pi_4, \cdots, \Pi_{n-m}\right) \tag{3.A.3b}$$

in which Π_1 expresses the variable a_1 as a non-dimensional parameter.

By way of illustration, consider the wall shear stress, τ_o, exerted by flow in a rough pipe. The variables influencing τ_o and, thereby, flow resistance are the pipe diameter, D, the pipe roughness, k, a characteristic flow velocity, U, and fluid properties (density, ρ, and kinematic viscosity, v); *i.e.*,

$$\tau_o = f(D, k, U, \rho, v) \tag{3.A.4}$$

The variables combined as a non-dimensional product are

$$\Pi = D^{x_1} k^{x_2} U^{x_3} \tau_o{}^{x_4} \rho^{x_5} v^{x_6} \tag{3.A.5}$$

As there are three fundamental dimensions, M, L, T, Equation (3.A.5) becomes

$$\Pi = [L]^{x_1} [L]^{x_2} [LT^{-1}]^{x_3} [ML^{-1}T^{-2}]^{x_4} [ML^{-3}]^{x_5} [L^2 T^{-1}]^{x_6} \tag{3.A.6}$$

or

$$\Pi = [L^{(x_1+x_2+x_3-x_4-3x_5+2x_6)} T^{(-x_3-2x_4-x_6)} M^{(x_4+x_5)}] \tag{3.A.7}$$

For the net exponents of L, T, and M to be zero,

$$x_1 + x_2 + x_3 - x_4 - 3x_5 + 2x_6 = 0$$
$$-x_3 - 2x_4 - x_6 = 0$$
$$x_4 + x_5 = 0$$

Because there are three more unknowns than equations, a unique solution of the three equations is impossible. In dimensional-analysis terms, this finding means that it is possible to form more than one set of non-dimensional parameters from the variables.

The Π theorem, as a mathematical theorem concerning the simultaneous solution of a set of linear equations, holds that only a limited number of solution combinations are independent from each other for a set of linear equations. In dimensional-analysis terms, only a limited number of non-dimensional parameters will be independent. As shown below, three such combinations result for the present illustration of flow through a pipe. The theory indicates that there are only $n - r$ linearly independent solutions to a set of m linear equations forming a matrix of n columns and m rows. The rank of the matrix, r, is the maximum order of non-zero determinants; $r \leq m$. In general, the number of independent non-dimensional parameters formed from the set of variables equals the number of linearly independent solutions for the equation set. In other words, the number of non-dimensional parameters equals the matrix rank, $n - m$.

The dimensional matrix of the exponent coefficients for variables associated with flow in a pipe (Equation (3.2.A.7)) has the following form:

	D	k	U	τ_o	ρ	v
L	1	1	1	−1	−3	2
T	0	0	−1	−2	0	−1
M	0	0	0	1	1	0

The rank of the matrix, $r = 3 = m$, because the matrix has at least one third-order determinant that does not equal zero. Therefore, the variables form $n - m = 6 - 3 = 3$ independent non-dimensional parameters. The parameters can be found by prescribing values to exponents of three of the dimensional variables resulting from Equation (3.2.A.7). In this regard, it is important to choose one variable from each of the variable categories (geometry, flow, fluid), thereby ensuring that each fundamental dimension (L, T, M) is assigned a value. Once the prescribed exponents are chosen, each of them are used to solve for the remaining unknown exponents. The exponents of the variables k, τ_o, and v are chosen for the present illustration. In some flows, there may be only one or two dimensions involved, and it may not be clear whether a parameter belongs to the flow or the geometry category; *e.g.*, in the analysis of buoyant jets in terms of volume and momentum fluxes.

When $x_2 = 1$, $x_4 = 0$, and $x_6 = 0$

$$x_1 + 1 + x_3 - 3x_5 = 0$$
$$x_3 = 0$$
$$x_5 = 0$$

so that $x_1 = -1$. The resulting dimensional parameter is $\Pi_k = k/D$, the relative roughness of the pipe.

When $x_2 = 0$, $x_4 = 1$, and $x_6 = 0$, the equations become

$$x_1 + x_3 - 1 - 3x_5 = 0$$
$$x_3 = -2$$
$$x_5 = -1$$

so that $x_1 = 0$. The resulting non-dimensional parameter is $\Pi_{\tau o} = \tau_o/\rho U^2$, a resistance coefficient.

When $x_2 = 0$, $x_4 = 0$, and $x_6 = 1$, the equations become

$$x_1 + x_3 - 3x_5 + 2 = 0$$
$$x_3 = -1$$
$$x_5 = 0$$

so that $x_1 = -1$. The resulting non-dimensional parameter is $\Pi_\nu = \nu/DU = 1/Re$.
Therefore, for wall shear stress or flow resistance in a pipe,

$$\frac{\tau_o}{\rho U^2} = \Phi_2\left(\frac{1}{Re}, \frac{k}{D}\right) \tag{3.A.8}$$

Note that by definition,

$$\tau_o = f\frac{\rho U^2}{2} \tag{3.A.9}$$

where f is the Darcy-Weisbach resistance coefficient and

$$H_L = \frac{\tau_o}{\gamma}\frac{L_p}{D} \tag{3.A.10}$$

Thus, the Darcy-Weisbach relationship appears as

$$H_L = f\frac{L_p}{D}\frac{U^2}{2g}. \tag{3.A.11}$$

With

$$f = \Phi_2\left(\frac{1}{Re}, \frac{k}{D}\right)$$

In summary, a host of dimensionless Π parameters can be identified from a list of variables. To establish a consistent and meaningful set of Π parameters, it is necessary to select m relevant or repeating variables containing the m fundamental dimensions. Each other variable is combined in turn with the repeating variables to form the dimensionless parameters. Consequently, none of the dimensionless quantities, Π_{n-m}, depends on more than $m + 1$ of the physical quantities, length, pressure, velocity, etc.

The following practical steps are involved in applying the Π theorem:

1. List all n physical quantities considered relevant to a process, and express them in terms of the fundamental dimensions. In this regard, a preliminary analysis of a process may help disclose the variables. Success with this step will depend on the experience of the modeler. If the process is not adequately understood so that the variables are not identified properly, then dimensional analysis may not be useful. Omission of a variable, or erroneous inclusion of a variable, may be evident if a fundamental dimension m (M, L, T) appears in only one variable listed. A list of physical quantities and their dimensions are given in Table 3.3.1.
2. Identify which variables are dependent (and of focal interest for the purpose of the experiment) and which are independent.
3. Note the number of fundamental dimensions, m.

4. Select m number of physical quantities as repeating variables. They must be selected such that –
    (i)   none is dimensionless;
    (ii)   no two should have the same dimensions;
    (iii)   together, the repeating variables should not form a Π parameter;
    (iv)   they must include all the fundamental dimensions involved; and,
    (v)   they should be chosen from different categories of variables (geometric, fluid properties, flow).

    Usually, repeating variables are the variables of focal interest. Sometimes, they are the variables most difficult to vary (e.g., for incompressible flows, fluid density and gravity are convenient repeating variables). Additionally, at least one repeating variable should be a representative length, although for some experiments (e.g., density flows) the variable may have to directly relate to the flow or fluid involved.
5. The other terms are then expressed as the product of the terms selected in step 3, each term raised to an unknown power, and one other term raised to an arbitrary power such as 1.
6. Solve for the unknown exponents in accordance with the requirement for dimensional homogeneity of the Π parameters.

As a further illustration of the application of the above principles, consider non-uniform, one-dimensional, flow in a wide open channel. The variables are water-surface slope, S_w, channel slope and roughness, S_o and k, a flow velocity and depth at a characteristic section, U and Y, fluid properties (density, ρ, kinematic viscosity, v, and surface tension, σ), and acceleration due to gravity, g. The nine variables, assembled with the water-surface slope S_w, as the dependent variable, are related functionally as

$$S_w = f(S_o, k, U, Y, \rho, v, \sigma, g) \tag{3.A.12}$$

As there are three fundamental dimensions for flow (M, L, T), the nine variables reduce to six independent, non-dimensional parameters, which can be identified from

$$\Pi = S_o{}^{x_1} k^{x_2} U^{x_3} Y^{x_4} S_w{}^{x_5} \rho^{x_6} v^{x_7} \sigma^{x_8} g^{x_9} \tag{3.2.A.13}$$

The parameters can be identified by prescribing the exponents of three repeating variables. For the present illustration the repeating variables are U, Y, and ρ. These variables contain all three fundamental dimensions of length, time, and mass. The procedure described above for flow through a pipe transforms Equation (3.2.A.7) into the following functional relationship between non-dimensional parameters:

$$S_w = \Phi\left(S_o, \frac{k}{Y}, \frac{v}{UY}, \frac{gY}{U^2}, \frac{\sigma}{\rho U^2 Y}\right) \tag{3.A.14}$$

which indicates a general functional dependence of water-surface slope on five non-dimensional parameters, channel slope, S_o, relative roughness, k/Y, Re, Fr, and We. These parameters can be identified also by directly working with the gradually varied flow equation;

$$\frac{dY}{dx} = \frac{S_o - S_f}{1 - Fr^2} \tag{3.A.15}$$

With x is downstream distance, and energy gradient, S_f, depends on relative roughness for fully turbulent flow, for which W_e exerts negligible influence.

Dimensional analysis results in a set of consistent non-dimensional parameters for describing a process, but this original set may not necessarily be the most meaningful set of parameters for describing the process. Indeed, they may not be all that useful in elucidating what is happening physically. It may be necessary to modify and regroup the variables in alternate sets until a set is found that facilitates clear explanation of the process.

A re-grouped set of parameters may be formed by choosing different relevant or repeating variables. Consider, for example, flow in an open channel with a loose bed. Dimensional analysis using U, Y, and ρ as repeating variables (as in the preceding open-channel flow example) results in a Reynolds number UY/ν, a Froude number, $U/(gY)^{0.5}$; relative roughness, Y/d, and density ratio ρ/ρ_s, with ρ_s = the density of the particles comprising the loose bed. These parameters are valid, but by no means as useful for describing flow resistance and sediment transport in loose-bed channels as are the parameters u_*d/ν, $\rho u_*^2/(\gamma_s d)$, Y/d, and ρ/ρ_s. These parameters result if g is replaced with the submerged specific weight of bed particles, $\gamma_s = g(\rho_s - \rho)$, and particle diameter, d, shear velocity, u_*, and density, ρ, are used as the repeating variables.

The appearance of recognized parameters in a parameter set may indicate the appropriateness of the set. For example, if a set of parameters describing a free-surface flow process includes the Froude and Reynolds numbers, the set likely will be effective in describing the process. Also, a useful set of parameters will contain parameters, such as ratios of forces whose physical significance is readily apparent. In this regard, for example, the parameter $\rho u_*^2/(\gamma_s d)$ mentioned above is a ratio of shear force exerted by flow relative to the submerged weight of a bed particle; its value expresses the likelihood of bed particles being moved by the flow.

NOTATION

A = area
c_p = specific heats of water at constant pressure
c_S = speed of sound in water
c_v = specific heats of water at constant volume
d = particle diameter; diameter of cylinder
F_D = drag force
F_L = lift force
f = Darcy-Weisbach resistance coefficient
g = gravitational acceleration
g_r = ratio of gravitational acceleration
k = thermal conductivity
L = a reference length
M = mass
p = pressure
p_o = reference pressure
p_v = vapor pressure
R_h = hydraulic radius
S_f = slope of the energy gradient
T = time

T_o = reference temperature
U = a reference velocity
$\forall$ = volume
Y = flow depth

Greek Symbols

β = expansion coefficient
Δ = increment
μ = dynamic viscosity
ν = kinematic viscosity
ρ = density
σ = coefficient of surface tension (water)
ω = characteristic frequency of oscillation

Subscripts

r = ratio
m = experiment (or hydraulic model) value
p = prototype value

Common dimensionless groups

$Re = \frac{UL}{\nu}$ = Reynolds number
$Fr = \frac{U}{\sqrt{gL}}$ = Froude number
$Ma = \frac{U}{a}$ = Mach number
$We = \frac{\rho U^2 L}{\sigma}$ = Weber number
$Eu = \frac{p - p_o}{\rho U^2}$ = Euler number
$Ca = \frac{p - p_v}{\rho U^2}$ = Cavitation number
$C_D, C_L = \frac{F_D, F_L}{0.5\rho U^2}$ = Drag or lift coefficient
$Pr = \frac{\mu C_p}{k}$ = Prandtl number
$EC = \frac{U^2}{C_p T_o}$ = Eckert number
$St = \frac{\omega L}{U}$ = Strouhal number
$Gr = \frac{\beta \Delta T g L^3 \rho^2}{\mu^2}$ = Grashof number

REFERENCES

Anwar, H.O., Weller, J.A., & Amphlett, M.B. (1978) Similarity of free-vortex at horizontal intake. *J Hydraul Res.*, 16(2), 95–105.

ASCE (1975) *Sedimentation Engineering. Manual on Engineering Practice No. 54*. New York, NY, ASCE Publications.

ASCE (2000) *Hydraulic Modeling: Concepts and Practice*. ASCE Manuals and Reports on Engineering Practice No.97, American Society of Civil Engineers, R. Ettema Editor, Reston, VA, ASCE Publications.

Boes, R.M. (2000) Scale effects in modelling two-phase stepped spillway flow. In: Minor, H.E., & Hager, W.H. (eds.) *Hydraulics of stepped spillways*. Rotterdam, Balkema. pp. 53–60.

Buckingham, E. (1921) Notes on the method of dimensions. *Philoso Mag.*, 6(42), 696–719.

Chanson, H. (2009) Turbulent air-water flows in hydraulic structures: Dynamic similarity and scale effects. *Environ Fluid Mech.*, 9(2), 125–142.

Chen, C.L. (1992) Power law of flow resistance in open channels: Manning's formula revisited, Channel flow resistance. In: Yen, B.C. (ed.) *Centennial of Manning's formula*. Littleton, CO, Water Resources Publications, Litteton, CO, 206–240.

Ettema, R., Kirkil, G., & Muste, M. (2006) Similitude of large-scale turbulence in experiments on local scour at cylinders. *J Hydraul Eng.*, ASCE, 132(1), 33–40.

Fuchs, H. & Hager, W.H. (2012) Scale effects of impulse wave run-up and run-over. *J Waterw, Port C Eng.*, ASCE, 138(4), 303–311.

Ghetti, A. & D'Alpaos, L. (1977) Effets des forces de capillarité et de viscosité dans les écoulements permanents examinées en modèle physique. In: *Proceedings of the 17th IAHR Congress*, 15–19 August, 1977. Baden-Baden, Germany, pp. 389–396.

Hager, W.H. (1994) Breitkroniger Überfall. *Wasser Energie Luft*, 86(11/12), 363–369.

Hager, W.H. & Bremen, R. (1989) Classical hydraulic jump: Sequent depths. *J Hydraul Res.*, 27(5), 565–585.

Heller, V. (2017) Self-similarity and Reynolds number invariance in Froude modelling. *J Hydraul Res.*, 55(3), 1–17.

Heller (2011) Scale effects in physical hydraulic engineering models. *J Hydraul Res.*, 49(3), 293–306.

Heller, V., Hager, W.H., & Minor, H.E. (2005) Ski jump hydraulics. *J Hydraul Eng.*, 131(5), 347–355.

Heller, V., Hager, W.H., & Minor, H.E. (2008) Scale effects in subaerial landslide generated impulse waves. *Exp Fluids.*, 44(5), 691–703.

Henderson F.M. (1966) *Open Channel Flow*. New York, NY, The Macmillan Company.

Hughes, S. (1993) *Physical Models and Laboratory Techniques in Coastal Engineering, Advanced Series on Ocean Engineering*. 7, Singapore, World Scientific Publishing.

Kobus, H. (1980) *Hydraulic Modeling, German Association for Water Research and Land Development, Bulletin* No 7, Berlin, Germany.

Kobus, H. (ed.), (1984) *Symposium on Scale Effects in Modeling Hydraulic Structures*. Esslingen, Germany, Technische Akademie Esslingen.

Langhaar, H.L. (1951) *Dimensional Analysis and Theory of Models*. New York, NY, John Wiley and Sons.

Lauber, G. & Hager, W.H. (1998) Experiments to dambreak wave: Horizontal channel. *J Hydraul Res.*, 36(3), 291–307.

Martins, R. (ed.), (1989) *Recent Advances in Physical Modeling*. Dordrecht, Netherlands, Kluwer Academic Publishers.

Miller, R.L. (1972) The role of surface tension in breaking waves. In: *Proceedings of the 13th Coastal Engineering Conference*, 10–14 July 1972, Vancouver, BC. ASCE, New York. 1, 433–449.

Möller, G., Meyer, A., Detert, M., & Boes, R.M. (2012) *Lufteintragsrate durch Einlaufwirbel— Modellfamilie nach Froude*. Wasserbau Symposium, Austria, Technical University of Graz. pp. 371–378.

Oliveto, G. & Hager, W.H. (2005) Further results to time-dependent local scour at bridge elements. *J Hydraul Eng.*, 131(2), 97–105.

Oumeraci, H. (1984) Scale effects in coastal hydraulic models. In: Kobus, H. (ed.) *IAHR Symposium on scale effects in modelling hydraulic structures*. Esslingen, Germany, Technische Akademie. 7(10), pp. 1–7.

Pfister, M. & Chanson, H. (2013) Scale effects in modelling two-phase air-water flows. In: Z. Wang (ed), *The Wise Find Pleasure in Water: Meandering Through Water Science and Engineering: Proceedings of the 2013 IAHR World Congress*, 8-13 September 2013, Chengdu, China. Curran Associates, Inc. pp. 4562–4571.

Pfister, M. & Hager, W.H. (2010) Chute aerators I: Air transport characteristics. *J Hydraul Eng.*, 136(6), 352–359.

Ranieri, G. (2007) The surf zone distortion of beach profiles in small-scale coastal models. *J Hydraul Res.*, 45(2), 261–269.

Reynolds, A.J. (1974) *Turbulent flows in engineering*. New York, NY, John Wiley and Sons.

Rouse, H., Siao, T.T., & Nagaratnam, S. (1959) Turbulence characteristics of the hydraulic jump. *Trans. ASCE.* 124, 926–966.

Rutschmann, P. (1988) Belüftungseinbauten in Schussrinnen. *VAW Mitteilung 97*. Vischer, D. (ed.) Zurich, Switzerland, ETH Zurich.

Schüttrumpf, H. & Oumeraci, H. (2005) Scale and model effects in crest level design. In: G. Viggosson (ed), *Proceedings of the 2nd Coastal Symposium, Höfn, Iceland*. pp. 1–12.

Schmocker, L. & Hager, W.H. (2009) Modelling dike breaching due to overtopping. *J Hydraul Res.*, 47(5), 585–597.

Sedov, L.I. (1959) *Similarity and Dimensional Methods in Mechanics*. New York, NY, Academic Press.

Shen, H.W. (ed.) (1990) *Movable Bed Physical Models*. Dordrecht, Netherlands, Kluwer Academic Publishers.

Skladnev, M.F. & Popov, I.Y. (1969) Studies of wave loads on concrete slope protections of earth dams. In: Battjes. J.A., & Bijker, E.W. (eds.) *Proceedings of the Symposium on Research on wave action, The Netherlands. Delft Hydraulics Laboratory.* 2(7), pp. 1–11.

Stagonas, D. (2010) Micro-modelling of wave fields. *PhD thesis*. University of Southampton, Southampton.

USBR (1981) *Hydraulic Laboratory Techniques*. U.S. Dept. of Interior, U.S. Bureau of Reclamation, Denver, CO.

Weichert, R. (2006) Bed morphology and stability of steep open channels. *VAW Mitteilung 192*, H.-E. Minor (ed.) ETH Zurich, Zurich, Switzerland.

Westrich, B. (1980) Air-tunnel models for hydraulic engineering. Section 14.4. In: Kobus, H. (ed.), *Hydraulic Modeling, German Association for Water Research and Land Development, Bulletin No 7*, Berlin, Germany.

White, F.M. (1986) *Fluid mechanics*. New York, NY, McGraw-Hill Book Company.

Chapter 4

Selection and Design of the Experiment

4.1 THE EXPERIMENTAL PROCESS

4.1.1 Introduction

There is a vast array of fluid flow systems that are important to us. They range from flows of fluids that are usually Newtonian such as air and water to flows of more complex fluids like pyroclastic flows, flows of molten plastics, and flows within the human body. Manmade systems are often quite simple when compared with natural ones, but both are generally too complex to fully comprehend with analytical or computational approaches alone. In order to better understand natural or manmade fluid flow systems or discover new features, experiments are often necessary. Such experiments are usually conducted in a laboratory by down-scaling their size through physical models of various scales. For complex flow systems, it is often necessary to study subcomponents of the systems by developing simplified experimental models that capture only the most important physical flow mechanisms. Special situations require the use of models which are larger than their prototypes; *e.g.*, models of flow within capillaries. At sites in the field, say a river reach, measurements must be made, at full scale in-situ. An example is the vast network of land, sea and space based weather stations around the globe.

Experiments are tests performed under known conditions with a measurement system to determine or demonstrate a principle or effect, test a hypothesis, or possibly collect data to address specific needs such as design evaluation, validation, or calibration. Experimental hydraulics is the scientific method based on dimensional analysis, similarity, and experimentation used for defining and quantifying the behavior of hydraulic systems or processes that cannot be satisfactorily treated using analytical or computational approaches. Experimental hydraulics continues to be needed, since only a handful of the hydraulic processes have been thoroughly solved analytically or computationally. Even with the increase in the reliability and popularity of Computational Fluid Dynamics (CFD), the need remains for experimental data to establish boundary and initial conditions and verify CFD results. As in any scientific field, experiments in hydraulics are performed for a variety of purposes. Table 4.1.1 lists common purposes for hydraulic experimentation. Common to all the purposes, including those not listed, is the necessity to follow a well-defined procedure for designing and conducting experiments.

Table 4.1.1 Purposes of hydraulic experiments (adapted from Stern *et al.*, 1999)

Purpose of Hydraulic Experiments	Example
1 Investigating a specific phenomenon	How do organized roughness elements affect frictional resistance in channels or on flow surfaces?
2 Substantiating a hypothesis	Is the derived force on a sluice gate accurate?
3 Obtaining benchmark data to establish a standard	What is the distribution of friction factors for new designs of corrugated metal pipe?
4 Calibrating instrumentation	What is the relation between pressure and voltage for a pressure transducer?
5 Evaluating alternative designs	Which spillway cross-section passes more flow for a given head?
6 Proving product reliability	To what extent is a proposed bridge pier design susceptible to scour?
7 Directly measuring key parameters	What is the discharge delivered by an irrigation canal?

There are physical and procedural components involved in the design of a hydraulic experiment. Physical components include facilities, measurement systems, operators and the measurement environment. These components are described in Section 4.2, Volume I. Procedural components entail measurement and data-processing operations typically grouped in customized procedures. These procedural aspects are described in Chapters 5 and 6, Volume I, respectively. Modern measurement systems often package sensors, data-acquisition systems, data-reduction procedures, and computer storage in an integrated hardware-software assembly that closely follows the steps of the experimental process. This section uses a typical "roadmap" to introduce the experimental process. Such a roadmap is a useful tool for the planning of experiments in hydraulics, because it helps assure, to the extent possible, that the experiments will be successful, accurate, and useful.

4.1.2 Experiment phases

Experiments typically comprise three phases: planning, conducting the experiment (execution), and data-processing. Figure 4.1.1 illustrates these phases and their sub-components. Additionally, several reference books on laboratory or field experiments discuss the three phases most experiments involve (*e.g.*, Yalin, 1971; Coleman & Steele, 1989; Doebelin, 1995).

Although the subcomponents of each phase have been organized in the order in which they are often completed, the order may vary as experimental processes often involve iteration. For example, goals and objectives may require modification if the initial uncertainty analysis shows that the original objectives cannot be achieved with available instrumentation.

The planning phase includes defining the experiment's goals, objectives and scope; conducting a dimensional analysis to isolate important variables, detect potential relations between defining parameters, and determine appropriate model scales; selecting appropriate facilities and equipment; and running an initial uncertainty analysis to determine if the facilities and equipment are adequate for achieving stated objectives. The most important part of planning in the experimental process is to establish a goal or set of

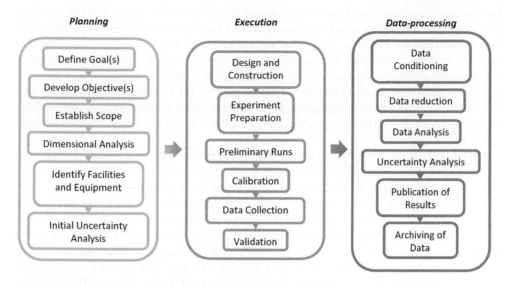

Figure 4.1.1 Diagram of the experimental process.

goals. In other words, what do you hope to achieve by the experiments? All experiments have goals, but an experiment is more likely to succeed if its goals are well defined.

A second, important step is to identify objectives, or steps that must be taken to realize the goals of the experiments. Goals and objectives should be clearly worded with precise detail. For instance, the statement, "Determine how the volume flow rate over a broad-crested unsuppressed weir is influenced by tail water depth" is much better than the overly broad statement, "Investigate weirs." Finally, once the goals and objectives are established, developing a scope is useful. The scope defines what elements are included and excluded from the experimental process. If the scope is too narrow, the experimenter may forget to collect important experimental data; but if the scope is too broad, the time or resources available to collect the data may be insufficient.

For both laboratory and field experiments, it is essential to first identify the variables most likely to influence relevant flow phenomena. The variables should be grouped as follows:

–   Targeted variables – the focal variables of an experiment that must be measured.
–   Independent and dependent variables – independent variables are those adjusted to observe their effects on other variables, whereas dependent variables change as a result of changes in independent variables.
–   Controlled variables – in experiments with multiple variables, it is often useful to adjust one variable at a time to observe its effect on the outcome of the experiment. Variables held constant or set to specific values while another variable is adjusted are called controlled variables.
–   Extraneous variables – these variables are not or cannot be controlled but may affect results.
–   Parameters – parameters are used for defining functional relations between multiple variables.

For example, an experiment may be performed in which the flow stage and the resulting discharge over a weir in a canal are measured for a range of canal flows to determine the functional relation between stage and discharge. Both the stage and discharge are targeted measured variables. In this case, stage is an independent variable and discharge is a dependent variable, because the effect that stage has on discharge is being observed. If wind-setup in the canal affects experimental results, the experiment can be done at a time when the wind speed is negligible, making wind speed a controlled variable. Water temperature is often an extraneous variable, because it is usually not controlled and can affect water viscosity.

Variables that influence the outcome of an experiment are most useful if organized into familiar dimensionless parameters. A thorough dimensional analysis reveals the physics of the problem at hand through simplified dimensionless numbers, and provides insight into the amount and range of data that should be collected. In developing dimensionless groups, it is important to distinguish independent and dependent variables so that the functional relations between the variables can still be observed in the results. The process of identifying relevant dimensionless parameter groups, while extremely important, is briefly discussed here in Chapter 3, Volume I because it is extensively discussed in introductory fluid mechanics texts (*e.g.*, Rouse, 1938; White, 2003). For laboratory models, dimensional analysis provides guidance for selecting appropriate scales or a means of relating model-scale results to the prototype. Selection of the experimental methodology (facility, equipment, instrumentation, procedures, etc.) that provides the necessary information in an economical and timely fashion should follow dimensional analysis.

Decisions on conducting experiments should be governed by the ability of the expected outcome to achieve the experiment objectives within allowable uncertainties. If an initial uncertainty analysis shows that the objectives cannot be achieved with the available instrumentation and facilities, modifications will be necessary or the experiment will have to be abandoned or goals redefined. Data quality assessment is a key part of the entire experimental procedure and includes several parts: determination of error sources, estimation of uncertainty, and documentation of the results. Strictly speaking, experiments require rigorous uncertainty assessment as part of the experiment process and documentation of results, so as to quantify the reliability of data produced.

Experiments normally begin with design and construction of the experiment setup. This initial phase of experiment preparation includes installation and testing of equipment and assembly of data acquisition hardware and software. A preliminary run or set of runs should then be performed to check the experimental design. Such a check provides a qualitative assessment that yields insight about leaks, important flow features, and other details that could impact the outcome of the experiment (*e.g.*, Tavoularis, 2005). Equipment and instrumentation must be calibrated or tested prior to data collection. For example, PIV experiments usually employ a calibration grid to assess imaging scales. Once calibration is completed, data can be collected and validated.

Data-processing is an eventual step in the experimental process. It includes filtering and proofing of data, and involves eliminating erroneous data. Data-reduction entails reorganizing data into usable forms, then calculating values of meaningful variables or parameters. Reduced data may then be analyzed to evaluate initial hypotheses, compare different models, or meet other purposes. Rigorous data analysis can be a difficult part of an experiment, because it requires insight into the correct theoretical relationships between parameters. Uncertainty analysis shows how errors introduced by instruments and methods may propagate through to the final results provided by an experiment. Such analysis is important for assessing the significance of the results from an experiment and revealing means to further improving the experimental process. An experiment is fully useful when accurately reported. Therefore, experiments require careful recording and communication of the activities described here. Throughout all phases it is important to review and cite relevant literature. Published literature, especially on similar prior experiments, can help guide the design and the performance of an experiment and can provide useful context for evaluating the results it produces.

4.1.3 Safety considerations

Risk to personal safety is inherent to many laboratory experiments. The experimenter must be mindful of eliminating safety risks. Doing so requires being familiar with the facilities and equipment used, identifying the potential risks associated with them. A partial list of potential risks is given in Table 4.1.2 for both laboratory and field experiments. For each new experiment, an exhaustive list of potential safety hazards should be developed and each hazard should be properly managed. All requirements for safety training should be met.

Table 4.1.2 Potential safety risks

Risk Type	Comment
Electrical	Note and mitigate all electrical hazards, including ungrounded equipment, equipment near water, and equipment that has the potential to overheat or overload.
Mechanical	Eliminate sharp edges. Clearly mark low elevation objects. Stabilize heavy objects.
Chemical	When working with chemicals, thoroughly consult and understand manufacturer material safety data sheets. Wear appropriate respiratory, eye and skin protection. Do not release chemicals into the environment without obtaining required approvals.
Traffic	If traffic poses a potential safety risk, wear brightly colored safety clothing. Arrange signs and cones to warn motorists.
Geotechnical	Avoid steep stream banks, especially when the soil is saturated. Collapsing banks are extremely dangerous.
Drowning	Be wary of fast or deep water. Be cognizant of submerged hazards like fallen trees or underwater currents.
Physical	When relevant, wear safety clothing, including hard hats and suitable footwear. Avoid excessive lifting which can cause back or other physical injuries. Make sure that all personnel have completed proper training. Know your limits and the limits of those working with you.
Optical	Wear proper eye protection when working with flying debris or lasers. Observe required precautions to prevent injury to others.

4.2 EXPERIMENTAL SETUP COMPONENTS

4.2.1 Introduction

This section describes the components of a typical experiment. As Figure 4.2.1 depicts, the components include the experimental facility, the measurement system, the operator of the system, and the operational environment (the surroundings of the experimental setup). The facility is any physical representation of a flow of interest, whether it is a fully-sized or scaled model of the flow. Examples of facilities are provided later in this section.

As the measurement system is at the core of the measurement process, this section describes typical components and configurations of such systems. The input to the system is an instrument that measures a property of the fluid in the facility. Instruments are an integral part of a system, and they are described in detail in Volume II. The general criteria for instrument selection are introduced in Section 4.4, Volume I. At any step during the experimental process, there are possible influences from various parts of the experimental setup that might affect the measured signal and the observed system response. System response implications are extremely important when evaluating the resulting output signal and are discussed in Section 4.2.4, Volume I. The last part of Section 4.2, Volume I discusses the measurement environment and its impact on experimental results. Additional aspects of experimental measurements are detailed in the remaining sections of Chapter 4 in Volume I.

4.2.2 Facilities

Facilities are replicas or idealizations of prototype flows and processes of interest. Common facilities include open channel flumes, scaled hydraulic structures, and wind tunnels or closed channels for studying boundary layers. Facilities generally have a test section with access to the flow, pumps or fans to recirculate the working fluid, and additional devices to condition and control the flow and the transported material. Often, hydraulic facilities must also be outfitted with large tanks for water storage and for maintaining steady conditions in the system. Flows must be replicated as accurately as possible in the test section of the facility while in the remaining areas of the facility the flow can be quite different from prototype conditions. For instance, the flow in the

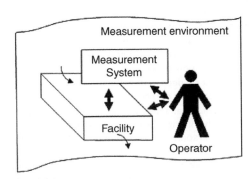

Figure 4.2.1 Components of an experimental setup.

recirculation pipe of a flume modeling an open-channel process can be a pressurized closed-conduit flow. Moreover, management of flow and handling of transported materials is often closely controlled by artificial means that are nonexistent in the prototype. For example, sediment and gravel may be continuously fed from a hopper into a model to represent sediment transport that occurs naturally in the prototype.

Flow and transported materials need to be continuously monitored during the experiments. Therefore, a suite of additional measurement devices are used to ensure that important prototype boundary conditions are accurately represented by the model; *e.g.*, orifices for flow measurements, erosion pins for bed form tracking, graduated scales for free surface and slope monitoring. Samples of hydraulic models are shown in Figure 4.2.2.

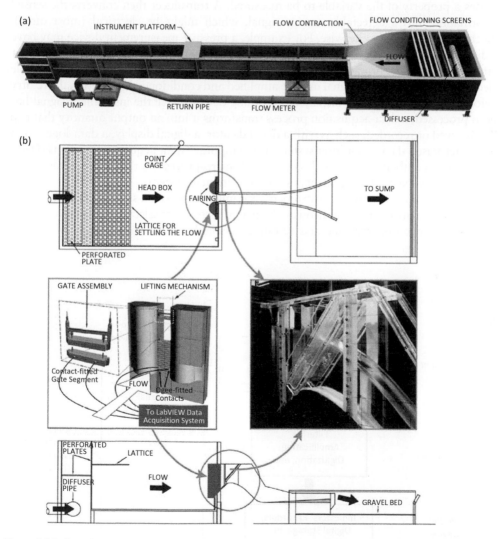

Figure 4.2.2 Sample experimental facilities: a) a multi-purpose open-channel flume; b) 1:24 scale spillway model for supporting the design of emergency stoplogs at a dam fish passage (Muste *et al.*, 2006).

Sections 5.2 and 5.5 in Volume I describe specific means of controlling the hydraulic conditions in models.

4.2.3 Measurement systems

Following selection of an appropriate facility, instrumentation and ancillary components must be selected and appropriate operational settings must be specified so that experimental data can be collected. These elements make up the measurement system (MS), as Figure 4.2.3 shows.

The measurement instrument (*e.g.*, pressure transducer, thermocouple) produces signals that represent the physical variable being measured. The instrument has a sensor that senses a property of the variable to be measured. A transducer then converts the sensed information into a detectable output signal, which might be electrical (most often), mechanical, optical or otherwise. For example, a pressure measurement device may have a diaphragm that is deflected when the pressure on it changes. A piezo-electric transducer attached to the diaphragm can convert the sensed pressure change into a measurable electric signal. The output signal may be amplified and conditioned by additional circuitry before it is transmitted to a display or recording device. Once the instrument signal has been processed, a data-acquisition process transforms it into an output quantity that can be observed or recorded, such as with a dial indicator, a digital display, a data-logger or a computer-based data acquisition system. Contemporary data acquisition systems (DAS) increasingly combine sensors with customized micro-processors that support tasks formerly done by external components of the measurement system.

DAS might include only one instrument or it might comprise a suite of coupled instruments that acquire data simultaneously. Note that instrument selection and measurement procedure depend on the time and resources available for the experiment. Moreover,

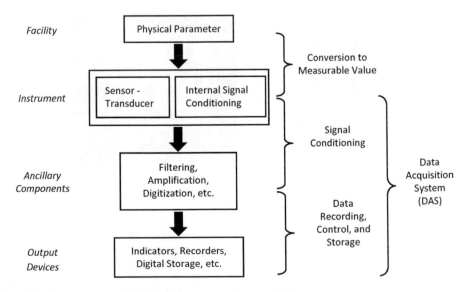

Figure 4.2.3 Component model of a Measurement System (MS).

sound decisions regarding the experiment can be made only after a dimensional analysis and an uncertainty analysis ensure that the selected instruments and ancillary components are suitable for measuring key parameters within allowable uncertainties.

Most contemporary DAS utilize electrical signals and computers for data collection, conditioning, and acquisition. Table 4.2.1 indicates common components and sub-components of such systems. As can be observed, a wide variety of customized components are associated with sensor-transducers, signal conditioning, and acquisition of data.

Table 4.2.1 Terminology for sensor-transducers, signal conditioning devices, and acquisition (adapted from Durst et al., 1976; Figliola & Beasley, 1991; Dally et al., 1993; Goldstein, 1996; Dunn, 2014)

Component	Role/Signal type	Specifications
Sensor-Transducers	Convert physical data into an electric signal/*Analog*	Sensors are devices that respond to a physical stimulus and convert it into a signal. Transducers are devices that change the signal into the desired quantity. Most often sensors and transducers are combined physically into one device that outputs an electrical signal (voltage or current). Sensor/transducer can be broadly organized based on the type of variables they sense: chemical, electrical, magnetic, mechanical, radiant, or thermal. Examples of physical principles for sensing include: electric, piezo-resistive, fluid mechanic, optic, photo-elastic, thermoelectric, or electrochemical.
Voltmeters	Measure static or quasi-static currents or voltages/*Analog*	The current in DAS circuitry can be time-continuous static or dynamic. Independent of whether the current is direct (DC) or alternating (AC), voltmeters can be used to measure the current as long as it does not change with time (static), or varies slowly with time (quasi-static).
Oscilloscopes	Rapidly measure voltage or current characteristics/*Analog, digital*	Oscilloscopes convert an input signal into a visual or recorded voltage-time series. Signals with frequencies of more than 1 GHz can be recorded and viewed with oscilloscopes.
Amplifiers	Increase the voltage of an input signal/*Analog*	Amplifiers may be single-ended (the input signal is referenced to the same ground as the output) or differential (the input and output signals do not share a common ground). Operational amplifiers (OpAmps) are integrated circuits that can be used to amplify a signal. OpAmps can also be used in filter circuits.
Root-mean-square (RMS) converters	Convert AC signals to DC values/*Analog, digital*	The RMS value of an alternating current (AC) is a voltage (also known as its *heating value*) equivalent to the direct current (DC) value that would be required to get the same heating effect.
Bridge circuits	Measure resistance, inductance, or capacitance/*Analog*	Bridges allow accurate measurement of subtle and fast changes in the input quantity. One of the most popular bridges is the Wheatstone bridge, which converts small changes in electrical resistance into measurable output voltages.

Table 4.2.1 (continued)

Component	Role/Signal type	Specifications
Frequency modulators (FM)	Convert variations in signal amplitude into variations in signal frequency/*Analog*	The transducer signal can be sinusoidal, transient or random. The signal is transposed onto a constant-amplitude, high frequency carrier signal such that the frequency of the carrier signal is altered to reflect the amplitude of the transducer signal. The process of separating the transducer signal from the carrier signal is known as demodulation.
Frequency filters	Remove frequencies from signals/*Analog, digital*	Low-pass filters transmit low-frequency signals and attenuate high-frequency signals. High-pass filters do the opposite. Band-pass and band-reject filters allow selection or elimination of a range of frequencies.
Data loggers	Record the signal from transducers/*Analog, digital*	Data-loggers are often autonomous devices that incorporate sensor-transducers, signal conditioning equipment and a data recorder. Data-loggers that record from multiple sensors may utilize a scanner to interrogate one sensor at a time.
Data Acquisition Systems (DAS)	Condition, digitize, and record the signal from multiple transducers/*Analog, digital*	Data acquisition systems are like data loggers, but are generally more sophisticated and more versatile. In the context of this book, they are microprocessor-based. Some data-acquisition systems provide signal conditioning prior to digitizing an incoming signal.
Time-measuring circuits	Handle the timing of various operations/*Analog, digital*	Time based measurements (frequency, time intervals) are typically made with digital electronic counters. Other time measurements may be done with binary counting units, gates, triggers, or liquid-crystal displays.
Optical and acoustic system components	Facilitate the control of optical or acoustic wave propagation/*Analog, electric*	Lens systems, prisms, polarizers, diffraction gratings, and photomultipliers are examples of optical components that direct and condition the light waves in the measurement volume of the sensor. Similarly, a variety of components are used to shape the form and direction of acoustic beams and pulses.
Signal processors	Extract various aspects of the DAS signals/*Analog, digital*	These are typically instrument specific devices that perform a sophisticated task. For example in laser-Doppler velocimetry, signal processors include correlators, spectrum analyzers, counters and frequency trackers. In some cases, the device that is best suited for an application must be selected from multiple options.
Data processors	Convert data to produce information/*Digital*	For example, Doppler signals appear as bursts of various temporal densities in the output signals. Data processors contain specific sampling algorithms that allow these bursts to be converted into continuous time measurements.

Data acquisition and storage can be accomplished with simple data-logging systems (data-loggers) or with more sophisticated data acquisition systems. While simple data-loggers might directly sample and record the output of a transducer at evenly spaced intervals, more sophisticated systems may pre-condition the data or employ intricate sampling algorithms. Thus, acquired data can be anything from the raw, unprocessed sensor output to signals that have been transformed by a variety of complex conditioning processes. The difference between a DAS and data-loggers are somewhat fuzzy, but there are two clear distinctions: (1) the data collected by a data-acquisition system is usually sent directly to the computer where it is processed, whereas data-loggers typically record data that must later be downloaded to a computer for processing, and (2) DAS are generally more versatile than typical data-loggers and may be capable of interfacing with many types of instruments. Because the components of DAS are electronic, mechanical, and computer-based in nature, it is useful to explain the terminology used for standard components, as illustrated in Table 4.2.2.

Table 4.2.2 Terminology for DAS components (adapted from Figliola & Beasley, 1991)

Component	Role/Signal type	Specifications
Signal conditioner	Cleans analog signals prior to sampling and digitization/*Analog, digital*	Performs offset, gain, and filtering on incoming analog signals. Anti-alias analog filters remove any frequency components of the input analog signal that exceed the Nyquist frequency prior to quantization (see Section 4.4.4.1 and 4.5.4 in Volume I).
Analog-to-Digital (A/D) Converter	Converts an analog signal to a digital signal/ *Analog, digital*	Digitizes the incoming analog signal into a binary number. Sometimes digital-to-analog (D/A) converters are used to convert digital numbers into analog voltages. Such devices may be used internally by the DAS controller to check the conversion accuracy of the A/D converter. They may also be used to control external analog devices.
Multiplexer/Scanner	A multiple-port switch that sequentially connects multiple analog inputs to a common A/D converter/*Analog, digital*	Each analog signal enters through an individual switch port that is sequentially connected to the A/D at a rate that is determined by the MS clock. The switch may be mechanical or digital. Mechanical switches are less prone to electrical noise but are inherently slower than digital ones. In some cases, a mechanical switch is a less expensive alternative to multiple transducers hence it might be included in the DAS configuration. Multiplexers are generally packaged in 8- or 16-channel modules. Modules can often be interconnected to expand the number of input ports. Multiplexing reduces the maximum sampling rate of the input signal.

Table 4.2.2 (continued)

Component	Role/Signal type	Specifications
Digital Input/Output (I/O) Devices	Acquire and record digital data/*Digital*	Most digital devices receive and transmit digital data at two standard voltage levels: 0 volts signifying a low signal or a binary 0, and 5 volts signifying a high signal or a binary 1. Low and high voltage levels vary depending on the type of computer logic. For some applications it is also important to become familiar with digital communication standards, such as RS-232C, IEEE-1394, and USB (Universal Serial Bus).
Master controller	Interface between operator and DAS/ *Analog, digital*	The controller sets sampling rates, the sequence of channels to be monitored, signal levels and triggers, filter settings, activation limits and supervisory alarms. Common controllers contain an arithmetic logic unit, memory, control circuits, and an internal clock. The arithmetic unit performs mathematical operations and makes decisions. The memory contains the digital data, as well as the software-driven and fixed execution instructions. The control circuits provide routing information and set up the functions of the arithmetic unit. The controller is driven by an executive program or operating system which supervises all tasks, compilations, and input/output from the master controller. A clock provides timing pulses for execution of each step.
Random-access Memory (RAM)	Temporary storage (buffer)/*Digital*	Acquired data are stored in digital buffers in RAM until they can be sent to a more permanent storage device. Many DASes permit double buffers (one buffer absorbs data from the A/D converter and the other transfers data to storage). Data buffers make real-time capabilities such as live display of data and limited mathematical processing possible.
Output devices and storage	Display and store data acquired by the system/ *Digital*	DASes may display output in a number of ways, including monitors and printers. The software driving a DAS can generally store data to hard disk drives, USB flash drives or optical drives like CDs or DVDs.

DASes generally contain the following subsystems: a) a signal conditioner; b) a multiplexer or scanner, c) an A/D converter; d) a controller; e) input/output displays and, f) data storage. Figure 4.2.4 outlines a measurement system that includes a two-channel acquisition system (*i.e.*, two variables are simultaneously measured). The sequence of subcomponents is not always as shown in this figure. Automated data-acquisition systems, containing their own microprocessors, are increasingly used for data acquisition. Their advantages lie in their unsupervised operation speed, data-handling capabilities, and small dimensions. Many modern systems are easily connected to computers via serial buses, Ethernet ports, or WiFi devices. A thorough understanding of the nature and form of all input and output signals in the successive steps of a measurement system is necessary for proper selection of various components and of the operational parameters in a data-acquisition system, such as is briefly described below. This description includes some of the components described in Tables 4.2.1 and 4.2.2.

The signal from a transducer needs to be appropriately conditioned prior to sampling and digital conversion. Many data-acquisition systems have built-in conditioning capabilities. For example, some systems have programmable amplifiers at their inputs. Most hydraulic experiments require collection of multiple variables or multi-component vector variables. In these cases, the system must be capable of collecting multiple channels, as illustrated by the two-channel system in Figure 4.2.4. The system may then include a multiplexer or scanner to sample all of the data channels in rapid succession so that only one A/D convertor is necessary. Moreover, since data must usually be collected at regular time intervals and in a particular sequence, a digital clock is used to time the scanning of the channels and the conversions.

The A/D converter transforms the analog signals obtained from each channel into binary numbers that can be transferred to a computer (or other device). The timing of the A/D conversion process is crucial and is closely controlled by the DAS Master controller. In contrast, personal computers, to which data are transferred, typically do not have tightly controlled timing, and are usually not synchronized with the data conversion process. Consequently, the data from the conversion process must be temporarily stored in a buffer until it can be displayed or transferred to permanent storage. Buffers are also important for data-acquisition systems that collect data much faster than can be permanently stored. In such cases, data can be sent to the buffer at a high rate until it is full and then slowly transferred to a more permanent storage location. The buffer may be dedicated random access memory (RAM) that is directly integrated into the acquisition device or it may be the RAM of a computer to which the device is connected. If the device stores data directly to the RAM of a computer without using the processor of the computer, it uses a technique called direct memory access (DMA). The master controller initiates and terminates the data acquisition process, controls the flow of data into and out of the DAS, and organizes the data for storage. In most modern systems, the acquisition system communicates directly with a computer which ultimately receives the data from the system. The computer then displays the results and stores the data to a more permanent storage device such as a hard drive, an optical drive (CD, hard disk drive - HDD), or a USB flash drive. In some cases, data may also be printed or plotted.

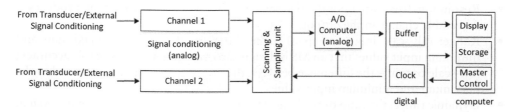

Figure 4.2.4 Signal routing in a two-channel DAS.

4.2.4 Measurement system response and characteristics

The response of a measurement system is defined by how the output of the system reacts to a change in the input. Since the concepts described here are relevant to both end-to-end measurement systems and to individual components, the term "measurement system" is expanded in this section to generically refer to either of these. Due to the nature of turbulent flows, most engineering fluid mechanics and hydraulic experiments are time-dependent, even if the mean flow is steady. Consequently, the accuracy of the experimental results depends on the measurement system's response to input signals (Figliola & Beasley, 1991). Equally important is the system sensitivity to temporal fluctuations of the input variables.

The operational characteristics of a measurement system are determined through static and dynamic calibrations (*e.g.*, Tavoularis, 2005). In a static calibration, the input to the system is varied and allowed to reach a static value and the corresponding output is recorded and compared to a simultaneous measurement from a second device with an output of known accuracy. For instance, a pressure transducer can be statically calibrated by applying different pressures across it and comparing the transducer output to the pressure measured with a liquid manometer. For systems with multiple inputs, it is necessary to hold all but one of the inputs constant while doing the calibration. Each input is varied to determine how it affects the output value. A static calibration is usually specified by an algebraic relation (*i.e.*, calibration curve), as, for example, is the linear relationship between the pressure input and the voltage output of a pressure transducer. Experimenters should be aware that even when a system is carefully calibrated, environmental and other factors may alter the static response by introducing a zero drift (a shift of the calibration curve) or a sensitivity drift (a change in the slope of the calibration curve). The important static performance characteristics for selecting a sensor, instrument, or measurement system are:

- Accuracy: the ability of a system to acquire unbiased data with low uncertainty (typically defined as a percentage of the full-scale reading).
- Sensitivity: defined by the slope of the relation between the output and the input of the system. A large change in the input of an instrument with low sensitivity does not cause a significant change in the output. Although the output of the sensor may be very accurate, that accuracy may be lost if the DAS is unable to discriminate small changes in the output because the system components have low sensitivity.
- Precision: reflects the ability of an instrument to reproduce repeated measurements of the same quantity with small scatter. This is not the same as accuracy because an instrument with good precision can consistently give the same measurement, but the measurement may deviate from the true value (bias).
- Resolution (spatio-temporal): the smallest change in the input that produces a detectable change in the output.
- Range: full-scale input (or span) – the difference between the maximum and minimum input values that an MS is designed to measure with acceptable accuracy; full-scale output – the difference between the output values that correspond to the maximum and minimum input values.
- Dynamic range: the ratio of the largest to smallest values of the input that a system can measure.

- Linearity: a measure of how well the relation between the output and input of an instrument can be represented by a least-squares straight line curve fit (expressed as a percentage of the output reading or of full-scale).
- Hysteresis: a measure of the dependence of the measurement-system output on the direction (ascending or descending) from which the input value was approached.

In practical situations, a measurement system rarely responds instantaneously to an input that varies in time. Therefore dynamic calibrations are needed. For analyzing the response of a system to signal fluctuations, it is common to approximate the relation between the input and the output of the system as an ordinary differential equation that contains time derivatives. Using the model shown in Figure 4.2.5, a measurement system can be conceptualized as a transformer of an input signal $x(t)$ into an output signal $y(t)$ dependent on a set of initial conditions $y(0)$. Although the relation between the input and output of a dynamic system can be very complex, the most common systems can be represented by a zero-order, first-order, or second-order differential equation.

The zero-order system is one in which the output of the system immediately adjusts to a change in the input. Essentially, a zero-order system behaves in the same fashion as a static system – the output is dependent only on the input. The output of a zero-order system can be represented by an equation like

$$y = Kx \qquad (4.2.1)$$

in which K is the static sensitivity of the system. Based on Equation (4.2.1), the output of the system is linearly dependent on the instantaneous input value and is unaffected by the dynamic aspects of the system. The more rapidly the input values of a system change, the less likely that the system can be treated as a zero-order system.

In a first-order system, a first-order differential equation defines the relation between the input and the output of the system:

$$\tau \frac{dy}{dt} + y = Kx \qquad (4.2.2)$$

When the input of a first-order system rapidly changes to a new value, the output of the system requires time to adjust to the change. In general, the variation of the output of the system is an exponential function that is dependent on the time constant (τ) of the system. The larger the time constant, the longer it takes for the system to adjust. An instantaneous change in the input from one static value to another is known as a step function. A good example of a first order system subjected to a step function change in

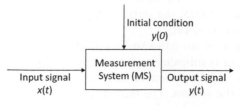

Figure 4.2.5 Illustration of measurement system response (modified from Figliola & Beasley, 1991).

input is when a thermocouple at room temperature is plunged into an ice bath. In such a system, this will result in an output response of:

$$\frac{y}{KA} = 1 - e^{-\frac{t}{\tau}} \tag{4.2.3}$$

where A is the amplitude of the step change and t is the time that has elapsed since the step change occurred. The response of a first-order system to a step change is illustrated in Figure 4.2.6a, which shows both the step function input and the resulting output. The time constant (τ) is important because it provides information about whether or not a system is fast enough to achieve the necessary dynamic response. When the elapsed time equals τ, 2τ, and 3τ after a step change in the input, the output reaches 63%, 87%, and 95% of its final value, respectively, as depicted in Figure 4.2.6a. Thus, one indicator of whether or not a system has adequate temporal response is if τ is much smaller than the required temporal resolution of the measured parameter.

A second order differential equation defines the relation between a second-order system output and input:

$$\frac{d^2y}{dt^2} + 2\zeta\omega_n\frac{dy}{dt} + \omega_n^2 y = K\omega_n^2 x \tag{4.2.4}$$

where ω_n is the undamped natural frequency of the system (in radians per second) and ζ is the damping ratio. When a step function of amplitude A is applied to a second order system, the response of the output is dependent on K, ω_n, and ζ as depicted in Figure 4.2.6b, which shows the system responses that result from a variety of damping ratios. A good example of a second-order system is a liquid manometer for measuring pressure in a turbulent flow. The behavior of the manometer is influenced by both damping and inertia of the manometer fluid. When a step change in pressure is applied to a manometer, the manometer may display oscillatory behavior due to the inertia of the liquid in the manometer. The oscillatory behavior observed in second-order systems is called ringing.

In addition to step functions, additional functions that can be applied to understand the response of dynamic systems include impulse functions, ramp functions, and frequency functions. Although it is beyond the scope of this text to discuss these functions here, more details can be found in other books about measurements in fluids such as Tavoularis (2005), Figliola and Beasley (1991), and Beckwith and Marangoni (1990). The frequency function is one of the more useful functions and can be used to analyze frequency response or the capability of the measurement system output to replicate the time evolution of a fluctuation in the input. For example, the frequency function can be used to assess how the ratio of the output and input amplitudes (or gain) of a system is affected by the frequency of the input signal. When the frequency of an input signal becomes higher than the frequency response of a particular system, the output amplitude can decrease to zero even if input amplitude remains constant.

The temporal response of an MS can be established with a dynamic calibration whereby the system input is subjected to a square-wave or frequency test (Tavoularis, 2005). The square-wave test represents a step function, and is simply a square wave applied to the input of the system. The output of the system can then be analyzed to determine the time constant of a first-order system or the natural frequency and

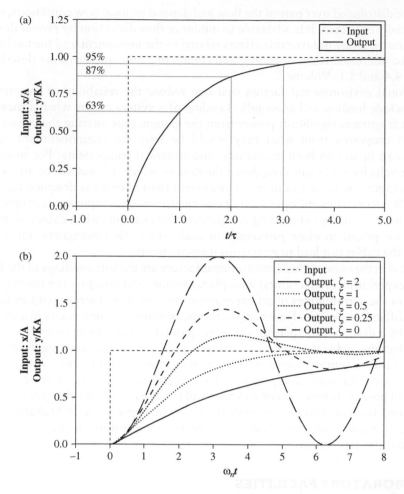

Figure 4.2.6 Dynamic response of a measurement system to a step function input: a) first-order (inertia); b) second-order (inertia and damping). (adapted from Figliola & Beasley, 1991).

damping ratio of a second-order system. The frequency test is done by applying a sinusoidal input to the system. The frequency of the input is varied to determine the frequency at which the ratio of the output and input amplitudes begins to decline. Both of these tests are easy to apply to electrical components and conditioners since it is easy to produce electrical square waves and sine waves of various frequencies. It is more difficult, but still possible to apply such tests to physical sensors like pressure and temperature transducers (Tavoularis, 2005).

4.2.5 Measurement environment

Any instrument inserted into a flow distorts the flow to some extent. Distortion effects may include blockage, streamline displacement, vortex shedding, and instability. Such effects are distributed over part of the flow and depend on local flow conditions, so they are not easily corrected. It is advisable to minimize flow distortion by proper design of the apparatus. Other unfavorable effects related to the instrument and the facility that need to be addressed when designing and conducting an experiment are described in Sections 4.4 and 5.1, Volume I.

Additional environmental factors that can reduce the reliability of experimental results include loading and cross-talk. Loading of a system occurs when a measuring component extracts significant power from the system, thus altering the values of the measured properties from what they would be were the component not present. Loading can be due to both mechanical and electrical components. For instance, a propeller velocity meter can slow down the flow in which it is placed, or an amplifier with low input resistance can draw more current than a sensor is designed to supply. Cross-talk occurs when the operation of one measurement component is coupled with the operation of another measuring component. For example, when cables to multiple sensors are placed in close proximity to each other, electromagnetic interference between the cables can lead to erroneous measurements.

Included in the measurement environment factors are the surroundings of the facility and the experimenters (operators) who plan, execute, and interpret the results. Subtle changes in the surroundings of an experiment may result in discrepancies in the data that are difficult to explain. For example, the measurement of air velocity in a jet can be affected by a draft in the experimental room. Similarly, a change in room temperature might affect the ambient pressure, viscosity, and density of the working fluid and the outputs of the measurement system by interference with the calibrations of the sensors, transducers or electronic components. To mitigate these effects, it is advisable to identify all potential environmental factors and to note even slight changes in environmental conditions (like room temperature) during an experiment. In hydraulic experiments, it is particularly important to record the temperature of the working fluid throughout the experiment.

4.3 LABORATORY FACILITIES

4.3.1 Introduction

Performing hydraulic experiments in a laboratory facility has the advantage that flow conditions are more readily controlled than in the field. This advantage enables more accurate investigations whereby independent variables can be controlled so as to reveal their influences on one or more dependent variables, and thereby processes and trends revealed. Typically, to fit within a laboratory facility, a model has to be smaller than its prototype, although in some situations the model may be larger than its prototype. In all situations, the experimenter must be aware of the similitude considerations associated with laboratory modeling.

There are two basic kinds of laboratory experiments:

1. The geometry of a prototype is reproduced in the laboratory (*e.g.*, flow over a spillway); and,
2. A subset of pertinent physical processes is reproduced in a representative, but not necessarily scaled, geometry (*e.g.*, a bouyant plume in a crossflow).

Similitude considerations described in Chapter 3, Volume I are strictly enforced for the first category of experiments (scaled models) where the results from the experiment are used to predict aspects of processes in the prototype. This section addresses considerations that pertain to the design and use of laboratory facilities irrespective of the type of experiment with emphasis on flumes and basins that are extensively used in hydraulic experiments. For specific applications in other areas of hydraulics many additional resources are available (*e.g.*, ASCE, 2000; Frostick, *et al.*, 2011; Hughes,1993; Tavoularis, 2005; USBR, 1981).

4.3.2 Flumes and basins

4.3.2.1 Dimensions

Hydraulic modeling and laboratory experiments commonly involve a size reduction of the flow process investigated. Additionally, sometimes the size reduction differs for vertical and horizontal dimensions of a prototype flow; such models are vertically distorted models. Such distortion is usually done to fit the model into limited laboratory space while still maintaining model depths that are necessary to satisfy physical or practical modeling constraints. In such flows, the bed roughness in the model must be adjusted to compensate for the effects of geometric distortion.

As discussed in Chapter 3, Volume I, similitude consideration of the physical processes associated with a flow situation commonly requires simultaneous satisfaction of more than one kinematic and/or dynamic similitude requirement (*e.g.*, concurrent Reynolds and Froude number similarity), and may require the use of unusual model fluids (see also Table 4.3.1). Consequently, it is generally best to choose a model scale that is as close as possible to the scale of the prototype yet enables the model to be tailored to an available laboratory facility.

For most water flow experiments, gravity governs flow behavior; hence, proper scaling requires Froude number similarity. As suggested above, it is usually impractical to satisfy both Froude and Reynolds number similarity requirements, and the resulting Reynolds number of the flow is often too small to permit accurate representation of the prototype boundary layer. Consequently, models of flows that should be turbulent may have uncharacteristically thick viscous sub-layers or may even be laminar. In order to create a small-scale turbulent sub-critical water flow, a mean velocity of about 0.1m/s with a depth of at least 0.1m is typically required; whereas for supercritical flows thinner water layers with higher velocities are sufficient. Boundary layers for waves

Table 4.3.1   Scaling factors for modeling open channel flows.

Property →	Re	Fr	Friction rough	Friction smooth
Scaling factor* →	$n_u = n_a^{-1}$	$n_u = \sqrt{n_a}$	$n_k = n_a$	$n_\delta = \sqrt{c_f}\,n_u^{-1}$

* Notations : n_u, n_a, n_k and n_δ represent the scaling factors for the velocity, water depth, roughness height and thickness of the viscous sublayer, respectively. c_f is the friction coefficient.

and currents are in fully developed status when the processes that govern fluid motion, including turbulence, are in equilibrium for the entire flow cross-section. For open channel flows, a fully developed state is realized when the boundary layer extends from the bed to the free-surface such that the stress profile and associated turbulence statistics are similar in the streamwise direction. This typically requires a flow development reach that is 50 to 100 times the water depth. These simple considerations imply that properly designed studies of fully developed flow in open channels should have depths of the order of tens of centimeters and lengths of tens of meters.

Provided that Froude-based scaling prevails in gravity driven free-surface experiments, the artifacts that result from insufficiently high Reynolds numbers need to be identified. As the bed friction coefficient for hydraulically smooth conditions decreases with increasing Reynolds number, Froude based scaling leads to overestimation of the bed friction for these cases. This can be compensated by using a distorted model in which the vertical scaling factor is smaller than the horizontal one. For the hydraulically rough case, the problem is less severe since the roughness height can also be adjusted to compensate for the overestimation of bed friction. Reducing the vertical scaling factor has the added advantage of producing a higher Reynolds number, a possible shift to the hydraulically rough regime, and better similitude of the vertical velocity profile with standard turbulent velocity profiles. An added practical advantage is the additional vertical space available for measurements and a relative reduction of the horizontal space needed for the model. Ideally, facilities should be evaluated for their scale effects before finalizing their design. Some of the most common causes of scale effects in hydraulic models are cavitation, friction, and surface tension (ASCE, 2000). The best solution for probing scale effects is to simultaneously use multiple model scales and then to inspect results obtained at the different scales for scale independence.

Frostick *et al.* (2011) provide a handy overview of the experimental facilities and ancillary instrumentation typically used in hydraulic modeling of riverine and coastal sediment transport. Facilities are classified as "standard", *i.e.*, widely used in various sizes or specially designed facilities (found only in a few laboratories). Table 4.3.2 summarizes specifications for standard hydraulic facilities.

4.3.2.2 Layout

Selecting an appropriate experimental arrangement requires consideration of scale requirements in conjunction with the space available for accommodating the geometrical layout of the problem to be investigated. The experimenter must determine whether the priority of the model is to simulate a sub-set of flow processes (*e.g.*, mixing) or the behavior of a complete flow system (*e.g.*, a spillway). In the latter case, the three-dimensionality of the flow through, for example, a water intake for hydropower turbine might require the reproduction of the entire undistorted geometry of the intake. On the other hand, if the flow in one of the horizontal dimensions is uniform or quasi-uniform, the flow can be modeled as a 2D-vertical flow in a flume configuration. This latter approach requires that the walls of the flume which confine flow in the third dimension do not affect the relevant processes of the remaining two dimensions. For example, an experiment on sediment bed forms in a straight flume normally requires

Table 4.3.2 Characteristics of typical hydraulic facilities (adapted from Frostick *et al.*, 2011)

Experimental facility	Essential features	Typical dimensions and scales
Flumes	Typical flumes include a pump, headbox, a sediment feeder, flow stabilization devices (e.g., screens, baffles, diffusers), a mobile sediment bed, a sediment trap and an outflow. Most of them can be adjusted for the slope.	• Width = 0.3-4m, Depth = 0.1-1m, Length = 10-150m • Optimal width/ depth ratio:6-10 • Structure blockage area < 1/6
Wave flumes	Wave flumes are similar to standard flumes, but they are also equipped with wave paddles and wave absorbers. Wave flumes can be built with flat or sloping beds.	• Width = 0.3-5m, Depth = 0.1-6m, Length = 10-324m • Recommended approach slope ≤ 1:20 (flat or sloped bed)
Wave basins	Wave basins are much larger than wave flumes and have fixed bed slopes. They are equipped similarly to wave flumes but often provided with mechanisms to create multi-dimensional currents.	• Width = 20m, Depth = 1m, Length = 30m • Scales: 1:10 to 1:100
Pipe flow apparatuses	Closed conduit flows of various cross-section (circular, ovoid, rectangular) for pressurized flow processes.	Diameter < 0.25 m, Length < 10m
Oscillating water tunnels (OWTs)	Closed ducts for wave boundary layer modeling. Tunnel dimensions should not affect the modeled process.	• Test section characteristics: Length = 5-10m, Width = 0.5-1m, Height = 0.5-1m

that the flume's side walls are smooth and do not significantly alter the vertical and longitudinal flow patterns responsible for the formation of the bedforms.

Models often are built within existing model basins or large flumes, or they can be built stand-alone having the required geometry in a manner that fits available floor space within a laboratory. In either case, the model's upstream and downstream boundaries should extend far enough to include features that significantly affect the flow at the points of interest in the model. Recommendations for decisions in this regard are suggested in ASCE (2000).

4.3.2.3 Flow boundary conditions

Fresh water is used to simulate water in most hydraulic models and experiments of water flow. Structural elements are often made of wood, metal, fiberglass, concrete or glass and plastic. Perspex or glass windows are often installed to allow optical access into the facility. Visibility of the flow is often an important consideration when building a model, because observing flow patterns is usually a critical modeling activity (ASCE, 2000). Fluid and structural elements in the model need to be designed to fully replicate the spatio-temporal flow boundary conditions of the prototype. This is a challenging task, both in terms of similitude analysis as well as from construction and operational perspectives. This section focuses on the structural and operational design considerations required during the planning stage of an experiment, with emphasis on the spatial continuity and distribution of the flow and sediment in the facility.

Most hydraulic models and experiment configurations have headboxes and tailboxes constructed from a variety of materials; notably framed lumber and plywood, concrete blocks, metal framing and sheet metal. Epoxy-coated, marine-grade plywood produces a durable product that is smooth enough to be used for planar flow surfaces. Fiberglass or urethane coatings are also used in many laboratories. For mobile-bed flumes, the dimensions and wall roughness have to be selected to ensure that sediment mobility is attained for the entire flow range.

A supply of water with sufficient energy head is required for every experiment. This is achieved by fitting the facility with a head tank (for creating a constant pressure in the facility) or with pumps (that recirculate the flow). The primary function of the inflow section of the flume is to establish a smooth flow that is evenly distributed across the flume inlet and does not contain large-scale disturbances. Flow straightening devices (*e.g.*, screens, guide vanes, baffles, etc.) are placed across the inlet for this purpose. The head box must be far enough upstream of the measurement section so that the flow is fully-developed when it reaches the measurement section.

Most tailboxes serve to collect the flow, and sometimes bed sediment, at the downstream end of a flume or model. They may be fitted with a control gate of some form so as to control the flow depth, or setting the tailwater level in the model or flume. A variety of so-called tailgates or other control structures can be used to control the tailwater level. The tailgate should be strong and easy to adjust under load, and it should be far enough downstream so that backwater effects do not affect the depth and velocity distribution at the measurement section.

While the headbox and tailbox control the inlet and outlet conditions of a flume or model, the material and configuration of the bed as well as the slope of the facility also play a critical role for hydraulic experiments. The bed of a flume or model may be smooth or rough and may have fixed or adjustable slope. Sediment transport experiments usually involve a mobile bed, but the mobile bed is generally situated above a fixed or adjustable sloping floor. Uniform flow conditions in an open channel require that the bed friction and the streamwise component of the gravitational force balance each other. For a fully-developed boundary layer in equilibrium, this balance results in a constant water depth over the entire length of the measurement area. In addition, if two-dimensional flow behavior is required within the measurement section, the aspect ratio (width/depth) of the flow cross section should be at least six (Nezu & Nakagawa, 1993).

The bathymetry and topography of a prototype site is customarily replicated in the model using reduced-scale digitized maps, especially for large models of rivers and large water bodies in which high accuracy of the bottom boundary is important. Fixed topographies can be replicated with concrete, foam, or other durable materials. Details about bed materials and construction can be found in ASCE (2000). Experiments with mobile beds are complex, as the bed of a river or estuary can be flat or have bedforms, depending on slight changes in the flow or fluid properties. Preparing the sand bed and performing the experiments requires careful consideration of the bed morphology and sediment transport processes as well as establishing detailed protocols for proper simulation of the most important sediment transport processes. More information about these aspects can be found in Frostick *et al.* (2011).

Bed roughness on the bottom boundary is modeled using the concept of equivalent sand roughness height introduced by Nikuradse (Frostick *et al.*, 2011). In the case of a movable sediment bed, the effective bed roughness depends on two factors: skin friction generated by grain roughness and form drag generated by pressure forces acting on bed forms. Moreover, the effective bed roughness for a given bed material size is not constant but depends on the flow conditions. There is an essential difference that needs to be accounted for in modeling of sediment transport processes that occur above rippled and plane sand dunes. For steady currents, correct modeling of the effective roughness in the hydraulically rough regime requires that the ratio of water depth to effective bed roughness is the same in the model and the prototype. For waves, the ratio of wave orbital excursion to effective bed roughness must be the same in the model and prototype. Generally, satisfying these criteria is not possible and corrections are necessary (Frostick *et al.*, 2011). While the flow dynamics of hydraulic models are often scaled based on Froude similitude, Froude scaling of sediment transport is usually inappropriate. The scaling laws for modeling the geometry of the bed and sediment transport processes are extensively described in (ASCE, 2000; Frostick *et al.*, 2011).

Boundary conditions for sediment models are also challenging. In tests where sediment transport is significant, the mobile bed may erode significantly during the tests unless sediment is fed into the system at the upstream end (see also Figure 4.3.1). Most often the sediment supply is achieved with a sediment feeder system in which the sediment is continuously introduced at the upstream end, but not recirculated, as illustrated in Figure 4.3.1b. To reduce the amount of sediment used in this type of facility, sediment that accumulates at the flume exit is often collected and returned to the feeder. For long-running experiments the sediment may be recirculated as a slurry. The capabilities and limitations of the feeding systems for sediment and gravel experiments are described in Frostick *et al.* (2011).

Additional provisions are needed both in the design as well as in operation of the facilities for minimizing the effects of flow-facility interactions. They include verification that wall interference with the flow (*e.g.*, due to uneven wall panels in the model) is negligible, that boundary conditions in the model and prototype are not dissimilar (*e.g.*, high turbulence at the flume entrance), and that there are not undesirable fluctuations or vibrations in the facility produced by pump or valve operations. Unwanted artifacts in laboratory flumes are often related to sharp corners and vertical walls. These in combination with the free-surface can give rise to secondary circulation cells which are sustained by turbulence anisotropy. Depending on the aspect ratio of the flume cross-section, secondary cells are usually much stronger in the laboratory than in the field (Nezu & Nakagawa, 1993).

4.3.2.4 Special facilities

The majority of natural hydraulic flows are free-surface flows, suggesting that most laboratory simulations will also be free-surface flows. However, for studying transport processes close to a wall where no influence of the free-surface is expected, the flow might be more easily conditioned in a closed system. In the absence of a free surface, Froude number no longer plays a role and Reynolds scaling is more easily achieved. For example, wave boundary layers and associated sediment transport can be simulated at full scale by forcing an oscillating flow within a rectangular duct. Oscillating water tunnels are often used to study oscillatory boundary layer flow over a sandy bed, as is caused by waves in coastal areas (Ribberink & Al-Salem, 1994). The advantages of using such flumes are that no further scaling is necessary regarding the sediment properties and that waves of arbitrarily large periods can be generated (Abad & Garcia, 2009).

Closed channels also have the advantage of optical accessibility from above; whereas free surfaces always have perturbations that distort the overhead view. Naturally, ice-covered flows can be studied in closed ducts if the dynamics of the ice layer are unimportant (Muste *et al.*, 2000). Finally, wind driven flows and wind-wave studies can be performed in a closed flume partially filled with water and partially filled with flowing air (Liberzon & Shemer, 2011).

Settling and flocculation processes as well as erosion processes also require special facilities. Particularly for the study of erosion and settling of fine sediments, special care needs to be taken to ensure that the fine material is easily suspended and the floc formation is gently entrained. For this purpose, a large settling column with the ability to induce specific turbulence levels by means of a moving grid can be used. Conditions in the column can be controlled to observe settling and flocculation processes at full scale, independent of the flow (Maggi *et al.*, 2002; Weitbrecht *et al.*, 2011). Similarly, annular or Couette flumes are suitable for the study of erosion of fine bed material without the influence of flow artifacts caused by inflow or outflow since the physical boundary conditions are periodic. Velocity gradients can be introduced by using a lid that rotates with respect to the flume (Crosato *et al.*, 2012). The rotational speed of both the lid and the flume can be adjusted to minimize secondary circulation. For flows in which the effects of the earth's rotation play a role, the model can be placed in a rotating tank to simulate the Coriolis force; the rotational speed can be adjusted for straightforward control of the magnitude of the effect (see for example Hopfinger *et al.*, 1982; Sous *et al.*, 2013). For other special facilities refer to section 5.4 in Volume I.

4.3.3 Flume assembly

Conventional hydraulic flumes are typically designed for handling water and sediment. Nearly all hydraulic models require an inlet reservoir with a baffle system to straighten and condition the flow and a tailwater to receive the flow. In closed conduit models, the inlet reservoir is often replaced by a pressure tank with a baffle system to condition the flow. Besides inlets/outlets, the flume assembly also includes some or all of the following features (Frostick *et al.*, 2011): pumps or head tanks for driving the flow, flow stabilization (conditioning) systems, systems that support sediment transport (recirculation or sediment feeding systems), and sediment traps. Typical layouts of flumes for sediment transport studies are shown in Figure 4.3.1 (see also the flumes described Section 5.4.1 in Volume I).

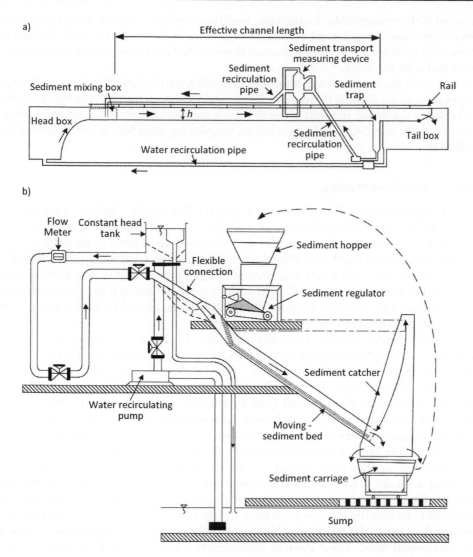

Figure 4.3.1 Flumes for studying flow and sediment transport: (a) closed-loop flume (adapted from Diplas, 1988); (b) head-tank driven flume (adapted from Bathurst *et al.*, 1984).

4.3.3.1 Flow circulation system

Water can be supplied to flumes in various ways, ranging from stand-alone flumes with their own (re)circulation pumps to constant-head lab-circulation systems that simultaneously serve many flumes. Many large facilities are equipped with a circulation system for the whole laboratory. Such a system has the advantage that water storage and pumping capacity is arranged centrally and that only one control system is needed to control the pressure for all connected flumes. The disadvantages are that water must be distributed throughout the laboratory and sediment or other transported materials need to be separated from the system in order to prevent pollution or accumulation of sediment in the storage tanks. In addition, facilities coupled to a single water circulation system also have the potential to interact when water demands are high or display strong time variations. Interdependence of flows in the facilities increases uncertainties associated with the flows.

Stand-alone flumes often have improved flexibility, particularly if the required range of operation of the flume is known in advance and local water supply storage is readily available for the flume. In stand-alone flumes, water and sediment (or other constituents) can easily be recirculated as a mixture, simplifying mass balance calculations for the facility.

4.3.3.2 Flow Conditioning

A significant factor that affects, to some degree, the flow quality (including the performance of flow measurement devices) in a facility is the distribution of the flow into important areas of the facility. Particular attention needs to be given to the entrance and exit of the flume to ensure that they do not influence the spatial-temporal flow distribution in the test section (see also Section 5.1 in Volume I). It is also important to ensure that flow measurements collected with meters installed in pipes that supply the facility are not influenced by irregular flow distributions caused by piping or channel configurations that precede or follow the meters.

Common sources of flow anomalies include: 1) elbows (in pipes) and bends (channels), 2) small aspect ratios (*i.e.*, width to depth ratios in channels), and 3) entrance and exit sections (channels, tunnels). Significant attention should be given to attaining a good flow distribution at flow meters set in the recirculating pipe of a flume when the adjacent pipe lengths are not long and straight. Several issues arise when design flow conditions do not satisfy the conditions needed for flow and velocity meters. Various types of meters are affected differently in flows that do not have fully developed boundary layers, and the effects are difficult to predict unless documented during earlier studies. A common difficulty when designing a model is to have enough space for installing the necessary length of straight pipe upstream and downstream of a flow meter. To address these issues, a variety of conditioning devices have been devised to form full developed approach flow at a flow meter if it is not possible to install the required length of pipe to assure fully-developed flow. These devices produce the same specific flow distribution, regardless of the flow pattern entering the flow conditioner. It is usually important to satisfactorily condition the flow without producing large static-pressure lossesin the flow. An example of a conditioning device for pipe flow downstream of a bend is illustrated in Figure 4.3.2.

4.4 INSTRUMENT SELECTION

4.4.1 Preliminary considerations

A typical hydraulic experiment entails measurements of physical properties of the working fluid such as density and viscosity, scalars such as temperature and pressure, and vectors and tensors such as velocities and strains. These measurements are

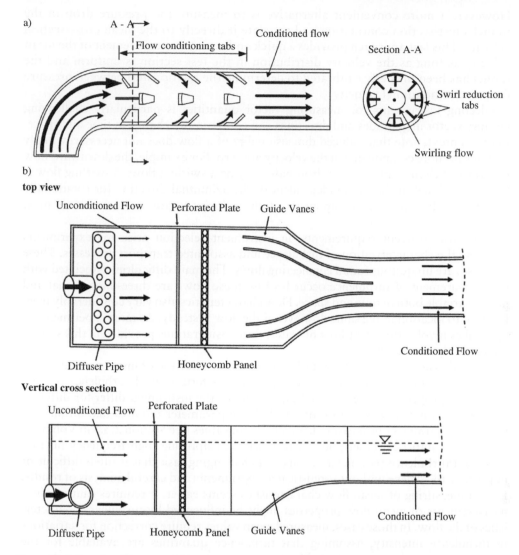

a)

A - A

Flow conditioning tabs

Conditioned flow

Section A-A

Swirl reduction tabs

Swirling flow

b)

top view

Unconditioned Flow   Perforated Plate   Guide Vanes

Conditioned Flow

Diffuser Pipe   Honeycomb Panel

Vertical cross section

Unconditioned Flow   Perforated Plate

Conditioned Flow

Diffuser Pipe   Honeycomb Panel   Guide Vanes

Figure 4.3.2 Sample of flow conditioning systems associated with: a) a flow meter set in a closed pipe (adapted from www.vortab.com); b) the entrance of an open channel flume.

used to determine bulk flow parameters (*e.g.*, discharge, Froude or Reynolds numbers) or to characterize temporal distributions of specific variables over a volume, area, line, or at a point in the flow (*e.g.*, recording a point velocity time series or tracking three-dimensional vortical structures near the flume bed). For each of the above situations, the selection of an instrument for measuring a specific variable must satisfy the purpose of the measurement. For example, if the purpose is to determine Reynolds number for a model test in a water tunnel, the mean velocity needed for the Reynolds number can be obtained using point velocities measured (with a Pitot or hot-wire anemometer, for example) over the tunnel cross-section.

However, a more convenient alternative is to measure the pressure drop in the tunnel's test-section contraction and to relate it directly to the mean cross-section velocity. This last approach provides a quick and accurate measurement of the mean velocity as long as the velocity distribution in the test section is uniform and the tunnel has been properly calibrated to determine the relationship between pressure drop and wind tunnel velocity.

Selecting instruments for measuring vector quantities is particularly challenging because vector magnitudes and directions can continuously change in time and space. In this context, note that reduced dimensionality of a flow does not necessarily imply that one or more components of the velocity are zero. For example, the downward flow in a vertical circular pipe (*i.e.*, a dropshaft) may be a swirling flow. A swirling flow is two-dimensional in that it is independent of the azimuthal direction (or rotationally symmetric), but all three components of the velocity vector are nonzero at most locations within the flow.

The most stringent requirements on instrument selection arise for experiments targeting the characterization of turbulence and associated transport processes. These processes are ubiquitous for all engineering flows. The main difficulties associated with the measurement of turbulence occur because these flows are three-dimensional and highly variable both in time and space. Flow characteristics also vary considerably from location to location in the same flow even if the flow is steady. As most experiments in hydraulics involve turbulent flow of water, the considerations provided in this section focus mainly on such flows.

As in all experimental work, in turbulent flows the selection of instruments depends on measurement objectives (*e.g.*, is complete characterization of turbulence or only mean flow properties). The required capabilities of the instrument differ for different objectives. Complete description of a turbulence parameter requires that all fluctuations of the parameter be measured, not just mean values (see Section 2.3.3 in Volume I, for detailed statistical description of turbulence). Capturing high frequency fluctuations of a turbulence parameter is usually a challenging task that is often difficult or impossible to achieve without sophisticated instruments and careful analysis of results. Even if measuring of mean flow characteristics seems easier, it requires caution since measurements of mean flow properties might be influenced by turbulence characteristics of the flow. In those cases, measured mean values require correction for variations in turbulence intensity, assuming that turbulence quantities are available for the correction.

Examples of instruments for measuring mean and turbulent flow variables are given in Table 4.4.1. Due to their variety and importance, instruments for measuring turbulent velocities are further detailed in Table 4.4.2. Volume II of the handbook provides extensive details on the capabilities of various categories of instruments to measure in turbulent flows. This section, describes the general guidelines for selection of the instruments in relationship with the scope of the measurement. It is important to bear in mind that each of the listed instruments may be best-suited for specific measurement tasks, hence all should be considered in the selection process. For example, if the targeted measurement is the streamwise mean velocity in a turbulent flow, a normal-sized Pitot probe will suffice without the need for a more expensive and complex alternatives such as a hot-wire anemometer.

Table 4.4.1 Sample instruments for the measurement of flow variables

Targeted variable	Variable type	Candidate Instruments
Wall shear stress	Mean	Preston tube, Stanton gauge, differential pressure gauges
	Fluctuations	Heat- and mass-transfer probes, hot-film, LDV
Pressure	Mean	Pitot probes
	Fluctuations	Pressure transducers - (microphones, piezoelectric devices)
Temperature	Mean	Resistance type, expansion type, thermocouples, pyrometers
	Fluctuations	Optical pyrometry, radiation detectors
Velocity	Mean	Pitot, propeller anemometers
	Fluctuations	See Table 4.4.2

Table 4.4.2 Sample instruments for measurement of turbulent velocities*

Intrusive	Up to 3-D, point measurements, velocity magnitude only	Hot wires, hot films, Pitot probes
Nonintrusive	Up to 3-D, point measurements	Laser-Doppler Velocimetry (LDV), Phase-Doppler Velocimetry (PDA), Acoustic-Doppler Velocimetry (ADV)
	Up to 3-D, line measurements	Light Detection And Ranging (LIDAR), SO(und) DAR, Acoustic-Doppler Current Profilers (ADCP)
	Up to 2-D, whole-field measurements	Particle Image Velocimetry (PIV), Particle Tracking Velocimetry (PTV), Laser-Speckle Velocimetry (LSV)
	3-D, volume measurements	Tomographic PIV, Stereo PIV, Doppler-Global Velocimetry (DGV), Holographic PIV

* Laser-, acoustic-, and image-based instruments are increasingly used to infer scalar properties such as concentration and size of the suspended fraction in multi-phase flows in addition to their capabilities to measure velocities.

4.4.2 Instrument spatial and temporal resolutions

Measurements with intrusive instruments imply that the instrument sensor is immersed at the location where the measurements are acquired. Each measurement is of finite duration and can be repeated in time and space at rates that depend on the temporal response of the instrument. The geometry of the sensor determines its spatial resolution while the frequency response determines its temporal resolution. Collectively, the spatial and temporal resolution of the instruments are critical in defining the sensitivity (response time), the precision (repeatability), and the accuracy (deviation of a measurement from its true value) of the instrument (Holman, 1989). Noninstrusive instruments do not have immersed probes and usually require flow tracers that are suspended in the flow to obtain measurements. Therefore, their spatio-temporal resolution depends on both instrument and tracer characteristics. The spatial resolution is defined by the geometry of the measurement volume and the physical and other properties of the tracer particles passing through it. The sampling frequency (the inverse of the sampling interval) is also dependent on instrument frequency response but may also be conditioned by particle (*i.e.*, size and shape) and flow (seeding concentration and homogeneity) properties. For example, the response of a LDV (operated in single-particle mode) is basically the time it takes for a particle to pass through its measurement volume, which can be on the order of milliseconds, implying a very high sampling frequency. However, if the seeding particle concentrations are low, the occurrence of particles within the measurement volume is random, so frequency of data points in the acquired dataset is dictated by the density of the tracers in the flow. As this data density might be inadequate for analyzing turbulence features, special sampling strategies are used as described in Section 4.5.4, Volume I.

Irrespective of their type, the outputs of most contemporary instruments are converted to digital signals. Signals in digital data-acquisition systems can be both analog continuous (prior to digitization) and discrete digital. Presently, 10-, 12-, 16-, and 32-bit A/D converters are commonly used for digitization. Using an A/D converter with greater bit precision allows the analog signal to be more accurately represented by its digital form. For example, for an A/D board with a 10 volt range, the resolution of a 10-bit conversion is 9.8×10^{-3} volts; while the resolution of a 32-bit conversion is 2.3×10^{-9} volts. However, there are trade-offs associated with selecting an A/D board with high digital resolution. First, a 32-bit digital signal requires four times as much storage space as an 8-bit digital signal. This is probably insignificant for most point measurements, but for multi-point and image-based instruments it can become important. Second, it generally takes significantly longer for high resolution A/D boards (*e.g.*, 32-bit) to accurately resolve a digital value from the analog input. This of course means that high resolution A/D boards may limit the temporal resolution of some instruments. For all A/D boards, high sampling rates can lead to degradation of measurement accuracy since there is less time for the A/D board to settle to its final value when sampling rates are high (Goldstein, 1996), but for high resolution boards the problem is compounded because it takes more time to achieve the higher bit resolution.

Spatial and temporal resolutions of instruments are often provided by the manufacturers along with other specifications. For intrusive instruments (*e.g.*, Pitot tubes, hotwire probes) spatial and temporal characteristics are related to size (*e.g.*, probe diameter, probe active area) and maximum sampling rate (*e.g.*, samples per second or Hz). The spatial resolution for non-intrusive instruments (*e.g.*, LDV, ADV, PIV, or ADCP), is less uniformly and thoroughly specified by manufacturers for several obvious reasons, depending on the instrument. For example, the measurement volume for LDVs is defined as the intersecting volumes of the sending and receiving optics, making the estimation of measurement volume size complex and in some cases application dependent. The same is true of ADV measurement volumes. The size of the measurement volume for PIV is established by the user during data processing as a function of image and tracer particle resolution, flow characteristics, and the speed of the camera. The measurement volume of ADCPs is variable with the depth due to the divergent angle between the beams of these multi-beam systems. For some newer (non-intrusive) instruments, the user may have several options to choose from to accommodate a range of temporal resolutions. Most non-intrusive instruments offer a great deal of flexibility to accommodate a wide range of experimental conditions. One consequence of this flexibility, however, is that accurate assessment of the spatial and temporal characteristics of the instruments might be quite complex. Successful application of such devices requires that the user be very familiar with the physics that control measurement volume geometry and instrument behavior.

4.4.3 Practical relationships for estimating turbulent flow scales

Length and time scales of a turbulent flow govern the selection of suitable instruments for measuring turbulent parameters in the flow. Thus, prior to selecting instruments, these flow scales should be estimated. Surprising as it sounds, estimates are attainable in typical hydraulic experiments by combining preliminary observations and simple measurements with theoretical knowledge and experience. Given the importance of the turbulence scales for hydraulics, and especially experimental hydraulics, complementary issues are presented in Chapter 2, Section 2.3.3.6 in Volume I and Chapter 6 (Sections 6.11 and 6.12 in Volume I). Presented below are basics of the physical background and practical relationships relevant for the selection of the data acquisition system.

Given the applied nature of this handbook, it is useful to summarize the main scales and practical relationships to quickly estimate them prior to the actual experiments. Consequently, Table 4.4.3 summarizes turbulence scales relevant to experimentation, Table 4.4.4 provides practical relations for estimating various turbulence properties used to estimate turbulence scales in various types of flows, and Table 4.4.5 provides practical relations for the length scales based on analytical procedures, empirical relations, and observations garnered through extensive measurements in common turbulent flows. The latter relations are based upon bulk flow variables (*e.g.*, flow geometry, bulk average velocity, and turbulence intensities) that are easy to measure or estimate for many flows of practical importance.

The notation of the variables in Tables 4.4.3 through 4.4.5 is based on Reynolds decomposition (Tennekes & Lumley, 1972, Pope, 2000). For example, the x component of the instantaneous local velocity can be written as $u = \bar{u} + u'$, where u' is the deviation of the instantaneous x component of the velocity from the time-averaged x component of the velocity, $\bar{u}$. Similar decomposition is used for all turbulence quantities. Dissipation rate is obviously useful for estimating Kolmogorov scales, but be aware that formulas given in the table are mostly for calculating global dissipation rate, and local dissipation rates are widely variable in nearly all shear flows. Still, the relations in the table can be used for obtaining rough initial estimates of relevant length scales. One hypothesis used to establish the relations in Tables 4.4.3 through 4.4.5 worth explaining is Taylor's hypothesis that velocity and scalar fields in some turbulent flows behave as though they are "frozen" (see also Section 2.3.6 in Volume I). That is, the velocity and scalar fields in a turbulent flow do not change much when they are convected downstream. Taylor's hypothesis states that:

$$\frac{\partial f}{\partial t} = U_c \frac{\partial f}{\partial x} \tag{4.4.1}$$

where f is any local variable in the flow and U_c is the velocity at which the flow is convected downstream. This hypothesis is very useful because many turbulence length

Table 4.4.3 Turbulence flow scales (see Sections 2.3.3.6 and 2.3.3.7 in Volume I for details)

Scales	Sizes	Purpose	Representation	
			Spatial	Temporal
External scales	Largest scales – associated with largest eddies	Used to define the Reynolds number (ratio of inertial and viscous forces)	Length scale, ℓ (labeled $\mathcal{L}$ in Section 2.3.3.6, Volume I) is determined by flow geometry (e.g., boundary layer thickness, jet or wake width, pipe diameter, open-channel flow depth).	The corresponding time scale is $\ell/\overline{U}$, where $\overline{U}$ is a mean characteristic velocity of the flow.
Integral scales	Large – macroscales	Integral scales represent the largest eddies containing the majority of the energy in the flow. They are characterizing large flow scales and are determined from measured data– not from the geometry of the flow.	Streamwise integral length scale, L_x: $$L_x = \int_0^\infty R_x(r)dr \quad (T.1)$$ $R_x(r)$ is the spatial correlation coefficient : $$R_x(r) = \frac{\overline{f'(x)\cdot f'(x+r)}}{\overline{f'^2}} \quad (T.2)$$ r is the spatial lag; f' is the turbulence fluctuation of any turbulence variable.	Integral time scale, T: $$T_x = \int_0^\infty R_x(\tau)d\tau \quad (T.3)$$ $R(\tau)$ is the time autocorrelation coefficient (ACF): $$R_x(\tau) = \frac{\overline{f'(t)\cdot f'(t+t)}}{\overline{f'^2}} \quad (T.4)$$ τ is the time lag and f' – fluctuation of any turbulence variable.
Taylor scales	Intermediate – microscales	Taylor scales represent small-scale eddies in the intermediate region of the energy spectrum responsible for transferring energy from the large structures to the small structures dissipating turbulent energy. Taylor scales characterize turbulent shear rates.	Taylor length scale, λ, is defined by the curvature of the ACF for $r = 0$ (ACF as defined in T.4). $$\left[\frac{d^2 R}{dr^2}\right]_{r=0} = -\frac{2}{\lambda^2} \quad (T.5)$$ The longitudinal microscale λ_x is defined by: $$\lambda_x^2 = \frac{\overline{u'^2}}{\overline{(\partial u/\partial x)^2}} \quad (T.6)$$	Temporal Taylor microscale, T_λ: $$T_\lambda = \frac{\lambda}{U} \quad (T.7)$$
Kolmogorov scales	Smallest turbulence scales – microscales (see also 2.3.3.7)	These represent the smallest eddies which dissipate turbulent energy. Kolmogorov length scales are typically on the order of tenths of a millimeter or less.	The Kolmogorov length scale, η, is defined as: $$\eta = (\nu^3/\overline{\varepsilon})^{1/4} \quad (T.8)$$ $\overline{\varepsilon}$ - dissipation rate per unit mass (see Equation (2.3.7) and Section 2.3.3.7 in Volume I);ν – fluid kinematic viscosity	The Kolmogorov time scale, T_η, is defined as: $$T_\eta = (\nu/\overline{\varepsilon})^{1/2} \quad (T.9)$$ There is also a Kolmogorov velocity scale, v_η: $$v_\eta = (\nu\overline{\varepsilon})^{1/4} \quad (T.10)$$

Table 4.4.4 Practical relations for estimating dissipation and turbulent kinetic energy (see also Section 2.3.3.1 in Volume I)

Variable	Application	Relation	Description
Mean energy dissipation ($\bar{\varepsilon}$)	Turbulent boundary layer	$\bar{\varepsilon} \approx \dfrac{U_1^3}{L_\delta}$	U_1 is the external or free-stream velocity of the boundary layer and L_δ is its thickness at a particular streamwise location. This is a global estimate of dissipation for a cross-section.
	Mixing layer	$\bar{\varepsilon} \approx \dfrac{(U_1 - U_0)^3}{L_\delta}$	Global estimate of dissipation for a cross section at the interface between two flows with velocities of U_1 (the higher speed flow) and U_0 (the lower speed flow) if the thickness of the mixing layer (L_δ) at the cross section can be estimated.
	Pipe flow	$\bar{\varepsilon} \approx \left(\dfrac{\bar{U}}{\rho}\right)\left(\dfrac{dP}{dx}\right)$	Global estimate of dissipation for fully developed pipe flow where $\bar{U}$ is the bulk average velocity of the flow, ρ is the fluid density, and dP/dx is the mean pressure drop per unit length of pipe.
	Open-channel flow	$\bar{\varepsilon} \approx \dfrac{u_*^3}{H}$	Global estimate of dissipation for fully developed open-channel flow where u_* is the friction (or shear) velocity and H is the flow depth.
	Open-channel flow	$\dfrac{\bar{\varepsilon}y}{u_*^3} = 9.8\left(\dfrac{y}{H}\right)^{1/2} exp\left(\dfrac{-3y}{H}\right)$	Local estimate of dissipation in a fully developed open-channel flow where y is the height above the bed (Nezu & Nakagawa, 1993).
	Isotropic turbulence	$\bar{\varepsilon} = \dfrac{15\,v\overline{u'^2}}{\lambda^2}$	Turbulence intensities and Reynolds stresses are isotropic; In other words, $\overline{u'^2} = \overline{v'^2} = \overline{w'^2}$ and $\overline{u'v'} = \overline{v'w'} = \overline{w'u'}$

Table 4.4.4 (continued)

Variable	Application	Relation	Description
	Isotropic, locally homogeneous turbulence	$$\bar{\varepsilon} = 15\nu \overline{\left(\frac{\partial u}{\partial x}\right)^2}$$	Based on the definition of the longitudinal microscale, this equation allows the local dissipation to be estimated based on measurements of the spatial gradient of the streamwise velocity.
	Isotropic, locally homogeneous turbulence	$$\bar{\varepsilon} = 15\frac{\nu}{\bar{U}^2} \overline{\left(\frac{\partial u}{\partial t}\right)^2}$$	This equation can be found by applying Taylor's frozen field hypothesis to the equation above. It permits estimation of local dissipation from a time series of longitudinal velocity data, something that can be measured with a single probe. $\bar{U}$ is the mean convective velocity of the flow.
	High Reynolds number flows	$$\bar{\varepsilon} \approx \frac{k^{3/2}}{\ell}$$	The assumption here is that the relevant velocity for calculating dissipation scales with the turbulent kinetic energy.
Turbulent Kinetic Energy (k)	High Reynolds number flows	$$\frac{k^{1/2}}{U_i} \approx 0.1$$	In many high Reynolds number flows, this equation can be used to provide a global estimate of turbulent kinetic energy (k), i.e., $k = \left(\overline{u'^2} + \overline{v'^2} + \overline{w'^2}\right)/2$. U_i is a bulk velocity that can be estimated (e.g., bulk average velocity of the flow).

Table 4.4.5 Practical relations for estimating turbulence flow scales

Scale	Application	Relation	Description
External scales			
			Estimate ℓ directly from the geometry of the flow (e.g., pipe diameter, flow depth, etc.).
Streamwise integral length scale	If the streamwise component of velocity is dominant	Directly determine T.1 in Table 4.4.3	For the spatial correlation coefficient in T.1, use two streamwise velocity components $u'(x)$ and $u'(x+r)$ with a streamwise lag distance r. $$R_x(r) = \frac{\overline{u'(x) \cdot u'(x+r)}}{u'^2}$$ This requires knowledge of the streamwise distribution of velocity. Taylor's frozen field hypothesis can be applied to obtain this from a single strategically located velocity series. $S(k_w)$ is the one-dimensional normalized spectral function of the u-fluctuations and k_w is the wave-number in the streamwise direction. The spectral function is defined as: $$S(k_w) = \frac{2}{\pi} \int_0^\infty R_x(r) \cos(k_w r)\,dr$$
	If the streamwise component of velocity is dominant	$$L_x = \frac{\pi}{2} \lim_{k_w \to 0} S(k_w)$$	
Taylor microscales	Isotropic turbulence	$$\frac{\lambda}{\ell} \sim \left(\frac{\nu}{\frac{(u'^2)^{1/2}}{\ell}\ell} \right)^{\frac{1}{2}} = Re_\ell^{-1/2}$$	The Reynolds number (Re_ℓ) is based on the scale of the flow.
	High Reynolds number	$$\frac{\lambda}{\ell} = \left(\frac{\nu}{k^{1/2}\ell} \right)^{\frac{1}{2}}$$	The turbulence Reynolds number is defined using k as the velocity scale.

Table 4.4.5 (continued)

Scale	Application	Relation	Description
Longitudinal Taylor microscale	Convective Flows- Spectrum Method	$$\frac{1}{\lambda^2} = \frac{\bar{\varepsilon}}{15 v \overline{u'^2}} = \int_0^\infty k_w^2 S(k_w)\,dk_w$$	This relation is valid for isotropic homogeneous turbulence. $S(k_w)$ is found using the cosine transform described above. Nezu and Nakagawa (1993) suggest that this method has been found to perform better than the other methods for computing the longitudinal Taylor microscale.
	Convective Flows- Probability Method	$$\lambda_x = \sqrt{\frac{\overline{u'^2}}{\overline{(\partial u/\partial x)^2}}}$$	Use this to find the longitudinal Taylor microscale from a time series of streamwise velocity measurements at a point. For a time series of data from a single probe, $\partial u/\partial x$ can be replaced with $(\partial u/\partial t)/\overline{U}$ based on Taylor's frozen field hypothesis for turbulence.
	Convective Flows- Zero-crossing Method	$$\lambda_x = \frac{\overline{U}}{\pi N}$$	For random variables, the Taylor microscale equals the average distance between zero-crossings of the fluctuating component. For velocity, this is true if measured distributions of u and $\partial u/\partial t$ are independent and have Gaussian distributions. N is the average frequency of zero-crossings of the measured velocity (when $u = 0$) and can be determined from a measured time series.
	Convective Flows - Autocorrelation	$$\left[\frac{d^2 R_x(r)}{dr^2}\right]_{r=0} = -\frac{2}{\lambda^2}$$	The Taylor microscale can be found from the autocorrelation function of the streamwise velocity component at two locations in space $R_x(r)$. The spatial Taylor scale is defined by the curvature of the autocorrelation coefficient at the origin ($r = 0$).
Kolmogorov microscales	High Reynolds	$$\frac{\eta}{\ell} = \left(\frac{v}{k^{1/2}\ell}\right)^{\frac{3}{4}}$$	The turbulence Reynolds number is defined using k as the velocity scale ($Re_\ell = k^{1/2}\ell/v$) and $k^{3/2}/\ell$ for the mean energy dissipation.

scales require knowledge of the spatial distribution of turbulence properties. Equation (4.4.1) affirms the merit of measuring the temporal distribution of a property with an instrument located at a single point in the flow, and then converting the distribution into a spatial distribution using the convective velocity, U_c. For Equation (4.4.1) to be accurate, the following criterion must be met:

$$\frac{\overline{u'^2}}{U_c^2} << 1$$

in which $\overline{u'^2}$ represents turbulence intensity of the velocity. Hence, Taylor's hypothesis is limited to low-intensity turbulent flows. A somewhat sophisticated way around this problem is to move the probe in a direction opposite to the flow, thereby artificially creating a high convective velocity, U_c. Such is the case when meteorological data are collected with instrumentation mounted on an airplane.

4.4.4 Selection of the instruments and settings for measurements in turbulent flows

The selection of instruments is an important step in designing an experiment, and involves several important considerations. The top rules-of-thumb for instrument selection include the following considerations:

1. Select the simplest appropriate instrument;
2. Absolute measurements are preferred to relative measurements;
3. Cumulative measurements are preferred to rate measurements for characterizing mean flow properties; and,
4. Accurate, sensitive, and precise instruments are preferred for both mean and turbulent flow characteristics (see term definitions in Section 4.2.4, Volume I).

Additional selection considerations include: the nature of the measured quantity (*e.g.*, single or multi-phase, scalar, vector, static, fluctuating), the targeted temporal and spatial flow scales (see Section 4.4.3 in Volume I), and satisfaction of uncertainty analysis constraints. The extensive list of desirable features for instrument selection is often drastically limited by the resources available for an experiment. Therefore the experimenter typically has to make trade-offs between goals, accuracy, and efficiency of the measurement system and protocols.

The present discussion does not cover all details about instrument selection, but instead discusses the salient aspects. Moreover, it focuses on instruments for measuring flow variables in turbulent flows. For measurements in such flows further distinction should be made depending on how turbulence is defined (Tropea *et al.*, 2007). In the classical RANS decomposition (see Section 2.3.1 in Volume I) time averaging applied to point measurements techniques might suffice to provide the needed information. For example, accurate measurement of Reynolds stresses in turbulent flows requires instruments that can capture Taylor time and spatial scales (see Table 4.4.3). These scales vary with the type of flow (*e.g.*, boundary layers, jets, flows around bluff bodies) and from location to location in the same flow. For finer flow descriptions, such as description of the turbulence structure functions (see Section 2.3.4 in Volume I), however, instruments and techniques that allow turbulence decomposition based on three-dimensional spatial averaging is required. Hence the velocity data supporting this approach have to be acquired with multicomponent, multipoint velocity measurement instruments such as optical planar velocimetry (see Section 3.7 in Volume II).

There are a number of general requirements that must be satisfied by instruments measuring turbulence:

Intrusive techniques

1. An instrument's measuring volume should be small and cause minimum disturbance to the flow. The instantaneous velocity distribution must be uniform in the region occupied by the element, or the detecting element should be smaller than the dimensions of the smallest eddies to be detected. In many cases, this means that the tip diameter of a probe should not exceed about 1 mm for laboratory measurements;

2. The instrument should sample at high rates (and have good frequency response) so that the output signal captures even the most rapid turbulence fluctuations. Sampling frequencies of up to 100Hz are typically required in water flows (5 KHz in low-speed air flows);

3. The instrument must be capable of recording small differences in the fluctuations (good sensitivity); these differences are often only a few percent of the mean value. It also should have a wide dynamic range;

4. The instrument must be stable, so that no noticeable change in its calibration occurs during the experiment; and,

5. The instrument should be sufficiently strong and rigid to exclude flow-induced vibrations.

Nonintrusive techniques (optical, acoustic, or electromagnetic)

Requirements 1 to 5, above, also apply to non-intrusive instruments. Additional requirements include:

6. Flow tracer particles (seeding) should be small enough to follow the smallest spatial scales of the turbulence yet large enough to be detected by the instrumentation; and,

7. Optical or acoustic access to the location of the measurement is required.

4.4.4.1 Attaining the required spatio-temporal resolution

Two basic requirements when selecting an instrument are that the probe size must be smaller than the size of the turbulent scales to be captured, and the probe must sample fast enough to capture the smallest turbulent fluctuations. Using practical relationships as those provided in Tables 4.4.3 – 4.4.5 or prior knowledge of the flow, the size of turbulent flow features (*e.g.*, wavelength λ and boundary-layer thickness, δ) can be estimated with sufficient degree of accuracy for providing information for instrument selection. It is important to consider not only the spatial resolution but also the temporal response (how fast the probe can capture the signal) of an instrument as will be discussed next.

Consider first a probe placed in a flow feature of wavelength λ and propagating with speed U_c, as illustrated in Figure 4.4.1a (Goldstein, 1996). There are two criteria to fulfill when selecting the probe. First, its size must be small enough to resolve the spatial

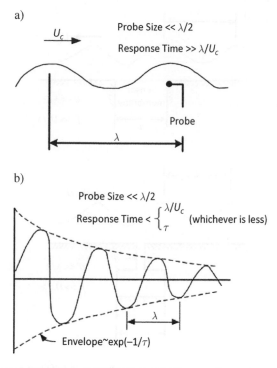

Figure 4.4.1 Spatial versus temporal resolution for: a) a "frozen" fluctuation; b) a decaying fluctuation (Copyright 1996 from *Fluid mechanics measurements* by Goldstein. Reproduced by permission of Taylor and Francis Group, LLC, a division of Informa plc).

extent of the feature; *i.e.*, the probe size must be less than $\lambda/2$. Second, since the probe is fixed in space it must have adequate frequency response; *i.e.*, the response time must be less than λ/U_c to ensure that the fixed probe does not sense a time-varying magnitude during the measurement. Now consider the more complex case of resolving a wave-like disturbance whose amplitude is decaying. The signal illustrated in Figure 4.4.1b displays an exponential decay, $a \sim \exp(-t/\tau)$. The original spatial requirement continues to hold (probe size $< \lambda/2$), but the response-time requirement is two-fold: the response time must be less than λ/U_c, or τ, whichever is less.

An example of a trade-off between temporal and spatial resolution that an experimenter may have to make is illustrated in Figure 4.4.2. In the example, the measurement goal is to determine fluctuating wall pressures having a characteristic wavelength of λ. Placing a pressure transducer flush with the wall ensures maximum frequency response. However, the diameter of the probe must be less than half the wavelength. Sensitive pressure sensors generally require fairly large diameters, but if a small effective sensor diameter is necessary to meet spatial resolution requirements, one option is to place the transducer in a chamber with a pinhole leading to the surface as shown in Figure 4.4.2b (Goldstein, 1996). Here spatial resolution is achieved at the expense of temporal response. The quantification of this trade-off is provided in Goldstein *et al.* (1996, p.27).

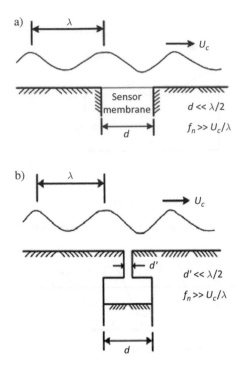

Figure 4.4.2 Trade-off between spatial and temporal resolutions in fulfilling: a) temporal resolution; b) spatial resolution (Copyright 1996 from *Fluid mechanics measurements* by Goldstein. Reproduced by permission of Taylor and Francis Group, LLC, a division of Informa plc).

4.4.4.2 Attaining the required frequency content for the data

In modern laboratory practice, turbulence measurements are rarely processed using a fully analog technique. This is due to the development of very efficient methods for analog-to-digital data conversion, followed by digital data processing (see Section 4.2.3 in Volume I). Usually, this means that the data are sampled at equally spaced time intervals, δt. It is necessary to distinguish the time between each sample (the sampling rate or sampling frequency) from the time required for acquiring individual samples (data points). The time for recording one data point is typically very short, and is sometimes assumed to be an instantaneous measurement. However, when measuring the mean and other statistical moments of a stationary signal it is required that each sample be recorded in less than the characteristic time of the highest frequency component of the signal. If the recording time of a sample point is too long, the result will be averaging or "slewing" of the signals in the high-frequency range, attenuating the maxima of the recorded signal magnitude (Goldstein, 1996).

For measurements of time-resolved flow features such as autocorrelations and spectra, the choice of time between samples, Δt, must be based on turbulence flow scales so that the fluctuation waveform (shape of the fluctuation) can be reconstructed as accurately as possible. Irrespective of the sampling method, if samples are collected at intervals of Δt, the sampling frequency is $f_s = 1/\Delta t$. As previously discussed, the highest frequency that is of importance is flow dependent. For example in an open-channel flow the maximum frequency can be approximated by $f_m = (50/\pi)(U/h)$ (Nezu & Nakagawa, 1993). Fulfilling the Nyquist criterion regarding sampling rate, requires that the sampling rate be more than twice the highest frequency contained in the signal, f_m (perhaps determined from a frequency power spectrum), i.e., $f_s > 2f_m$. The corresponding sampling interval of the selected instrument should then be less than $1/2f_m$.

4.4.4.3 Summary for instrument selection in turbulent flows

The choice of the device for turbulence measurements depends on the measurement environment of the application. General criteria include stability, compactness, sensitivity, and response to high frequencies. In order to choose the most appropriate instrument, estimate in advance the scales for turbulence, time and length scales as well as the total sampling time required for reconstructing the original signal from the measured data. Table 4.4.6 summarizes the steps involved in selecting the instruments and settings of the instruments for conducting measurements in turbulent flows. For illustration purposes, Table 4.4.7 provides an example of specifications that can be compared to select an appropriate instrument for velocity measurements in turbulent flows.

Table 4.4.6 Steps for selecting appropriate instruments for turbulence measurements

Step	Description	Comments
1. Estimate scales in the flow	Integral length scales, Taylor microscales (mixing length), Kolomogoroff velocity, time, and length scales, rate of viscous dissipation.	Do this prior to selecting the instrument and before data acquisition.
2. Review the dynamic content of the signals	Using the scales obtained in step 1, create the waveforms, and assess their distribution throughout the flow (along vertical and spanwise directions). The dynamic content of the signal must be considered to select a suitable sampling type, sampling frequency (Nyquist criterion) and sampling time (see next section).	Use Table 4.4.5 for assessing scales if limited or no information is available for an experimental setup.
3. Select instruments with appropriate spatial and temporal resolutions	• The probe size/measurement volume should be always smaller than the size of the structure to be measured • The duration for acquiring individual samples (data points) should be smaller than the characteristic time of the higher frequencies in the flow • Sample spacing should be smaller than the time scale of the structure to be measured	• Small sample duration for statistical analysis (especially higher order moments) • Fast sampling rate for time resolved measurements (e.g., correlations, spectra)

Table 4.4.7 Sample of specification comparison for supporting the selection of a velocimeter

Instrument specification	Particle Image Velocimetry (PIV)	Laser-Doppler Velocimetry (LDV)	Constant-temperature Anemometry (CTA)
Measurement principle	optical	optical	thermal
Measurement volume	global	local	local
Intrusiveness	non-intrusive	non-intrusive	intrusive
Tracer particles	yes	yes	no
Output function	linear	linear	non-linear
Calibration	no	no	yes
Sampling type	random in space	random in time	equally sampled
Frequency response	poor (in most cases)	good	excellent
Flow reversal	yes (cross correlation)	yes (Bragg shifting)	yes (flying hot-wire film)
Conditional sampling	yes	yes	yes
Bias and correction	yes	yes	yes

4.5 FROM SIGNALS TO DATA

4.5.1 Signal classification

Hydraulic experimental data are usually collected for finite durations as the signals are time-dependent. Consequently, the signal discussions in this section focus primarily on the treatment of time-series data. However, much of the presented discussion can be extended to spatial dimensions by simply replacing time with space in the analysis. Such a procedure is extensively discussed in Section 2.3.2, Volume I, whereby spatial averaging of rough-bed open-channel flows is considered in addition to time averaging.

Time series data acquired through experiments can be broadly classified as deterministic or nondeterministic (Bendat & Piersol, 2000). Deterministic data are those that can be described by a mathematical relationship. Non-deterministic data usually derive from random signals having patterns that cannot be predicted and must be described in terms of probability statements and statistical parameters rather than by explicit relationships. Given that turbulent flows dominate experimental hydraulics, most of the measured signals are nondeterministic or random (*e.g.*, a velocity fluctuation). While truly deterministic dynamic signals are not practically encountered in experimental hydraulics, many fundamental aspects of the physical processes of the investigated flows can be treated as deterministic for analytical purposes (*e.g.*, wave investigations). The distinction between deterministic and random signals is somewhat imprecise in practice, since any deterministic physical process modeled in an experimental facility can be contaminated by unknown factors so that it is not strictly deterministic. In such cases, the deterministic signal definition is often extended to include situations in which the influences of random perturbations on the measured variable are minimal.

A classification of deterministic signals is provided in Figure 4.5.1. This figure shows frequency power spectra widely used to illustrate the partitioning of a dynamic signal with respect to temporal scales of the fluctuating eddies in the flow. It also includes steady (static) signals fully characterized by their amplitude as they do not vary in time. All other signals illustrated in the figure are dynamic and can be fully characterized by their amplitude and frequency power spectra. Details about the nature of various dynamic signals can be found in Bendat and Piersol (2000). More details on the implementation of spectral analysis to acquired data can be found in Section 6.12, Volume I.

The vast majority of hydraulic flows, including steady and unsteady flows, are turbulent with the flow variables randomly varying in time and space. Such randomness is obvious from the visualization of the turbulent structures occurring near the bed of an open-channel flow and the fluctuating quantities depicted in Figure 4.5.2. The area near the channel bed contains self-organized structures moving away or toward the bed (*i.e.*, coherent structures) as illustrated in Figure 4.5.2a. These continuously changing structures are advected downstream by the mean flow. A point-velocity instrument measuring in this flow produces time series of fluctuating quantities that are depicted in Figure 4.5.2b. The appearance of the signals depends on the measurement location, *i.e.*, larger and less frequent fluctuations near the bed and smaller but more intense away from the wall. These complex structures and their interactions result in recorded signals that cannot be described using deterministic methods. Therefore, statistical tools are used for the analysis of such random (stochastic) processes to subsequently develop insights about the flow mechanics.

The statistical functions used to describe and analyze random signals (*i.e.*, turbulent fluctuations of flow variables) include: the mean value, the probability density function (pdf), and the energy spectrum as a function of the frequency or wavenumber. These statistical functions are described in Section 2.3 and 2.4, Volume I. Figure 4.5.3 illustrates some of those characteristics for a stationary turbulent flow. Figure 4.5.3a

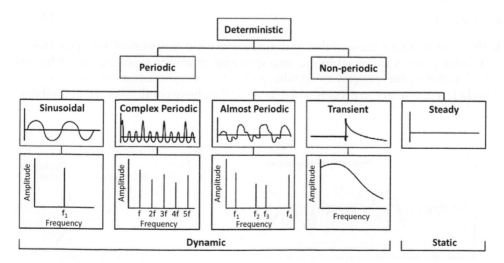

Figure 4.5.1 Classification of deterministic signals including their associated power spectra (adapted from Bendat & Piersol, 2000).

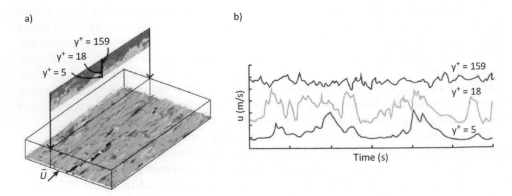

Figure 4.5.2 Visualization of the vorticity associated with near-wall random structures in turbulent channel flow and associated traces of the velocities at various distances from the bed.

replicates a raw signal while Figure 4.5.3b exemplifies a processed signal from the same flow. The generic spectrum plotted in Figure 4.5.3b reflects the contribution of the turbulence eddy frequencies or wavenumbers to the total flow energy. The conversion of frequency to a wavenumber-based representation of the power spectrum assumes that eddy size can be computed from a measured frequency (see more details in Section 5.2.4, Volume I). The random process illustrated in Figure 4.5.2 is just one of several types of random processes encountered in hydraulic engineering as illustrated in Figure 4.5.4.

For stationary processes, whereby the statistical functions are time-invariant, the signal describing the turbulent quantities can be conceptualized using Reynolds decomposition: *i.e.*, the instantaneous value of a fluctuating quantity is expressed as a mean value and a fluctuation about the mean, as illustrated in Figure 4.5.3a. The Reynolds decomposition for a turbulent fluctuation is mathematically expressed as:

$$f(t) = \bar{f} + f' \tag{4.5.1}$$

If the time-averaged mean value and autocorrelation function of the data from a stationary process are equal to the corresponding ensemble averaged values for any set of samples, the process is ergodic (only stationary random processes can be ergodic). A stationary process is non-ergodic if the above conditions are not fulfilled.

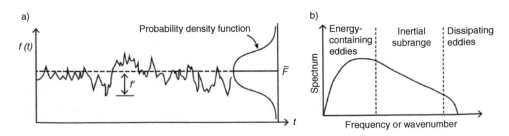

Figure 4.5.3 Time series of a signal, pdf, and associated energy spectrum.

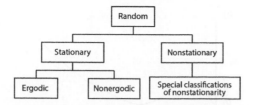

Figure 4.5.4 Classification of random signals (republished with permission of John Wiley and Sons Inc, from Bendat & Piersol, 2000; permission conveyed through Copyright Clearance Center, Inc.).

Often stationary random processes are ergodic. Therefore, stationary random processes can be fully characterized by a single observed time history record. Analytical procedures for assessing whether or not a flow is stationary are presented in Section 6.12.1, Volume I.

If the statistical characteristics of a process vary in time it is called nonstationary. Nonstationary random signals are associated with processes that have changing boundary conditions (*e.g.*, turbulence downstream of a sluice gate that is closing) or oscillations in the flow (*e.g.*, vortex shedding). There are no general analysis methodologies to describe all types of nonstationary random processes from individual sample records. Fortunately, for measurement and analysis purposes it is often possible to treat nonstationary data as piecewise stationary. Due to the complexity of nonstationary data, some more considerations are provided in Section 6.12.1, Volume I. A detailed discussion of nonstationary processes is provided in Bendat and Piersol (2000). Random signals may be discrete or continuous. A discrete random variable takes on only distinct values. For example, there are only two possible outcomes when a coin is flipped. A continuous random variable may take any value within a continuum of values. Random processes may involve multiple random variables (see Section 6.2.1 in Volume I, Bendat & Piersol, 2000 for more details).

The signals can be classified based on their acquisition type and format; *i.e.*, analog, discrete time, or digital, as illustrated in Figure 4.5.5. Analog signals are continuous in time and magnitude (Figure 4.5.5a). For example, a conventional dial gage can be read at any time, and the indicator can point to any value within the range of the instrument, including values that are between gauge scale marks. A discrete time signal reveals the magnitude of the signal only at discrete points in time (Figure 4.5.5b). A discrete time signal results from sampling of a continuous variable at finite time intervals. The variable can still take on a full range of magnitudes, but is only recorded at discrete times. Since information is available only at discrete times, an assumption must be made about the behavior of the variable between samples. If the sampling rate is sufficiently high, the signal may be assumed to be constant over the period of time between samples. It is obvious from the visualizations in Figure 4.5.5 that if a waveform is sampled with reduced time between samples, the produced discrete signal better represents the original signal.

Digital signals are produced with the computer-based data acquisition and processing as described in the next section. A digital signal has two important characteristics:

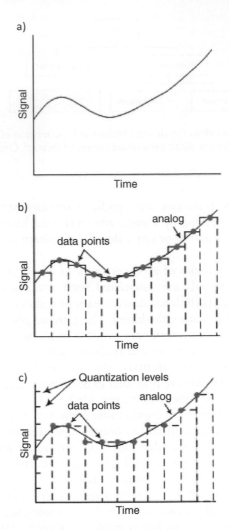

Figure 4.5.5 Signals from a data-acquisition system: a) analog; b) discrete time; c) digital (from Figliola & Beasley, *Theory and Design for Mechanical Measurements*. Copyright © 1991 by John Wiley Sons, Inc. Reprinted by permission of John Wiley & Sons, Inc.).

1. It is sampled at specific times, like discrete time signals; and,
2. Its magnitude is limited to a set of discrete values determined at each point in time by a process called quantization. Quantization assigns a single number to represent a range of magnitudes for the continuous signal (Figure 4.5.5c).

4.5.2 Signal digitization

Contemporary data-acquisition systems transform analog data into digital numbers for transmission and processing. Sampling an analog signal to produce a digital signal is done with an analog-to-digital (A/D) converter. Modern A/D converters are solid-state devices (semi-conductors) that convert an analog voltage signal to binary representation. The binary system is a system in which data are represented as a set of ones or zeros; such a system is necessary because digital circuits in contemporary electronics can only take on one of two states (either on or off). The smallest unit of information in the binary system is the bit. A bit is either 1 or 0 and may represent either logical (yes-no) or base-2 numeric information.

A word is a combination of bits that represents a number. The number of bits in the word determines the maximum size of the number that the word can represent. A word with M bits can represent any integer from 0 to 2^M-1. For example, a 2-bit word can represent four possible integers (*i.e.*, 00, 01, 10, and 11 represent 0, 1, 2, and 3, respectively), and a word with 32 bits can represent integers from 0 to 4,294,967,295. An M-bit A/D device converts an analog signal into a digital one, with each conversion resulting in a data point with a binary resolution of M bits. Typical binary resolutions of modern A/D converters are 8, 12, 16, 18, and 32 bits. Most A/D converters have voltage inputs. Typical voltage ranges for A/D converters are 0 to 10 V (unipolar) and ± 5 V (bipolar), but there are many available voltage ranges for modern A/D converters.

Figure 4.5.6 provides a schematic of the digitization process. This figure visualizes three primary sources of error intrinsic to any A/D converter operation: quantization, saturation, and conversion error. For the 2-bit converter shown in the figure, any input between 0 and 1 volt results in a digital output of 0. So the limited resolution of the converter can bring about an error in the analog-to-digital conversion. The difference between the analog input and the resulting digital value assigned by an A/D converter is called quantization error and is more significant when measuring small voltages or using a low resolution A/D converter. The quantization error for a 2-bit A/D converter with a full scale range of 0 to 4 V is $\pm 1/2$ of the binary output resolution $= \pm 1/2 \ (4 \ V/2^2)$ $= \pm 0.5$ V. Similarly, a 10 V 8-bit A/D converter has a quantization error of ± 0.020 V, and a 10 V 12-bit system has a quantization error of ± 0.0012 V.

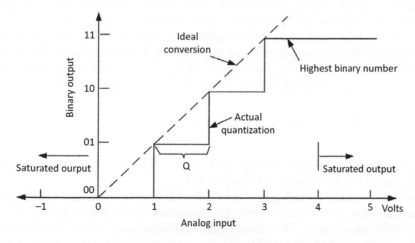

Figure 4.5.6 Digitization process illustrated by a 2-bit A/D converter with a full-scale input range of 0 to 4 V (from Figliola & Beasley, *Theory and Design for Mechanical Measurements*. Copyright © 1991 by John Wiley Sons, Inc. Reprinted by permission of John Wiley & Sons, Inc.).

An A/D converter also has a limited range of analog voltages that it can convert. If the input voltage is outside of the conversion range, the A/D converter output becomes saturated and no longer adjusts to changes in the input level. For the example in Figure 4.5.6, an input below 0 V results in a binary output of 00 and an input above 4 V results in an output of 11. The saturation error is defined as the difference between the input analog signal level and the equivalent digital value assigned by the converter. Saturation errors can be avoided by quantizing only signals that are within the limits of the A/D converter. This can be accomplished either by selecting a converter with a sufficiently wide sampling range or by conditioning the analog signal before quantization.

A/D converters are subject to additional conversion errors such as linearity, sensitivity, zero, precision (repeatability), and hysteresis (the previous measurement condition affects the current measurement). Typically, hysteresis is a result of inertia or resistance in the acquisition process produced by such factors as elastic deformation, magnetic, or thermal effects. Factors that contribute to conversion errors include converter settling time, signal noise during analog sampling, temperature effects, and device excitation power-setting effects. An inherent uncertainty associated with a conversion of signal from analog to digital is± 1/2 of the binary output resolution, hence use of a higher resolution A/D converter is a first step in enhancing the output data quality.

4.5.3 Representation of continuous and discrete data

An ideal (zero-order) measurement system has constant parameters and is linear over its range of measurement (Bendat & Piersol, 2000). This idealization is assumed if the response time of the measurement system is negligible and the system responds linearly over some limited range of inputs. The A/D converter of an ideal measurement system produces a continuous, deterministic signal input expressed by a continuous function f (t) sampled over a time interval. However, such a system does not exist and the results are samples of discrete data. Assuming the samples are uniformly spaced in time, N samples at time intervals of Δt provide a discrete time signal $f(r\Delta t)$ with $r = 0,1,2,...$, N-1 (i.e., $r\Delta t$ is the time step). Using the sampling convolution theory applied to a set of discrete measurements conceptualized as impulses equal to the value of $f(t)$ outputted by the DAS at each time step $r\Delta t$, a relationship between $f(t)$ and $f(r\Delta t)$ can be obtained. This transformation is described by the series (Figliola & Besley, 1991)

$$f(r\Delta t) = f(t) \sum_{r=0}^{N-1} \delta(t - r\Delta t) \qquad r = 0, 1, 2, ..., N \qquad (4.5.2)$$

where $\delta(t - r\Delta t)$ is the delayed unit impulse function. $N\Delta t = t_{N-1} - t_0$ is the duration of the sampling period for the continuous function $f(t)$.

In most experiments involving real-world flows, system inputs are random signals (whether stationary or nonstationary). The measurement system converts the inputs into discrete measured data. The subset of data collected over a finite time interval is a sample (or realization) of all possible outcomes of a particular experiment (a.k.a. the sample space or population). A collection of samples forms an ensemble. The points belonging to a sample space can be grouped into events of various kinds. For example, if the sample space is defined as the daily precipitation amounts over a city, an event could be the rainfall that occurred on a particular day. All possible events that might occur constitute a completely additive class of sets, and a probability measure may be assigned to each event (Bendat & Piersol, 2000). The assigned probabilities permit the investigator to make generalizations about the complete sample space (population) using information obtained from random samples as will be discussed in Section 6.2.1, Volume I.

Table 4.5.1 summarizes the statistical functions most commonly used as descriptive properties of random signals. In this table, the notation for Reynolds decomposition are: f = instantaneous variable, $\overline{f}$ = time averaged variable and f' = fluctuation. When applying these functions to measured data, the experimenter should be aware that the raw signals inherently include unwanted random noise produced by the MS components and protocols. Therefore, an inspection of the outcomes of the statistical analysis applied to the raw data is needed to verify if the obtained statistical results are those expected. If they are not, there may be indications that the data are affected by anomalies in the measurement system.

Table 4.5.1 specifies the role played by statistical functions for evaluating several aspects of the data acquisition and processing. Table 4.5.2 contains basic mathematical expressions for computing statistical properties of continuous and discrete signals and data. Relationships are developed for a generic variable denoted by f. Additional statistical functions and considerations for continuous and discrete datasets are described in Section 6.3.1, Volume I, and by Bendat and Piersol (2000) and McCuen (1985). The relevance of presenting the summary of statistical functions in this chapter is to substantiate the format of some of the outputs of the modern DAS whereby estimation of data quality is provided to users as a diagnostic with computation embedded in the firmware. More discussions on the assumptions and implementation details for application of the statistical functions to data processing are provided in Section 6.12, Volume I.

4.5.4 Selection of an optimum data-acquisition system

Data-acquisition systems have relied almost exclusively on analog components to condition, observe, and store data, but contemporary systems tend to be digital. The

Table 4.5.1 Basic descriptive properties for random signals

	Statistical functions	Main roles in signal and MS analysis
1	Mean statistical moments	Central tendency and dispersion of the random signal
2	Probability density functions	Evaluation of normality, detection of data acquisition errors, indication of nonlinear effects, analysis of extreme values
3	Autocorrelation functions	Detection of periodicities, detection of noise and interference, identification of spatial and temporal features in the signal
4	Spectral density functions	Determination of MS properties from input and output data, prediction of MS properties, identification of energy and noise sources, filtering specifications

Table 4.5.2 Selected statistical functions for characterizing signals and data

Function	Continuous signal	Discrete signal
Probability density function (pdf)	$p(f) = \frac{dP(f)}{df}$ where $P(f) = Prob[f(k) \leq f]$	$p(f) = \lim_{N\to\infty} \sum_{i=1}^{k} \frac{n_i}{N\Delta f}; \sum_{i=1}^{k} n_i = N$ N - total samples, k is the number of intervals spanning entire range of f; n_i = number of samples between f_i and $f_i + \Delta f$
Mean (Average)	$\bar{f} = \int_{-\infty}^{\infty} f p(f)\, df$ For ergodic random data $\bar{f} = \lim_{T\to\infty} \frac{1}{T} \int_{0}^{T} f(t)\, dt$ T: Sampling Time	$\bar{f} = \lim_{N\to\infty} \frac{1}{N} \sum_{i=1}^{N} f(i\Delta t) = \lim_{N\to\infty} \frac{1}{N} \sum_{i=1}^{N} f_i$
Mean square value	$\overline{f^2} = \int_{-\infty}^{\infty} f^2 p(f)\, df$	$\overline{f^2} = \frac{1}{N-1} \sum_{i=1}^{N} (f_i)^2$
Variance (2nd moment of the pdf)	$\overline{f'^2} = \int_{-\infty}^{\infty} \left(f - \bar{f}\right)^2 p(f)\, df$	$\overline{f'^2} = \frac{1}{N-1} \sum_{i=1}^{N} \left(f_i - \bar{f}\right)^2$
n^{th} order moment of pdf	$\overline{f'^n} = \int_{-\infty}^{\infty} \left(f - \bar{f}\right)^n p(f)\, df$	$\overline{f'^n} = \frac{1}{N} \sum_{i=1}^{N} \left(f_i - \bar{f}\right)^n$
Root-mean square (rms)	$f_{rms} = \sqrt{\overline{f^2}}$	$f_{rms} = \sqrt{\overline{f^2}}$

Table 4.5.2 (continued)

Function	Continuous signal	Discrete signal
Standard deviation	$\sigma_f = \sqrt{\overline{f'^2}}$	$\sigma_f = \sqrt{\overline{f'^2}}$
Autocovariance function	For ergodic random data $$C_{ff}(\tau) = \lim_{T\to\infty}\frac{1}{T}\int_0^T f'(t)f'(t+\tau)dt$$ where $f' = (f - \bar{f})$	$$C_{ff}(\tau = k\Delta t) = \frac{1}{N-k}\sum_{i=1}^{N} f'_i f'_{i+k}$$
Cross-covariance function	For ergodic random data $$C_{fg}(\tau) = \lim_{T\to\infty}\frac{1}{T}\int_0^T f'(t)g'(t+\tau)dt$$ where $f' = (f - \bar{f})$ and $g' = (g - \bar{g})$	$$C_{fg}(\tau = k\Delta t) = \frac{1}{N-k}\sum_{i=1}^{N} f'_i g'_{i+k}$$
Auto-correlation coefficient	$$R_f(\tau) = \frac{C_{ff}(\tau)}{\overline{f'^2}}$$	$$R_f(\tau = k\Delta t) = \frac{C_{ff}(\tau = k\Delta t)}{\overline{f'^2}}$$
Cross-correlation coefficient	$$R_{fg}(\tau) = \frac{C_{fg}(\tau)}{\sqrt{\overline{f'^2}}\sqrt{\overline{g'^2}}}$$	$$R_{fg}(\tau = k\Delta t) = \frac{C_{fg}(k)}{\sqrt{\overline{f'^2}}\sqrt{\overline{g'^2}}}$$
Fourier transform	$$F(w) = \int_0^T f(t)e^{-i2\pi wt}dt$$	$$F\left(w_k = \frac{k}{N\Delta t}\right) = \Delta t\sum_{n=0}^{N-1} f_n exp\left[\frac{-i2\pi kn}{N}\right]$$
One-sided autospectral density function	$$G_{ff}(w) = \frac{2}{T}|F(w)|^2$$	$$G_{ff}\left(w_k = \frac{k}{N\Delta t}\right) = \frac{2}{N\Delta t}|F(w_k)|^2$$

earlier systems included analog transducers such as thermistors and piezometers to produce the signals; analog outputs like gages, voltmeters, and oscilloscopes to observe the signals; and analog devices like magnetic tape or plotters to record the signals. Analog instruments were also used to perform mathematical operations on signals (summation, multiplication, division and integration, mean and root-mean-square values). Details about many analog components can be found in Dally *et al.* (1993) and Goldstein (1996). Although much of the conditioning, observation, and recording of measured signals continues to be done with analog instruments, the overwhelming trend is to directly transform raw signals into discrete-time series and to then digitally condition and record the discrete data. This trend will continue as the cost and capabilities of digital circuitry improve.

With the miniaturization and advancement of digital electronics in recent decades, analog instruments have increasingly been combined with digital circuitry adding new capabilities to both engineering analysis and process control. Contemporary measurement systems are often provided by manufacturers as "optimum" configurations, leaving few decisions to users in terms of modifying system components. Moreover, the components are also "optimized" to automatically execute hardwired operations with limited user input. However, the experimentalist must still be familiar with the operating principles of the components, the impacts of component settings, and the transfer of information between components in order to assess the significance and uncertainty of the measured data.

This section provides basic information about the operation of several common devices and procedures used for data acquisition. This information is useful for selecting the appropriate data-acquisition system, and specific instrument to meet specific experimental goals. Section 5.2 in Volume I describes complementary operations of which the experimenter has full control, such that the data-acquisition system can be adapted to specific flow conditions in the experiments.

Amplification: offset and gain adjustments. Amplification is usually done to analog signals (Dally *et al.*, 1993) to take advantage of the full range of a digital converter or an MS component with range limitations. Many sensors have low sensitivity and produce very low currents or voltages in response to changes in the physical property that they measure. Furthermore, sensor outputs may also have an offset. For example, a sensor may have an output range that spans only 1.0 V and is offset by 5 V. The output of such a sensor varies from 4.5 to 5.5 V over its measurement range. A/D converters, on the other hand, often require input voltages with standard ranges of 0 to 5 V or 0 to 10 V. An amplifier can be used as illustrated in Figure 4.5.7a to amplify, attenuate, or offset the instrument output to make it compatible with the input range of the A/D converter. Amplifiers are widely used in signal processing systems, and it is important to be aware of noise and potential filtering effects that they may introduce to the signal. For signal conditioning purposes, amplification is done primarily to analog signals, but a digital signal may also be amplified; to extend its transmission distance in a communication cable, for instance.

Filtering. The signals in a data-acquisition system are prone to be affected by noise and interference. Noise is generated by various internal (*e.g.*, thermal, optical, and shot noise) and external influences (*e.g.*, variations in the power supply). It is usually random and cannot be known precisely, and so is best analyzed statistically. Many common sources of noise (*e.g.*, thermal noise) have a constant power spectral density and a Gaussian distribution and are referred to as Gaussian white noise. In general, it is easier to statistically analyze white noise than other forms of noise. The relative significance of noise may be characterized by a signal-to-noise ratio (SNR) defined as the ratio of the average value of the signal and the standard deviation of the noise (Tavoularis, 2005). Noise is sometimes caused by interference from a nearby process or piece of equipment. For example, a large AC motor can produce electromagnetic noise in a measurement system. Interference produces undesirable deterministic effects on the measured value because of extraneous variables. One form is that of a sinusoidal wave superimposed onto a signal. Interference can also be associated with internal (e.g., thermoelectric voltages, ground loops) or external sources. Some forms of interference can be removed by improved electromagnetic shielding or elimination of interference sources. Some effects of interference can be eliminated by filtering. For example, if the line voltages have a carrier frequency of 60 Hz, a frequency that might be found in signals because of electromagnetic interference, a band-reject filter may be used to reduce the impact of the 60 Hz noise on measurement results.

There are a wide variety of analog filter designs (Figliola & Beasley, 1991; Dally et al., 1993). Filters significantly attenuate a frequency or range of frequencies (called stopbands) while the remaining frequencies are much less attenuated (called passbands) as illustrated in Figure 4.5.7b. Digital filters are applied to a time series following A/D conversion of the analog signals. For many applications, digital filters are superior to analog filters since they are more versatile and can better preserve the shape of the original waveform (Tavoularis, 2005). Moreover, digital filters can be applied in both the frequency domain and the time domain. Implementation of filtering for sample datasets is provided in Section 6.11, Volume I.

Sampling. It is obvious in Figure 4.5.7c that there is a difference in the information contained in analog and digital signals. There are two key concerns when converting signals from analog to digital: achieving adequate conversion accuracy (e.g., selecting a digitizer with sufficient resolution);and, sampling in such a way that the digital data adequately represent important statistical characteristics of the sampled signal. Sampling protocol addresses the following issues: how to sample, how fast to sample and how many samples should be collected. Because of the wide range of scales in turbulent flows, sampling rates and periods can be quite intensive. For example, Goldstein (1996) explains that for a turbulent boundary layer in an air flow with a Reynolds number Re = 10^6 and a free stream velocity of 15 m/s, 50-60 thousand measurements per second may be required for 30 to 60 seconds to adequately represent a particular flow parameter; this amounts to as many as 4 million measurements per sample. Requirements like these can strain the limits of some measurement systems, so it is highly useful to determine minimum sampling requirements.

Fingerson et al. (1993) distinguish between three sampling methods:

1. Sampling at uniformly spaced time intervals;
2. Sampling at non-uniformly spaced time intervals; and,
3. Random sampling.

The first two methods can be applied when using analog instruments with continuous outputs, while the latter must be employed for instruments that collect data at random times (e.g., for LDV there is no control of the times when particles arrive at the measurement volume of the instrument), as is illustrated in Figure 4.5.7d. A common method adopted for uniformly spaced sampling is the "sample and hold" method whereby the value sampled at any given instant is retained until another measurement is taken. This method is necessary when using a multi-channel system simultaneously sampling multiple instruments. For instruments acquiring data with random sampling, post-processing corrections are sometimes necessary to reconstruct the original signals (Fingerson *et al.*, 1993).

The maximum rate at which a data-acquisition system can sample must be high enough to meet the sampling rate requirements of the experiment. To illustrate the importance of sampling rate, consider Figure 4.5.7e, where variation of a 10 Hz sine wave is plotted against time (Figliola & Beasley, 1991). Suppose this sine wave is uniformly sampled with time increments of Δt, corresponding to a sampling rate of $1/\Delta t$. As shown in the figure for sampling rates of 100Hz, 27 Hz, and 12 Hz, selection of sampling rate has a significant effect on the ability to accurately reconstruct the original 10 Hz signal. Specifically, as sampling rate decreases, the amount of information collected per unit time decreases, until the frequency of the reconstructed signal is less than that of the original (clearly shown by the last plot of Figure 4.5.7e).

The sampling theorem known as the Nyquist criterion on sampling rate ensures that the frequency of the original signal can be reconstructed from the sampled signal; it requires that the original signal be sampled at a rate that is more than twice the highest frequency contained in the signal. Denoting the maximum frequency in the measured signal as f_m, the Nyquist criterion requires $f_s > 2f_m$, in which f_s is the sampling frequency. When the Nyquist criterion is not satisfied, it is impossible to reconstruct the analog signal using the acquired time series.

Furthermore, if the analog signal is not sampled at a high enough rate, high-frequency components of the signal are misconstrued as lower-frequency components. The misinterpretation of the analog signal caused by these false frequencies is called "aliasing." An illustration of the effects of aliasing on an unfiltered signal is shown in Figure 4.5.8a in which the Fourier transform used to represent the sampled signal folds the frequencies higher than $f_s/2$ back into the power spectrum (Goldstein, 1996); the higher frequencies appear as lower frequencies in the power spectrum. Aliasing is an inherent consequence of the discrete sampling process and it cannot be removed by digital filtering or other digital data processing. To prevent aliasing, the signal must either be sampled at a rate that is higher than the maximum frequency of the signal or low-pass analog filtering must be applied before discretization.

When sampling a signal of unknown frequency content, the potential of aliasing is always a concern (Figliola & Beasley, 1991). If only a limited frequency range is of interest for a particular flow, the analog signal must be low-pass filtered to eliminate frequencies above the maximum frequency of interest prior to sampling; this prevents aliasing associated with the higher frequencies. If all frequencies are of interest and the maximum frequency of the signal is unknown, the maximum measureable frequency is limited to half the sampling rate by the Nyquist criterion. For all cases in which the signal may contain frequencies above $f_s/2$, the signal must be low-pass filtered with a cutoff frequency at or below $f_s/2$ to avoid aliasing. More details on the filtering schemes and their effect, including the extensively used Butterworth filter, are provided in Section 6.11.2, Volume I.

Sufficiently high sampling rates ensure that the frequency signature of a signal is captured. However, in order to guarantee that the analog signal is fully represented by the digital signal, it is also necessary to capture the amplitude signature of the signal.

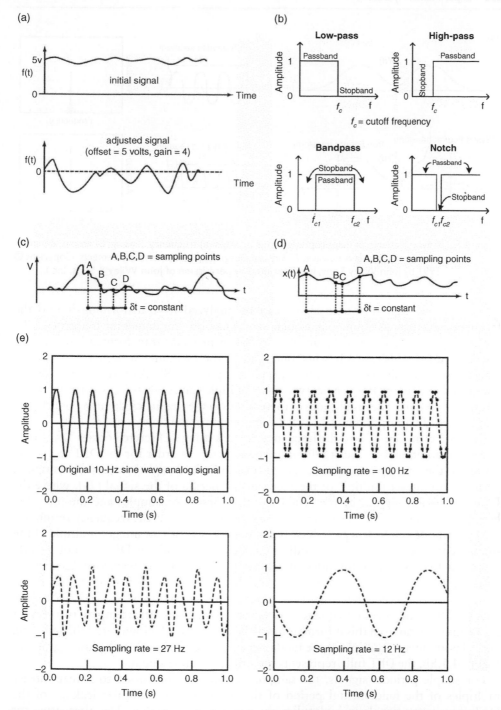

Figure 4.5.7 Selected signal conditioning operations: a) offset-gain-amplification; b) filtering (Figliola & Beasley, 1991); c) "sample and hold" sampling (Fingerson *et al.*, 1993); d) random sampling (Fingerson *et al.*, 1993); e) effect of sampling rate (from Figliola & Beasley, *Theory and Design for Mechanical Measurements*. Copyright © 1991 by John Wiley Sons, Inc. Reprinted by permission of John Wiley & Sons, Inc.).

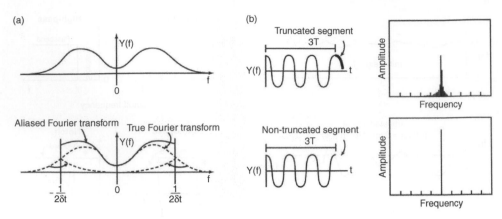

Figure 4.5.8 Adverse effects of inappropriate sampling rates: a) frequency aliasing; b) amplitude ambiguity (from Figliola & Beasley, *Theory and Design for Mechanical Measurements*. Copyright © 1991 by John Wiley Sons, Inc. Reprinted by permission of John Wiley & Sons, Inc.).

The amplitude signature of a signal can be analyzed by transforming it into the frequency domain. A simple sinusoidal signal has one fundamental frequency- like the analog signal shown in Figure 4.5.7e. When properly transformed into the frequency domain, this signal has just one spike at the fundamental frequency – such a transformation is shown by the lower graph of Figure 4.5.8b. A more complex signal, composed of multiple frequencies, will have multiple spikes when transformed into the frequency domain. The amplitudes of these spikes in the frequency domain relate to the amplitudes of corresponding sine wave components in the time domain.

Data are transformed into the frequency domain using a Discrete Fourier Transform (DFT). In addition to Nyquist criterion requirements, in order for the DFT to exactly represent the original analog signal, the signal must be sampled so that the sampling period is an integer multiple of the fundamental period of the signal (T_1), where the sampling period is given by the product of the number of samples and the time step between samples $(N{\cdot}\Delta t)$. The inverse of the sampling period is the frequency resolution of the DFT: $\Delta f = 1/N{\cdot}\Delta t = f_s$. If $N{\cdot}\Delta t$ is not coincident with an integer multiple of the fundamental period T_1 of a periodic signal $y(t)$, the resulting DFT cannot exactly represent the spectral amplitudes of the sampled continuous waveform as demonstrated in the upper graph in Figure 4.5.8b. This figure shows that when the sampling period is a non-integer multiple of the fundamental period, there is "leakage" of the energy of the fundamental frequency to adjacent frequencies. An attempt to reconstruct the original signal from this information will be unsuccessful. However, if the sampling period is an integer multiple of the fundamental period as shown in the lower graph of Figure 4.5.8b, the DFT fully represents the original analog signal.

For simple periodic signals, the sampling period can be varied to realize integer multiples of the fundamental period of the signal and to minimize leakage of the DFT, but many signals in hydraulics are much more complex. The alternative for these complex signals, which may be aperiodic or nondeterministic, is to choose long sampling periods which results in high frequency resolution (Δf becomes very small). Leakage is inversely proportional to frequency resolution, and so selecting longer sampling periods reduces leakage and improves the accuracy of spectral amplitudes. Selecting a sampling period of sufficient length can be accomplished by trial and error, increasing the sampling period until the frequency domain representation becomes independent of sample size. This, of course, requires that the sample set be stationary.

In summary, during the selection of a data-acquisition system careful attention should be given to the following considerations:

1. Applying gain-offset-amplification procedures to utilize the full input range of the converter by accounting for the variation range of the measured signal and the resolution of the A/D converter;
2. Documenting (*e.g.*, calibrations, instrument specifications) system and sensor characteristics as well as quantization, saturation, and conversion (hysteresis, gain, linearity, and drift) errors; and,
3. Identifying and estimating sources of noise and interference from internal and external sources and filtering the analog signal to eliminate unwanted frequencies in the signals, and making sure that the important content of the dynamic analog signal is accurately captured by the digitized experimental data by appropriately sampling the signals.

The information presented in Sections 4.2, 4.4 and 4.5 in Volume I show that the configuration and settings of a data-acquisition system can considerably affect the perceived outcomes of experiments. Therefore it is important to pair the experimental needs with the specifications of the selected data-acquisition system. The main differences between a targeted (ideal) system and the typical actual system are listed

Table 4.5.3 Departure of actual data-acquisition systems from ideal systems

Ideal System	Actual System	Procedures/Countermeasures
Output signal is proportional to measured quantity.	Output signal follows a static or dynamic transfer function.	Conduct calibration and verify interference with extraneous variables.
Measurements are conducted at precisely defined points, along a line, or over a surface.	The spatial resolution of the sensors is finite. Measuring volumes can be affected by velocity gradients and other interference.	Estimate the scales of the flow and compare with the instrument spatial and temporal resolution. Select an instrument with satisfactory resolution or correct measurements appropriately.
Output signal has no frequency distortion.	MS specifications and sampling protocol alter the frequency of the perceived input.	Check system type and follow sampling protocols (Nyquist criterion, alias frequency, amplitude ambiguity).
No noise in output signal.	Unavoidable noise produced by components and interference.	Determine signal-to-noise ratio and if unacceptable remove sources of interference or apply filters.
No interactions between the probe and the flow or the facility.	Large dimensions of sensors and proximity to the boundaries affect measurements.	Apply corrections (calibrations) and follow instrument specific guidelines.
Output signal is directly related to measured variable.	Output signal may be affected by external variables (*e.g.*, room temperature and pressure).	Apply filtering, compensation, or analytical correction for external effects.

in Table 4.5.3. Irrespective of the devices comprising a data-acquisition system, the experimenter should be well familiarized with the flux of information within the system so that collected data can be objectively interpreted and, if needed, corrected with post-processing procedures.

REFERENCES

Abad, J. D., & Garcia, M.H. (2009) Experiments in a high-amplitude Kinoshita meandering channel: 1. Implications of bend orientation on mean and turbulent flow structure. [Online] Water Resour Res., [Accessed 4th March 2017] 45, W02401. Available from: doi:10.1029/2008WR007016.

ASCE (2000) *Hydraulic Modeling: Concepts and Practice*, ASCE Manuals and Reports on Engineering Practice No.97, American Society of Civil Engineers, R. Ettema (ed.), Reston, VA, ASCE Publications.

Bathurst, J.C., Cao, H.H., & Graf, W.H. (1984) *The Data from the EPFL Study on Hydraulics and Sediment Transport in a Steep Flume*. Lausanne, Switzerland, Ecole Polytechnique Federale de Lausanne.

Beckwith, T.G., & Marangoni, R.D. (1990) *Mechanical Measurements, 4th Edition*. Reading, Addison-Wesley.

Bendat, J.S., & Piersol, A.G. (2000) *Random Data: Analysis and Measurement Procedures, 3rd Edition*. New York, John Wiley & Sons.

Coleman, H., & Steele, W.G., Jr. (1989) *Experimentation and Uncertainty Analysis for Engineers*. New York, John Wiley & Sons.

Crosato, A., Desta, F.B., Cornelisse, J., Schuurman, F., & Uijttewaal, W.S.J. (2012) Experimental and numerical findings on the long-term evolution of migrating alternate bars in alluvial channels. [Online] *Water Resour Res.*, [Accessed 4th March 2017] 48, W06524. Available from: doi:10.1029/

Dally, J.W., Riley, W.F., & MCConnell, K.G. (1993) *Instrumentation for Engineering Measurements, 2nd Edition*. Hoboken, NJ, John Wiley & Sons.

Diplas, P. (1988) Characteristics of self-formed straight channels. *J Hydraul Eng.*, 116(5), 707–728.

Doebelin, E. (1995) *Engineering Experimentation: Planning, Execution, Reporting*. New York, McGraw Hill.

Dunn, P.F. (2014) *Measurement and Analysis for Engineering and Science, 3rd Edition*. New York, CRC Press, Taylor & Francis Group.

Durst, F., Melling, A., & Whitelaw, J.H. (1976) *Principles and Practice of Laser-Doppler Anemometry*. New York, Academic Press.

Figliola, R.S., & Beasley, E.B. (1991) *Theory and Design for Mechanical Measurements. John Wiley & Sons.* 185–258.

Fingerson, L.M., Adrian, R.J., Menon, R.K., Kaufman, S.L., & Naqwi, A.A. (1993) *Data analysis, Laser Doppler Velocimetry and particle image velocimetry*. TSI Short Course Text.

Frostick, L.E., McLelland, S.J., & Mercer, T.G. Editors (2011) *Users Guide to Physical Modelling and Experimentation*. New York, CRC Press, Taylor & Francis Group.

Goldstein, R.J. (1996) *Fluid Mechanics Measurements, 2nd Edition*. Bristol, PA, Taylor & Francis.

Holman, J.P. (1989) *Experimental Methods for Engineers*. New York, McGraw Hill.

Hopfinger, E.J., Browand, F.K., & Gagne, Y. (1982) Turbulence and waves in a rotating tank. [Online] *J Fluid Mech.*, [Accessed 4th March 2017] 125, 505–534. Available from: doi:10.1017/S0022112082003462.

Hughes, S.A. (1993) Physical models and laboratory techniques in coastal engineering. *Adv Ser Ocean Eng., (7)*. Singapore, World Scientific.

Liberzon, D., & Shemer, L. (2011) Experimental study of the initial stages of wind waves' spatial evolution. [Online] *J Fluid Mech.*, [Accessed 4th March 2017] 681, 462–498. Available from: doi:10.1017/jfm.2011.208

Maggi, F., Winterwerp, J.C., Cornelisse, J.M., Fontijn, H.L., & van Kesteren, W.G.M. (2002) A settling column to study turbulence-induced flocculation of cohesive sediment. In: Wahl, T.L., Pugh, C.A., Oberg, K.A., & Vermeyen, T.B. (eds.), *ASCE Proceedings of the Conference on Hydraulic Measurements & Experimental Methods*. Estes Park, Colorado, USA. Paper 93.

McCuen, R.H. (1985) *Statistical Methods for Engineers*. Englewood Cliffs, New Jersey, Prentice-Hall.

Muste, M., Braileanu, F., & Ettema, R. (2000) Flow and sediment transport measurements in a simulated ice-covered channel. *Water Resour Res.*, 36(9), 2711–2720.

Muste, M., Haug, P., Troy, L., Weber, L., & Hay, D. (2006) Model Study for the Design of Emergency Stoplogs Deployed in a Complex Flow Field. *Part 1: Experimental Results: Proceedings ASCE World Environmental & Water Resources Congress*. Omaha, NE.

Nezu, I., & Nakagawa, H. (1993) *Turbulence in Open-Channel Flows*. Balkema, A.A., Rotterdam, Netherlands.

Pope, B.P. (2000) *Turbulent Flows*. Cambridge, UK, Cambridge University Press.

Ribberink, J.S., & Al-Salem, A.A. (1994) Sediment transport in oscillatory boundary layers in cases of rippled beds and sheet flow. [Online] *J Geophys Res.*, [Accessed 4th March 2017] 99 (C6), 12707–12727. Available from: doi:10.1029/94JC00380.

Rouse, H. (1938) *Fluid Mechanics for Hydraulic Engineers*. New York, Dover Publications.

Sous, D., Sommeria, J., & Boyer, D. (2013) Friction law and turbulent properties in a laboratory Ekman boundary layer. *Phys Fluids*, 25(4), 046602-1–046602-18.

Stern, F., Muste, M., Beninati, M.L., & Eichinger, W.E. (1999) *Summary of Experimental Uncertainty Assessment Methodology with Example*. Iowa Institute of Hydraulic Research, The University of Iowa, Iowa City, IA. Report number: 406.

Tavoularis, S. (2005) *Measurement in Fluid Mechanics*. New York, Cambridge University Press.

Tennekes, N., & Lumley, J.L. (1972) *A First Course in Turbulence*. Cambridge, MA, The MIT press.

Tropea, C., Yarin, A.L., & Foss, J.F. (eds.) (2007). *Handbook of Experimental Fluid Mechanics* Germany, Springer-Verlag Berlin Heidelberg.

USBR (1981). *Hydraulic Laboratory Techniques*. U.S. Dept. of Interior, U.S. Bureau of Reclamation, Denver, CO.

Weitbrecht, J., Seoul, D-G., Negretti, E., Detert, M., Kuhn, G., & Jirka, G.H. (2011). PIV measurements in environmental flows: Recent experiences at the Institute for Hydromechanics in Karlsruhe, *J Hydro-environ. Res.*, 5(4), 231–245.,

White, F. (2003) *Fluid Mechanics, 5th Edition*. New York, McGraw Hill.

Yalin, M.S. (1971) *Theory of hydraulic models*. London, Macmillan Press.

Experiment Execution

5.1 INSTRUMENT-FLOW AND FACILITY-FLOW INTERACTIONS

5.1.1 Introduction

All commonly used measurement techniques are indirect in that they rely on physical interpretation of a measured surrogate quantity. For example, when pressure is measured with a membrane-based transducer, the deflection of the membrane is sensed and the resulting output is interpreted as being due to a change in pressure. Interactions between the flow, the facility, and the measurement instrument affect the physical relation between the quantity that is being measured and the output of the instrument. Because such interactions often govern instrument selection, a thorough understanding of relevant interactions is necessary if instrument outputs are to be correctly interpreted.

For a proper analysis of results, old and new measurement technologies alike require that users understand how flow and measurement instrument interact to produce an output signal – especially in turbulent flows and in flows with strong spatial or temporal velocity gradients. Table 5.1.1 lists commonly used instruments for the measurement of velocity, pressure, and concentration and their corresponding concerns.

5.1.2 Types of Interactions

To examine instrument-flow and instrument-facility interaction, velocity measurements may be considered for illustration purposes. Most commonly used velocity measurement techniques fall into one of two categories:

1. Intrusive instruments that directly measure a flow property that can be related to flow velocity; and,
2. Non-intrusive instruments that track the motion of particles or tracers suspended in a flow.

Instruments that fall within the first category include traditionally used devices like Prandtl tubes, propeller velocity meters, and thermal anemometers. These instruments rely on a physical relationship between velocity and pressure or heat transfer. They are usually inserted at a location in the flow, directly affecting pressure and velocity distributions at that location. For most devices of this type, the physical relation between the scalar and the velocity is nonlinear, making effects of turbulence and velocity gradients on interpretation of results more complex.

Table 5.1.1 Flow interaction concerns related to common instruments

Instrument	Measurement	Flow Interaction Considerations
Prandtl Tube	Velocity	Measurements in spatial gradients are biased high. Walls influence velocity measurements. Turbulence affects sensor output.
Laser-Doppler Anemometer	Velocity	Measurements in spatial gradients and in unsteady flows are biased. Turbulence moments are biased.
Particle Image Velocimetry	Velocity	Measurements in velocity gradients are biased.
Acoustic-Doppler Velocimeter	Velocity	Strong turbulence impacts signal correlation. Measurements in strong velocity gradients have poor signal correlation. The presence of the probe in the flow affects the adjacent velocity field.
Hot-film Anemometer	Velocity	Relies on non-linear physical relation between convective heat transfer and flow velocity. Particulates deposited on the wires affect accuracy.
Static-Pressure Taps	Pressure	Geometry of pressure tap influences pressure measurement error (Tavoularis, 2005).
Suction Sampler	Concentration	Sampling velocity influences observed concentration. Non-linear concentration gradients bias measurements.
Acoustic Sediment Profiler	Concentration	Concentration gradients bias measurements. Turbulence levels influence signal calibration.

Contemporary laser-, acoustic-, and image-based instruments are less intrusive, relying on optical or acoustic detection of small particles suspended in the flow. Techniques such as Laser-Doppler Velocimetry (LDV), Particle Image Velocimetry (PIV), Acoustic-Doppler Velocimetry (ADV), and Acoustic-Doppler Current Profilers (ADCP) indirectly measure flow velocity by measuring velocities of suspended tracer particles existing in or added into the flow. These techniques are partly or completely non-intrusive and do not greatly affect the flow, though the presence of tracer particles and immersion of the transducer may have a small effect. They are based on the assumption that measured tracer particle velocities accurately reflect the flow velocity. Obviously, particle mass influences particle inertia, and particles with a small diameter and a density that matches the density of the fluid carrying them follow the velocity and acceleration of the flow much more faithfully. Sediment samplers are subject to similar constraints. Even when tracer particles follow the flow exactly, particle-based velocity and concentration measurement techniques are subject to biases that result from the finite size of the measurement volume and spatial and temporal gradients in the flow.

In the following sections, some of these biases are introduced, though not exhaustively discussed for all instruments and the errors associated with them. Rather, the examples are intended to alert the experimenter to the importance of completely understanding the relation between a measured parameter and the physical quantity that it represents. In many cases, the bias may be viewed as arising from sampling that is not random, and/or not representative of the intended population (see the general discussion of population and sample in Chapter 6, Volume I).

The sensors of all intrusive instruments and some non-intrusive instruments (*e.g.*, ADVs and ADCPs) must be immersed in the flow to collect data. Sensors immersed in the flow disturb the flow and are a potential source of uncertainty. The extent to which the flow is disturbed depends on sensor design and geometry and on the characteristics of the flow field. Consideration of flow disturbance by non-intrusive instruments is particularly important when the measurement volume is in close proximity to the sensor. For example, a typical ADCP probe is a cylinder with a diameter of about 25 cm and requires an immersion depth of about 5 cm. Mueller *et al.* (2007) and Muste *et al.* (2010) found that the ADCP can significantly disturb the flow in the vicinity of the sensors. The extent of the disturbed region varies with the flow velocity, with impacts of flow disturbance being worse for high-velocity and shallow-water measurements (Muste *et al.*, 2010). Mueller *et al.* (2007) estimated velocity biases of more than 25% at a distance of 5 cm from the transducer head for some scenarios. This type of information can help to assess instrument suitability or at least an instrument's region of applicability for any particular application.

Issues associated with instrument interaction with the experiment facility are common for both intrusive and non-intrusive instruments. Model or prototype boundaries produce strong gradients of velocity and other variables, and can alter the expected operation of an instrument (see Section 5.1.4 in Volume I, for example). Non-intrusive instruments, such as laser- and acoustic-based instruments, are prone to the effect of backscatter from facility boundaries. For example, velocities in the millimeter-thin boundary-layer region adjacent to flume walls are difficult to measure with PIV because of reflections and glare caused by the walls (Adrian, 1991). Similarly, for some regions of an ADCP velocity profile, parasitic side-lobes of acoustic pulses generated by the ADCP impinge on the bed and banks of the channel, produce reflections that interfere with valid velocity measurements, and result in randomly biased velocities that must be discarded (Simpson, 2001).

5.1.3 Measurements in flows with velocity gradients

A common experiment task involves measuring flow velocity in a location subject to a substantial velocity gradient. An instrument with a measurement volume of finite size does not necessarily yield an average over the measurement volume. For example, in PIV, displacements of groups of suspended particles are measured to determine flow velocities. To do this, PIV images are subdivided into small sub-windows, which are the measurement volumes. During the time between two consecutive images, particles within each sub-window are subject to small displacements, depending on the instantaneous velocity of each particle. An averaged displacement is determined for all particles within each sub-window by measuring a correlation between their positions in the first and second images, and this is then divided by the time between images to compute an averaged velocity that will be assigned to that sub-window. If a sub-window is located in a position with a strong spatial gradient, the particles within the sub-window may be moving at significantly different velocities. In addition to a spatial averaging effect, faster moving particles enter and leave the sub-window at a faster rate, such that correlations between these particles will not be as strong as correlations between slower moving particles. Consequently, PIV velocity measurements are biased low when steps are not taken to mitigate the effect of the velocity gradient.

LDV measurements face a similar problem though LDV measures the velocities of individual or groups of particles, rather than a correlation between groups of particles. If it is assumed that the distribution of suspended tracer particles in a flow is uniformly distributed, and that the LDV measurement volume is located within a region with a large velocity gradient, then the flux of fluid will be higher across the part of the measurement volume with the higher velocities. Consequently, the flux of particles that crosses the higher-velocity part of the measurement volume will also be higher. This will result in more velocity measurements in the high-velocity region than in the low-velocity region, and if individual velocity measurements are averaged to determine an average velocity for the measurement volume, the result will be biased high. There are averaging techniques that have been developed to compensate for this bias (Tropea *et al.*, 2007), but this is another example of a bias that results from misinterpretation of the physical relation between a flow parameter and the associated measurement.

Measurements of point suspended sediment concentration also face similar issues. Traditionally, suction tubes have been used to measure the suspended sediment concentration profile. These tubes must isokinetically sample the sediment concentration at the measurement point, *i.e.*, the velocity at the inlet of the suction sampler must be the same as the velocity of the flow in which it is sampling. Otherwise, the suspended sediment measurement will be biased because the sediment, which is usually denser than the fluid, will not follow the same streamlines as the fluid. Then, the measured sediment concentration will be higher or lower than the actual sediment concentration, depending on the difference between the sampling velocity and the flow velocity. In some cases, there is no practical way to isokinetically sample the flow - in oscillatory flows, for example. In such cases, corrections must be made to suspended sediment samples to account for the differences in inertia between the fluid and solid particles. Bosman *et al.* (1987) provide guidance for making such corrections.

Contemporary sediment measurement techniques like ultrasonic samplers must also be carefully used to avoid spatial biases. In strong spatial gradients, measured sediment concentration may be biased high because of the nonlinear relation between acoustic pressure and sediment concentration. In addition, it is best to calibrate these transducers in situ, because even when average concentrations remain the same, changes in turbulence levels and water properties affect the output of the transducer.

5.1.4 Influence of turbulence levels

Turbulence levels may alter values of velocity, pressure, and other mean flow characteristics measured with certain instruments. When performing such measurements, in addition to selecting an instrument that has adequate frequency response and a sufficiently small measurement volume, it is important to identify how the output of the instrument might be biased by flow turbulence. Averaged parameters, whether the averages are done by the instrument or during post-processing, can be particularly problematic. For example, when LDV is applied in a turbulent flow, homogeneously distributed tracer particles arrive more frequently during high velocity fluctuations than during low velocity fluctuations. A simple (uniformly weighted) arithmetic average of the sampled velocities will produce an average velocity that is biased high because the samples are not equally distributed in time (Adrian & Yao, 1987).

Care is also required to properly collect turbulence measurements with acoustic devices. An ADV measures the Doppler velocity of groups of tracer particles moving through its measurement volume. Good instrument performance requires a sufficiently large number of particles within the measurement volume to provide a strong reflection of the acoustic pulse transmitted by the ADV. The ADV determines how well the reflected pulse correlates with the transmitted pulse to estimate the average velocity within the volume over the integration period. If the measurement volume is in a strong shear flow or if turbulence levels are high, the correlation will be poor because not all of the particles within the volume are moving at the same velocity over the integration period. In such cases, it is likely that there is bias in the resulting velocity measurement. Moreover, due to particle scattering characteristics, all acoustic measurements contain noise. ADVs compensate for this noise by collecting data at high frequencies (*e.g.*, 250 Hz), which are then averaged and reported at a much lower frequency (*e.g.*, 25 Hz). Since the noise is random, spatial and temporal averaging of the velocities reduces the effect of noise in the mean characteristics, but quantities based on higher-order moments, such as turbulent kinetic energy (TKE) may be biased high because velocities are squared prior to averaging so that noise spikes of opposite sign do not cancel (Garcia *et al.*, 2005). If the energy level of the noise is known or can be estimated, corrections can be made to the TKE. Similar correction methods are necessary if ADCPs are used to provide estimates of higher-order turbulence properties (Stacey *et al.*, 1999).

Example 5.1.1: Measurements with Pitot tubes.   A Prandtl (or Pitot-static) tube is used to indirectly measure velocity by converting velocity head into stagnation pressure (Figure 5.1.1a). The Prandtl tube depicted in this figure is a blunt-nosed cylindrical device with a vertical support tube. The device is directed into the flow and is aligned with a streamline, and so is suitable only for unidirectional flow with known flow direction. The local velocity and static pressure of the undisturbed flow at this location are respectively denoted as u and p_0. The pressure sensed at the nose of the Prandtl tube (at location *1*) is the total or stagnation pressure, p_t. The static pressure is sensed with static pressure ports usually located on the circumference of the Prandtl tube at a section downstream of the nose (at location *2*).

Accurate conversion of the measured pressure difference into velocity is subject to the influences of velocity gradients, measurement location and turbulence effects. The measurement of velocity in a uniform, laminar flow with a Prandtl tube provides an example of how measurements even in simple flows require careful analysis of inter-action between the flow and the measurement instrument. The static pressure, p_0, is

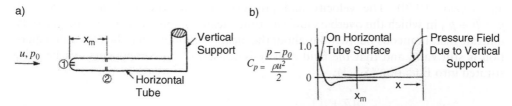

Figure 5.1.1 Prandtl tube (Goldstein, 1996): a) probe configuration; b) optimal location for static pressure ports operated in a uniform, laminar flow (copyright 1996 From *Fluid mechanics measurements* by Goldstein. Reproduced by permission of Taylor and Francis Group, LLC, a division of Informa plc).

related to the stagnation pressure, p_t, through the Bernoulli Equation, provided viscous effects are negligible,

$$p_t = p_0 + \frac{\rho u^2}{2} \qquad (5.1.1)$$

in which ρ is the fluid density. Equation (5.1.1) can be rearranged to give u if the difference between p_0 and p_t is measured:

$$u = \sqrt{\frac{2(p_t - p_0)}{\rho}}. \qquad (5.1.2)$$

The pressure observed along the side of the Prandtl tube is influenced by interaction between the probe and the flow. Probe-flow effects are depicted in terms of the dimensionless pressure coefficient, C_p, in Figure 5.1.1b, which shows how the nose and support of the Prandtl tube affect the pressure (C_p) along the tube. C_p has a value of 1 when the observed pressure (p) equals p_t, and a value of 0 when $p = p_0$. Pressures near the nose of the Prandtl tube are elevated since the solid tube acts as a bluff body. Farther downstream, acceleration around the nose of the tube causes the pressure to drop below p_0. The pressure rises asymptotically to p_0 as distance (x) downstream of the nose increases. The support for the device also acts as a bluff body, and pressures upstream of the support are elevated. The effects of the support decrease with distance upstream of the support. The optimal distance, x_m, to place the static pressure port depends on both interaction of the flow with the nose and interaction of the flow with the tube support. At the location where these two effects cancel each other, the measured pressure will be closest to the static pressure of the undisturbed flow, and is where the static pressure ports should be placed.

Interactions between the flow, facility and the instrument have three effects on the stagnation pressure measurement that should be considered:

1. The stagnation pressure sampled over time is affected by the turbulence level of the local flow,
2. Velocity measured with the probe is biased high in strong velocity gradients, and
3. Velocity bias near the boundary facility where the strong velocity gradients cause asymmetrical flow around the tube.

The effect of turbulence on the velocity measurement can be explored by decomposing velocities and pressures into time mean and fluctuating components, as shown by Arndt and Ippen (1970). The velocity and pressure are replaced with $u = \bar{u} + u'$ and $p = \bar{p} + p'$, in which the overbars indicate time mean components and prime symbols indicate instantaneous fluctuations about the mean. Assuming that the Bernoulli equation is still valid and that the fluid density is constant, the components can be substituted into Equation (5.1.1):

$$\overline{p_t} + p_t' = \overline{p_0} + p_0' + \frac{\rho(\bar{u} + u')^2}{2} \qquad (5.1.3)$$

Expanding the velocity term and then time-averaging the equation yields:

$$\overline{u}\left(1 + \frac{\overline{u'^2}}{\overline{u}^2}\right)^{1/2} = \sqrt{\frac{2(\overline{p_t} - \overline{p_0})}{\rho}}$$ (5.1.4)

Thus, if difference between the time-mean stagnation and static pressures is used to calculate the time mean velocity, the mean velocity will be over-predicted because of the non-linear contribution from velocity fluctuations about the mean. However, turbulence effects may be negligible if $\sqrt{\overline{u'^2}/\overline{u}^2}$ is less than about 10% (Tavoularis & Szymczak, 1989).

Prandtl-tube measurements in a velocity gradient (as shown in Figure 5.1.2a)produce measured velocities that are biased high. The bias has two causes:

1.   Streamlines are deflected toward the lower velocity part of the velocity profile by the presence of the probe (downwards in Figure 5.1.2.a), and,
2.   Because of the nonlinear relation between velocity and pressure, integration of the stagnation pressure across the Pitot opening necessarily results in an average pressure that is higher than the centerline pressure.

The second effect has a small influence on the velocity measurement (Chue, 1975). Traditionally the velocity bias has been treated as a shift in the effective centerline of the probe as shown in Figure 5.1.2a. The centerline shift has been estimated to be linearly related to the probe diameter by multiple studies (MacMillan, 1956; Tavoularis & Szymczak, 1989) and is ~ 0.15D, in which D is the tube diameter.

When a Prandtl tube is near a boundary, the velocity measurement may be affected by the presence of the boundary. This effect is demonstrated in Figure 5.1.2b, which shows that when the Prandtl tube is within two diameters of a boundary, the velocity measurement may be biased by as much as 2%. Moreover, without considering this and other influences that the flow and the probe have on each other, seemingly reliable Prandtl tube measurements could be incorrect by several percent. For precision measurements, it is clearly beneficial to calibrate these tubes to avoid biases caused by interaction between the flow, the instrument, and proximate boundaries. This example

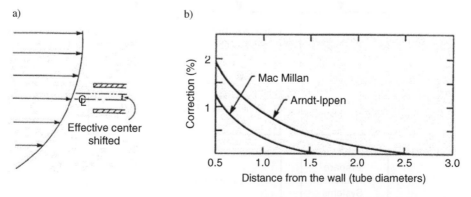

Figure 5.1.2 Various effects of flow interactions in measurements with Pitot tubes. a) shift of the measured velocity location due to velocity gradient effect; b) corrections of the measurement velocities near boundaries (copyright 1996 From *Fluid mechanics measurements* by Goldstein. Reproduced by permission of Taylor and Francis Group, LLC, a division of Informa plc).

also shows the importance of carefully considering instrument-flow interaction for all instruments measuring near boundaries.

5.2 CONDUCTING THE EXPERIMENT

5.2.1 Activities

Once the planning and design of an experiment is complete, the instrumentation selected and calibrated, and the experiment apparatus is built, experimentation can begin. Preliminary experiments are recommended as a means of debugging experimental procedures, gaining an appreciation of the information the experiment will yield, and identifying aspects of the experiment that need adjustment.

Debugging runs sometimes are necessary to determine and fix unanticipated problems. Of special importance in this stage is to determine if there are errors involved in the design, construction, and execution of the experiment. Debugging may involve observing measured variables and parameters and through balance checks consisting of application of basic physical conservation laws (mass, momentum, or energy) to the main quantities governing the experiment. These balance checks are strategically repeated in subsequent execution of the experiment to make sure that the desired qualities of flow performance and data acquisition are maintained throughout the experimental program.

The essential activities during the performance of the experiment directly relate to data acquisition. Figure 5.2.1 outlines them in step-wise sequence, and indicates that a

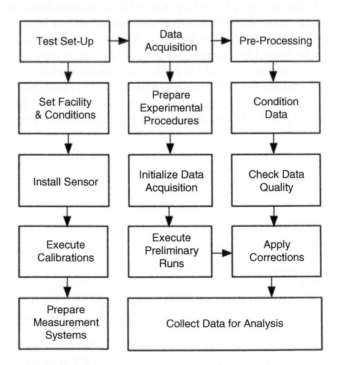

Figure 5.2.1 Diagram of experiment execution

reliable procedure should be in place for controlling the measurements and associated conditions of facility or experiment operation.

5.2.2 Instrument Calibration

Measurement systems (MS) used to collect data in an experiment should be verified by way of a calibration process. The main purpose of calibration is to identify the accuracy of the measurement system or of a measurement protocol. When a measurement system or instrument is calibrated it is compared with some standard with presumably well-substantiated accuracy. The standard can be a unit of a dimension (*e.g.*, mass, length, second), a piece of equipment that the operator trusts, or a well-accepted technique known to produce a reliable value. Calibration of a measurement protocol is the tuning of the model to reproduce a single known event. This event may be a single instantaneous occurrence, such as a particular flood discharge, or it may be a prolonged event, such as a complete flood hydrograph.

Calibration of instruments is typically achieved by comparing the instrument's output with a primary standard, a secondary standard, or a known input source. The metrologic definition of a primary standard is made with respect to a reference unit (Figliola & Beasley, 1991). Some primary standards are physical objects such as the standard kilogram, which is a bar of platinum-iridium kept at the International Bureau of Weights and Measures in France (*e.g.*, Holman, 1989). Other primary standards are arbitrarily defined but accepted values that can be more easily reproduced in laboratories. These definitions are enforced by international agreements. For example, the meter is defined as the distance light travels in a vacuum in $1/299,792,458$ of a second (Figliola & Beasley, 1991). In the present context, a primary standard is an established standard like those set by the National Bureau of Standards; it is believed that primary standards do not change with time.

A secondary standard is a second measurement system with a known accuracy. To calibrate a system to its full potential using a secondary standard, the secondary standard must have a higher accuracy than the system being calibrated. The accuracy of the calibrated system can be no greater than the accuracy of the standard used to calibrate it. For instance, a magnetic flow meter can be calibrated with a weir, but the accuracy of the magnetic flow meter cannot be expected to exceed the accuracy of the weir used to calibrate it, even if it is determined that the flow meter is a more sensitive device.

A known input source is essentially a direct measurement that can be used to calibrate devices in situ – especially for devices that may not maintain their calibrations well. For example, electrical pressure transducers are sometimes calibrated by subjecting them to a range of water depths, measuring each depth with a piezometer and plotting the depth against transducer output. Why not just use the piezometer to measure pressure all the time? While a piezometer can provide an accurate measure of average pressure, the temporal resolution of the pressure transducer might be much greater than that of the piezometer, hence the capability of the piezometer to measure pressure fluctuations.

Calibration of velocimeters is usually performed in a towing tank, where the speed of a carriage forms a primary standard. For depth gauges, calibration is usually performed by altering the relative heights of the gauge to follow various stages of a still water surface through a range of positions covering those in the experiment. Calibration of instruments that include electronic circuitry for acquisition of high frequency dynamics signals is more complex, requiring the application of a known input value to the selected measurement system for the purpose of observing the system output. Constant known values sequentially applied to the instrument to be calibrated produce relationships for the magnitude of the signal over the range of the instrument operation (static calibration). A short pulse applied to the MS tests its ability to react to a change in the input value (dynamic calibration). These types of calibration are required for hot-wire or hot-film (thermal anemometry) probes. As thermal-anemometer probes are sensitive to the change in temperature of the working fluid, calibration needs to be repeated for the range of temperature variation that is expected during the measurements to be conducted. More information about the complex calibrations can be found in Figliola and Beasley (1991).

The calibration of a hydraulic model so as to reproduce the prototype behavior does not ensure that the model will reproduce different or future events. However, by reproducing a known event, the model gains the confidence of its operators. Model calibration is typically conducted in conjunction with verification and validation. Verification ensures that the model behaves properly; in other words, the model obeys basic physical laws, such as conservation of mass, momentum, and energy without losses that might occur due to imperfect construction or operation of the model. Instrumentation should be included in the model verification to the extent possible. Validation consists of ensuring that the model reproduces the relevant physical processes that affect the performance of the structure or device being tested. Calibration, verification and validation of the model will lend the greatest degree of confidence to the results during the testing phase of the experiment.

5.2.3 Establishing the measurement procedure

The sound conduct of an experiment requires that there be a clear procedure for taking measurements. The procedure entails so-called set points that should be chosen carefully such that the experiment will provide consistent information about the targeted results over the full range of expected values. Set points form a series of values for the controlled variables selected such that they cover uniformly the range of variation for both independent and dependent variables. The spacing between the set points is established by choosing the values of the controlled variables that will be 'set' (Coleman & Steele, 1999). These test points should be chosen according to the nature of the relationship between the dependent and controlled variables. In many situations, general relationships are known beforehand. For example, if the relationship is nonlinear, then the spacing of the controlled variables will be unequal over the range of interest. For linear dependencies, fewer points are needed for establishing these relationships.

Sometimes, though not for all experiments, randomization of the order of the data acquisition can be useful (Coleman & Steele, 1999). A random rather than a sequential order of the test points may be chosen to neutralize the effect of extraneous factors such as noise (random variations in the environmental conditions) and interference (undesirable deterministic effects on the measured variables). For example, noise can be introduced in the circuitry of the experimental apparatus at a frequency of 50- 60 Hz (from electric power) and at 100-120 Hz by fluorescent lighting. Interference can be produced by electromagnetic effects produced by nearby motors, transmitters, and electric transformers. In addition to these perturbing factors, human activities (variable proficiency during the work day), hysteresis effects (measurements varying not only with the measured quantity but also with the direction of the measured value over its range), and drift in facility conditions and instrumentation can also affect the measured variables. Randomization breaks up potential trends developed by gradually and slowly changing the independent variables making any measured value truly independent from the previous one. Practical means to attain randomization can be applied to independent, dependent and extraneous variables as well as to various operations included in the experiment acquisition (Figliola & Beasley, 1991).

A sequential or progressive series of experiments, which is often the more convenient, can on the other hand be useful for delineating trends in an experiment's outcomes, and adjusting the experiment so as to capitalize on information being revealed by an experiment. At times it is found useful to include additional experiments so as to gain further data and insight into a process or the performance of, say, a physical model for a hydraulic structure.

5.2.4 Data acquisition for turbulent flows

The recommendations discussed in Section 4.5.4, Volume I, guide the experimentalist in the design stage when decisions are made on selecting the instrument and the ancillary components of the data-acquisition system (DAS) for a particular measurement situation. After DAS selection, the experimenter has to set the system to meet the particular needs of the experiment. The configuration and strategy for sampling the data should be optimized to make the recorded digital signals representative of the actual physical process (*e.g.*, flow turbulence) by minimizing error during sampling. Optimization of the sampling strategy depends on the flow conditions and includes provisions to acquire the data with sufficiently high sampling frequency and spatial resolution (see also Section 4.4.4 in Volume I) and to ensure an adequate sampling duration for obtaining a representative record. In addition, in some cases, the instrument response should be verified and adjusted to minimize the effects of spatial and time averaging (filtering). The errors introduced by an improper setting of the sampling parameters are not easily detected in the output of the DAS but can considerably affect the accuracy of the results and the experimental inferences.

The main concerns in setting the DAS for measurements in turbulent flows are related to the appropriate selection of the sampling rate and the sampling duration. The settings need attention for the measurement of all the variables affected by the influence of turbulence (notably, velocity, pressure, and shear stress). The following guidelines are strictly valid for turbulent flows whose statistical properties are stationary, which are the most commonly measured water flows. For other situations, additional considerations need to be developed.

5.2.4.1 Sampling rate

In cases where time-resolved characteristics are of interest, the selected instrumentation should accordingly be capable of sustaining sampling rates required to reconstruct time-resolved quantities of interest (*e.g.*, power spectra). Some instruments can accommodate a range of sampling rates enabling the user to adopt an optimum sampling rate satisfying the Nyquist criterion (see Section 4.4.4.2 in Volume I) and efficiently using the computer storage which is another important aspect of managing the experiments. The latter aspect is illustrated by considering a data acquisition scheme aiming at time-resolved analysis of a boundary-layer air-model flow with a Reynolds number, $Re = 10^6$, and a free-stream velocity of 15 m/s. This flow situation requires (via the Nyquist criterion) 50-60 thousand readings/second for 30-60 seconds, or a total of 4 million points. With a 16-bit digitizer in the DAS, this will require 8 Mbytes of storage for each measurement location.

There are several aspects to consider when establishing the strategies for efficient storage of the data. When the aim of the measurement is to generate the high frequency portion of the power spectrum, long sampling durations are not necessary; likewise, high sampling rates are not necessary when generating the low-frequency portion of the spectrum (Goldstein, 1996). The sampling rate has little effect on estimates of statistical moments (*e.g.*, mean value, standard deviation, turbulence intensities) for stationary flows as long as the sampling duration is appropriately selected. Consider for example measurements of a time series for a flow variable measured at a point in a steady turbulent flow. If the sampling rate is chosen to resolve the smallest time scales in the flow, consecutive measurements with high sampling rate are generally not statistically independent (*e.g.*, successive velocity measurements will be similar to each other hence being strongly correlated). Velocities become statistically independent only for sampling intervals (the inverse of the sampling rates) longer than about twice the integral time scale, $2\,T$ (see Table 4.4.3, Chapter 4 in Volume I). Using this approach will ensure that none of the samples are correlated, reducing the standard deviation of the sample population to almost minimum level hence minimizing the number of samples needed. Thus, for estimating the mean, variance, and higher order moments, long periods of low resolution data are typically superior to short periods of high resolution data. This also has implications for confidence intervals as conventional confidence interval formulae assume that sampling is random. A general discussion of sampling and confidence intervals is given in Section 6.4, Volume I.

For time-resolved measurements, the chosen instrument should be capable of sampling the relevant variables in turbulent flows at a frequency that is at least twice the highest frequency of interest contained in the actual signal (see Section 4.4.4.1 in Volume I). When the DAS does not have this capability or the operator overlooks to properly set the sampling rate, the highest frequencies contained in the signal are mis-represented by the DAS, leading to aliasing (see Figure 4.5.8 in Volume I). Aliasing refers to distortions that result when the digital signal reconstructed from samples is different from the real world analog data due to the use of an incorrect sampling frequency. Consequently, it is recommended that for cases where the signal may contain frequencies above $f_s/2$ (whereby f_s is the sampling frequency), the analog signal must be low-pass filtered by analog means with a cutoff frequency at or below $f_s/2$ to avoid aliasing.

5.2.4.2 Sampling duration

The issue of sampling duration (also termed as sampling period or exposure time) for the measurements of variables is often overlooked in the data acquisition process despite its critical importance for accurate quantification of turbulent flow (*i.e.*, mean and higher-order statistical moments). Selecting an appropriate sampling duration requires a compromise between accuracy of the results and the cost (or effort) associated with the acquisition of relevant flow characteristics. The number of sampling points in the sample population needs to be sufficiently large to build a representative sample population. For example, Goldstein (1996) states that 3000–5000 (statistically independent) data points would probably suffice for determining the mean characteristics of a stationary turbulent flow. This assessment is made with consideration of a given uncertainty level in the estimation of the flow characteristics as discussed below.

For instruments that have adjustable sampling rates, the needed number of points per sample can be evaluated by collecting several datasets using the same sampling duration but different sampling rates. By gradually increasing the sampling rates, a larger number of data points (N) are recorded in each dataset. Comparison of the statistical moments obtained with such datasets will allow the experimentalist to decide the appropriate sampling rate for a given sampling duration (Goldstein, 1996). For instruments that have a fixed sampling rate, a practical procedure to attain reliable datasets is to repeat the data acquisition with increased sampling durations until the desired statistical moments become stable (Muste et al., 2004). This trial-and error approach is quite efficient in ensuring reliable results (even for higher-order moments) but requires additional tests as will be subsequently discussed.

The estimation of the appropriate sampling duration for measurements in turbulence can also be made with analytical approaches (e.g., Nezu & Nakagawa, 1993; Tennekes & Lumley, 1972; Garcia et al., 2004). Relevant for the present context are the semi-empirical approaches using estimates of the integral time scales (T) in turbulent boundary layers (Equation (T.3) in Table 4.4.3 in Volume I). A rough estimate of the integral time scale in boundary layer can be

$$T \cong \delta_x/U \qquad\qquad (5.2.1)$$

where δ_x is the thickness of the boundary layer and U is the mean (or freestream) velocity in the boundary layer. Estimates for the boundary thickness can be based on the geometry of the flow (e.g., measurement of the extent of the flow). An estimate of the mean velocity in the boundary layer can be easily obtained, say, through a simple visualization test whereby a floating object carried by the flow is timed for a short distance. Both estimates can be quickly obtained prior to the production measurements.

According to Garcia et al. (2004), the time required to appropriately sample one of the larger turbulent structures in a flow is on the order of five integral time scales (5 T). Soulsby (1980) suggests that measuring 30 large structures may be enough to capture reliable statistical moments for the flow, indicating that a sampling period of 150 T is sufficient. Nezu and Nakagawa (1993) recommend that at least 100 burst events should be recorded for accurately capturing mean flow features in open-channel flows. This requirement implies that for an experiment in open-channel flows the sampling duration (T_s) can be approximated by

$$T_s \approx 100T = 100 \cdot C \cdot \frac{U_{max}}{h} \qquad\qquad (5.2.2)$$

where T is the integral time scale of the flow at the measurement location, U_{max} is the maximum bulk velocity, h is the flow depth, and C is a constant between 1.5 and 3.0.

Tennekes and Lumley (1972) proposed an alternative procedure for estimation of the sampling duration using the integral time scale, T, as an independent variable. In addition to those presented above, this procedure allows a consideration of an allowable uncertainty associated with the fact that in reality the data is always acquired over a finite sampling duration. If it is assumed that the measurement goal is to obtain the average value of a fluctuating variable, $f(t) = f' + \bar{f}$, with a targeted degree of accuracy, e, then the required sampling duration, T_s, can be obtained as

$$T_s = \frac{2T\overline{f'^2}}{e^2 \; \bar{f}^2} \tag{5.2.3}$$

with the allowable error, e, approximated by

$$e = \frac{\sqrt{(\bar{f}_T - \bar{f})^2}}{\bar{f}} \tag{5.2.4}$$

in which $\bar{f}$ is the true mean and $\bar{f}_T$ is the (sample) mean computed from a finite integration time (equal to the sampling duration). The basis of Equation (5.2.3) in standard statistical sampling theory is pointed out in Chapter 6.4, Volume I. Figure 5.2.2 illustrates relationships between T and T_s for obtaining results within $e = 0.025$, i.e., 2.5%. These relationships vary with the flow turbulence levels ($\sqrt{\overline{f'^2}/\bar{f}^2}$). Similar diagrams can be obtained for other levels of accuracy for the estimation of the results (Muste et al., 2004). Lenschow et al. (1994) used Equation

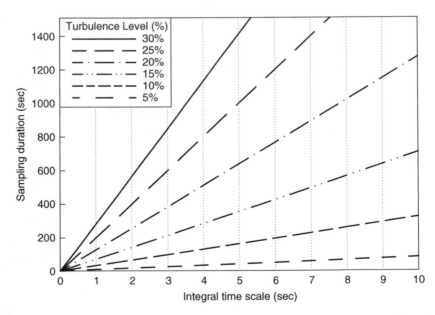

Figure 5.2.2 Relationship between the integral time scale and the total sampling time for various turbulence levels at 2.5% error in the estimation of the mean fluctuating quantity (from Muste et al., 2004. Reprinted by permission of the publisher Taylor and Francis Ltd., www.tandfonline.com).

(5.2.3) for determining the T_s needed to estimate the fourth-order moments for a specified value of e. They found that, for the normalized error variance of the moments to be smaller than 5%, T_s/T would have to be greater than 100. Using the same approach, Bendat and Piersol (1993) found that for estimating the power spectrum with errors of less than 10% would require a sampling duration exceeding 500 T.

Another approach for the diagnostic analysis is to perform bootstrap analyses of sample data. The analyses outcomes define the uncertainty of each statistical parameter using subsamples of different lengths from long records. The optimum sampling time is selected when the parameter uncertainty is smaller than a threshold value. For correlated samples, a Moving Block Bootstrap (Garcia et al., 2005) should be performed. The Moving Block bootstrap technique, commonly used in econometrics, has been validated by Garcia et al. (2005) to provide a good approximation of the confidence intervals of statistical parameters and functions, intervals which cannot be estimated using other methodologies because of their computational complexity and reliance on the correlation structure in the signal. More details on bootstrapping and its applications is given in Section 6.5, Volume I.

Even though the sampling duration can be obtained using the analytical considerations presented above, it is good practice for the experimentalist to conduct an "in-situ" verification of the analytically-derived results for the choice of an experimental parameter before taking the production measurements. One such trial-and-error verification evaluates the running average for each statistic to be computed (e.g., mean, variance, covariance, skewness, etc.) using a long-duration measurement collected in the region with the highest turbulence in the flow. For example, in a natural stream the measurements should be acquired in the deep and high velocity area of a stream where the largest low-frequency turbulence structures are expected. Subsequently, estimates of each statistic are calculated using an increasing number of data points from the long-duration sample. These estimates are compared to identify when they do not appreciably change as the number of points in the sample is increased (Muste et al., 2004). In other words, the sampling duration for which calculated statistics become independent of the duration is selected as the acquisition time. This approach also ensures that sample statistical calculations are not dependent upon the low-frequency components of a flow. Results from such a diagnostic analysis is provided in the example below.

While the discussions above are related to steady turbulent flows, a number of widely studied hydraulic phenomena are unsteady and include periodic (e.g., waves) and nonperiodic (e.g., river hydrographs) flows. For these types of flows, turbulence characteristics must be accurately captured within short time intervals (during which the flow can be considered quasi stationary for practical purposes) and there is a need for extended the testing time to obtain enough information to derive the time-evolving features of the unsteady process. For example, reliable estimation of the load/response due to tidal exposure requires extended experiment durations (from 3 to 12 hours) and repetitions of the modeling simulations with different wave time series (Frostick et al., 2011). Obtaining time-resolved discharge during the propagation of the storm in a river may take anywhere from one day to weeks of continuous measurements depending on the intensity and duration of the storm (Muste & Lee, 2013).

Example 5.2.1:   Evaluation of the sampling duration for river measurements.   This first part of this example illustrates the use of Equation (5.2.2) for estimating the sampling

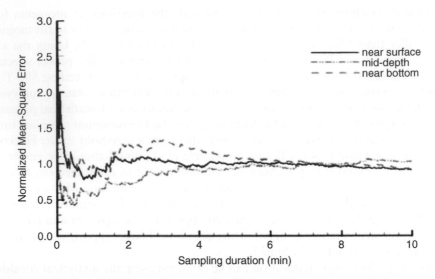

Figure 5.2.3 Normalized mean-square error for the time-averaged streamwise velocity collected at three different vertical locations over the depth in a river

duration required to obtain mean values for the streamwise velocity in a reach of the Mississippi river. Assume that the values for the relevant variables needed in the equation at the time of the measurements are: mean velocity, $U = 1$ m/s; turbulence intensity in the streamwise direction, $\sqrt{u'^2} = 0.23$ m/s (near the bed), integral time scale = 2.7 seconds. Using Equation (5.2.2) with an allowable error (e) of 2.5% leads to a sampling time of 460 seconds (7.66 minutes) as can be observed in Figure 5.2.2. For higher moments in the statistics, the averaging time will generally be longer. Assuming turbulence measurements in a smaller stream, say with the water depth of 1m and a mean flow velocity of 0.5 m/s, the integral time scale of the flow is estimated to be $T = 2$s. The sampling duration required to limit the relative mean-square error (e) of the mean flow velocity to less than 10% of the ensemble variance is estimated to be $T_s > 40$s (Garcia *et al.*, 2004).

The second part of the example provides results of a running-average analysis conducted for measurements collected at the same field site. The long-duration measurements were acquired for 16 minutes using fixed-ADCP deployment. Velocities acquired in three points over the vertical (with bin 2 being the highest location) were subject to running-average analyses. Visual inspection of the results, plotted in Figure 5.2.3, suggests that the normalized mean square error applied to the running averages of the mean streamwise velocity do not vary more than 5% for sampling durations longer than 7 minutes. This value is in good agreement with the 7.66 minutes above-reported result obtained with Equation (5.2.2).

5.2.5 Preliminary runs

An important task prior to taking extensive measurements is a check that the selected instrument is properly set to characterize the targeted flow features. To do so, the experimentalist should be familiar with the appearance and significance of the signals produced by the data-acquisition system as they start producing signals/data. This requires knowledge of both the flow to be studied and the selected data-acquisition system. The foremost prerequisite to collecting reliable measurements is to broadly understand the flow processes in the experimental facility and identify factors that might interfere with the experiment. Even when using simple instruments in a simple facility, problems can arise if data are collected without this understanding (Goldstein, 1996). Such knowledge is even more vital when using modern instruments like LDVs or ADVs, which will produce data regardless of their relevance. The experimenter must decide, based on knowledge of the experimental system and characteristics of the signal, whether the data is usable or not. Examples of data outputs are briefly discussed below to illustrate this important aspect of conducting experiments.

Assume that velocity measurements are acquired with a one-dimensional Laser-Doppler Velocimeter (LDV). The signal output caused by light-scattering particles in the fluid passing through the LDV measurement volume is a series of short, random bursts, as illustrated in Figure 5.2.4a. The experimenter must verify that the output signal has certain expected characteristics to ensure that the LDV signal processor can convert each burst in Doppler frequency into a valid and accurate velocity measurement (Durst *et al.*, 1976). To this end, the experimenter must assess several operational aspects of the measurement procedure, including the amount of seeding in the facility, the amount of light received by the photodetector (which is also dependent on levels of ambient illumination), and the settings of filters applied to the analog output signal.

Next, assume that the sampling rates on the data-acquisition system are properly set, and after successfully performing the above checks, the trace of a turbulent velocity component looks like that shown in Figure 5.2.4b. The instrument records fluctuations in the velocity about the mean velocity at that location in the flow. The fluctuations are caused by eddies of various sizes being advected past the stationary probe. From Taylor's frozen-field hypothesis (see Sections 2.3.6 and 4.4.3 in Volume I), the eddy distribution in a turbulent flow field does not change instantaneously, and hence the analysis of the velocity signal measured with a stationary probe reveals information about the spatial distribution of the turbulent eddies. For example, the frequencies embedded in the velocity time series can be related to eddy time scales, *i.e.*, the

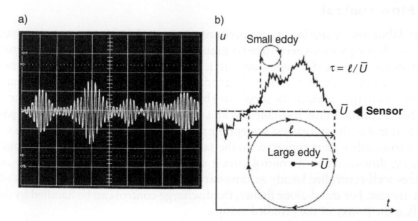

Figure 5.2.4 Signals and data observed by the instrument operator: a) LDV raw signal; b) the processed (final) data acquired with optimal settings

frequency of a fluctuation depends on the size of the turbulent eddy that triggers the fluctuation and the velocity at which the eddy is advected past the measurement volume (Goldstein, 1996). The size, l, of an eddy can be determined from frequency as $l = 2\pi\overline{U}/\omega$ where $\overline{U}$ is the convective velocity of the flow, and ω is a characteristic eddy frequency (the inverse of the eddy time scale, τ). A wavenumber, $k = \frac{\omega}{\overline{U}} = \frac{1}{l}$, can also be defined, and is inversely related to eddy size (See also Section 2.3.3.5 in Volume I).

Turbulent flows consist of a wide range of eddy sizes, which contain all of the energy of the flow (only one eddy size is featured in Figure 5.2.4b), and it is worthwhile to define an eddy size that is characteristic of a particular turbulent flow. One very useful "average" eddy size is defined by the integral length scale. Taylor hypothesized that the duration over which the velocity signal is correlated with itself (T) is a useful measure of the time scale of a typical eddy (Tennekes & Lumley, 1972). The integral time scale T is found by integrating the autocorrelation function of the velocity time series, and can be used to find an integral length scale (L) using the relation: $T = L/\overline{U}$.

As previously remarked, an important task prior to collecting final measurements (while debugging the experiment) is a check if the system is properly set to capture the targeted flow features. To address this task, the experimenter has to extract additional information from the velocity time series and check if the information is in agreement with the expected features obtained analytically (or with literature results) for the specific flow and measurement location. Using such analytical relationships in conjunction with the acquired data, the experimenter can assess if the final results will fulfill the goals of the study. Section 4.4.3 in Volume I provides practical relationships for finding spatial and temporal flow scales using bulk flow parameters and information about the flow (*e.g.*, a rough estimate of turbulence intensity). These estimates may be compared to preliminary measured data. Agreement between preliminary estimates and experimental data substantiates the current configuration and settings of the data-acquisition system, allowing extensive further measurements to be initiated with confidence. Otherwise, adjustments to the system or its settings may be necessary before proceeding with further data acquisition.

5.2.6 Flow control

As many laboratory experiments are performed in simplified and relatively small geometries, the flow conditions need to be continuously checked during data acquisition and between different runs (for reproducibility). Related main concerns are maintaining constant flow rate and water levels in the model. A constant water head tank guarantees a constant discharge provided the supply line between the head tank and the flume inflow section has a constant flow resistance. With sufficiently large piping, an overcapacity is created that makes the system less susceptible to small changes in resistance or to varying demands from other users connected to the same circuit. Integrating an electromagnetic or acoustic flow-meter in a control circuit with the valve feeding the flume usually guarantees well-controlled steady or time-varying conditions as response times of such a circuit are fast. For stand-alone flumes, the discharge control can be handled by directly controlling the power of the installed pumps.

The control of water levels is more complicated and Froude number dependent. Equilibrium depth and critical depth are the first considerations in deciding at which location the level needs to be controlled. Straight subcritical flows are governed by the downstream water level, super-critical flows by the upstream water level. Actively controlling the water level with subcritical flows is problematic as surface disturbances travel as gravity waves through the flume, introducing large and varying time lags in the control cycle. Backwater curves resulting from the downstream weir can be avoided by tilting the bed to establish an equilibrium depth that matches the discharge and bed resistance. This requires a long flume, particularly with large depths and smooth beds.

Irrespective of the bottom boundary type (with or without bedforms for channel beds covered by sand) the overall quality of the flow in open channel flows is reflected by the water depth over the measurement area. In some cases, a uniform flow where the bed slope is parallel with the free-surface slope is an essential aspect, while producing and checking such a flow may require considerable effort. In cases of a fixed (horizontal) slope, simple estimates of the water depth variations can be made using the Belanger equation (Bélanger, 1849). It is also helpful for determining the length scales over which obstacles like weirs or roughness transitions influence the flow. For erodible beds, transport processes can be sensitive to small deviations from flow uniformity. A straightforward approach for erodible beds would start with a horizontal bed, allowing the system to develop toward equilibrium, during which time the erodible bed will develop toward its equilibrium slope. This however may require large amounts of sediment and long running times. Use of an adjustable tilting flume may alleviate this problem. At the beginning of the experiment the anticipated bed slope can be set, with minor adjustments needed to match the eventual bed friction as established by the details of the morphology at the levels of grain size, bed forms or channel shape.

In studying the interaction of a flow with a structure or a surface, the velocity profile should be regular and predictable, or at least well characterized, preferably an equilibrium logarithmic profile (Nezu & Nakagawa, 1993). As the latter is established in a rather delicate process where turbulence transfers momentum over the vertical in a diffusive manner, acquisition of velocity profiles (vertical and horizontal) within and in the vicinity of the measurement section needs to be considered. If, at the experimental facility inlet, large-scale swirling structures or surface-wave disturbances develop that are not representative of the modeled flow, then means to dissipate these extraneous flow features, such as screens or other flow-conditioning devices, should be installed upstream of the working region to dissipate these features and reduce their effect on the studied problem.

The temperature of the water used in the experiment is highly affected by its storage conditions. Small volumes stored in the laboratory fluctuate with the room temperature and hence storage in large containers is preferred for limiting the temperature variations. Temperature differences between the water and the laboratory atmosphere can affect an experiment at low-flow conditions due to temperature induced density currents. The combination of a low temperature and a limited exposure to light reduces the possible growth of algae and is therefore beneficial for the water quality.

5.2.7 Quality control

The success of an experiment also depends on the skill, experience, and background preparation of the experimenter (Tavoularis, 2005). An experienced investigator continuously assesses the system to identify:

1. Possible defects (*e.g.*, leaks, vibrations, etc.);
2. Formation of unwanted separation, recirculation, or pulsation of the flow;
3. Presence of impurities in the water or within the measurement system components (*e.g.*, connecting tubes); and,
4. Possible interactions between instrument and flow, and instrument and facility.

The vast majority of the qualitative assessments are visual but the experimenter should also cautiously use other senses during the execution of the experiment, *i.e.*, touch the apparatus to check for vibration, listen to the flow, and generally use experience to note any undesirable effects that might adversely affect experimental results or complicate the data analysis. In parallel, the experimenter should maintain a detailed, chronologically arranged, scientific journal of all conditions, actions, and results obtained. These logs should contain a list of all acquired measurements recorded during each experiment, the settings for different instruments, reference values for controlled variables, and personal observations concerning the experiment (including difficulties, malfunctions of equipment, and additional notes to help with the interpretations of the results). All information should be recorded tidily and detailed enough so that it can be reviewed by other members of the research team and possibly to be replicated at a later time.

Immediately following the acquisition of the data, a preliminary assessment of the acquired data is recommended as a complement to the qualitative assessments conducted during the experiments. Such quantitative assessments include pre-processing of selected data samples to verify the statistical parameters (first and second statistical moments) that control the adequacy of the size of the samples collected for subsequent more detailed data analysis. Where relevant, checks on power spectra might be made to examine flow quality through agreement with known results, such as Kolmogorov's $-5/3^{rd}$ law for the inertial range (see Sections 2.3.3.5 and 6.12.4 in Volume I) or the adequacy of instrument settings by identifying a noise floor (where the power spectral density is constant over a range of frequencies). The results of such checks may motivate adjustments of the flow facility, the instrument settings, or the use of alternative protocols for the data production runs.

Once data acquisition is deemed adequate, some experiments may need to be repeated to obtain data for uncertainty analysis. The experimental process can readily generate uncertainties. Errors and uncertainties (error estimates) are typically classified as bias and precision based on the way they affect the final results. Bias errors are especially difficult to estimate as they are embedded in the experimental processes in various forms. Frequently, measurement errors are the only ones considered. This is unfortunate, because in many situations errors generated by instrument interactions with flow and facility (Section 5.1 in Volume I) as well as in the measurement process (Chapter 4 in Volume I) and data processing (Chapter 6 in Volume I) can be larger than the errors induced by the measurement itself. Repeatability tests demonstrate the capability of the MS to produce similar results under similar forcing conditions.

In discussing uncertainty in measurements, Coleman and Steele (1999) distinguish between repetition of an experiment in its common sense and replication as a distinct concept. Replication is defined as a repetition carried out in a very specific manner to obtain: zeroth order, first order, and Nth order replication levels. This distinction between different levels of replication is summarized in Table 5.2.1. While

Table 5.2.1 Levels of replication for a time-wise experiment

Replication level	Measurement process	Relevance
Zeroth order	one experiment, multiple measurements, same instrumentation	Indicative of the scatter of the experiment
First order	multiple experiments, multiple measurements, same instrumentation	Indicative of the scatter due to environmental factors
N-th order	multiple experiments, multiple measurements, multiple instrumentation of the same type	Indicative of all sources of uncertainties in the experiment

replication of the experiments typically captures random sources of uncertainties, the evaluation of the sources of uncertainties that are fixed and embedded in the execution of the measurements (bias sources of uncertainty) is a much more complex task.

Similar to random errors, bias errors are generated throughout the measurement process. Most bias errors are generated in the data acquisition phase by the use of inadequately calibrated instruments or the influence of unwanted effects active during the experiments. A special type of bias error is conceptual (*i.e.*, error that might stand between concept and the actual measurement), generated both during the test design through idealizations (assumptions) in the interpretation of equations, and/or use of equations that are incomplete and do not incorporate all significant factors. During the experiment execution, conceptual errors occur by not measuring variables which are assumed (Moffat, 1988). There seems to be no limit to the mistakes that can be made in assigning significance to what has been measured. For illustration purposes, Table 5.2.2 highlights potential bias sources in open-channel flows with suspended sediment transport. The table lists for completeness bias error sources generated in all the phases of the experimental process along with the conceptual bias error sources. The methodology for assessment of experimental uncertainties is presented in Chapter 7, Volume I. Notable for the present context is that robust uncertainty analyses are based on the assumption that the bias errors are corrected before or immediately after acquiring the data. Uncorrected bias uncertainties are not supposed to be part of the uncertainty analysis.

5.2.8 Record keeping

When conducting experiments it is imperative that complete records be kept throughout the testing period. Questions may subsequently arise about how data were collected, what supplemental information is available, and whether the data need to be analyzed differently. During the planning phase, documentation of the results of dimensional analyses and of the preliminary uncertainty analyses are needed. During the experimentation phase, the data for instrument calibrations, data-collection sheets, and diagrams showing the design of the experimental setup should be retained. During the data-processing phase, adjustments to measured data, details about data analysis, and uncertainty analysis should be similarly recorded. In addition, references and personal communications, potential unaccounted sources of

Table 5.2.2 Illustration of the bias error sources in suspended-sediment experiments (from Muste, 2002. Reprinted by permission of the publisher Taylor and Francis Ltd., www.tandfonline.com).

Error Type	Error Source	Error Removal
Conceptual bias	– sediment and water considered as mixtures – use of clear water equations for the mixture	– use two-phase flow approach in design, instrumentation selection, execution and processing of the results – use separate equations for the interacting phases in the mixture
Experiment Design & Monitoring	– Bedform presence – Change of fluid characteristics – Spatial/temporal flow non-uniformity – Non-equilibrium sediment transport – Non-equivalent flows	– use starved-bed or capacity condition – monitor temperature and correct for viscosity change – use aspect ratio (b/h) > 6 – set test section at 160 hydraulic radii from flume entrance – measure bed and free-surface slopes accurately – increase duration of the experiment – select proper sediment feeding system – use precise flow definitions
MS & Data Acquisition	– Data acquisition design – Instrumentation calibration – Flow-sensor interaction – Lack of flow-phase discrimination – Insufficient spatial and temporal resolution	– use multiple methods for documenting same flow variables – gradually change flow parameters, one parameter at a time – use best available information (inter-laboratory comparison, manufacturer specifications, publications) – use non-intrusive techniques – select non-intrusive instruments capable of distinguishing flow phases (LDV, PIV) – select adequate instrumentation for measurement of flow and sediment characteristics
Data Reduction & Analysis	– Data sampling & filtering – Statistical analysis & curve fits – Non-uniform analysis methods	– correlate spatio-temporal scales with data acquisition parameters – use multiple points with large data samples at each location – analyze regression residuals – adopt standard procedures for data reduction and data analysis

error, and brief narratives and observations related to the experiment should be duly noted.

Keeping a laboratory notebook is useful for recording daily activities, observations, and procedures. Annotating entries with dates can help decipher the exact sequence of events that occurred during an experiment long after details of the experiment have been forgotten. It is also beneficial to prepare standard summary sheets for important points before beginning an experiment. Subsequently, when the experiments are underway, data should be entered into a datasheet. These data often include information like temperature, flow settings, and names of data-acquisition files. As contemporary records often are electronic, records should always be backed-up. At times, informal records may have legal ramifications, as explained by Tavoularis (2005), who discusses the benefits of recording a variety of information on the experiments as they unfold to prevent ethical rule infringements such as intellectual property violation.

Once data are acquired, considering appropriate approaches to synthetically present the results is a necessary step. The exclusive use of large numerical tables is not always recommended as the scatter of the data likely obscures relationships between the variables. Usually, graphic and mathematical representation of the data and the derived relationships are used in addition to the tabular data representation (Coleman & Steele, 1999). An experimental investigation report should contain at minimum the following tabular information for each result: a) the value (single or average), b) the total uncertainty and individual uncertainty components specified with confidence level. Often mathematical expressions are derived or proposed to capture succinctly the relationships between variables. This equation allows interpolation between the discrete data points in regions where measurements are not available. The graphical presentation of the data should include uncertainty intervals around the measured results indicative of all the sources of errors involved in the measurement process.

Example 5.2.2   Sample measurements acquired in turbulent open-channel flows.   For exemplification purposes, a sample of typical measurements of turbulence characteristics in an open-channel flow over a transitionally-rough bed acquired with Particle Image Velocimetry (PIV) are provided below (Bigillon et al., 2006). The experimental conditions for the reported experiments are provided in Table 5.2.3. Quantitative flow visualizations were made using a commercially-available PIV system in a vertical plane 0.04m-long by 0.04m-high illuminated by a laser located at the center of the channel (see Figure 5.2.5a). For each test, a series of 1,000 pairs of images was recorded and analyzed. The acquired images were processed with the software associated with the

Table 5.2.3 Experimental conditions for the sample results provided in Figure 5.2.5 (Bigillon et al., 2006)

Runs	h (m)	u_* (m/s)	$\overline{U}$ (m/s)	$k_s^+ = u_* k_s/v$	Re = Uh/v
01 cw	0.025	0.018	0.25	7.59	4,490
02 cw	0.03	0.019	0.28	8.19	5,871
03 cw	0.037	0.022	0.34	9.66	8,980
04 cw	0.044	0.024	0.37	10.34	11,915

h is flow depth, u_* is shear velocity, $\overline{U}$ is the cross-sectional (bulk)channel velocity, k_s^+ is roughness Reynolds number, k_s is roughness height, v is fluid kinematic viscosity, Re is Reynolds number.

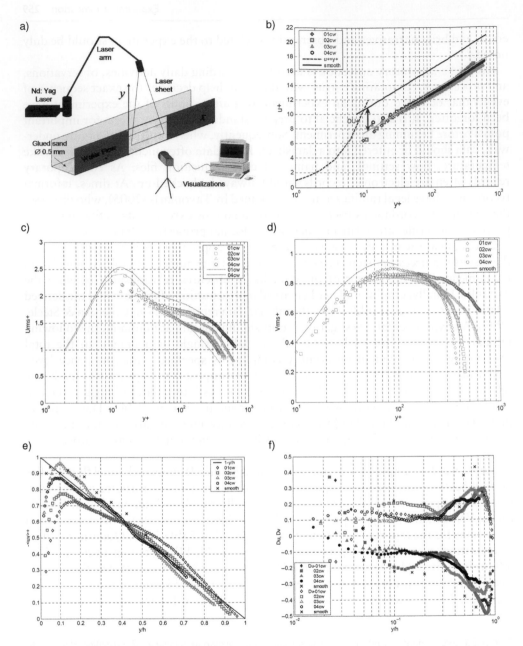

Figure 5.2.5 Illustration of PIV data collected in a steady open-channel flow (Bigillon *et al.*, 2006, with permission of Springer): a) experimental arrangement; b) vertical distribution of the mean streamwise velocity. Continuous line corresponds to smooth-wall situation (Nezu & Nakagawa, 1993); c) streamwise turbulence intensity distribution. Continuous lines are obtained using the relationships proposed by Nezu and Nakagawa (1993) for the 01cw and 04cw flows; d) wall-normal turbulence intensity distribution. Theoretical curve for a smooth bed is taken from Nakagawa and Nezu (1977); e) Reynolds shear-stress distribution. The data for the smooth wall case are taken from Antonia and Krogstad (2001); f) turbulent diffusion coefficient distribution for the streamwise (full symbols) and wall-normal directions (open symbols). Data for smooth walls are taken from Nakagawa and Nezu (1977)

PIV system. The processed PIV data yield instantaneous vector fields for the whole field of view for each pair of images analyzed. From those data other quantities characterizing the average properties of the flow as well as turbulent characteristics were computed.

Sample results from these experiments are provided in Figure 5.2.5. A first observation on this figure is in regard to the high density of the data points and smoothness of the distribution, which reflects the capabilities of the PIV system and the good quality of the experiment, respectively. The PIV could not measure, however, close to the bed, i.e., $y^+ < 10$, due to the lack of quality image data in the near wall region. The qualitative analysis of the results reveals that they are very similar to those obtained in previous studies for smooth walls (see provided references in Figure 5.2.5). They also reveal typical features of the turbulence characteristics in an open channel with both smooth and rough walls. Most notable, the plots in Figure 5.2.5b substantiate the three distinct areas of the velocity profile (i.e., viscous sublayer, logarithmic region and outer layer described in Section 2.4.2.3, Volume I. A shift in the velocity, DU^+, is obvious and is attributed to the presence of a rough wall. The turbulence intensity profiles plotted in Figures 5.2.5c and 5.2.5d illustrate that the maximum turbulence activity occurs for $10 < y^+ < 20$. The Reynolds stress distribution collapses well in the logarithmic region (i.e., $y/h > 0.2$) on the expected linear dependency with $1 - y/h$. Finally, the turbulent diffusion factor profiles displayed in Figure 5.2.5f display small, close to zero, values in the logarithmic region suggesting that probability distributions of fluctuating velocities are close to the normal distribution. Toward the wall, the values of the diffusivity are larger due to the strong and highly intermittent turbulent events in that region. The latter plots show more variability in the trends compared with the previous, indicating that the higher statistical moments are more difficult to capture, even if we use the most advanced instruments.

5.3 FIELD EXPERIMENTS

5.3.1 Introduction

Field measurements and experiments have a long history in hydraulic and geophysical research and are critical for understanding flows in nature as well as interrelations between physical, biological, and chemical processes. They have been a basic tool in hydraulics before the advent of laboratory studies and have since remained an important tool for scientific investigations at the prototype scale. Moreover, field measurements are required for water resources management and to assess the impact of hydrological and hydraulic flow features on existing or planned projects.

The vast majority of field measurements pertain to monitoring activities conducted for a specific purpose, such as determining stage-discharge curves at a gauging station, generating digital elevation models, assessing floodplain characteristics through remote sensing, identifying bed material characteristics, monitoring of bed levels to assess morphological development and/or stream navigability, etc. Although field measurements are often obtained with sophisticated and advanced instrumentation and experimental methods for each of the above purposes, they are not easy to assemble to conduct integrative studies for several reasons. First, field measurements are often administered by specialized agencies that monitor various aspects of water resources management and measurements carried out by different agencies are often not connected among themselves despite that they are carried out over the same geographic areas. Second, the resulting datasets are large, produced in various formats, and stored in diverse repositories. Nonetheless, data from field measurements at particular sites are often available over longer time periods and can thus provide valuable information and input for scientific studies as well as for the validation of theories and hypotheses.

Fewer field experiments are specifically designed to address open scientific questions that cannot be adequately addressed in laboratory studies, or to provide prototype data for further analyses as well as the validation of scientific theories, numerical models, and physical scale models. An advantage of field experiments is that physical processes can be directly investigated in the natural environment so that the scale distortion of important parameters and boundary conditions is eliminated. This is of particular importance for environmental and ecohydraulic studies as many physical, chemical, biological/microbiological, and ecological boundary conditions cannot be adequately reproduced in laboratory studies. Thus conditions can differ dramatically in the field and laboratory due to complex feedback loops between abiotic and biotic factors so that it is not clear if such laboratory experiments are representative for the natural environment (*e.g.* Bouma *et al.*, 2007, Cameron *et al.*, 2013).

The distinct difference in observing process variables and their natural interactions in field experiments compared to laboratory experiments is relinquishing control over key parameters such as discharge and water depth. This complicates not only data interpretation and analysis, but also the planning of field experiments at particular sites, as the desired experimental conditions (*e.g.*, experiments aiming at the investigation of flood scenarios and/or flood wave propagation) may prevail only during a brief time window. Consequently, components of laboratory facilities have been implemented in natural-scale environments to allow for limited control of the boundary conditions (*e.g.*, Wilcock *et al.*, 2008). Examples include the Wienfluss in Austria (*e.g.*, Meixner & Rauch, 2004), the Outdoor Stream Lab at the University of Minnesota in the USA (*e.g.*, Legleiter & Overstreet, 2014), and the Environmental Fluid Dynamics Lab in the Field in Germany and Italy (*e.g.*, Sukhodolov, 2015; Sukhodolov & Uijttewaal, 2010).

Technological advances have made feasible a wide variety of field experiments. Accurate data on river hydrodynamics may be acquired using acoustic methods such as acoustic Doppler velocimetry and acoustic Doppler current profilers (*e.g.*, Holmes & Garcia, 2008; Demers *et al.*, 2012), Particle Image Velocimetry (*e.g.*, Cameron, 2013) and Large Scale Particle Image Velocimetry (*e.g.*, Fujita & Muste, 2011). Novel techniques and instrumentation for determining important parameters have emerged (*e.g.*, Aberle *et al.*, 2012; Muste *et al.*, 2016). For example, advances in remote sensing allow data such as bathymetry, water surface levels, roughness, and vegetation parameters using terrestrial, airborne or even satellite-based methods to be collected for river reaches at varying spatial resolutions (see Volume II). These complex and comprehensive measurements can be carried out repeatedly over time, enabling the documentation of processes over a range of temporal and spatial scales, from local to reach and even catchment scales (e.g. Marcus & Fonstad, 2010).

At times the complexity of field experiments and measurements require detailed planning, taking into account many different aspects such as cost, available instrumentation and manpower, spatial and temporal scales and last but not least, health and safety aspects. These aspects are discussed in the following sections with general considerations to illustrate that each field experiment is unique and cannot be repeated for exactly the same conditions.

5.3.2 Planning

As with laboratory experiments, a key requirement for successful field experiments is a clear formulation of the main research questions. The corresponding objectives define the parameters that must be determined while the investigative approach depends on the phenomena of interest (*e.g.*, turbulence characteristics, waves, transport processes, flow-biota interaction), the associated spatial (micro to global) and temporal scale (*e.g.*, turbulence measurement or long term monitoring), the environment (*e.g.*, lentic or lotic freshwater), and the required boundary conditions (*e.g.*, hydrologic, hydraulic, geo-morphological, ecological, etc.). Equally important for planning is a review of the applicability of the available instruments. The applicability is not limited to their accuracy but also to safety and logistical aspects (*i.e.*, power availability, manpower, auxiliary field installations, and reassurance that the instrumentation is ready to use when it is needed). These aspects need special attention.

As field instrumentation is often shared by many users, additional checks on instrument operability are critical before deployment in the field. Malfunctioning, incomplete or damaged instruments can endanger the success of the experimental campaign, especially if the campaign is carried out in remote places or on short notice. If instruments are shared between many users it must be ensured that the required instruments are available for the planned measurements. This may be achieved through an internal booking system. Another important aspect that needs to be taken into account in the planning stage is to clarify the availability of external sources if these are required for the measurements. Relying for example on vessel or airborne based measurements carried out by third parties requires that these resources will be available as desired. This applies also to the availability of adequate transportation for human resources and instruments to and from the field site. Overall, careful planning of field experiments includes realistic budgeting and timing. Budgeting for field experiments is significantly different from budgeting for laboratory experiments and includes non-scientific expenses such as logistics (*e.g.*, accommodations, transportation, food), instrumentation (*e.g.*, rent and operation), resources (*e.g.*, human resources, special installations, special equipment, external services), expendables etc. The time schedule must include preparation before departure; travel time to and from the field; installation of the setup in the field, including potential calibration of instruments; execution of the experiments, including appropriate considerations of operation safety (see Section 5.3.4 in Volume I); and the packing of the instruments and clean-up of the site following completion of the experiments.

The selected experimental site must be well-suited to address the research questions of interest and should be as representative as possible for the process being investigated. Consequently, the identification and selection of an appropriate field site can be a time-consuming process. Detailed guidance on scientific selection criteria cannot be offered here due to the wide variety of possible study objectives (*e.g.*, hydrodynamics, morphody-namics, ecohydraulics). Given the extensive effort and costs involved, field research is commonly carried out within an interdisciplinary framework, thereby possibly increasing the level of complexity of the experiments (*e.g.*, Marion *et al.*, 2014). A practical factor in selecting a field site is accessibility to the site. Site selection also requires consideration of downstream features that can potentially affect the experiments or its boundary conditions (*e.g.*, confluences, bed and bank protection, bends, weirs). The site selection should also be accompanied by a thorough background-check if supplementary data from the literature, official bodies or other institutions are available. Probably the best example in this context is the identification of the closest gauging station in order to have background information on the hydrological regime and discharge.

5.3.3 Execution

As indicated before, each field experiment is unique and thorough and detailed documentation is therefore crucial. The technical aspects of data acquisition, execution, and processing covered for laboratory experiments are all valid for field experiments as well. The most important distinction between conducting laboratory and field experiments lies mainly in the safety considerations, and they will be separately discussed in the next section. Field experiments require heightened and constant attention of all field team participants during data acquisition, and comprehensive records of experimental details like the timing of events that might affect the process under investigation are a necessity.

A field log should document both the measured parameters and peculiarities at the experimental site. The log information should not rely on the instruments to gather all necessary information, but should complement the instruments with details about the measurements at the beginning and end of data sampling, including relevant boundary conditions. In general, all log-entries need to have a time stamp so that the occurrence of events can be linked to the measured data. Field experimentation often includes sampling, and the nature of the samples can vary quite significantly (*e.g.*, bed material samples or water samples). Each sample needs to be documented in the log and adequately labeled, stored and carefully transported so that its usefulness is maximized. Additionally logs should include observations of phenomena or unexpected processes that occur during the execution of the measurements and which may affect the measurements (*e.g.*, passing of a boat when measuring velocities) as well as problems encountered before and during the measurements.

Logs can take various forms (written, audio, photographic, or video recordings). In terms of spatial coverage and representation of detail, photographs and video recordings are especially valuable. They need to be acquired as often as possible and labeled and documented appropriately in order to relate them to particular sites and events (both in time and space). Logging tools must remain reliable in case of sudden changes in weather conditions. Therefore, selection of adequate documentation equipment is critical (*e.g.*, blotting pads, waterproof pens). Upon return from the experiments it is advisable to immediately review the log and soon thereafter prepare and file a campaign report that can support the analysis of the data at a later stage.

5.3.4 Health and safety aspects

A field experiment or measurement can only be successful if the safety of the participants, public, and the environment is guaranteed. The safety requirements of a field experiment are directly dependent on the purpose of the field measurements, and it is not possible to provide a specific safety protocol that covers all field experiments. Nonetheless, some commonly relevant field safety considerations are provided below.

5.3.4.1 Health and safety of participants

The health and safety of the participants are paramount when carrying out field experiments. Thus, the planning of field activities requires an appropriate risk assessment so that all participants are aware of potential hazards and their liabilities in order to mitigate potentially dangerous situations. Many safety issues can be addressed before departing for the field by preparing field trip policy forms and information sheets; some of these are often available from employers. Such forms should at least cover safety principles in vehicles and in the field, responsibilities of the participants to the team, required safety gear, as well as medical and emergency contact information (*e.g.*, Whitmeyer & Mogk, 2013). Local customs and methods for contacting emergency service providers may vary from country to country and should be clarified before departing to the field for international field campaigns. It is the responsibility of the team leader to check that each participant is adequately insured and informed about his/her liability.

The vehicles used for transport to and from the field must be adequate for the trip (*e.g.*, appropriate number of seats and space for safe transport of the equipment, four-wheel-drive, snow-chains) and should contain safety equipment as well as vehicle repair kits. It is generally advisable to use official institution vehicles and to have at least two authorized drivers in the team. Safe transport after field experiments should also be guaranteed as field work can be rather exhausting. Driving long distances after completing extensive measurements is potentially hazardous and in such cases the arrangement of adequate local accommodations is recommended. Moreover, all drivers should be made aware of their liabilities as insurance of official vehicles may be significantly different from insurance of private vehicles. In case of an accident, the procedures of the corresponding institution must be followed.

Before traveling to the field site, safe accessibility to the site must be ensured. This includes accessibility through difficult terrain as well as obtaining the right to access private or public property. Moreover, potential hazards at the experimental site must be identified. Such hazards are not necessarily related to the experimental procedures but may also be associated with wildlife, flora, weather conditions, etc. For example, carrying out experiments in areas with poisonous plants and/or dangerous or poisonous animals (*e.g.*, crocodiles, spiders, snakes, bears) represents a significant hazard and may influence experimental procedures. Similarly, weather-conditions may change on short notice imposing threats to human lives as well as to the instruments. Field conditions are unpredictable and therefore emergency procedures should be planned and contingency plans should be available.

Field work should never be carried out alone so that in case of critical situations, accidents or injuries, help is directly available and further help can be called. It is important that a sufficient number of team members have first-aid experience. Portable first-aid kits should be standard field equipment and more complete first-aid gear should be available in the vehicles. Adequate communication equipment is also essential for field experiments. Mobile phones are an option, but there may be limited cell coverage at the field site. In remote areas, other options include two-way radios with sufficient range and satellite phones. It is strongly advised that colleagues or other people not involved in the experiments be informed about the experiments as an additional safeguard.

Adequate protective clothing and safety equipment are also indispensable. Examples include life vests for vessel-based measurements or for measurements in deep or fast flowing water, warning devices and adequate safety clothing for measurements in the presence of automotive traffic (*e.g.*, measurements from a bridge). It is also advisable that the field-team members have training on how to react in emergency situations; *e.g.*, when a small boat capsizes or when someone loses his/her footing in a river while wearing a wader. For more sophisticated measurements, for example when diving or using chemicals, the responsible team members must have appropriate certified training.

5.3.4.2 Health and safety of the public

Field experiments are generally carried out in an environment which is accessible to the public so that adequate health and safety procedures are required to avoid harm to the public. The required level of health and safety measures depends on the complexity of the experiments. A first step clarifies whether special permits are required to conduct the experiments; *i.e.*, if the site is generally accessible and if the desired measurements can be carried out. For example, simple stream gauging usually does not require special health and safety procedures, but in experiments in which the flow rate of a river reach is controlled so that the water depth may change rapidly and may impose a threat, measures must be taken to insure that nobody is in the vicinity of the water at the measurement site as well as in other river reaches which are affected by the experiment. Another example is the use of tracers in experiments. Although special permits are usually required to conduct such tests, it should also be considered that if visible tracers are used (*e.g.*, fluorescent dye tracers), the public may be alarmed if no information about the experiments has been provided beforehand. Similarly, the use of potentially harmful instrumentation should be reviewed when planning the experiments. Laser-based measurements for example may need to strictly adhere to health and safety regulations so that the public is not endangered.

Required permits should be obtained well in advance of experiments to prevent delays associated with processing of the permits. Moreover, through the application process the legality of the required instruments and planned experimental methods can be verified. Last but not least, carrying out experiments in the public space can attract the interest of the public. Thus the field-team should be prepared to answer questions from the public and also consider that interactions with the public can affect the time schedule of the experiments.

5.3.4.3 Protection of the instrumentation

Instrumentation used in field experiments is often unique and expensive. Thus, it is important to treat all instruments with care and to protect them adequately during transport to and from the field site and during the measurements so that they will not be damaged or lost. The instruments need to be adequately secured and protected from hazards during the experiments (*e.g.*, potential damage from driftwood, vessels, wildlife). If instruments are deployed for a long time period and cannot be constantly supervised, they need to be protected against vandalism and theft. The security of instrumentation is important since it often cannot be adequately insured, and may be difficult or impossible to replace if damaged or lost.

5.3.4.4 *Consideration of the environment*

During field experiments, the environment should be treated with respect. The presence of the field team, installation of instrumentation, and the measurements themselves may have consequences for the environment. Examples are the removal of vegetation, the preparation of the experimental site, destructive sampling (*e.g.*, bed samples, collection and sacrifice of plants), observations of animals such as fish, pollution through the use of tracers, pollution of the experimental site, etc. In the best case, there will be no evidence of the experiments at the field site after they have been completed. This means that the site will be left in its original state, and that the stresses imposed on the ecosystem and the environment during the measurements have been within natural limits. If it is not possible to leave no trace of the experiments, the environmental impact of the measurements should be minimized, remaining within ethical and regulatory limits (*e.g.*, Farnsworth & Rosovsky, 1993). Otherwise, the experiments must be modified or abandoned.

Applicable special permits must be obtained prior to collecting measurements. Measurements in parks or on public land often require special permits, and private land owners may be reluctant to allow access to their land. Therefore it is wise to formulate a memorandum of understanding that details access to the site and the potential effects of the measurements on the local environment and infrastructure. In any case, no waste and pollution should be left behind, gates and fences should be left as they have been found, wild and range animals should not be disturbed or harmed, and hazards resulting from smoking or other fire dangers should be mitigated (*e.g.*, Whitmeyer & Mogk, 2013).

5.4 COMPLEX EXPERIMENTS

5.4.1 Sediment transport

5.4.1.1 *Introduction*

An important category of experiments concerns the interaction of water flow and boundary sediment which can produce varied phenomena associated with sediment transport. Three categories of phenomena are of interest:

1. Flow transport of sediment particles as both bed load and suspended sediment load;
2. Various bed forms, ranging from small ripples, through dunes and anti-dunes, to bars and pool riffle systems (Yalin, 1972). Such bed forms typically move along the flow direction with a velocity smaller than that of the individual grains and the flow itself; and,
3. At larger spatial and temporal scales, channel cross-section and plan form geometries change, resulting in features like meanders and braids.

The limited spatial and temporal conditions available in laboratories restrict the scope of the experiments allowing the investigation of one or a subset of components of the natural sediment-transport processes. These processes include the motion of individual grains, the formation and dynamics of bed forms, and river plan form changes. In all investigations, a choice has to be made as to what sediment is most representative. The most important decisions on the sediment characteristics are related to the scaled mass density and size (or size distribution) of sediment grains, as they determine the mobility of a sediment in a flow. An additional scaling complexity is the cohesiveness of bed sediment as it affects the behavior of very fine sediment in water, notably in the sediment floc formation and settling as well as in determining the maximum angle of repose. Moreover cohesive behavior imposes a sediment-size limit when downscaling sediment size for an experiment. These sediment properties must be related to the flow properties using dimensionless numbers such as the Shields number, Rouse number and particle Reynolds number (see Section 2.5 in Volume I).

As laboratory flumes are generally much shorter than river reaches, representative equilibrium conditions can only be attained when sediment is fed at the upstream side of the flume. In some situations it is convenient to collect sediment at the downstream end of a flume, then recirculate it back to flume's upstream end as illustrated in Figure 4.3.1, Volume I. For experiments involving unsteady transport of sediment, whereby sediment accumulates or erodes from a simulated reach, it may be necessary to feed sediment into the flume but not recirculate it through the flume. The sediment flux out of the model can be determined by either intermittently weighing the recirculated amount or continuously measuring the concentration in the sediment return circuit (Muste, 2002). In case recirculation is not possible, feeding a known flux of sediment and trapping the sediment at the downstream end can be a solution (Yossef & de Vriend, 2010). LiDAR scans (described in Section 4.2, Volume II) of the model bed are useful means to control the sediment accumulation or erosion in the model and to apply the needed operational corrections.

5.4.1.2 Experimental flow complexity

Experimental techniques for determining suspended particle concentration. Instrumentation for characterizing suspended sediment is described in Section 5.3 in Volume II. The simplest way to determine particle concentration is by taking a sample and counting the number of sediment particles per unit of volume after a sieving, drying, weighing procedure. Using suction tubes, local iso-kinetic sampling can be done with a vertical resolution of the order of 10^{-2} m. As solid particles in a fluid scatter light, this phenomenon can be exploited by relating the extinction of laser light over a well-defined pathway to a concentration, labeled as the optical back scatter technique. The scattering properties are a function of particle size, so calibration with specific particle classes is necessary.

Similarly the scattering of ultrasound by sediment particles can also provide information on the local sediment concentration. With proper calibration, acoustic Doppler velocimeters, including profilers can be used to obtain the concentration from the signal intensity whereas the Doppler shift of the scattered sound wave is a measure of particle velocity (Holdaway *et al.*, 1999). Only using a particle image velocimetry technique (PIV) can the individual particles be made visible, counted and tracked (Kiger & Pan, 2000).

Sediment-turbulence interaction. For the interaction of suspended sediment grains with a turbulent flow, particles need to be picked up by the flow and transported in the water column. The settling velocity should therefore be small in comparison with the friction velocity, resulting in a small Rouse parameter. In this case, particle motion can be observed in combination with the fluid motion. The remaining parameter is the particle response time in relation to the time scale of the turbulence, determining whether the inertia of the particle allows it to follow the fluctuations in the flow. Using laser-Doppler techniques or whole-field PIV (see Sections 3.6 and 3.7 in Volume II), the particle velocities and fluid velocities can be obtained simultaneously but separately as associated with the flow phases (*e.g.*, Durst, 1982; Kiger & Pan, 2000). Sample PIV measurements with discrimination of the phases are illustrated in Figure 5.4.1 (Muste *et al.*, 2005). A two-phase approach to channel flows with suspended sediment may allow directly measurement of the Reynolds flux $\overline{u_i' c'}$ in Equation (2.5.34), Volume I that cannot be obtained with any of the previously developed techniques (see Figure 5.4.1b). The above type of two-phase flow measurements are currently limited to low concentrations in which the flow phases can be easily discriminated (Kiger & Pan, 2000; Muste *et al.*, 2005). Another experimental approach using the same type of laser-based instrumentation selects the suspended material such that its refractive index matches that of the fluid (Budwig, 1994; Cui & Adrian 1997). Such experiments are useful for the development of two-phase flow models in which the motion of the solid and fluid phases are modeled with their mutual interactions.

Bed forms. The design of experiments focused on bed forms requires flume lengths of several tens of meters to allow for the development of the characteristic length of the largest features (Crosato *et al.*, 2011). As the transport of sediment depends on particle size, the characteristics of the bed forms for a mobile bed will change locally in accordance with the (selective) transport and local bed shape. Bed-form characteristics can be determined by tracking the surface of the bed, mechanically, optically or acoustically. Measuring surface-bed composition without disturbing the bed is best done optically, imaging the bed surface and determining the granulometry by image analysis (Detert & Weitbrecht, 2012). After finishing the experiment, careful removal of the bed allows for further analysis of the bed stratigraphy.

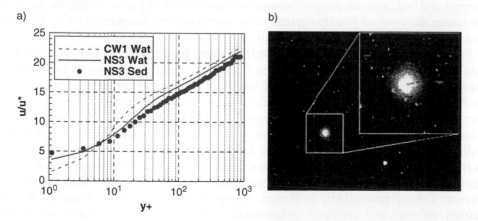

Figure 5.4.1 PIV measurements in sediment-laden channel flows using two-phase flow approach: a) streamwise velocity profiles for sand (NS3Sed), water (NS3Wat) and clear water equivalent flow (CW1Wat); b) instantaneous velocity vectors in the vicinity of a sand particle (Muste *et al.*, 2005)

Plan-form changes. For larger-scale morphodynamics at the reach scale, the erosion and accretion processes at the banks often play an important role. The transverse motion of the channel should therefore be made possible by using a wide flume in which a relatively narrow initial channel is excavated, which in time evolves into a self-shaping geometry. Many of the features found in natural rivers can be reproduced on a very small scale (*e.g.*, van Dijk *et al.*, 2013). With the appropriate sediment composition, channel slope and discharge, the characteristic patterns associated with braiding, anabranching, meandering and pattern changes like chute cut-offs and avulsions could be qualitatively replicated in hydraulic models on the order of 10 to 500m^2. When a more quantitative interpretation of the processes is required (*e.g.*, in engineering applications), the proper representation of turbulence and flow separation needs to be replicated in the physical model (*e.g.*, Friedkin, 1945). Accordingly, the dimensions of the models must be increased. Plan-form changes can easily be visualized by ensuring that the flowing water has a color that contrasts with the sediment. A more detailed analysis can be provided by a 3D mapping of the bed morphology using laser, acoustic probe, or stereo-photogrammetric techniques (Butler *et al.*, 2002). Applying these techniques during flow conditions is challenging as the free surface is a strong reflector of light and sound whereas refraction changes the direction of light and sound that do penetrate.

5.4.2 Gravity Currents

5.4.2.1 Introduction

Gravity currents, also called density or buoyancy currents, are buoyancy-driven flows moving as a result of density differences primarily in the horizontal direction. The density difference between the involved fluids can be caused by temperature or by dissolved (salinity) or suspended material (turbidity currents, snow avalanches). Turbidity currents are a special type of gravity current with material exchange between current and bed, whereas density currents are characterized by no or negligible material exchange. Reviews of gravity currents have been given by Simpson (1997), Meiburg and Kneller (2010) and Ungarish (2010).

Examples of gravity currents include freshwater fronts merging with sea water in estuaries, sediment-laden flows entering reservoirs and lakes, plumes of warm moist air emitted by cooling towers, and the interflow of water of different temperatures. Such currents also occur in other fluids besides water: *e.g.*, outflows from thunderstorms, as well as base surges formed from gases and solids from volcanic eruptions or snow avalanches. Figure 5.4.2a shows an atmospheric gravity current entraining sediment in Khartoum, Sudan, caused by the collapse of a thunderstorm. Gravity currents are also of practical importance in mining (spread of dense and potentially explosive gases), for natural hazards (pyroclastic flows, submarine cable breakage), aircraft safety, atmospheric pollution, oil spillage on water, discharge of power station effluent and reservoir sedimentation (Simpson, 1997). Sometimes water flows are used to model density currents in other fluids. Figure 5.4.2b shows images of one of the first experimental investigations of an atmospheric gravity current in a water channel by Schmidt (1911).

Figure 5.4.2 (a) gravity current in the atmosphere in Khartoum, Sudan (http://www.techeblog.com/) and (b) experimental investigation of an atmospheric gravity current in a laboratory water channel by Schmidt (1911) with increasing Reynolds number from i to vi.

5.4.2.2 Experimental flow complexity

Gravity currents are commonly investigated with saline or a suspended particle-water mixture (denser fluid) intruding into fresh water (lighter fluid), and sometimes using water flows of different temperatures. Early studies investigated the advancement of the front and the current height (Keulegan, 1957). These parameters are now relatively well understood and recent studies (Gerber *et al.*, 2011; Fragoso *et al.*, 2013; Nogueira *et al.*, 2013) concentrate mainly on the entrainment rate, mixing processes and turbulent features in the most active areas of the developing gravity currents. Most measurements are based on optical visualization methods where the resolution of the density field is important for the quantification of the extent of mixing (Fragoso *et al.*, 2013). Sophisticated measurement methods allow for the investigation of both velocity and density fields simultaneously (Ramaprabhu & Andrews, 2003; Martin & Garcia, 2009; Gerber *et al.*, 2011).

Gravity currents are mainly investigated with one of the following three methods/ facilities:

1. *Lock-release method.* A barrier separating a fixed quantity of denser fluid from less dense fluid is suddenly removed such that the denser fluid intrudes into the less dense one. Relevant parameters affecting the gravity current in this method include the initial lock height, the density difference between the two fluids, the lock length and the surface roughness. Three phases can be observed namely (i) the slumping phase where the less dense fluid moves backwards over the released fluid, is reflected at the back wall, and then eventually catches up with the head, (ii) a self-similar phase during which the current moves under a balance between buoyancy and inertial forces, and (iii) the viscous phase where the movement results as a balance between buoyancy and viscous forces (Keulegan, 1957; Huppert & Simpson, 1980; Simpson, 1997; Kneller *et al.*, 1999; Adduce *et al.*, 2012; Fragoso *et al.*, 2013; Nogueira *et al.*, 2013). Less common are studies where the fluid behind the barrier is less dense (fresh water) propagating on a denser (saline water) fluid in order to investigate oil spills (Hoult, 1972; Simpson, 1997).

2. *Arrested gravity current method.* The water level in a channel is controlled with an overflow weir or a similar structure (Simpson, 1969). The denser water is fed through the bottom corner in front of this structure. The water in the channel

flows over the denser fluid and the weir. The head of the current is then quasi steady and the gravity current is approximated by a logarithmic velocity profile resulting in small mixing due to small velocity close to the channel floor (as in a tide driven salt water wedge propagating upstream in a fresh water river). A uniform velocity profile (as experienced by a turbidity current) and considerably larger entrainment and mixing can be generated if the floor of the flume is a moving conveyor belt that extends to the front of the current head. In this method the entire (steady) head can be recorded at once over time and the Kelvin-Helmholtz billows above the front of the dense fluid are not disturbed by lobes and clefts (Simpson, 1997; Parsons & Garcia, 1998; Martin & Garcia, 2009).

3. *Constant feed (constant flux) method*. In this method, the denser water is prepared in a mixing tank or overhead reservoir, and is then constantly fed into the test section of a fresh water tank until it reaches a downstream tank with a constant water level controlled by a weir or pump. In this method the entire current can be recorded at once over time (Parker *et al.*, 1987; Ramaprabhu & Andrews, 2003; Choux *et al.*, 2005; Gerber *et al.*, 2011).

The previous paragraphs consider turbulent gravity currents caused by small density differences in water flow propagating over a horizontal surface with consideration of the Boussinesq approximation (see Section 2.4.3.2 in Volume I). As mentioned above, density currents occur in other fluids, and have been the subject of laboratory investigation. For example, experiments on viscous gravity currents, such as propagating lava, were conducted by Huppert (1982), and were further discussed by Simpson (1997) and Huppert (2006). Also gravity currents propagating on an inclined boundary (snow avalanches, turbidity currents) behave differently from gravity currents moving over a horizontal surface, as addressed by Britter and Linden (1980) and Alavian (1986). In Parker *et al.* (1987) the boundary was inclined and additionally covered with silt in order to investigate the sediment exchange between the bottom and turbidity current. Simpson (1997) describes many further special cases such as gravity currents meeting obstacles, high density ratios, propagation over a porous media, intrusion into ambient stratified or turbulent fluids and gravity currents on a rotating framework.

5.4.3 Flow through vegetation

5.4.3.1 Introduction

Flow-vegetation interaction is a rapidly developing research area due to the significance of vegetation for both ecological and engineering applications. Despite significant research work in recent years (see summaries in *e.g.*, Nikora, 2010; Folkard, 2011; Nepf, 2012; Aberle & Järvelä, 2013), many research questions remain unanswered. Given the complexity of the flow disturbance due to the presence of vegetation, many of these questions have to be tackled by means of laboratory and field experiments.

Plants and plant communities in a flow are exposed to flow-induced forces in the form of viscous drag, pressure drag and lift, as the water is forced to move around stems, branches, leaves, and patches of vegetation. The proportions of the hydrodynamic forces depend not only on flow parameters, but also on plant characteristics such as biomechanical properties, topology, age, seasonality, foliage, volumetric and areal vegetation porosities, vegetation density, and patchiness (*e.g.*, Nikora, 2010; Nepf, 2012; Aberle & Järvelä, 2013).

Vegetation has often been modeled in a simplified way using rigid cylinders without considering the actual behavior of natural plants. While rigid cylinders may be used to appropriately mimic rigid plant-stems, such models are inappropriate for modeling the hydraulic resistance of flexible natural vegetation. This is due to resistance imposed by foliage (*e.g.*, Jalonen *et al.*, 2013) and the ability of plants to bend and become more streamlined with the flow (*e.g.*, Nikora, 2010; Nepf, 2012), as visualized in Figure 5.4.3 for an isolated natural foliated willow twig exposed to flow velocities ranging from still water (0 m/s) to 0.89 m/s. Streamlining occurs when flow forces exceed the reaction forces of flexible plants and is thus governed by plant biomechanical properties (Luhar & Nepf, 2011).

Vegetation characteristics significantly affect flow conditions across a large variety of spatial scales ranging from the stem scale through the leaf, canopy, depth, to the reach scale. This means that flow-vegetation interaction is characterized by a complex multi-scale flow pattern (Nikora, 2010). For example, experimental results obtained with a single isolated plant cannot be directly up-scaled to the patch or canopy scale, as wakes generated by upstream elements alter the approach flow conditions for individual plants within a patch or the canopy. The complexity of the problem entails determining both hydraulic and plant parameters in experimental studies so that flow-induced and plant-reaction forces can be parameterized at the plant scale; *i.e.*, for individual plants, as well as for plant communities at the canopy, patch or patch mosaic scale (*e.g.*, Nikora, 2010; Luhar & Nepf, 2013).

Another important aspect to be considered relates to field and laboratory investigations of flow-vegetation interaction. Field experimental studies have the advantage that plants grow in their natural environment (*e.g.*, Sukhodolov & Sukhodolova, 2010; Cameron *et al.*, 2013) but are rare due to heterogeneous boundary conditions that are difficult to control (*e.g.*, discharge, flow velocity, water level), challenges in using

Figure 5.4.3 Frontal projected area and side view of a submerged willow twig exposed to different flow velocities. The percentages indicate the proportion of the projected area compared to the no-flow case (Figure from Aberle & Järvelä, 2013 with //www.informaworld.com. Reprinted by permission of the publisher, Taylor & Francis Ltd, http://www.tandfonline.com; data from Schoneboom, 2011).

sophisticated measurement instruments in the field, and safety concerns. Laboratory studies, on the other hand, can be carried out with controlled boundary conditions, sophisticated and sensitive measurement techniques, and both natural and artificial vegetation elements (*e.g.*, Thomas *et al.*, 2014, Banerjee *et al.*, 2015). The scaling of biomechanical properties of natural vegetation is not yet resolved so that corresponding experiments in the laboratory have been carried out with prototype aquatic vegetation (*e.g.*, Siniscalchi & Nikora, 2012), trees in towing tanks (*e.g.*, Whittaker *et al.*, 2013; Jalonen & Järvelä, 2014), plant parts (*e.g.*, Västilä *et al.*, 2013), or artificial objects with a simplified geometry and structure such as rigid cylinders (see Aberle & Järvelä, 2013).

In general, laboratory experiments with natural vegetation are difficult due to the limited size of flumes or experimental channels, and the altered habitat conditions for the plants which in turn may affect plant biomechanical properties. An option to determine plant conditions is the monitoring of physiological stresses, but such experimental aspects add complexity and require support from biologists or ecologists. The use of artificial plants helps to overcome difficulties with plant nursing and can eliminate the variability of individual plants within patches or the vegetation canopy, but choosing artificial vegetation only on the basis of geometrical similarity with the prototype vegetation does not guarantee adequate modeling of biomechanical properties and the resistance behavior of natural plants (Järvelä, 2006).

5.4.3.2 Experimental complexity

Vegetation characterization. The most common parameters for describing individual plant elements in hydraulic engineering applications have been the stem diameter, plant height, and the vegetation density. A more complete description of natural vegetation would require information on plant morphology, topology, foliage, biomass, and biomechanical properties. These properties can be determined using methods used in ecology and forestry (*e.g.*, Frostick *et al.*, 2011) but are often not adequately reported in publications. Moreover, geometric information of vegetation in dry conditions (*e.g.*, riparian vegetation) may be obtained through remote sensing techniques allowing for a detailed description of such properties even over depth (*e.g.*, Antonarakis *et al.*, 2010). Determining biomechanical properties of natural vegetation requires harvesting the plants so that they cannot be tested again in subsequent experiments. The applicability and the potential influence of the test procedure for the determination of plant characteristics should be considered, as the cutting of a plant can trigger physiological reactions that may modify plant tissue characteristics.

Plant deformation and frontal projected area (as illustrated in Figure 5.4.3) can be quantified on the basis of photogrammetric techniques (*e.g.*, Statzner *et al.*, 2006). Care needs to be taken in the case of flexible elements as the deformation can alter the elements' distance to the camera which in turn can result in biased estimates when applying image analysis (Sagnes, 2010). Plant deformation is also a dynamic process so that sequences of pictures may be required for the appropriate characterization of plant deformation which may also be used to determine plant motion (Cameron *et al.*, 2013).

Hydrodynamic forces. The hydrodynamic forces acting on vegetation elements can be measured with drag force sensors (see Section 6.6 in Volume II). These sensors may be installed in a false bottom below the bed so that they do not disturb the flow. In case the measurement system is located above the water surface, the plants with which the measurements are carried out can no longer be attached to the flume bed. This may not be a problem for rigid elements in emergent conditions but imposes some experimental peculiarities for submerged conditions, flexible plants or plant parts, and buoyant plants as components of the measurement system are exposed to the flow. If drag force measurements on plants are not possible, the accurate estimation of the vegetative drag is only possible by assuming that friction losses at the bed are negligible. This, however, depends on the vegetation density (*e.g.*, Schoneboom *et al.*, 2010).

Hydrodynamic parameters. The experimental quantification of flow hydrodynamics in experiments with vegetation is challenging due to the aforementioned spatial flow heterogeneity and scale dependency. In general, the turbulent flow field may be measured with common methods and, in case of submerged conditions, flow velocities can easily be measured above the canopy. Within the canopy, plants or plant parts may interfere with the measurements (*e.g.*, leaves, bending vegetation) thus complicating the measurements. For a detailed analysis of the dynamics of flow-vegetation interaction, the measurement of plant movement, drag forces, and flow velocities should be synchronized thus enabling cross-correlation analyses (*e.g.*, Siniscalchi & Nikora, 2012; Cameron *et al.*, 2013).

A theoretical framework for investigating flow through vegetation is provided by the double-averaging methodology (DAM) (*e.g.*, Nikora *et al.*, 2007; Nikora & Rowinski, 2008, and Section 2.3.2 in Volume I). The DAM framework allows the upscaling of physical interactions and mass transfer processes, as well as for considerations of mobile-boundary conditions (Nikora *et al.*, 2013) such as flexible vegetation.

5.4.4 Aerated Flows

5.4.4.1 Introduction

Aerated or bubbly flows occur when air bubbles become distributed in a water body, either by air entrainment at the air/water interface or by direct injection from natural or artificial sources at the fluid boundaries. From a chemistry perspective, the technical term aeration refers to the increase in gas concentration (usually of oxygen) in the water resulting from dissolution out of the bubbles. Likewise, gas stripping (removal of dissolved gas from the water) may occur when the bubbles are deficient in gasses dissolved in the water.

From a physics perspective, bubbly flows are examples of multiphase flow, and the lower density of the gas bubbles compared to that of the displaced water results in a local upward, buoyant force on the water from the bubbles. When the region of bubbly flow is narrow, the collective effect of the buoyancy is to drive an upward plume flow with entrainment at the plume edges; such flows are termed bubble plumes. When the aerated flow is more widely distributed, entrainment from the sides of the bubble swarm is limited, and the buoyant force does not result in net fluid flow of the water; idealized examples of this type of flow are bubble columns. Reviews on aspects of bubbly flows can be found in Mudde (2005), Balachandar and Eaton (2010), and Kiger and Duncan (2012).

Aerated flows are a very common scene in the natural aquatic environment, such as in rivers, lakes, reservoirs, and the ocean. Examples of aerated flows resulting from entrainment at the air-water interface include hydraulic jumps, white water, breaking waves, aeration at dams and weirs, and droplet breakup in water falls. Bubble plumes can result from natural gas emissions at submerged seeps or by direct injection through point or line-source diffusers, as in reservoir aeration.

Measurements in aerated flows generally include liquid and bubble velocities and bubble void fraction and size distribution. The measurement tools include non-intrusive image-based techniques and nearly non-intrusive probes of small size, such as conductivity probes, fiber-optic probes, and more bulky probes such as impedance-based probes. Image-based methods obtain bubble size and void fraction by image analysis and correlate consecutive images to obtain bubble velocities. Their application is limited, however, to low void-fraction flows where visual access to the region of interest is permitted with minimal overlap of bubble signatures in the images. At higher void fraction, velocity and void-fraction measurements may still be permitted, but size distribution becomes limited by bubble overlap. Point-based probes measure void fraction and bubble size by means of signal processing and may also measure bubble or water velocity, though usually in only one dimension along the probe axis. They have the advantage that they can be applied at any point in the flow field for any value of void fraction. The impedance-based probes are limited by their large sample volume (with length scales typically of order 25 mm), hence, the conductivity and fiber optic probes are perhaps the best tools for bubble size-distribution measurements if a good spatial resolution (typically of the order of millimeters or less) is desired, even though they are point-measurement tools (Lamarre & Melville, 1992).

The general rule on what tools to choose is as follows: Use imaging methods if the bubble concentration is sufficiently low (*i.e.*, few bubbles overlap in each image). Otherwise, use imaging methods for bubble and water velocity determination if possible, and use conductivity or fiber optic probes for bubble size distributions as well as bubble and fluid velocities.

5.4.4.2 Experimental flow complexity

Aerated flows are commonly investigated with air as the working gas. The flows are generally characterized by their void fraction and the resulting bubble and fluid motion; however, the presence of the bubbles complicates design of experiments and instrumentation. For Eulerian measurements (*e.g.*, hotwire, LDV, conductivity probe, or optical probe) the continuous measurement of the water properties are interrupted whenever a bubble passes through the sample volume (Lance & Bataille, 1991; Harteveld *et al.*, 2005; Rensen *et al.*, 2005). For volume (*e.g.*, PIV) or Lagrangian (*e.g.*, PTV) measurements based on non-intrusive optical methods, foreground bubbles may interrupt the viewing path to the region of interest (Ryu *et al.*, 2005; Seol *et al.*, 2007; Seol *et al.*, 2009). Despite these challenges, standard approaches to void fraction and bubble and fluid velocity field measurements are well developed as described below. Recent experiments focus on properties of the induced flow field and the flow agitation caused by the bubbles.

When designing aerated flow experiments, two important aspects relate to the experimental scale: the geometric scale of the bubbles and the length scale of the bubbly flow. As with all multiphase flow experiments, it is generally not possible to scale the bubble size by the same geometric factor as the rest of the model. For aerated flows, this is due to the fixed interfacial tension between air and water and the resulting stable bubble sizes. This fact is alleviated to a certain extent when the slip velocity is used as the governing parameter for the physics analysis because terminal rise velocity is fairly constant over a wide range of bubble sizes in the millimeter diameter range (Clift *et al.*, 1978). The result, however, is that bubbles in scaled laboratory experiments are generally larger than the corresponding bubbles in the prototype.

As introduced in Section 2.5.3 in Volume I, by adding the bubble diameter or slip velocity to the list of governing variables, it is usually possible to formulate a characteristic length scale of the bubble motion. In the case of bubble plumes, this length scale, called D, is given by the buoyancy flux, B (L^4/T^3), divided by the cube of the slip velocity, u_s^3 (L^3/T^3) (Bombardelli *et al.*, 2007). When comparing results between experiments or between the model and prototype, it is important to compare similar locations scaled by the characteristic bubble flow scale. Data collapse may depend on scaling by this length-scale parameter, and the shape and size of boundary conditions are generally important over length scales comparable to this scaling parameter (Bombardelli *et al.*, 2007; Bryant *et al.*, 2009).

Aerated flows are generated in the laboratory generally by one of the following experimental approaches:

Free-surface air entrainment. Many of the natural examples of aeration can be simulated in the laboratory using scaled versions of the prototype generation dynamics. Breaking waves, flow over weirs, and hydraulic jumps are common examples. Because the geometric dimensions and flow forces are usually smaller than in the field, the resulting air bubble size distribution may differ from the prototype or the selected model scale. This is common throughout multiphase flow experimentation, as discussed above, and the size distribution in these experiments is largely out of the control of the modeler.

Point and porous plate injectors. For bubble plume experiments, bubbles can be introduced either from an open orifice or from a porous injection stone. Point injection from the end of a pipe or nozzle generally results in the widest and most non-uniform size distribution owing to the bubble break-up process of a gas jet into water. An example of a specialized nozzle injection is given in Zhang and Zhu (2013). Air injection by pushing air through a porous stone or plate can result in more uniform bubble size distributions. Examples include the plume experiments by Socolofsky and Adams (2005), Bryant *et al.* (2009) and Hugi (1993).

Bubble columns. In most environmental applications of aerated flow there is either a background mean flow or the aerated region itself generates a mean flow. Bubble columns can be used as an artificial laboratory method to isolate the flow dynamics of the bubbles by removing the background flow. Bubble columns are also common in industrial applications. In an idealized bubble column, the bubble distribution is uniform throughout a narrow and tall laboratory tank. As long as localized bubble aggregation does not occur, entrainment and plume dynamics are prevented by the bounding sidewalls and recirculation cells are inhibited by the uniform void fraction and narrow facility aspect ratio. Such flows are difficult to achieve and require a careful bubble injection system, usually consisting of an array of micro tubes or hypodermic needles uniformly distributed over the full bottom of the column. A good example and review of related work is presented in Riboux *et al.* (2010).

5.4.4.3 Measurement approaches

Due to the presence of the bubbles, the instrumentation for bubbly flow measurement is generally not commercially available, or at least studies cannot utilize turn-key systems. That has forced researchers to build their instruments and analysis software in-house. Although the situation has improved in recent years, commercial systems are mostly limited to image-based techniques. For conductivity-based or fiber-optic-based systems, although they are indeed quite simple to design and build, there are few commercial alternatives. Researchers may have to build their own system or may be able to purchase custom systems through individual contact with researchers who have reported their work using such systems. These two main approaches (image- and probe-based systems) for flow velocities, bubble void fraction, and bubble size distributions are discussed in the following:

Imaging methods. Imaging methods are perhaps the first method researchers may attempt in a modern fluid mechanics laboratory since they are usually an extension from the popular particle image velocimetry (PIV) described in Chapter 3.7, Volume II. Details about PIV can be found in Raffel *et al.* (2007) and free software for PIV image analysis can be downloaded from Mori and Chang (2003).

Since bubble velocities are different from liquid velocities, directly applying PIV in a bubbly flow will generally result in only the bubble velocities being obtained, since their diameters (order of mm) are usually much greater than that of seeding particles (order of 10 µm) while image brightness is proportional to diameter squared. In some cases, the laser power could be adjusted to obtain vectors for both the seeding particles and the bubbles. In that case, a post-processing technique is needed to distinguish the bubble velocity vectors from the water velocity field (Seol & Socolofsky, 2008). More commonly, PIV is applied to images where bubbles and tracer particles have already been separated into separate images. One may remove bubble images either by image post processing (Seol *et al.*, 2007) or by the use of fluorescent seeding to separate bubbles from particle images during image capture (Seol & Socolofsky, 2008). In both cases, the laser power needs to be carefully adjusted so that the phases can be distinguished. A high dynamic-range camera (10 bits or above) is required for quality results.

PIV methods will work only for low bubble-concentration flows, defined as bubble images not having too much overlap (such as bubble plumes generated by a low air flow rate). For highly aerated flows, bubble image velocimetry (BIV) may be applied. The BIV technique, introduced by Ryu *et al.* (2005), acquires backlit images without the need of a laser. Bubbly flows are illuminated from behind by a uniform light source, such as an LED panel, while a high-speed camera captures shadow textures created by gas-liquid interfaces, including gas bubbles and liquid droplets. By limiting the camera depth of field (DOF) through adjusting the camera aperture and distance from its focal plane, objects appear sharp only within the DOF (typically a few centimeters) because they are in focus and have a higher weight in correlation analysis so velocities are based on bubbles/droplet images within that thin layer (analogous to a light sheet in PIV). Because of the high void fraction, the region of interest must be near the boundary of the bubbly flow region. The technique has been applied to a wide range of violent flows, including hydraulic jumps as shown in Figure 5.4.4a. Interestingly, BIV works in the highly aerated region while PIV can be applied in the continuous phase (water in this case) region – combining them may result in the entire flow field. Note that the BIV-

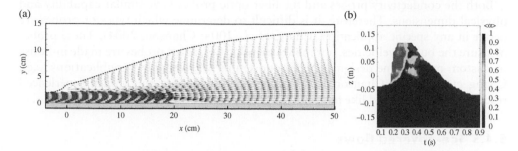

Figure 5.4.4 (a) Mean velocity field of a hydraulic jump (phase independent, for Froude number 5.0) measured by BIV (blue vectors) and PIV (red vectors) in Lin *et al.* (2012). (b) Mean void fraction measured at the center of the first splash-up roller of a plunging breaking wave in Lim *et al.*, 2015, with permission of Springer.

measured bubble velocities are, as shown in Lin *et al.* (2012) for hydraulic jumps, lower than the water velocities. The BIV-measured velocities are more representative under the condition where the inertia force is much greater than the buoyancy force such as in many overtopping flows (Ryu & Chang, 2008).

Conductivity Probes and Fiber Optic Probes. The conductivity probes and fiber-optic probes are perhaps the most commonly employed tools for bubble sizing and velocity determination in bubbly flows. These probes are generally applied in highly aerated flows where imaging methods do not work well. The conductivity probes have been used in the measurements of bubble velocities and chord lengths for decades (Chanson, 1997; Chanson, 2004). These probes typically have dual tips, made of platinum inside a stainless steel enclosure and have a diameter ranging from 100μm to a few mm. By changing conductivity when encountering a phase change, the probes obtain time-series electronic signals that contain phase information. Using two tips separated by a small gap (typically a few mm), the probes are able to measure the bubble residence time and void fraction (percentage of gas phase in the time series) and, through cross correlation, bubble velocities in the direction of the two tips. That in turn allows the calculation of the corresponding bubble chord lengths. The technique has been proven effective and accurate. The probes do not measure bubble diameters directly but through statistical analysis, and determining an appropriate gap between the dual tips is flow dependent.

Fiber-optic probes are relatively newer. They are very similar to the conductivity probes except the fiber-optic probes detect phase changes based on reflective index changes (Chang *et al.*, 2003). Similar to the conductivity probes, most fiber-optic probes also have dual tips that enable the measurements of void fraction, bubble velocities, and bubble chord lengths. Examples of the development of fiber-optic probes can be found in Chang *et al.* (2003), Chang *et al.* (2004) and Lim *et al.* (2008). Application of fiber-optic probes on bubbly-flow measurements can be found in Blenkinsopp and Chaplin (2007), Rojas and Loewen (2007), Lim *et al.* (2015), Lim *et al.* (2008) and Zhang *et al.* (2014). An example of the mean void-fraction field of a plunging breaking wave measured using fiber optic probes is shown in Figure 5.4.4b.

Both the conductivity probes and the fiber-optic probes have similar capability and physical dimensions. Therefore, it is difficult to determine which types of probes are better in any specific application (Chang *et al.*, 2004; Chanson, 2004). These probes measure the bubble velocities, not the water velocities. Most probes are made in-house or custom-made; some design sketches and setups are available in publications (*e.g.*, Chang *et al.*, 2003; Chanson, 1997). It may be necessary to contact experienced researchers if one would like to build or acquire such probes to study bubbly flows.

5.4.5 Ice-covered flows

5.4.5.1 Introduction

Flow situations involving ice are direct extensions of modeling free-surface flows, with the added complexities of the flow boundaries imposed by ice covers, and the consequences of ice-piece drift and accumulation. Further complexities arise when thermal and strength processes need to be simulated. Special laboratory facilities may be needed for replicating some of these processes. Often, substantial fieldwork has to be conducted, because some processes are not readily replicated in laboratories.

Common situations of laboratory investigation and hydraulic modeling involve free-surface flows with ice as a floating solid boundary retarding water flow, or ice occurring as solid pieces being conveyed and accumulated by water flow. Such flows primarily concern patterns in the model ice accumulations, velocity profiles of water flow, and possibly how they interactively affect ice movement and accumulation. The usual factors associated with water flow have to be considered in models aimed at studying ice effects on water flow: *i.e.*, water momentum, gravitational forces, the viscous and surface-tension properties of ice, and boundary resistance. When investigating or simulating ice movement and accumulation, additional forces are associated with ice momentum, hydrodynamic forces exerted against ice pieces, friction between ice pieces as well as between ice and other solid boundaries. These forces are normally handled in terms of the same parameters and similitude relationships used for simulating free-surface flows in channels of fixed or mobile (sediment) boundaries. Also to be considered are ice processes and their effects on coastal processes. Wind-driven ride-up of ice on beaches, wave damping by ice, ice-flow drift and accumulation are some of the issues faced in coastal waters.

The strength and deformation behavior of ice and accumulations of ice pieces, such as form ice covers or ice jams, are determined by geometric and material factors. Depending on the combination of these factors, the strength and deformation behavior can be relatively simple or complicated to formulate and replicate in the laboratory. At best they can be reproduced or simulated only within a fairly narrow range of length scales or facility extents and sometimes require special materials and additives, and procedures, for modifying the strength and material properties of ice. Thermal factors, such as freezing, freeze-bonding and the frictional behavior of ice pieces, can be difficult to simulate, because they involve processes at the molecular scale (notably the phase change of water) that cannot be further reduced in scale.

Useful references for additional general information about ice formation, ice transport, and ice strength behavior and their effects on water flow are Ashton (1989), Prowse (1993) and Beltaos (1995, 2013).

5.4.5.2 Laboratory Investigation

Practical problems can arise in experiments involving ice. Low air and water temperature may impede the use of some instruments, and ice accretion on instrumentation may hamper instrument use. Then, as mentioned above, special refrigerated laboratory facilities may be needed for conducting experiments in controlled circumstances. Such facilities can be expensive in construction and daily operation.

Simulation of ice processes often is not straightforward. Ice covers influence water velocity magnitudes and distribution, water stage, stream conveyance capacity, channel morphology and sediment transport. These influences occur due to variation of boundary shear stresses with and without ice cover as an upper boundary, and thus are influenced by ice roughness and water temperature near ice covers. Development of an upper boundary increases the flow resistance, and has been observed to decrease the shear stress on the river bed.

As rates of bed sediment transport vary non-linearly with bed shear stress, lower rates of bed material transport and suspended sediment fluxes are observed during the stable ice cover period. Nevertheless, significantly higher shear stress values on the lower and upper boundaries followed by higher concentration of suspended material can be observed, particularly during the break-up periods when the ice cover condition varies enormously and rapidly due to fragmentation and removal of the ice cover.

The temporal thickening of an ice cover, due to formation of hanging ice dams or the mechanical thickening process of "shoving", involves unsteady processes that may alter flow-related stresses by accelerating the flow through cross sections of changing area. This in turn may lead to local scour of channel beds during the freeze-up or stable ice-cover period. Obstruction to the flow due to the motion and deposition of ice floes in narrower parts of the river during the formation of ice jams can alter the flow stage and flow velocity drastically. Ice-jam formation and release is one of the main causes of riverine inundation and overbank flooding upstream and downstream of the jam location, respectively. Wind at times is a major driver of ice motion, especially in coastal regions. Creative design of experiments is needed to simulate wind-driven ice motion. These processes, and others described in the references given above, pose substantial challenges to laboratory experiments.

5.4.5.3 Field investigation

Ice-related processes often necessitate in-situ field investigations to determine river behavior under ice cover. This is particularly the case during the most dynamic stages of the river ice such as the freeze-up and break-up periods when the ice cover condition drastically affects the flow characteristics. There are significant difficulties to be overcome in obtaining reliable field measurements, because of the hazardous nature of the ice and river during the ice cover period – safety is a major concern for field work on rivers in winter. This concern prompts the imperative of developing and using new experimental techniques for ice covered rivers. New approaches are needed for continuous autonomous recording of flow and sediment transport under varying ice-cover conditions, and indeed of ice formation itself.

Fieldwork may necessitate installation of monitoring equipment on the bed below the ice cover for continuous autonomous data recording. Sample measurements using this approach are provided in Figure 5.4.5a. The alternative is to drill holes through the ice for data collection, but sampling in discrete, interval time steps - regardless of their adequacy – may only represent a restrictive condition. This method does not allow obtaining information on river and ice dynamics at all ice-cover stages.

Figure 5.4.5. Measurements in ice-covered flows: (a) Difference in water velocity profiles during different ice stages compared to open water condition (modified Zare *et al.*, 2016), (b) deployment of acoustic instruments through a cut in the ice cover; slush ice is present at the opening.

Study site accessibility in remote northern areas is often limited. In such cases, access may be constrained to aerial options, which are costly and limited in terms of instruments and crew transportation. This is even more critical in remote, wide rivers or lake and ocean studies, where helicopters must land on the ice surface. In this condition, the experiment must be timed properly in order to ascertain that the cover is thick and strong enough to land a helicopter. On the other hand, ice-cover presence acts as an obstacle that limits the progress of the experiments in many aspects. The cover usually must be drilled or cut, so that the instruments or probes can be deployed under the cover. Overcoming this task is a technical challenge when the cover is thick and rigid enough to provide a safe landing ground. If the instruments are oversized, like sediment traps or samplers, large holes must be cut into the ice to provide enough space to operate them. Cumbersome and heavy instruments like power ice augers or chain saws, both equipped with long blades, are necessary in this case. Sometimes the ice cover must be cut in more than one attempt. First, the upper part of the cover, as thick as the length of the blade, and then the lower, should be cut and removed. The cut ice must also be removed from the hole to open the water. Big ice chunks may be removed using winches and ice screws, or they can be cut in smaller pieces for removal by hand, which consumes more time and energy (Figure 5.4.5b). It is usually not possible to transport and operate heavy equipment, such as winches and cranes, and thus hand techniques are required.

In some cases, there is a significant amount of slush ice under the ice cover (*e.g.*, Zare *et al.*, 2016). This sometimes completely stops the experiment since the slush ice floats on the water and fills and blocks the opening. Due to its high buoyancy and density, slush ice in some cases acts like a rigid cover. At low temperatures, sampling holes may be rapidly covered by new ice formation. Thus, frequent site preparation may be needed. It is very important to emphasize that the dynamic and unstable condition of ice cover, during almost all river ice stages, raises the importance of safety considerations. Some stages, such as ice cover break-up, are especially hazardous; therefore, the feasibility of field investigations is only limited to remote observations. Hence, collection of consistent data that covers all river ice stages is not practically achievable using the traditional through-ice data acquisition technique.

5.5 INTERACTION OF EXPERIMENTS WITH NUMERICAL MODELING

5.5.1 Introduction

Experimentation in hydraulics emerged at the end of the 19th century in conjunction with the design of large water-related construction projects (Hager, 2015). This branch of experimentation, also known as physical or hydraulic modeling, investigates design and operation issues in hydraulic engineering. Physical models have the potential to replicate many features of a complicated flow. In many instances there is little recourse other than hydraulic modeling to make design and operational decisions for expensive and complex hydraulic works (Ettema, 2000). Physical modeling can also be used to better understand generic physical processes (*e.g.*, sediment transport, flow through vegetation). The first basic experimental research in open-channel turbulence started in the 1950s, triggered by the advent of hot-wire anemometry (Nezu, 2005). After 1980, even more powerful optical and acoustical techniques have made experimental studies in open-channel turbulence much less arduous, permitting detailed investigations of not only basic two-dimensional (2D) uniform flows, but also, more recently, unsteady and three-dimensional (3D) channel flows.

Hydraulic experiments allow insights into phenomena that are not yet described analytically or are so inadequately understood that they are inaccessible to numerical modeling. Moreover, laboratory or field experimentation provides much needed measurements to validate and calibrate numerical models. In turn, the validated numerical models can be used not only for extending the range of observations obtained from experiments but also for predicting the consequences of extreme conditions that cannot be measured in the field or laboratory because of safety reasons or operational complexities. Similar to all investigative tools in hydraulic research, physical experimentation has limitations too. A summary of the advantages and disadvantages of experimental hydraulics as compiled by Frostick *et al.* (2011) is provided in Table 5.5.1.

Today, numerical models are increasingly used in place of physical models. Physical and numerical models both have their strengths and weaknesses and their merits must be compared to the benefits of theoretical analysis and measurements made in the field (Frostick *et al.*, 2011). A single tool cannot adequately reproduce complex hydraulic processes and thus replace the others in all cases. Combining tools can add value, but also cost and time. In most cases the boundary conditions for a hydraulic model come from field measurements or a regional numerical model. Compromises are made with the amount of data or information that is available from the field and with the validity of the models for specific ranges of flows. When tools are used in parallel to model or test the same process, there are often differences in the results, but it is not always

Table 5.5.1 Advantages and disadvantages of hydraulic experimentation (Frostick *et al.*, 2011)

Advantages	Disadvantages
Enable discovery of new processes within phenomena not yet described or understood	Down-scaled models are affected by scale effects
Employ less simplifying assumptions compared to analytical and numerical models	Impossible to simulate all relevant process variables in correct relationship to each other
Validate theoretical hypotheses	Difficult to accurately simulate flow boundaries and forcing conditions
Can exceed the ranges of field measured data	Experiments are expensive to build and maintain
Enable insights into process aspects that were never observed in the prototype	
Provide immediate insights into processes that in turn allow the study to be better focused and reduce the planned testing	

possible to decide which tool is more trustworthy. A new approach increasingly considered in today's investigations entails combining field and physical modeling results with numerical simulations in a complementary, coordinated manner as described next.

5.5.2 Composite modeling

The term composite modeling was introduced by the Hydralab consortium to define the integrated and balanced use of field measurements, physical models and numerical models (Frostick *et al.*, 2011). The links between these knowledge building blocks and theoretical analysis are shown in Figure 5.5.1. Each block links with the other three in a two-way flow of information and feedback. For example, field experiments inform theoretical analysis, while the results of the analysis suggest further field experiments. Physical model data are used to calibrate and validate numerical models while theoretical analysis and numerical model results often lead to further physical experimentation and additional field-data collection to understand complex features in real-life processes. Moreover, numerical models can be used in the design of physical models with the former being used to decide on the physical model scale, location and density of various instruments and the sampling approaches used in the experiments.

The hydraulic community envisions that the use of composite modeling can result in (Frostick *et al.*, 2011): a) better simulation of the relevant prototype processes; b)

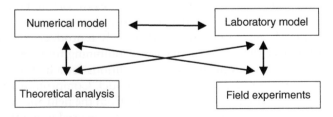

Figure 5.5.1 Knowledge-building blocks supporting hydraulic investigations (Frostick *et al.*, 2011)

improved modeling infrastructure; c) reduced uncertainty and cost of the studies; and, d) building knowledge-production technology for future investigations in hydraulics. To date, there is an increased number of situations that can be satisfactorily modeled. Frostick *et al.* (2011) expand on various aspects of composite modeling including: case studies, good modeling practices, selection of composite modeling hypotheses, organizing a composite modeling experiment and substantiating the pitfalls and traps of this investigative approach.

Example 5.5.1:   Synergy between experiments and numerical simulations.   The interaction between experimental and numerical studies is illustrated by a study conducted at the Iowa Institute of Hydraulic Research (IIHR) - Hydroscience & Engineering (Lyons *et al.*, 2007). The goal of the study was to provide design guidance for a vertical segmented gate (stoplog) to be used for emergency situations. The overall view of the 1:24-scale model is provided in Figure 4.2.2, Volume I. The gate segments are deployed sequentially under flowing-water conditions by a permanent hoist assembly and a lifting beam as illustrated in Figure 5.5.2a. Closing such systems gave rise to problems due to the complex interplay between the hydraulic downpull and uplift forces acting on the stoplogs. Moreover, gate closure might lead to vibrations or strong pressure waves around the individual stoplogs.

The relevant forces are indicated in Figure 5.5.2a. The position of the gate segment in the gate slot and selected hydrodynamic forces acting on the stoplog can be observed by the operator in real-time using customized data-acquisition software (see Figure 5.5.2b). The model tests provide the friction forces, upwards and downwards forces in the cables, net horizontal force, and dry and submerged weight on the stoplog at any location during the travel. Visual observations of the stoplog vibrations can also be observed using the monitoring software, as illustrated in Figure 5.5.2c. The drag force on the individual stoplog at selected locations is calculated with the numerical model (see Figure 5.5.2d and 5.5.2e).

Experimental results show an acceptable gate and lifting beam design with respect to dry and submerged weights, cable tensions, hydraulic conditions, and hoist requirements, and measurements agree well with numerical results with respect to vertical forces for several tested gate configurations. To a degree, this validates the forces orthogonal to the gate segments obtained from numerical calculations. Final gate design was based on combining the results of physical and numerical models. The required submerged weight for closure, dictating allowable gate volume and weight, was determined from experimental values for hydraulic force and numerical friction forces.

Comparison of the numerical and experimental results revealed capabilities and limitations of each methodology. Experimental results revealed a trend for lower friction forces deeper in the water column. The friction data, however, exhibit scatter, likely due to the friction force being determined by subtracting two cable-tension time series. Due to cable oscillation and deflection under flow, cable tensions are difficult to record at identical elevations in the upward and downward directions, leading to uncertainty. The agreement between experimental and numerical data was satisfactory for travel tests but poor for closure tests. The numerical model, however, was invaluable when analyzing flow fields around each geometric design and guided subsequent tests on the experimental model, ultimately arriving at the final design.

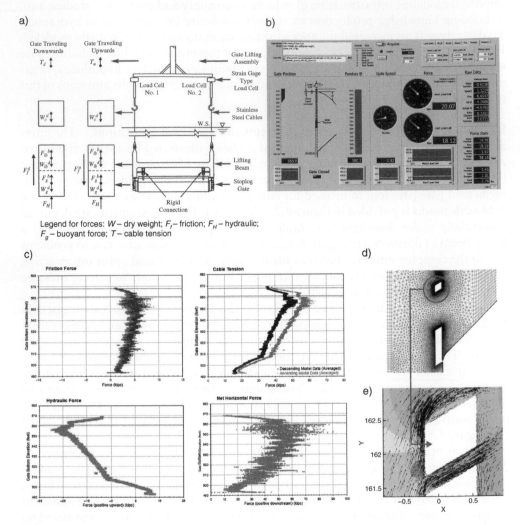

Figure 5.5.2 Interaction between experimental and numerical modeling (Lyons et al., 2007): a) hydrodynamic forces acting on the gate segment; b) experiment software graphical user interface; c) Sample of force variation during a downward-upward travel of the gate slot; d)computational mesh; e) results of numerical simulations.

REFERENCES

Aberle, J. & Järvelä, J. (2013) Flow resistance of emergent rigid and flexible floodplain vegetation. *J Hydraul Res.*, *51*(1), 33–45.

Aberle, J., Nikora, V., & Rennie, C.D. (2012) Evolution of environmental hydraulic instrumentation and experimental methods. In: Jordaan, J.M. & Bell, A. (eds.) *Hydraulic Structures Equipment and Water Data Acquisition System, Encyclopedia of Life Support Systems (EOLSS), Developed under the Auspices of the UNESCO.* Oxford, UK, Eolss Publishers. [Online] Available from: http://www.eolss.net.

Adduce, C., Sciortino, G., & Proietti, S. (2012) Gravity currents produced by lock exchanges: Experiments and simulations with a two-layer shallow-water model with entrainment. *J Hydraul Eng. – ASCE*, *138* (2), 111–121.

Adrian, R.J. (1991) Particle-imaging techniques for experimental fluid mechanics. *Annu Rev Fluid Mech.*, *23*, 261–304.

Adrian, R.J. & Yao, C.S. (1987) Power spectra of fluid velocities measured by laser Doppler velocimetry. *Experiments in Fluids*, *5*, 17–28.

Alavian, V. (1986) Behavior of density currents on an incline. *J Hydraul Eng. ASCE*, *112* (1), 27–42.

Antonarakis, A.S., Richards, K.S., Brasington, J., & Muller, E. (2010) Determining leaf area index and leafy tree roughness using terrestrial laser scanning. *Water Resources Research*. [Online] *46* (W06510), doi:10.1029/2009WR008318.

Antonia, R. & Krogstad, P. (2001) Turbulent structure in boundary layers over different types of surface roughness. *Fluid Dynamics* Res., *28*, 139–157.

Arndt, R. & Ippen, A. (1970) Turbulence measurements in liquids using an improved total pressure probe. *J Hydraul* Res., *8* (2), 131–158.

Ashton, G. (1989) *River & Lake Ice Engineering*, Water Resources Publication LLC, Highlands Ranch, CO, USA

Balachandar, S. & Eaton, J.K. (2010) Turbulent dispersed multiphase flow. *Annu Rev Fluid Mech.*, *42*, 111–133.

Banerjee, T., Muste, M., & Katul, G. (2015) Flume experiments on wind induced flow in static water bodies in the presence of vegetation. *Advances in Water Resources 76 (2)*, 11–28.

Bélanger, J.B. (1849) *Notes sur le cours d_hydraulique, session 1849–1850, 222p.Mém. Ecole Nat. Ponts et Chaussées*. Paris, France. (In French).

Beltaos, S. (ed.) (1995) *River Ice Jams*. Water Resource Publications, Littleton, CO.

Beltaos, S. (ed.) (2013) *River Ice Formation*. Canadian Geophysical Union, Montreal, Canada.

Bendat, J. & Piersol, A. (1993) *Engineering Applications of Correlation and Spectral Analysis, 2nd edition*. New York, Wiley.

Bigillon, F., Nino, Y., & Garcia M.H. (2006) Measurements of turbulence characteristics in an open-channel flow over a transitionally-rough bed using particle image velocimetry. *Experiments in Fluids*, *41*(6), doi:10.1007/s00348-006-0201-2, 857–867.

Blenkinsopp, C.E. & Chaplin, J.R. (2007) Void fraction measurements in breaking waves. *Proc. R. Soc. A.*, *0*(2088), 3151–3170.

Blom, A., Ribberink, J.S., & de Vriend, H.J. (2003) Vertical sorting in bed forms: Flume experiments with a natural and a trimodal sediment mixture. *Water Resour Res.*, *39*(2), doi:10.1029/2001WR001088.

Bosman, J.J., Van der Velden, E.T.J.M., & Hulsbergen, C.H. (1987) Sediment concentration measurement by transverse suction. *Coas Eng.*, *11*, 353–370.

Bouma, T.J., van Duren, L.A., Temmerman, S., Claverie, T., Blanco-Garcia, A., Ysebaert, T., & Herman, P.M.J. (2007) Spatial flow and sedimentation patterns within patches of epibenthic structures: Combining field, flume and modelling experiments. *Continental Shelf Res.*, *27*, 1020–1045.

Britter, R.E. & Linden, P.F. (1980) The motion of the front of a gravity current travelling down an incline. *J Fluid Mech.*, *99* (3), 531–543.

Bryant, D.B., Seol, D.G., & Socolofsky, S.A. (2009) Quantification of turbulence properties in bubble plumes using vortex identification methods. *Phys Fluids*, *21*(7), paper 075101.

Budwig, R. (1994) Refractive index matching methods for liquid flow investigations. *Experiments in Fluids*, *17*(5), 350–355.

Butler, J., Lane, S., Chandler, J., & Porfiri, E. (2002) Through-water close range digital photogrammetry in flume and field environments. *The Photogrammetric Record*, *17*(99), doi:10.1111/0031-868X.00196.

Cameron, S.M., Nikora, V.I., Albayrak, I., Miler, O., Stewart, M., & Siniscalchi, F. (2013) Interactions between aquatic plants and turbulent flow: A field study using stereoscopic PIV. *J Fluid Mech.*, *732*, 345–372.

Chang, K.A., Lim, H.J., & Su, C.B. (2003) Fiber optic reflectometer for velocity and fraction ratio measurements in multiphase flows. *Rev Sci Instrum.*, *74*(7), 3559–3565.

Chang, K.A., Lim, H.J., & Su, C.B. (2004) Reply to "Comment on 'Fiber optic reflectometer for velocity and fraction ratio measurements in multiphase flows'" [Rev Sci Instrum., 74, 3559 (2003)]. *Rev Sci Instrum.*, *75*(1), 286–286, doi.org/10.1063/1.1634361.

Chanson, H. (1997) Air bubble entrainment in open channels: Flow structure and bubble size distributions. *Int. J. Multiph. Flow*, *23*(1), 193–203.

Chanson, H. (2004) Comments on "Fiber optic reflectometer for velocity and fraction ratio measurements in multiphase flows" [Rev Sci Instrum., 74, 3559 (2003)]. *Rev Sci Instrum.*, *75* (1), 284–285, doi: 10.1063/1.1634360

Choux, C.M.A., Baas, J.H., McCaffrey, W.D., & Haughton, P.D.W. (2005) Comparison of spatio-temporal evolution of experimental particulate gravity flows at two different initial concentrations, based on velocity, grain size and density data. *Sedimentary Geol.*, *179* (1–2), 49–69.

Chue, S. (1975) Pressure probes for fluid measurement. *Progress in Aerosp Sci.*, *16*(2), 147–223.

Clift, R., Grace, J., & Weber, M.E. (1978) *Bubbles, Drops, and Particles*. Dover Publications, Inc., Mineola, New York.

Coleman, H.W. & Steele, W.G. (1999) *Experimentation and Uncertainty Analysis for Engineers*. New York, John Wiley & Sons.

Crosato, A., E. Mosselman, F. Beidmariam Desta, & Uijttewaal, W.S.J. (2011) Experimental and numerical evidence for intrinsic nonmigrating bars in alluvial channels. *Water Resour Res.*,47, W03511, doi:10.1029/2010WR009714.

Cui, M.M. & Adrian, R.J. (1997) Refractive index matching and marking methods for highly concentrated solid–liquid flows. *Experiments in Fluids*, *22*(3), 261–264.

Demers, S., Buffin-Bélanger, T., & Roy, A.G. (2013) Macroturbulent coherent structures in an ice-covered river flow using a pulse-coherent acoustic Doppler profiler. *Earth Surf Process and Landforms*, *38* (9), 937–946.

Detert, M. & Weitbrecht, V. (2012) Automatic object detection to analyze the geometry of gravel grains – a free stand-alone tool. In: Muños, R.M. (ed.) *River Flow 2012*. Taylor & Francis Group, London, ISBN 978-0-415-62129-8, pp. 595–600

Durst, F. (1982) Combined measurements of particle velocities, size distributions, and concentrations. *J Fluids Eng.*, *104*(3), 284–296, doi:10.1115/1.3241834.

Durst, F., Melling, A., & Whitelaw, J.H. (1976) *Principles and practice of Laser-Doppler Anemometry*. London, UK, Academic Press.

Ettema, R. (ed. and lead author) (2000) *Hydraulic Modeling: Concepts and Practice*. Reston, VA, ASCE Publications.

Farnsworth, E.J. & Rosovsky, J. (1993) The ethics of ecological field experimentation. *Conserv Biol.*, *7* (3), 463–472.

Figliola, R.S. & Beasley, D.E. (1991) *Theory and Design for Mechanical Measurements*. New York, John Wiley & Sons.

Folkard, A.M. (2011) Vegetated flows in their environmental context: a review. *Eng Comput Mech.*, *164* (EM1), 3–24.

Fragoso, A.T., Patterson, M.D., & Wettlaufer, J.S. (2013) Mixing in gravity currents. *J Fluid Mech.*, *734* (R2), R2-1-R2-10.

Friedkin, J.F. (1945) *A Laboratory Study of the Meandering of Alluvial Rivers*. Waterw. Exp. St., U.S. Army Eng., Vicksburg, Miss.

Frostick, L.E., McLelland, S.J., & Mercer, T.G. (2011) *Users Guide to Physical Modeling and Experimentation*. Balkema, Leiden, The Netherlands, CRC Press.

Frostick, L., Thomas, R.E., Johnson, M.F., Rice, S.P., & McLelland, S. (2014) *Users Guide to Ecohydraulic Modelling and Experimentation.* IAHR Design Manual. Balkema, Leiden, The Netherlands, CRC Press.

Fujita, I. & Muste, M. (2011) Preface to the special issue on image velocimetry. *J Hydro-environment Res.,* 5 (4), 213.

Garcia, C.M., Cantero, M.I., Jackson, P.R., & Garcia, M.H. (2004) *Characterization of the flow turbulence using water velocity signals recorded by acoustic Doppler velocimeters.* Civil Engineering Studies-Hydraulic Engineering Series, University of Illinois at Urbana-Champaign. Report number: 75.

Garcia, C.M., Cantero, M.I., Nino, Y., & Garcia, M.H. (2005) Turbulence measurements with acoustic Doppler velocimeters. *J Hydraul Eng.,131* (12), 1062–1073.

Gerber, G., Diedericks, G., & Basson, G.R. (2011) Particle image velocimetry measurements and numerical modeling of a saline density current. *J Hydraul Eng. – ASCE, 137* (3), 333–342.

Goldstein, R.J. (1996) *Fluid Mechanics Measurements, 2nd edition.* Washington, DC, Taylor & Francis.

Hager, W.H. (2015) *Hydraulicians in the USA 1800-200,* CRC Press, New York, N.Y.

Harteveld, W.K., Mudde, R.F., & Van den Akker, H.E.A. (2005) Estimation of turbulence power spectra for bubbly flows from Laser Doppler Anemometry signals. *Chem Eng Sci.,* 60(22), 6160–6168.

Holdaway, G.P., Thorne, P.D., Flatt, D., Jones, S.E., & Prandle, D. (1999) Comparison between ADCP and transmissometer measurements of suspended sediment concentration. *Continental Shelf Res.,19*(3) 421–441.

Holman, J.P. (1989) *Experimental Methods for Engineers.* New York, McGraw Hill.

Holmes, R. & Garcia, M. (2008) Flow over bedforms in a large sand-bed river: A field investigation. *J Hydraul Res.,* 46 (3), 322–333.

Hoult, D.P. (1972) Oil spreading on the sea. *Annu Rev Fluid Mech.,* 4, 341–368.

Hugi, C. (1993) *Modelluntersuchungen von blasenstrahlen fuer die seebelueftung, Ph.D. Thesis,* Inst. Hydromechanics and Water Resources, ETH, Zurich.

Huppert, H.E. (1982) The propagation of two-dimensional and axisymmetric viscous gravity currents over a rigid horizontal surface. *J Fluid Mech.,* 121, 43–58.

Huppert, H.E. (2006) Gravity currents: a personal perspective *Journal of Fluid Mechanics, 554,* 299–322.

Huppert, H.E. & Simpson, J.E. (1980) The slumping of gravity currents. *J Fluid Mech.,* 99 (4), 785–799.

Jalonen, J. & Järvelä, J. (2014) Estimation of drag forces caused by natural woody vegetation of different scales. *J Hydrodynamics Serial B,* 26 (4), 608–623.

Jalonen, J., Järvelä, J., & Aberle, J. (2013) Leaf area index as vegetation density measure for hydraulic analyses. *J Hydraul Eng.,* 139 (5), 461–469.

Järvelä, J. (2006) Vegetative flow resistance: Characterization of woody plants for modeling applications. In: *Proc. World Environmental and Water Resources Congress, Omaha.* Papers on CD-Rom, ASCE, USA.

Keulegan, G.H. (1957) An experimental study of the motion of saline water from locks into fresh water channels. *U.S. National Bureau of Standards Research Report,* 5168.

Kiger, K.T. & Pan, C., (2000) PIV technique for the simultaneous measurement of dilute two-phase flows, *J. Fluids Eng.* 122(4), 811-818 doi:10.1115/1.1314864

Kiger, K.T. & Duncan, J.H. (2012) Air-entrainment mechanisms in plunging jets and breaking waves. *Annu Rev Fluid Mech.,* 44, 563–596

Kneller, B.C., Bennett, S.J., & McCaffrey, W.D. (1999) Velocity structure, turbulence and fluid stresses in experimental gravity currents. *J Geophys Res.,* 104 (C3), 5381–5391.

Lance, M. & Bataille, J. (1991) Turbulence in the liquid-phase of a uniform bubbly air water-flow. *J Fluid Mech.,* 222, 95–118.

Lamarre, E. & Melville, W.K. (1992) Instrumentation for the measurement of void-fraction in breaking waves - laboratory and field results. *IEEE J. Ocean Eng.,17*(2), 204–215.

Legleiter, C.J. & Overstreet, B.T. (2014) Retrieving river attributes from remotely sensed data: An experimental evaluation based on field spectroscopy at the Outdoor StreamLab. *River Res Appl., 30* (6), 671–684.

Lenschow, D.H., Mann, J., & Kristensen, L. (1994) How long is long enough when measuring fluxes and other turbulence statistics? *J Atmos Oceanic Technol 11*, 661–673.

Lim, H.J., Chang, K.A., Huang, Z.C., & Na, B. (2015) Experimental study on plunging breaking waves in deep water. *J. Geophys. Res. Oceans, 120*(3), 2007–2049.

Lim, H.J., Chang, K.A., Su, C.B., & Chen, C.Y. (2008) Bubble velocity, diameter, and void fraction measurements in a multiphase flow using fiber optic reflectometer. *Rev Sci Instrum., 79*(12), 11.

Lin, C., Hsieh, S.C., Lin, I.J., Chang, K.A., & Raikar, R.V. (2012) Flow property and self-similarity in steady hydraulic jumps. *Exp. Fluids, 53*(5), 1591–1616.

Luhar, M. & Nepf, H.M. (2011) Flow-induced reconfiguration of buoyant and flexible aquatic vegetation. *Limnol Oceanography, 56* (6), 2003–2017.

Luhar, M. & Nepf, H.M. (2013) From the blade scale to the reach scale: A characterization of aquatic vegetative drag. *Adv Water Resour., 51* (1), 305–316.

Lyons, T, Muste, M., Weber, L., Politano, M., Carrica, P., & Hay, D. (2007) Experimental and Numerical Study for the Design of Emergency Stoplogs Deployed in a Complex Flow Field. *Proceedings Hydropower XV Conference*. Chattanooga, TN.

MacMillan, F. (1956) *Experiments on Pitot-Tubes in Shear Flow*. Aeronautical Research Council, London, UK. Reports and memoranda number: 3028.

Marcus, W.A. & Fonstad, M.A. (2010) Remote sensing of rivers: The emergence of a subdiscipline in the river sciences. *Earth Surface Process and Landforms*. [Online] Available from: doi:10.1002/esp.2094.

Marion, A., Nikora, V., Puijalon, S., Bouma, T., Koll, K., Ballio, F., Tait, S., Zaramel-la, M., Sukhodolov, A.N., O'Hare, M., Wharton, G., Aberle, J., Tregnaghi, M., Da-vies, P., Nepf, H., Parker, G., & Statzner, B. (2014) Hydrodynamics and Ecology: The critical role of interfaces in biophysical interactions. *J Hydraul Res*. [Online] Available from: http://dx.doi.org/10.1080/00221686.2014.968887.

Martin, J.E. & Garcia, M.H. (2009) Combined PIV/PLIF measurements of a steady density current front. *Experiments in Fluids, 46* (2), 265–276.

Meiburg, E. & Kneller, B. (2010) Turbidity currents and their deposits. *Annu Rev Fluid Mech., 42*, 135–156.

Meixner, H. & Rauch, H.P. (2004) Projekt "Neuer Wienfluss"– Messungen an der ingenieur-biologischen Versuchsstrecke. In: Florineth, F. (ed.) *Endbericht 1998–2003*. Universität für Bodenkultur, Vienna, Austria.

Moffat, R.J. (1988) Describing the uncertainties in experimental results. *Experimental Thermal and Fluid Sci., 1*, 3–17.

Mori, N. & Chang, K.-A. (2003) Introduction to MPIV (http://www.oceanwave.jp/softwares/mpiv - last access May 2015.).

Mudde, R.F. (2005) Gravity-driven bubbly flows. *Annu. Rev. Fluid Mech.*, Annual Reviews, Palo Alto, 393–423.

Mueller, D.S., Abad, J.D., Garcia, C.M., Gartner, J.W., Garcia, M.H., & Oberg, K.A. (2007) Errors in acoustic Doppler profiler velocity measurements caused by flow disturbance. *J Hydraul Eng., 133* (12), 1411–1420.

Muste, M. (2002) Sources of bias errors in flume experiments on suspended-sediment transport. *Journal of Hydraulic Research,40* (6), 695–708.

Muste, M., Yu, K., Pratt, T., & Abraham, D. (2004) Practical aspects of ADCP data use for quantification of mean river flow characteristics: Part II: fixed-vessel measurements. *Flow Meas Instr., 15* (1), 7–28.

Muste, M., Yu, K., Fujita, I., & Ettema, R. (2005) Traditional versus two-phase perspective on turbulent channel flows with suspended sediment. *Water Resour Res.*, *41* (10), W10402, 22.

Muste, M., Kim, D., & Gonzalez-Castro, J.A. (2010) Near-transducer errors in ADCP measurements: experimental findings. *Journal of Hydraulic Engineering,136* (5), 275–289.

Muste, M. & Lee, K. (2013) Evaluation of hysteretic behavior in streamflow rating curves: *Proceedings 2013 IAHR Congress, Chengdu, September 8-13*. China, Tsinghua University Press.

Muste, M., Baranya, S., Tsubaki, R., Kim, D., Ho, H-C., Tsai, H-W. and Law, D. (2016). Acoustic Mapping Velocimetry. *Water Resources Research*, *52*, doi:10.1002/2015WR018354

Nakagawa, H. & Nezu, I. (1977) Prediction of the contributions to the Reynolds stress from bursting events in open-channel flows. *J Fluid Mech.*, *80*, 99–128.

Nepf, H.M. (2012) Hydrodynamics of vegetated channels. *J Hydraul Res.*, *50* (3), 262–279.

Nezu, I. & Nakagawa, H. (1993) *Turbulence in Open-Channel Flows*. Balkema, A.A., Rotterdam, Netherlands.

Nezu, I. (2005) Open-channel flow turbulence and its research prospect in the 21st century. *J Hydraul Eng.*, *131* (4), 229–246.

Nikora, V. (2010) Hydrodynamics of aquatic ecosystems: An interface between Ecology, Biomechanics and Environmental Fluid Mechanics. *River Res Appl.*, *26* (4), 367–384.

Nikora, V.I. & Rowinski, P.M. (2008) Rough-bed flows in geophysical, environmental, and engineering systems: Double-averaging approach and its applications. *Acta Geophys*, *56* (3), 529–533.

Nikora, V., Ballio, F., Coleman, S., & Pokrajac, D. (2013) Spatially averaged flows over mobile rough beds: Definitions, averaging theorems, and conservation equations. *J Hydraul Eng.*, *139* (8), 803–811.

Nikora, V., McEwan, I., McLean, S., Coleman, S., Pokrajac, D., & Walters, R. (2007) Double-averaging concept for rough-bed open-channel and overland flows: Theoretical background. *J Hydraul Eng.*, *133* (8), 873–883.

Nogueira, H.I.S., Adduce, C., Alves, E., & Franca, M.J. (2013) Dynamics of the head of gravity currents. *Environmental Fluid Mech.*, *14* (2), 519–540.

Orrú, C., Chavarrías, V., Uijttewaal, W.S.J., & Blom, A. (2014) Image analysis for measuring the size stratification in sand–gravel laboratory experiments. *Earth Surface Dynamics*, *2*, 217–232

Parker, G., Garcia, M., Fukushima, Y., & Yu, W. (1987) Experiments on turbidity currents over an erodible bed. *J Hydraul Res.*, *25* (1), 123–147.

Parsons, J.D. & Garcia, M.H. (1998) Similarity of gravity current fronts. *Physics of Fluids*, *10* (12), 3209–3213.

Prowse, T.D. (1993) Suspended sediment concentration during river ice breakup. *Canadian J Civil Eng.*, *20*, 872–875.

Raffel, M., Willert, C.E., Werely, S., & Kompenhans, J. (2007) *Particle Image Velocimetry: A Practical Guide*. Springer.

Ramaprabhu, P. & Andrews, M.J. (2003) Simultaneous measurements of velocity and density in buoyancy-driven mixing. *Experiments in Fluids*, *34* (1), 98–106.

Rensen, J., Luther, S., & Lohse, D. (2005) The effect of bubbles on developed turbulence. *J Fluid Mech.*, *538*, 153–187.

Riboux, G., Risso, F., & Legendre, D. (2010) Experimental characterization of the agitation generated by bubbles rising at high Reynolds number. *J Fluid Mech.*, *643*, 509–539.

Ryu, Y. & Chang, K.A. (2008) Green water void fraction due to breaking wave impinging and overtopping. *Exp. Fluids*, *45* (5), 883–898.

Ryu, Y., Chang, K. A., & Lim, H. J. (2005). Use of bubble image velocimetry for measurement of plunging wave impinging on structure and associated greenwater,*Meas. Sci. Technol.*, *16*(10), 1945–1953.

Sagnes, P. (2010) Using multiple scales to estimate the projected frontal surface area of complex three-dimensional shapes such as flexible freshwater macrophytes at different flow conditions. *Limnol Oceanography: Methods, 8*, 474–483.

Schmidt, W. (1911) Zur Mechanik der Böen. *Z Meteorol., 28*, 355–362.

Schoneboom, T. (2011) *Widerstand flexibler Vegetation und Sohlenwiderstand in durchströmten Bewuchsfeldern.* Mitt. Leichtweiß-Institut für Wasserbau No *175*, Braunschweig, Technische Universität Braunschweig, Germany, in German.

Schoneboom, T., Aberle, J., & Dittrich, A. (2010) Hydraulic resistance of vegetated flows: Contribution of bed shear stress and vegetative drag to total hydraulic resistance. In: Dittrich, A., Koll, K., Aberle, J., & Geisenhainer, P. (eds.) *Proceedings of the International Conference on Fluvial Hydraulics River Flow 2010*, September 8–10 2010, Braunschweig, Germany. Karlsruhe, BAW. pp. 269–276.

Seol, D.-G., Bhaumik, T., Bergmann, C., & Socolofsky, S.A. (2007) Particle image velocimetry measurements of the mean flow characteristics in a bubble plume. *J. Eng. Mech., 133*(6), 665–676.

Seol, D.-G., Bryant, D.B., & Socolofsky, S.A. (2009) Measurement of behavioral properties of entrained ambient water in a stratified bubble plume. *J Hydraul Eng.-ASCE, 135* (11), 665–676

Seol, D.-G. & Socolofsky, S.A. (2008) Vector post-processing algorithm for phase discrimination of two-phase PIV. *Exp Fluids, 45*(2), 223–239.

Rojas, G. & Loewen, M.R. (2007) Fiber-optic probe measurements of void fraction and bubble size distributions beneath breaking waves. *Exp. Fluids, 43*(6), 895–906.

Simpson, J.E. (1969) A comparison between laboratory and atmospheric density currents. *Quart J Royal Meteorol Soc., 95* (406), 758–765.

Simpson, J.E. (1997) *Gravity Currents – In the Environment and the Laboratory.* Cambridge, Cambridge University Press.

Simpson, M.R. (2001) *Discharge Measurements Using a Broad-Band Acoustic Doppler Current Profiler.* U.S. Geological Survey, Open-File Report 01-.Sacramento, CA.

Siniscalchi, F. & Nikora, V.I. (2012) Flow-plant interactions in open-channel flows: A comparative analysis of five freshwater plant species. *Water Resources Research.* [online] *48* (5), W05503. Available from: doi:10.1029/2011WR011557.

Socolofsky, S.A. & Adams, E.E. (2005) Role of slip velocity in the behavior of stratified multiphase plumes. *J Hydraul Eng.-ASCE, 131*(4), 273–282.

Soulsby, R.L. (1980) Selecting record length and digitization rate for near-bed turbulence measurements. *J Phys Oceanography, 10*, 208–219.

Stacey, M.T., Monismith, S.G., & Burau, J.R. (1999) Observations of turbulence in partially stratified estuary. *J Phys Oceanography, 29*, 1950–1970.

Statzner, B., Lamoroux, N., Nikora, V., & Sagnes, P. (2006) The debate about drag and reconfiguration of freshwater macrophytes: Comparing results obtained by three recently discussed approaches. *Freshwater Biol., 51* (11), 2173–2183.

Sukhodolov, A. & Sukhodolova, T. (2010) Case Study: Effect of submerged aquatic plants on turbulence structure in a lowland river. *J Hydraul Eng., 136* (7), 434–446.

Sukhodolov, A. N. (2015). Field-based research in fluvial hydraulics: potential, paradigms and challenges. *J Hydraul Res, 53* (1), 1–19.

Sukhodolov, A. N. & Uijttewaal W. S. J. (2010). Assessment of a river reach for environmental fluid dynamics studies. *J Hydraul Eng-ASCE, 136* (11), 880–888.

Tavoularis, S. (2005) *Measurement in Fluid Mechanics.* New York, Cambridge University Press.

Tavoularis, S. & Szymczak, M. (1989) Displacement effect of square-ended Pitot tubes in shear flows. *Experiments in Fluids, 7*, 33–37.

Tennekes, H. & Lumley, J.L. (1972) *A First Course in Turbulence.* Cambridge, MA, The MIT press.

Thomas, R.E., Johnson, M.F., Frostick, L.E., Parsons, D.R., Bouma, T.J., Dijkstra, J.T., Eiff, O., Gobert, S., Henry, P.Y., Kemp, P., McLelland, S.J., Moulin, F.Y., Myrhaug, D., Neyts, A., Paul, M., Penning, W.E., Puijalon, S., Rice, S.P., Stanica, A., Tagliapietra, D., Tal, M., Tørum, A., & Vousdoukas, M.I. (2014) Physical modelling of water, fauna and flora: Knowledge gaps, avenues for future research and infrastructural needs. *J Hydraul Res., 52* (3), 311–325.

Tropea, C., Yarin, A., & Foss, J. (eds.) (2007) *Springer Handbook of Experimental Fluid Mechanics.* Berlin, Springer. p. 1557.

Ungarish, M. (2010) *An Introduction to Gravity Currents and Intrusions.* Boca Raton, CRC Press.

van Dijk, W.M., van de Lageweg, W.I., & Kleinhans, M.G. (2013) Formation of a cohesive floodplain in a dynamic experimental meandering river. *Earth Surf Process Landforms, 38,* 1550–1565. doi: 10.1002/esp.3400

Västilä, K., Järvelä, J., & Aberle, J. (2013) Characteristic reference areas for estimating flow resistance of natural foliated vegetation. *J Hydrology, 492,* 49–60.

Whitmeyer, S.J. & Mogk, D.W. (2013) Safety and liability issues related to field trips and field courses. *Eos, 94* (40), 349–351.

Whittaker, P., Wilson, C., Aberle, J., Rauch, H.P., & Xavier, P. (2013) A drag force model to incorporate the reconfiguration of full-scale riparian trees under hydrodynamic loading. *J Hydraul.Res., 51* (5), 569–580.

Wilcock, P.R., Orr, C.H., & Marr, J.D.G. (2008) The need for full-scale experiments in river science. *Eos, 89* (1), 6–6.

Yalin, M.S., (1972) *Mechanics of Sediment Transport.* Pergamon Press, University of Minnesota.

Yossef, M. & de Vriend, H., (2010) Sediment exchange between a river and its groyne fields: Mobile-bed experiment. *J Hydraul Eng., 136* (9), 610–625.

Zare, S.G.A., Moore, S.A., Rennie, C.D., Seidou, O., Ahmari, H., & Malenchak, J. (2016) Boundary shear stress in an ice covered river during break-up. *J Hydraul Eng.,142*(4), 04015065.

Zhang, W.M., Liu, M.N., Zhu, D.Z., & Rajaratnam, N. (2014) Mean and turbulent bubble velocities in free hydraulic jumps for small to intermediate froude numbers. *J Hydraul Eng.-ASCE,* 140(11), 9.

Zhang, W.M. & Zhu, D.Z. (2013) Bubble characteristics of air-water bubbly jets in crossflow. *Int J Multiphas Flow, 55,* 156–171.

Chapter 6

Data Analysis

6.1 INTRODUCTION

Hydraulic experiments are performed with different aims, and hence the appropriate data analysis will differ. In a scale-model experiment, where a single value (or a series of values without any attempt to relate them) is to be obtained, then a simple uncertainty analysis (discussed in Chapter 7) may suffice. In experiments where the physical phenomena are well understood, and an empirical coefficient and/or exponent relating dependent and independent variables, *e.g.*, a discharge coefficient for the flow over a spillway, additional data analysis, such as fitting to an accepted functional form, will be required. In the most complex case, where the physics of the flow and related transport phenomena are poorly understood, and especially where measurements are overlain with high levels of noise and/or spatio-temporal variability, analyses assume a more exploratory aspect, and inferences must be more limited and cautious. Much of this chapter focuses on this last category since it not only offers the most challenges and scope for analysis, but many problems of engineering and scientific importance fall into this category.

Traditional statistical inference theory provides the conceptual foundation for data analysis, and will be relied on heavily in the following. Readers are assumed to have a general familiarity with elementary probability and statistics (*e.g.*, at a level similar to Walpole *et al.*, 2007 or Mendenhall & Sincich, 1995; Kottegoda & Rosso, 1997, among many others), but a brief review of basic concepts is given and references will be made throughout to the more specialized statistical literature. The present chapter aims to encourage a judicious application of statistics to observations in hydraulics, which however entails a careful consideration of the physical phenomena. Two general principles are stressed throughout: i) physics supercedes statistics, or because statistical significance does not imply physical or engineering significance, the latter is cautiously given precedence if a conflict between the two is found, and ii) the simpler or the more parsimonious model is preferred (sometimes termed the principle of Occam's razor), provided no well-founded physical principle is violated. The limitations of statistics applied blindly without physical constraints must be stressed; purely statistical models, *e.g.*, forecasting based on past values as is commonly performed in statistical hydrology, can be useful in their limited context, but will not be dealt with here. Even if formal statistics are not evaluated, consideration of concepts such as population, model, statistical significance, may clarify not only the data analysis but also influence the prior steps of experimental design and choice of experimental techniques, procedure, and parameters. The most sophisticated data analysis techniques may alleviate but not entirely compensate for the limitations of the available data.

For the most part, the discussion follows the traditional so-called frequentist approach to statistics, as this will be most familiar to the typical reader. The major alternative, the Bayesian approach, has received much attention in the hydrological literature, and so is introduced as potentially useful in dealing with hydraulic data. Much of the emphasis will be placed on basic tools that can be applied to a wide variety of data, and highlighting the relation to the physical hydraulic context and the subtleties in interpretations that are often missing in purely statistical accounts. A special feature of the chapter is the use of examples, some quite detailed, to illustrate the concepts discussed and to provoke thinking about issues often overlooked. Due to the interests of the authors, the examples mainly involve turbulent open channel flows and sediment transport in such flows, though the techniques should be much more broadly applicable. More advanced or more specialized techniques, such as those from the machine-learning literature, or from the specialist turbulence literature, such as proper orthogonal decomposition are dealt with more briefly.

As in other domains, computational advances in both software and hardware have also had significant impacts on statistical data analysis. The broad availability of software tools, especially those with extensive graphic capabilities, has brought quite sophisticated data analysis techniques within the reach of researchers. Many of the extended examples presented in this chapter were analyzed using the R software system (Maindonald & Braun, 2013), which is freely available for multiple platforms, and it will be generally assumed that statistical analyses will be performed with similar software. These advances have also permitted a much more computationally oriented approach to data analysis. Whereas traditional approaches relied on restrictive assumptions to obtain simple analytical results, more recent approaches, such as bootstrapping, or those based on Bayesian concepts or machine learning make fewer assumptions and instead rely heavily on data-driven computationally intensive simulation or algorithms. Whether the latter approaches by themselves lead to a better understanding of the underlying physical phenomena remains debatable, but that their influence on statistical practice and prediction is undeniable. New measurement technologies based on acoustic or optical principles, especially when coupled with computer control, can produce large amounts of data that motivate new approaches to their handling and analysis. Many of the methods presented in this chapter, but especially machine learning (Section 6.10, Volume I) and extraction of coherent structures in turbulent flows (Section 6.14, Volume I), may be equally applied to processing of both experimental and numerical simulation results.

Fundamental concepts and terminology are first reviewed in Section 6.2, Volume I, followed by a discussion of descriptive statistics and exploratory data analysis and conditioning in Section 6.3, Volume I. General concepts of statistical inference are introduced in Section 6.4, Volume I, with a brief exposition of bootstrapping in Section 6.5, Volume I. The central topic of regression is treated in Section 6.6, Volume I, with elementary ideas from Bayesian analysis covered in Section 6.7, Volume I. Illustrative regression examples are given in Section 6.8, Volume I. More specialized techniques in classification, machine learning, filtering and time series analysis, spatial interpolation, and eduction of large-scale structures in turbulent flows are successively considered in Sections 6.9–6.14, Volume I.

6.2 BASIC CONCEPTS, TERMINOLOGY, AND NOTATION IN PROBABILITY AND STATISTICS

Fundamental concepts and the attendant terminology that will recur in later sections are here reviewed. An effort has been made to use a clear and consistent notation, following wherever possible that typically found in the statistics or hydraulics literature and in the other chapters of the book. For simplicity and convenience, this is not always practical, and the same symbol is sometimes used with different meanings in different contexts, $e.g.$, f may denote a probability density function, or a friction factor, or a response in regression model. As turbulent flows are frequently discussed, the traditional notation for the Reynolds decomposition of an instantaneous fluctuating quantity, u, is generally followed, namely, $u = \bar{u} + u'$, where $\bar{u}$, and u' denote respectively the time-averaged contribution and the randomly varying departure from the time-averaged mean.

6.2.1 Randomness, sampling, population, and homogeneity

Statistics is applied to a quantity thought to vary randomly, $i.e.$, in an 'unpredictable' manner, to some degree, and hence is a random variable, conventionally denoted as X. The random variable may also exhibit some deterministic variation or trend. X may be discrete, $e.g.$, taking only integer values (or even qualitative categories, such as success or failure rather than numerical values), or, as is more typical in the hydraulics context, may be continuous, taking real values, though possibly over a limited range. Examples include the instantaneous velocity at a point in a steady turbulent flow shown in Figure 6.2.1a (further details in Section 6.12.7, Volume I) where the random variation occurs about a constant mean, or the variation of bankfull width with bankfull discharge shown in Figure 6.2.1b (further details in Section 6.8.2, Volume I) where the random variation overlies what is thought to be a deterministic trend.

In practice, the statistic is evaluated from measurements or results of numerical 'experiments' assumed to be sampled from a population representing all possible outcomes of a specific measurement or experiment. For a meaningful statistic, such as a (population) mean, the population should be homogeneous, in both a physical as well as a more narrow purely statistical sense. The physical criterion for homogeneity will be problem-specific and may involve scaling or other transformation, and may not be straightforward to identify. The statistical aspect of homogeneity is conceptually clearer in implying that random samples from the same population should result in statistics that are not statistically different. Statistical homogeneity does not necessarily imply physical homogeneity. The two examples in Figure 6.2.1 illustrate the variation in sample statistics depending on the sample chosen. Any sample statistic, such as the sample mean, or a sample regression coefficient, is $also$ a random variable, as it is a function of random variables. In Figure 6.2.1a, the sample means based on different 5-second samples vary randomly about a constant mean though much less so than the original signal. Similarly, in Figure 6.2.1b, three different regression curves with different regression parameters are obtained from three different samples over different regions. Whether the samples were drawn from the same population (despite the

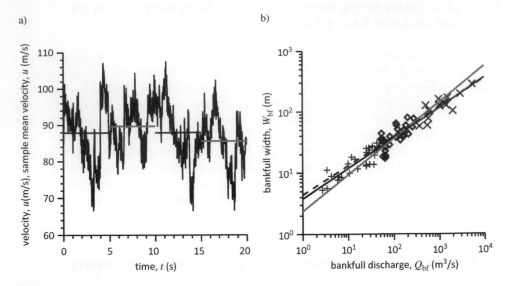

Figure 6.2.1 Examples of quantities modeled as random variables: a) the instantaneous velocity component, u, at a point over a period of time in a steady turbulent flow (data of Schemper & Admiraal, 2002), and different sample means, $\bar{u}$, indicated by horizontal solid lines over the range of the different samples, b) the variation of the bankfull width, W_{bf}, with bankfull discharge, Q_{bf}, and different power-law regression curves (again solid lines) from different samples (data of Parker et al., 2007).

different sample statistics values) is a typical question in data analysis that this chapter attempts to address.

Example 6.2.1 Consider measurements of time-averaged velocity, $\bar{u}$, at a point at a given distance from a smooth bed on the centerline of a wide channel in a steady uniform open-channel flow. For multiple depths, h, and slopes, S_0, questions regarding population may be raised. If the study were interested solely in depth-averaged velocities, $\langle u \rangle_h$, would the observed values of $\langle u \rangle_h$ be drawn from the same population? If not, would some scaled quantity, $\langle u \rangle_h / u_s$, where u_s is an appropriate velocity scale? If so, what is the appropriate u_s ? If u_s is chosen as the shear velocity, $u_s = u_* = \sqrt{ghS_0}$, g being the acceleration due to gravity, then $\langle u \rangle_h / u_*$ is directly related to a friction factor, which for a smooth bed is known to vary (deterministically) with a Reynolds number, such as $u_* h / v$, where v is the fluid kinematic viscosity. Thus, $\langle u \rangle_h / u_*$ might be considered as being drawn from the same population provided $u_* h / v$ is approximately the same. If the study were focused on the velocity profile, i.e., $\bar{u}$, as a function of the normal distance, y, from the bed, similar questions arise. Is $\bar{u}$ observed at a given y for flows of different h and S_0 drawn from the same population? If not, would some scaled version, $\bar{u} / u_s$, be sufficient, or would some scaled y/y_s (where y_s is an appropriate length scale) also be necessary? A conventional answer, invoking the law-of-the-wall model (and so assuming points sufficiently close to the bed), would be that the scaled quantity, $\bar{u} / u_*$, is drawn from the same population provided that measurements are made at $y/(v/u_*)$, which of course is local Reynolds number, and $y_s \equiv (v/u_*)$ is the appropriate length scale.

6.2.2 Probability and conditional probability; distributions, quantiles, moments, and expectations

From the traditional frequentist perspective, the probability of an outcome, A, of an experiment or measurement is its (limiting) relative frequency. Thus if an experiment is repeated identically(and hence independently) N times, and A occurs n times, then the probability of the occurrence of A, $P(A) = p_A = n/N$, in the limit of large $N \to \infty$. The limiting value p_A must be a non-negative (dimensionless) number, ranging from 0 and 1, and the probability of all possible outcomes must sum (or integrate) to 1. The conditional probability, denoted as $P(A|B)$, refers to the probability of A occurring with a qualifying constraint, B, which potentially could affect the probability of A. As an example, A might be the categorical event that a riprap blanket has failed, or the quantitative continuous event that the maximum scour depth, y_{sc}, has exceeded a specified value, y_{sc0}, i.e., $y_{sc} > y_{sc0}$, while B might be the continuous quantitative event that the approach cross-sectionally averaged velocity, $\langle U \rangle$, has exceeded a value, U_0. Thus, $P(A|B)$ may be expressed as $P(\text{riprap failure} \mid \langle U \rangle > U_0)$. If $P(A|B) = P(A)$, i.e., the probability of A is unaffected by the occurrence of B, then A is said to be statistically independent of B.

Standard theoretical distributions with a limited number of parameters have been traditionally used to aid in evaluating probabilities associated with a random variable, X. The cumulative distribution function (cdf), $F(x_p)$, is directly related to the probability of an event, $X < x_p$, by

$$P(X < x_p) = F(x_p) \tag{6.2.1}$$

The inverse of the cdf is the quantile function, with the term quantiles referring to values of x_p corresponding to values of $F(x_p)$ at regular intervals. Special cases of quantiles include quartiles (with an interval of 1/4) or percentiles (with an interval of 0.01 or 1%). For example, the condition $F(x_{75}) = 0.75$ defines x_{75} as the third quartile or the 75th percentile. The median, x_{50}, is the second quartile or the 50th percentile.

The characteristic features of different distributions tend to be more apparent in the probability density function (pdf, for continuous random variables), $f(x) = dF/dx$. Unlike F, which as a probability must be dimensionless, f has the inverse dimensions of x. The combined quantity, $f(x)dx = dF$, is interpreted as the probability that the value of x falls within the infinitesimal interval, $(x, x + dx)$, so that

$$P(x < X < x + dx) = f(x)\, dx \quad \Rightarrow \quad P(X < x_p) = \int_{-\infty}^{x_p} f(x)\, dx = F(x_p) \tag{6.2.2}$$

The expectation value, $E[g(X)]$, of a function, $g(X)$, is defined in terms of $f(x)$ as

$$E[g(x)] = \int_{-\infty}^{\infty} g(x)\, f(x) dx \tag{6.2.3}$$

and so is interpreted as a (probability-weighted) average of $g(X)$. The moments of the distribution, i.e., $E[X^m]$, $m = 1, 2, ...$, or the central moments, $E[(X - \mu_x)^m]$ are often used to characterize the distribution. The first moment of the distribution is the mean

$\mu_x = E(X)$. The second central (*i.e.*, relative to the mean) moment is the variance, $\sigma_x^2 = \text{var}(X) = E[(X - \mu_x)^2]$, while the normalized third central moment, $\gamma = E[\{(X - \mu_x)/\sigma_x\}^3]$ is the skewness. The normalized fourth central moment, $m_4 = E[\{(X - \mu_x)/\sigma_x\}^4]$, is related to the kurtosis, $K = m_4 - 3$ (sometimes K is defined as m_4, and $m_4 - 3$ is termed the *excess* kurtosis, the qualifier excess expressing the kurtosis relative to that of a normal distribution for which $m_4 = 3$). Being non-normalized, both μ_x and σ_x have the dimensions of x. The dimensionless coefficient of variation, σ_x/μ_x, is sometimes used as a normalized standard deviation, provided μ_x is not close to zero. In Bayesian practice, the reciprocal of the variance, $1/\sigma_x^2$, is termed the precision. Whereas μ_x gives the centroid or central tendency, and σ_x the 'width' of the distribution, the higher than second central moments are providing additional information regarding the tails of the distribution, *i.e.*, the range of x that is far from the mean.

A large variety of pdf's may be found in the statistical literature. Attention is restricted here to the normal (and the related lognormal), the (Student's) t, and the chi-squared (χ^2, and the related gamma) distributions. The well-known normal distribution is defined entirely by its mean, μ_x and its variance, σ_x^2, and can be expressed either in a 'raw' form, $\mathcal{N}$, or in a standardized form, $\mathcal{N}_{std}$:

$$\mathcal{N}(x; \mu_x, \sigma_x) = \frac{1}{\sigma_x \sqrt{2\pi}} \exp\left[-\frac{1}{2}\left(\frac{x - \mu_x}{\sigma_x}\right)^2\right], \quad \text{or} \quad \mathcal{N}_{std}(z; 0, 1) = \frac{1}{\sqrt{2\pi}} \exp\left[-\frac{z^2}{2}\right]$$

(6.2.4)

In its standard form, the original random variable, X, is re-scaled as the standard normal deviate, $Z = (X - \mu_x)/\sigma_x$, resulting in the *standard* normal distribution with mean 0 and variance 1. The variable x or z is often omitted as implicit. From a physical point of view, this rescaling can also be interpreted as resulting from a dimensional analysis reducing the original three (possibly dimensional) variables to a single dimensionless group. The skewness and the kurtosis for the normal distribution are both zero. The t distribution arises in sample mean and related statistics, and is characterized by a single parameter, n_{df}, usually termed the degrees of freedom, which in statistical inference refers to the number of independent 'pieces of information' available for a statistical estimate. The cumulative and density distributions of the standard normal and the t distributions are compared in Figure 6.2.2a.

In contrast to the normal (and the t) distribution, the lognormal and the chi-squared (χ^2), are restricted to positive or non-negative arguments and are positively skewed, *i.e.*, with longer tails on the right than on the left. The random variable, X, follows a lognormal distribution when its logarithm, $\ln(X)$, follows a normal distribution. The lognormal model is well known in hydraulics for its use in describing sediment size distributions, but also arises where log transformation of variables are taken. It is also characterized by two parameters, the mean $\mu_{\ln(x)}$ and the variance, $\sigma_{\ln(x)}^2$, both dimensionless though, as in modeling sediment size distributions, a possibly dimensional geometric mean, μ_{gx}, is often defined such that $\mu_{\ln(x)} = \ln\left(\mu_{gx}\right)$ to facilitate physical interpretation. Like the t-distribution, the χ^2-distribution is characterized by a single parameter, the degrees of freedom, n_{df}, and arises in sample statistics of variances

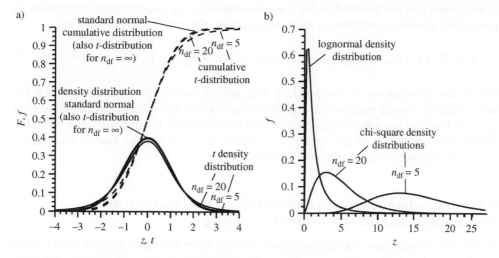

Figure 6.2.2 a) Cumulative and density distributions for the normal and the *t* distributions, b) density distributions for the lognormal and the chi-squared distributions.

(including power spectra as discussed in Section 6.12, Volume I). The (standard, $\mu_{\ln(x)} = 0$ and $\sigma^2_{\ln(x)} = 1$) lognormal and the χ^2 density distributions are shown in Figure 6.2.2b. The χ^2-distribution is a special case of the standard two-parameter gamma distribution, which is frequently encountered in Bayesian analysis. The random variable, X, follows an inverse gamma distribution when its reciprocal, $1/X$, follows a gamma distribution. Further details of these distributions are given in Appendix 6.A.1.

Practical problems often involve multiple random variables, $X_k, k = 1, ..., M$, which are dealt with through a joint density distribution, $f(x_1, ..., x_M)$, rather than the univariate distribution, $f(x)$, discussed so far. The basic concepts for the latter also apply to the former. The case where X_k are all statistically independent is important, e.g., in dealing with random sampling, and implies that $f(x_1, ..., x_M)$ can be factorized in a simple manner,

$$f(x_1, ..., x_M) = f_1(x_1)...f_M(x_M) \tag{6.2.5}$$

where $f_i(x_i)$ denotes a possibly individually different univariate pdf. The often assumed case of jointly uncorrelated (and hence independent) normal distribution (so each f_i denote a normal distribution, each with its own mean, μ_{x_k}, and variance, $\sigma^2_{x_k}$) can be expressed as

$$f(x_1, ..., x_M) = \mathcal{N}\left(\mu_{x_1}, \sigma^2_{x_1}\right)...\mathcal{N}\left(\mu_{x_M}, \sigma^2_{x_M}\right) = \frac{1}{C_M}\exp\left[-\frac{1}{2}\sum_k \left(\frac{x_k - \mu_{x_k}}{\sigma_{x_k}}\right)^2\right] \tag{6.2.6}$$

where the normalizing constant, $C_M = (2\pi)^{M/2}(\sigma_{x_1}...\sigma_{x_M})$ ensures a unit overall integral.

Moments can be obtained as for the univariate case by appropriate integration. In particular, for two (continuous) random variables, the covariance, defined as

$$\text{cov}(X, Y) = \sigma_{xy} = E[(X - \mu_x)(Y - \mu_y)] = \int (x - \mu_x)(y - \mu_y)f(x, y) \, dxdy \quad (6.2.7)$$

which has dimensions of the product XY, and measures the extent to which the two variables change together. If the two variables are statistically independent, then $\sigma_{xy} = 0$ but the converse is not necessarily true (though is true for joint normally distributed variables). In the special case where X and Y are the same variable, the covariance becomes identical to the variance.

6.2.3 Sampling statistics and distributions, statistical independence, and conditional sampling

In practice, population characteristics are inferred from a sample, usually a very small fraction of the entire population. From the frequentist viewpoint, sample statistics, such as the sample mean, are random variables characterizing the population, and must be related probabilistically to the corresponding population characteristic, such as the population mean. As random variables, a sample statistic may be assumed to follow a standard (sampling) probability distribution, such as the normal or the chi-squared distribution or a non-standard empirical distribution. Further, standard statistical tests assume random sampling, implying statistically independent outcomes. Reference will often be made to measurements that are independent identically distributed (iid), i.e., randomly sampled from the same population. More specifically, the measurements may not only be independent but taken from the same *normally* distributed population, $\text{nid}(\mu_x, \sigma_x^2)$. The sampling in typical hydraulic measurements (notably in time series) is not necessarily random therefore restricting what can be statistically inferred from the acquired datasets. Even if the sampling is random, the sample size may not be sufficient to justify a strong inference due to a possible large uncertainty in the measurements.

Example 6.2.2: The specific issue of choosing an adequate sampling duration in dealing with a turbulent-flow time series is covered at greater length in Section 5.2.4, Volume I. Here the point emphasized concerns the possible non-random nature of the sampling. For specificity, consider the measurement of the time series of the instantaneous velocity, u, at a point in a steady turbulent flow. If the sampling rate is chosen to resolve the smallest time scales in the flow, then consecutive measurements of u will generally not be statistically independent, as they will be highly (serially) correlated, i.e., successive velocities will be close to the preceding velocity. In the absence of statistical independence, the sampling is non-random, and hence any statistical inference result assuming random sampling is questionable. The same point would also be made for the less common case where spatial averages are taken in a spatially homogeneous direction.

Example 6.2.3: Consider measurements of sediment transport rate at two field sites. At one site, 100 data points are available for different depths and slopes but for a given sediment size, while at the other site, only 5 data points are available for different depths and slopes but for a different sediment size. Various statistics-related questions may be raised, including those related to the population and appropriate scaling and the sufficient number of samples, but highlighted here is the issue related to random sampling. If the data analysis is aimed at developing a general sediment transport formula that is independent of site, is each of the measurement point collected at the two sites statistically independent? If the population is restricted, *e.g.*, if the data analysis is aimed at developing a sediment rating curve for each site separately, then the answer might well be different.

For a specified population, a non-random sample will result in biased statistics that do not accurately characterize the entire population. This type of bias is termed sampling bias. The same sample may however be considered random for a more restricted sub-population. Because of the more limited control over conditions, sampling in field experiments may be prone to being non-random. For example, measurements during extreme events may not be feasible due to safety concerns, but provided the analyst is aware of the data limitations (*i.e.*, the sub-population) then statistical results may still be justified but possibly less useful.

In some refined analyses, the statistics of sub-populations may be of interest because they are considered more homogeneous in certain respects than the entire population. Such an analysis is performed in conditional sampling whereby the samples are obtained randomly _but_ satisfying an explicit condition. The analysis of such-obtained sub-population results in conditional statistics. In a sense, all practical flow measurements could be considered as applying conditional sampling since experiments are usually performed for specific conditions (*e.g.*, at a specified point under uniform-flow over a non-erodible smooth beds). More narrowly, the acquired data are partitioned according to a specified classification scheme, the definition of which itself may be open to question (see the following examples).

Example 6.2.4:   Consider a dataset on sediment transport rates obtained in laboratory and in field conditions, covering a range of flows and sediment characteristics for determining a sediment transport function. One possible approach considers the entire dataset as being homogeneous (assuming some appropriate scaling or choice of dimensionless parameters) and so describable by a single function. Alternative approaches can be based on conditional sampling by partitioning the data according to various criteria, such as sediment size range (sand or gravel), or data source (laboratory or field), or bed form regime (lower-regime or upper-regime). The motivation for these alternatives is presumably that the different classes differ substantially, either in mechanics or in scale, and so more reliable statistics or models would be developed if these differences were explicitly recognized in a model. The classification criterion can be open to question. For example, sand and gravel streams might be classified based on the median sediment size being less than 2 mm (sand) or greater than 2 mm (gravel), but the value of 2 mm is purely conventional without necessarily representing any physical basis in the particular application.

Example 6.2.5:   Consider measurements of a phenomenon with a strong deterministic component. The deterministic structure is often but not necessarily periodic in time, *e.g.*, diurnal variations in sewer flows, an oscillatory turbulent boundary layer, or turbulent vortex shedding flow around a bluff body. For some purposes, simple time-averaged statistics may provide useful information, but they ignore the strong time-varying feature, and thereby render more difficult an interpretation of the higher-order statistics, such as the variance, which includes the effect of the deterministic time-varying feature. An alternative conditional sampling approach partitions the data according to the phase of the periodic structure (e.g., see the study of a vortex-shedding flow in Lyn et al., 1995). In that study, conditional statistics were evaluated only for data corresponding to the same phase which are considered more homogeneous than all data at quite different phases lumped together. The definition of phase may itself not be trivial, and may require additional measurements.

Example 6.2.6:   The search for coherent structures in wall-bounded turbulent flows has also made use of conditional sampling (e.g., Antonia, 1981). The quadrant analysis is perhaps the most widely known approach (Wallace et al., 1972; Willmarth & Lu, 1972). Because the mean Reynolds shear stress, $-\overline{u'v'}$, is of prime concern in shear-flow turbulence, the instantaneous contributions to the mean are sorted according to the sign of the individual fluctuating velocity components (i.e., u', v'). For example, measurements satisfying the condition, $u' > 0, v' > 0$ are classed in quadrant 1, while those satisfying the condition, $u' < 0, v' > 0$ are classed in quadrant 2, and so on. Statistics based on quadrant 2 observations are thought to characterize 'ejection' events, while those based on quadrant 4 observations ($u' > 0, v' < 0$) characterize 'sweep' events. The combined effect of ejections and sweeps is thought to dominate the mean Reynolds shear stress particularly for those events that exceed a certain magnitude termed the hole size, $H = -u'v'/\left(\sqrt{\overline{u'^2}}\sqrt{\overline{v'^2}}\right)$. Hydraulic applications using this conditionally sampled quadrant analysis include Lyn (1987), Balachandar and Bhuiyan (2007), and Hurther et al. (2007) in studies of open-channel flows with roughness, and Dwivedi et al. (2010) in a study of hydrodynamic forces during sediment entrainment. The topic of detecting (or educing) coherent structures in turbulent flows is covered in greater detail in Section 6.14, Volume I.

6.2.4 Bias, variance, and formulating estimators

In addition to the sampling bias discussed in the preceding section, another type of bias may be identified. The bias, $\mathcal{B}$, of a statistical sample point estimator, $\hat{\Theta}$, is the difference between its mean or expected value, $E(\hat{\Theta})$, and the value of the population parameter, θ. An unbiased estimator, i.e., one with zero bias, is usually but not always preferred. An example is the conventional estimator, $s_x^2 = \sum_i (x_i - \overline{x})^2/(N-1)$, for the variance, σ_x^2, where $\overline{x}$ is the sample mean (in the following, unless otherwise specified, results deal with N randomly sampled observations, x_i). The divisor, $N - 1$, is preferred to the more intuitive, $s_x^2 = \sum_i (x_i - \overline{x})^2/N$, because its use leads to zero bias. For large N, the intuitive estimate (see Chapter 2.3.3 or Table 4.5.2 of Volume I) becomes unbiased. Interestingly, s_x is *not* an unbiased estimator of the standard deviation, σ_x (Panik, 2005). Bias is however not the sole performance criterion for an estimator; a small or minimum variance is also sought so that most estimates will be 'close' to the population parameter to be estimated. Thus, some bias may be tolerated if the advantage in a smaller variance were sufficiently large. This may also be understood in terms of the decomposition of the expected mean squared error, $E[(\hat{\Theta} - \theta)^2]$, into the sum of $\mathcal{B}^2$ and the variance of $\hat{\Theta}$, $\text{var}(\hat{\Theta})$ (Panik, 2005; Hastie et al., 2009). As such, minimizing the mean squared error may involve a tradeoff between bias and variance in the choice of an estimator.

Various general approaches to formulating estimators are available (Mendenhall & Sincich, 1995; Panik, 2005). The method of moments provides a simple and direct means of formulating the basic statistics by equating population moments to sample moments, $\sum_i x_i^m / N$. Another general approach is based on the concept of likelihood, defined as the joint pdf of the observations, and assumed to depend on population parameters, θ_k. The corresponding sample parameters, $\hat{\theta}_k$, are chosen so as to maximize the likelihood (or often the log of the likelihood). The maximum likelihood (ML) estimator is thus formulated to result in the greatest 'likelihood' of the observations. In simple cases, the ML estimator(s) can be obtained analytically, but in general the estimation requires a numerical solution to find the values of $\hat{\theta}_k$ associated with the ML. The ML estimator arises in the discussion of Bayesian inference (Section 6.7, Volume I) and in logistic classification (Section 6.9, Volume I). Neither the method of moments nor the ML method necessarily leads to an unbiased estimator. According to both methods, the estimator for the variance of a normal distribution is the biased $\sum_i (x_i - \bar{x})^2 / N$.

Finally, although mainly of theoretical use, two results should be mentioned, the law of large numbers and the central limit theorem (Panik, 2005). A precise account will not be given here, but suffice it to say that both give support to the frequent assumptions made with regards to behavior of large-sample statistics. As $N \to \infty$, for random samples, the sample mean approaches the population mean, and the distribution of the sample mean approaches the normal distribution. Much of classical statistical inference relies on assumptions of normality, which may only be justified for large N, and many conventional results assumed large N.

6.3 DESCRIPTIVE STATISTICS AND EXPLORATORY DATA ANALYSIS

Inferential statistics, which is the focus of most of this chapter, aims to infer population parameters from sample observations and expresses them probabilistically, e.g., in terms of a confidence interval, as the sample is random. Much of the experimental hydraulics literature uses descriptive or diagnostic statistics whereby the results are not expressed probabilistically, and so are restricted to characterizing the sample rather than the population. More recently, the broader field of exploratory data analysis made popular by Tukey (1977), has been concerned with summarizing basic statistics, checking statistical assumptions, identifying outliers and patterns, and emphasizing graphical displays of the manipulated data to facilitate interpretation.

6.3.1 Histograms, kernel density estimates, sample moment statistics, and quantiles

As the density distribution is central in the description of random phenomena, its determination is a starting point for characterizing a sample. The density distribution can be estimated through various procedures. The most common is the histogram (Scott, 2015), in which the range of values of x is divided into intervals, also termed bins, typically but not necessarily of equal width, Δx_{bin}. Subsequently, the number of occurrences, $n_{bin,i}$, with values of x in each of the i-th interval, $(x_{bin,i}, x_{bin,i} + \Delta x_{bin})$, is counted. A histogram ordinate is plotted as a function of x. Three types of histogram ordinates are commonly found: i) the raw count, $n_{bin,i}$, ii) the relative frequency or probability mass, $n_{bin,i}/N$, and iii) the probability density, $(n_{bin,i}/N)/\Delta x_{bin}$, which integrates to unity. The last option is chosen in Figure 6.3.1 as it affords a more consistent comparison with density distributions. In the construction of a histogram, the bin width, Δx_{bin}, is chosen as a compromise between bias and variance (or resolution and smoothness). Various prescriptions for estimating the bin width are available. Here only the Freedman-Diaconis rule (Scott, 2015) is mentioned that states that $\Delta x_{bin} = 2(IQR_x)/N^{1/3}$, where $IQR_x = x_{75} - x_{25}$ is the interquartile range (*i.e.*, the difference between the first and the third quartile). Such rules assume independent observations and so should not be applied blindly to time series data that are serially correlated. In Figure 6.3.1, the histograms were constructed from the time series of a point velocity in a turbulent channel flow illustrated in Figure 6.2.1 using the

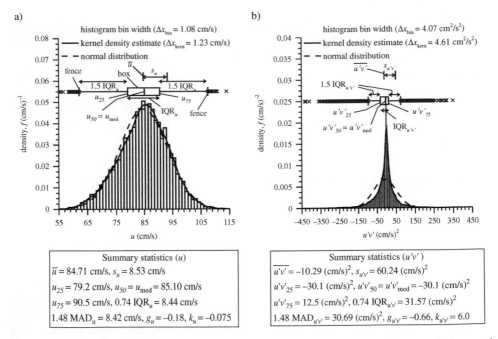

Summary statistics (u)
$\bar{u} = 84.71$ cm/s, $s_u = 8.53$ cm/s
$u_{25} = 79.2$ cm/s, $u_{50} = u_{med} = 85.10$ cm/s
$u_{75} = 90.5$ cm/s, $0.74\ IQR_u = 8.44$ cm/s
$1.48\ MAD_u = 8.42$ cm/s, $g_u = -0.18$, $k_u = -0.075$

Summary statistics ($u'v'$)
$\overline{u'v'} = -10.29$ (cm/s)^2, $s_{u'v'} = 60.24$ (cm/s)^2
$u'v'_{25} = -30.1$ (cm/s)^2, $u'v'_{50} = u'v'_{med} = -30.1$ (cm/s)^2
$u'v'_{75} = 12.5$ (cm/s)^2, $0.74\ IQR_{u'v'} = 31.57$ (cm/s)^2
$1.48\ MAD_{u'v'} = 30.69$ (cm/s)^2, $g_{u'v'} = -0.66$, $k_{u'v'} = 6.0$

Figure 6.3.1 Histogram and density estimates (obtained with "hist" and "density" R routines) along with the summary statistics for time series displayed in Figure 6.2.1: a) horizontal velocity component, u, and, b) corresponding (negative) Reynolds shear stress component, $u'v'$, from measurements in a turbulent channel flow.

Freedman-Diaconis rule. The results are reasonably smooth and unimodal (just one distribution peak) for both the horizontal velocity component, u, as well as the (negative) Reynolds shear stress component, $u'v'$.

The use of finite bin widths results in a histogram that exhibits jumps in density estimates for each bin as can be seen in Figure 6.3.1a. A continuous density distribution estimate can be obtained through the use of kernel functions (Scott, 2015). The estimated (univariate) density distribution, $\hat{f}(x)$, is obtained as the sum of a chosen kernel function, $K_h(x, x_i; \Delta x_{\text{kern}})$:

$$\hat{f}(x) = \frac{1}{N}\sum_i K_h(x, x_i; \Delta x_{\text{kern}}), \qquad (6.3.1)$$

where K_h is non-negative, symmetric about each data point, x_i, integrates to unity, and is characterized by a bandwidth, Δx_{kern}. A common choice for K_h is the normal distribution, i.e., $\mathcal{N}(x_i, \Delta x_{\text{kern}})$. Similar to Δx_{bin} in histograms, Δx_{kern} is chosen as a compromise between bias and variance. Rules similar to the Freedman-Diaconis can be applied for selecting Δx_{kern} (typically $\propto N^{-1/5}$ rather than $\propto N^{-1/3}$), but these again assume random (independent) sampling.

Kernel density estimates based on a normal distribution kernel with the specified band widths are compared with the corresponding histograms in Figure 6.3.1. For both presumably unimodal distributions, the kernel density estimates agree well with the histograms, which is not surprising since bandwidths were chosen comparable to bin widths. Also plotted in Figure 6.3.1 are the normal density distributions with the same sample means and variances. While the observed u -distribution may be considered to follow closely a normal distribution, the $u'v'$ -distribution illustrated in Figure 6.3.1b clearly does not (see also Lu & Willmarth, 1973). The latter displays a much more peaky behavior and long 'heavy' tails (for neither observed distribution is the skewness pronounced), with implications not least for the measurement of Reynolds shear stresses.

The density distribution contains much information, but not all of this information is of equal practical importance. Moreover, the tail of the distribution is associated with larger uncertainties, being less reliably estimated due to fewer observations. The characteristics of a distribution that are often of most practical interest may be gleaned from its lower-order moments. These include the mean, $\bar{x}$, or the median, x_{50}, and the range (or dispersion or variability), such as the standard deviation, s_x, the interquartile range, IQR_x, or the median absolute deviation, MAD_x. The sample (arithmetic) mean or the first moment is $\bar{x} = \sum x_i/N$, while the median(x) or second quartile, x_{50}, is defined such that $P(X < x_{50}) = P(X > x_{50}) = 1/2$. The median is sometimes preferred as a *robust* estimator of central tendency in being much less sensitive than $\bar{x}$ to the presence of outliers (see discussion below in Section 6.3.3, Volume I). The sample standard deviation is estimated as $s_x = \left[\sum_i (x - \bar{x})^2/(N-1)\right]^{1/2}$. In practice, when $N \gg 1$, the standard deviation (and the variance) may be evaluated with N rather than $N - 1$ as divisor as in Chapters 2 and 4 of Volume I. The interquartile range, IQR_x, and the median absolute deviation, $\text{MAD}_x = \text{median}(|x - x_{50}|)$ are considered more robust estimators of the distribution range or variation than s_x. For a normal distribution, $\sigma_x = 0.74\,\text{IQR}_x = 1.48\,\text{MAD}_x$.

Estimators for skewness and kurtosis may be also formulated but their normalization and the approach for bias correction may vary slightly. For larger values of N, however, the various estimators give for practical purposes the same result (Joanes & Gill, 1998). Using for illustration the results shown in Figure 6.3.1, the intuitively appealing but biased skewness and kurtosis estimators are

$$g_x = \frac{\sum (x_i - \bar{x})^3/N}{s_x^3}, \quad k_x = \frac{\sum (x_i - \bar{x})^4/N}{s_x^4} - 3. \tag{6.3.2}$$

For the nearly normal u-distribution, $s_u \approx 0.74 \text{ IQR}_u \approx 1.48 \text{ MAD}_u$, and both g_u and k_u are reasonably close to zero. In contrast, the distinctly non-normal though still roughly symmetric $u'v'$-distribution characterized by $0.74 \text{ IQR}_{u'v'} \approx 1.48 \text{ MAD}_{u'v'}$, both of which are notably smaller than $s_{u'v'}$, hence more accurately characterizing the narrow $u'v'$-distribution. A high value of $k_{u'v'}$ is indicative of the effect of peakiness and heavy tails of the $u'v'$-distribution.

When two random variables, X and Y, possibly correlated, are measured pair wise, the sample covariance, s_{xy}, and the sample correlation coefficient, r_{xy}, are estimated with Equation (6.2.7):

$$s_{xy} = \frac{1}{N-1} \sum_{i=1}^{N} (x_i - \bar{x})(y_i - \bar{y}), \quad \text{and} \quad r_{xy} = \frac{s_{xy}}{s_x s_y} \tag{6.3.3}$$

As in the case of the sample variance (which is obtained if $X = Y$), the $N - 1$ divisor is chosen to obtain an unbiased estimate. The sample covariance with dimensions of the product XY is extensively used in experimental hydraulics (e.g., the Reynolds shear stress), arising in regression theory (Section 6.6.2, Volume I), the analysis of time series (Section 6.12.2, Volume I), and in kriging theory for spatial interpolation (Section 6.13, Volume I). The sample correlation coefficient measures the linear relationship between the two random variables.

Also shown in Figure 6.3.1 are box-and-whiskers plots (drawn horizontally) to emphasize graphically the relationship to the density distribution (another example of box-and-whiskers plot drawn vertically will be seen in Section 6.13.2, Volume I). The vertical lines of the box, i.e., the sides and the internal line, indicate the location of the first, the third, and the second (the median) quartiles. Two additional vertical lines, sometimes termed inner fences, one located 1.5 IQR to the left of the lower quartile and the other located 1.5 IQR to the right of the upper quartile, are intended to span (most of) the range of observations. Another graphical tool to investigate the distributions of two random variables is the quantile-quantile (Q-Q) plot. This plot relates the estimated quantile of one distribution, which may be either empirically obtained or derived from a model, and the corresponding quantile of the other distribution. One common form of the Q-Q plot is illustrated in Figure 6.3.2 where quantiles, z_{obs}, of an observed distribution are compared with quantiles, z, of a standard normal distribution (sometimes referred to as a normal probability plot). z_{obs} are estimated using a plotting position formula, which is a standard procedure to estimate probabilities of ordered data (e.g., Kottegoda & Rosso, 1997; McCuen, 2003). If the observations follow a normal distribution, then $z_{\text{obs}} = z$, and the data points should fall on or close to the straight equality line. As might be expected from the preceding discussion of

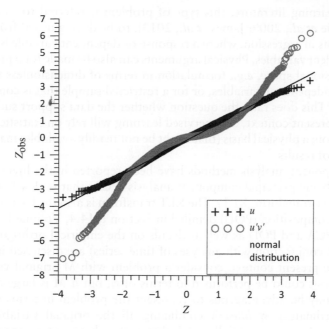

Figure 6.3.2 Quantile-quantile plot (quantiles obtained using the "qqnorm" function in R) of observed quantiles for the u- and the $u'v'$-time series data of Figure 6.3.1a and Figure 6.3.1b compared to that of a normal distribution.

histograms, the u-quantiles follow closely those of the normal distribution, with only slight deviations at the far tails, $z > 3$, for which the plotting positions are in any case associated with greater uncertainty. In contrast, the $u'v'$-quantiles exhibit noticeable deviations from the equality line, even for relatively small z, with a typical shape associated with heavy-tailed distributions. As will be discussed below, normal probability plots are also used in examining residuals in regression analyses.

6.3.2 Unsupervised learning: principal component analysis and clustering

Any sample statistic estimated without regard for the relation to the corresponding population parameter may be regarded as a descriptive statistic. Thus, for a time series, simply estimating the autocorrelation or the power spectrum (described in Section 6.12, Volume I) may be viewed as exploratory data analysis aimed at characterizing the sample rather than the population. In this section, two types of analysis, often applied at a preliminary stage, which are somewhat different from common statistics, are discussed, where simplification of the available sample data is sought rather than any relationship between dependent and independent variables, such as in regression (described in Section 6.6, Volume I). The simplification may lie in a reduction in the dimensionality of the input data (independent variables) by a projection onto a lower dimensional space, termed component analysis, or in a reduction of the number of observations to be considered in the analysis by grouping, termed cluster analysis. In the machine-learning literature, this type of problem is referred to as unsupervised learning (Hastie *et al.*, 2009; James *et al.*, 2013), to be distinguished from supervised learning, such as in regression, where a response or dependent variable is to be related to the independent variables. Physical arguments can also be made for a projection onto a lower dimensional space, *e.g.*, formulation in terms of dimensionless variables will lead to fewer independent variables, or for a restricted sample that is considered more homogeneous. This does raise the question whether the data support such an analysis choice. In the present context, unsupervised learning will rely on statistical criteria for the simplification; a physical basis (that might be not readily available) maybe sought in interpretation of results.

Several component analysis methods have been reported in the literature with the most commonbeing principal component analysis (PCA) that is closely related to the Karhunen–Loève transform (KLT). The KLT transform is also the basis for the proper orthogonal decomposition (POD) detailed in Section 6.14.4, Volume I. The relationship between PCA and POD as well as details on the empirical orthogonal functions (EOFs, used in conjunction with analyses of time series) are discussed in Liang *et al.* (2002). For the present context, consider a problem with M original variables, r_i. A statistical analysis could be performed in terms of r_i, but if M is large and the r_i are correlated, it may be advantageous to represent the problem in terms of alternative variables (coordinates) by *linearly* combining all the original variables. The new coordinates are chosen sequentially, such that each subsequent component selection maximizes the range of values or the variance of the observations (appropriately centered and scaled) and that direction is perpendicular (*i.e.*, orthogonal) to the preceding direction. The directions so obtained are termed the principal component directions or simply the principal components (PCs). Further details of PCA can be found in Hastie *et al.* (2009) and James *et al.* (2013). Tayfur *et al.* (2013) applied PCA in formulating a sediment transport function.

A somewhat artificial but simple, two-dimensional (*i.e.*, $M = 2$), case is provided by the input data for the bankfull-width problem for regression relationships previously illustrated in Figure 6.2.1b. For this case, the logarithms of the bankfull discharge and median sediment diameter, $r_1 = \log(Q_{bf})$ and $r_2 = \log(d_{50})$ are the input data as plotted in Figure 6.3.3. The two PCs (coordinates), $\tilde{r}_1 = \log(Q_{bf}^{0.99}/d_{50}^{0.11})$ and $\tilde{r}_2 = -\log(Q_{bf}^{0.11}d_{50}^{0.99})$, obtained with the PCA procedure, are also shown in the figure. Both $\tilde{r}_1$ and $\tilde{r}_2$ are *linear* combinations of the original r_1 and r_2, with $\tilde{r}_1$ in the direction where the input data vary the most, and $\tilde{r}_2$ perpendicular to $\tilde{r}_1$. For more complicated cases with $M > 2$, $\tilde{r}_2$ would be chosen such that the remaining variance is maximized in the $\tilde{r}_2$-direction. The line represented by $\tilde{r}_1$ is a lower (one)-dimensional projection of the original two-dimensional data, and the hope of PCA would be that instead of a model relation, $W_{bf} = g(r_1, r_2)$, the simpler $W_{bf} = g_*(\tilde{r}_1)$ may suffice. In more complicated problems with large M, the input data can be approximated more parsimoniously by the first few, say $\tilde{M}$, PCs, which might be advantageous in subsequent statistical analysis.

The extent to which the original data might be approximated by the first $\tilde{M}$ PCs is usually quantified by the fraction of the total variance of the input data accounted for by these PCs. For the case in Figure 6.3.3, the fraction of the total variance of the input data accounted for by $\tilde{r}_1$ is 0.94. While this may sound impressive, it is actually little

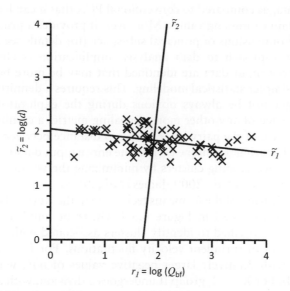

Figure 6.3.3 Principal components analysis (using the"prcomp" function in R) applied to the dataset illustrated in Figure 6.2.1b. In this case, the $M = 2$ original variables are: $r_1 = \log(Q_{bf})$ and $r_2 = \log(d_{50})$ while the principal components are $\tilde{r}_1 = \log(Q_{bf}^{0.99}/d_{50}^{0.11})$ and $\tilde{r}_2 = -\log(Q_{bf}^{0.11}d_{50}^{0.99})$.

improvement over the original coordinate, r_1, as $\tilde{r}_1$ is quite close to r_1 for this particular case. Because the underlying principle of PCA involves a maximization of variance, and the problem may involve variables of different dimensions and units, PCA can be sensitive to scaling. It is usually recommended to scale each input to unit variance. For the present problem, the log transformation does deal to some extent with the different scales of the variables; however, the scaling of the original log transformed variables was not performed in obtaining the results of Figure 6.3.3.

When conducting PCA, the assumptions of a *linear* decomposition and the importance of large variance of the variables involved must be borne in mind, as otherwise the PCs may be of little use. Neither assumption is necessarily valid in typical hydraulics applications. Even if the assumptions are thought reasonable, the physical interpretation of the PCs remains problematic. In the example of Figure 6.3.3, a physical interpretation of $\tilde{r}_1$ is not readily apparent. Thomson and Emery (2014) note regarding the closely related EOFs that "no direct physical or mathematical relationship necessarily exists between the statistical EOFs and any related dynamical modes". Practical questions such as appropriate scaling, transformations (such as a log transformation), as well as selection of features or relevant original variables need careful consideration, but this is not unique to PCA. PCA does deal with highly correlated variables (as illustrated by r_1 and r_2 in Figure 6.3.3), and as a technique in exploratory data analysis may motivate fruitful lines of enquiry, especially for large M. Hastie *et al.* (2009) have emphasized the relationship of learning (both supervised and unsupervised) to density estimation. Tipping and Bishop (1999) and Roweis (1998) proposed a probabilistic formulation of PCA making it amenable to a Bayesian treatment. The probabilistic

PCA has the advantages compared to conventional PCA that it can be applied to input data having uncertain or missing values. Moreover, it provides a principled identification of the optimal dimensions of principal sub-space (for details see Bishop, 2006).

The other main approach to data analysis simplification is clustering whereby smaller subsets of the input data are identified that may be more homogeneous and hence may permit simpler statistical modeling. This requires a definition of a similarity metric, which might not be always obvious during the exploratory stage of data analysis. In the absence of any other more appealing metric, a common conventional choice is d_E^2, the square of the pairwise (Euclidean) distance between data points. The popular K-means clustering technique divides the entire input data into a pre-specified number, K_c, of non-overlapping clusters by minimizing the within-cluster sum of d_E^2 over all K_c clusters (Hastie $et\ al.$, 2009; James $et\ al.$, 2013).

Results of an application of the K-means technique to the Reynolds shear stress data of Figure 6.3.1b are provided in Figure 6.3.4. Different numbers of clusters, $i.e.$, $K_c = 2$, 3, and 4 were specified to identify clusters associated with quadrant analysis (recall Example 6.2.6) of turbulent velocity fluctuations. For $K_c = 2$, one cluster (I) consists of points with relatively larger negative values of $u'v'$, with the remaining points in cluster (II). For $K_c = 3$, group II undergoes a division, with a new group (IIa) being identified as those characterized by larger positive values of $u'v'$, while, for $K_c = 4$, group I is further refined, with groups Ia and Ib being made distinct. The identified clusters resemble the conceptual picture of quadrant analysis, emphasizing events in the second and fourth quadrants (ejections and sweeps) that corresponds to group I. The clusters corresponding to group IIa are less prominent corresponding to the first and third quadrant events (inward and outward interactions). For reference, curves defining a quadrant-analysis hole (hole size, $H = 1$) have been added. As the choice of hole size in quadrant analysis remains somewhat arbitrary (see Balachandar & Bhuiyan, 2007), it is interesting that a purely statistical criterion may suffice to determine an effective hole size. Based solely on the Reynolds stress measurements, $i.e.$, $u'v'$, the technique cannot distinguish between second and fourth (or first and third) quadrants, and so the identified clusters (for $K_c > 2$) are essentially those with relatively large negative $u'v'$, relatively large positive $u'v'$, and the remainder, but of course the boundaries were determined by the K-means technique.

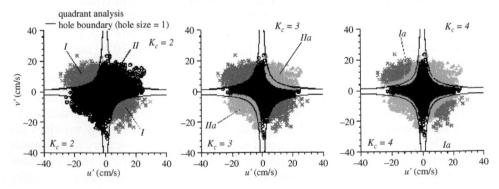

Figure 6.3.4 Results of K-means clustering (using the "kmeans" function in R) on Reynolds stress measurements illustrated in Figure 6.3.1b for different pre-specified number (K_c) of clusters.

A multitude of clustering techniques may be found in the literature (see Jain, 2010 for a review), but as Hastie *et al.* (2009) has commented, "Specifying an appropriate dissimilarity measure is far more important ... than choice of clustering algorithm". Clustering and component analyses may also be viewed as complementary (Ding & He, 2004) as for input data having N measurements (rows) in M-dimensions (columns), the cluster analysis groups rows while the component analysis groups columns.

6.3.3 Data conditioning (or validation): outlier detection and data replacement

The term data conditioning is used here to refer to a wide range of operations applied to acquired data, often prior to final data analysis or to drawing a statistical conclusion. This could include operations to ensure consistency in units or timing or coordinate axes of measurements. In later sections, filtering (Section 6.11, Volume I) and spatial interpolation (Section 6.13, Volume I), both of which could be also considered as data conditioning in the broadest sense, will be discussed. Bendat and Piersol (2000), which is oriented toward time series analysis (to be covered in Sections 6.11 and 6.12, Volume I) listed as two broad aims of what they term data validation: the detection of i) non-stationary (and non-Gaussian or non-normal) behavior, and ii) statistical anomalies, *i.e.*, outliers. Identifying non-stationary behavior will be dealt with in Section 6.12.1, Volume I; this subsection deals with identification of outliers, which are observations that differ markedly from the bulk of the sample, and hence considered *not* representative of the population of interest.

Outliers, also termed spikes or spurious observations, may result from a variety of sources including instrument malfunction, operator error, or factors of the measurement environment that are not considered in the experimental design. In some accounts, a rare event that is nevertheless a member of the population of interest is also termed an outlier (*e.g.*, Mendenhall & Sincich, 1995); this is *not* followed here. A general statistical account of outliers is found in Barnett and Lewis (1988), in the hydrological context in McCuen (2003), and in the standards document (ASTM, 2016), with focus on small sample sizes that might be especially affected by outliers. Detecting an outlier involves defining an acceptance region or range of values in terms of critical values, such that if an observation falls outside of this region it would be deemed an outlier. Almost invariably, the critical values are defined in terms of a 'standardized' variable or test statistic, $\zeta_x = (x_* - M_x)/R_x$ (in the multivariate case, this might be the equation of an elliptical region or equivalent) where x_* is the candidate outlier, M_x is a location parameter (such as the $\bar{x}$, or x_{50} or possibly the value closest to x_*), and R_x is a range parameter (such as s_x or IQR$_x$ or MAD$_x$), and M_x and R_x may be evaluated based on the sample including or excluding x_*. The acceptance region might then be specified as $(\zeta_x)_L < \zeta_x < (\zeta_x)_U$, where $(\zeta_x)_L$ and $(\zeta_x)_U$ would be the critical values (preferably expressed at some significance level as will be discussed below in Section 6.4, Volume I). If the distribution of ζ_x is assumed symmetric as in many applications, the rejection (non-acceptance) region might be expressed as $|\zeta_x| > (\zeta_x)_{\text{crit}}$. Formal statistical testing as in Barnett and Lewis (1988) usually requires a precise specification of hypotheses and hence a pre-specification of the number of outliers to be examined.

A common statistical rule of thumb (Mendenhall & Sincich, 1995; sometimes termed the Tukey rule as in ASTM, 2016) is formulated in terms of the box-and-whiskers plot, namely the acceptance region is defined by the inner fences as plotted in Figure 6.3.1. This can be viewed as a choice of $M_x = x_{50}$, and $R_x = \text{IQR}_x$, and critical value of ≈ 2 for an approximately symmetric distribution. For the one-dimensional standard normal distribution, this corresponds to $(\zeta_x)_{\text{crit}} \approx 2/0.75 = 2.7$, and so outliers would be associated with a (two-sided) probability of 0.007. Measured values outside of the inner fences, which might be flagged as possible outliers, have been plotted in Figure 6.3.1. A large number of points, especially in the $u'v'$ data lie outside the inner fences, but few if any are sufficiently egregious as outliers to warrant their exclusion, and so such simple rule of thumb should be used cautiously.

A well-known outlier-detection scheme is based on Chauvenet's criterion (McCuen, 2003; Tavoularis, 2005; Coleman & Steele, 2009), which chooses $M_x = \bar{x}$ and $R_x = s_x$, reflecting a normal-distribution assumption, with a critical value, $(\zeta_x)_{\text{crit}} = z_{1-(1/4N)}$, depending on the number of (independent) observations, N. This is appealing as more extreme values of the population are likely to be sampled as N increases. Chauvenet's criterion thus specifies the rejection region as $|z| = |(x - \bar{x})/s_x| > z_{1-(1/4N)}$. For $N = 1000$, the rejection region is $|z| > z_{0.99975} = 3.48$, while for $N = 10$, $|z| > z_{0.975} = 1.96$. Wahl (2003) used Chauvenet's criterion in a procedure for despiking acoustic Doppler velocimetry (ADV) signals. Barnett and Lewis (1988) however pointed out that the basis of the Chauvenet's criterion would suggest that it may reject too many good data. A similar criterion, examined by Goring and Nikora (2002) and Cea $et\ al.$ (2007), is the Universal Threshold, which defines the outlier region as $|z| > \sqrt{2\ln(N)}$, yielding somewhat higher critical values compared to Chauvenet's criterion. Such criteria were originally intended for application to independent observations and their application to high-sampling rate data such as those shown in Figure 6.3.1 is questionable. For the data of Figure 6.3.1, $N \approx 9000$ and the measurements were taken over a period of ≈ 180 s, but, based on the argument that independent samples are only taken when separated by twice the integral time scale (see Chapter 5.2.4), N_{eff} may be as low as 60, assuming an integral time scale of ≈ 1.5 s (see Section 6.12.7 in Volume I). Despite the questionable basis for the use of N rather than N_{eff}, its use may mitigate the issue of rejecting good data as the larger value of N enlarges the acceptance region. It is also usually specified that Chauvenet's criterion be applied only once (McCuen, 2003; Wahl, 2003; Coleman & Steele, 2009), presumably because of the underlying hypotheses, and hence this restriction should apply also to the use of Universal Threshold. The well-known approach of Goring and Nikora (2002) applied to ADV spiking does involve an iterative successive application of the Universal Threshold which might be also questionable in this regard.

Additional context-specific information or measurements can be used in outlier detection. In the ADV literature (see also the discussion in Section 3.2.4, Volume II), the Goring-Nikora (2002) approach to ADV despiking formulates a multivariate problem in which the sampled velocity signal, u_i, and its first and second differences, $\Delta u_i = (u_{i+1} - u_{i-1})/2$ and $\Delta^2 u = (\Delta u_{i+1} - \Delta u_{i-1})/2$ (which might be interpreted as time derivatives), were included in the analysis. In this multivariate problem, the acceptance region is defined as an ellipsoid (Wahl, 2003) or as in the original study the projections of the ellipsoid in the three planes. Wahl (2003) went further by also including all three components of velocity, though each component seems to have been considered independently. In a similar vein, Cea *et al.* (2007) proposed a scheme involving all three velocity components but not the first and second differences. Islam and Zhu (2013) also used u and Δu, but their detection scheme involved estimation of a two-dimensional kernel density (see Section 6.3.1, Volume I for a discussion of *uni-variate* kernel density estimators). Multivariate approaches bring additional complexities in the choice of variables and their evaluation (*e.g.*, Δu can be performed using central differences as in Goring and Nikora (2002), or forward or backward differences (Islam & Zhu, 2013), as well as the requirement for data replacement to allow re-evaluating of Δu in iterative procedures), possible scaling issues, and the definition of a multi-dimensional acceptance region.

Particle image velocimetry (PIV; discussed further in Raffel *et al.*, 2007; Adrian & Westerweel, 2011, as well as in Section 3.7.1, Volume II) also must deal with spurious velocity vectors resulting from initial image processing. PIV differs from ADV in being a planar field technique rather than a point technique and outlier detection focuses on spatial rather than temporal statistics. In the local scheme of Westerweel and Scarano (2005; see also Raffel *et al.*, 2007 and Adrian & Westerweel, 2011), attention is restricted to a small neighborhood of the examination point, (x_*, y_*). A test statistic similar to the one discussed above, with $M_u = u'_{50}$ and $R_u = \text{MAD}_{u'}$, is formulated where u'_{50} and $\text{MAD}_{u'}$ are the median and the MAD of the local sample *excluding* the point being tested. The local nature of the statistic test is important because PIV is frequently applied to spatially inhomogeneous flows. In contrast, ADV measurements at a point in a steady flow generate a stationary time series, and a local test statistic is less important.

The above engineering approaches to outlier detection may be criticized from a statistical viewpoint in that they do not specify a significance level (see the discussion of hypothesis testing below in Section 6.4, Volume I), and so lack a quantitative indication of the 'reliability' or 'stringency' of the criterion. The outlier detection schemes discussed above also address a specific type of data inhomogeneity; other noise sources may need to be dealt with by other means. For ADV applications, the treatment of the non-spike noise is discussed in Garcia *et al.* (2005), Doroudian *et al.* (2010), Khorsandi *et al.* (2012) as well as in Section 3.2.4, Volume II. The problem of appropriate sampling discussed in Section 6.2.3, Volume I should be reiterated here as inappropriate representation of the population by the acquired sample might still lead to incorrect analysis results not because of the presence of the outliers but rather because of flawed sampling.

Example 6.3.1: Despiking of ADV signals has received much attention (see also Section 3.2.4.3 in Volume II), and is viewed here as an application of outlier detection. A 20-s segment of the contaminated record of Goring and Nikora (2002) is shown in Figure 6.3.5a, where the spikes, *i.e.*, large signal excursions at some points leading to negative values, are evident. The total record of ≈ 170 s with the sampling frequency of 200 Hz, leads to a total number of data points of $N = 33995$.

Only results of three univariate (based on only the velocity) outlier-detection schemes are presented in this example. The first scheme is the max/min scheme studied by Cea *et al.* (2007) and adopted in Doroudian *et al.* (2010). It uses a simple test statistic based on the mean ($\bar{u}$) and the standard deviation (s_u) evaluated from the *entire* original record, so $\zeta = (u - \bar{u})/s_u$. The critical value was chosen as $\sqrt{2 \ln N}$ (with the dubious

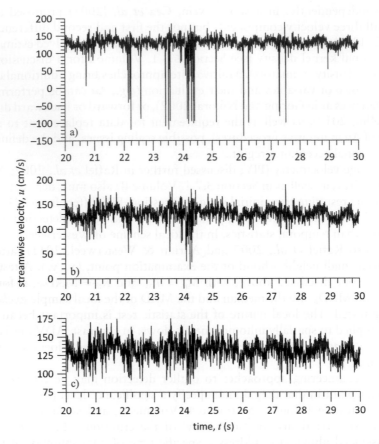

statistic	original data	Cea et al. (2007) max/min scheme	Hampel's rule scheme	K-means clustering
mean, $\bar{u}$ (cm/s)	127.5	129.1	129.6	129.1
median, u_{50} (cm/s)	129.7	129.8	129.9	129.8
standard deviation, s_u (cm/s)	23.9	15.33*	13.75*	15.20*
1.48×median absolute deviation, 1.48 MAD$_u$ (cm/s)	13.63	13.43	13.31	13.42
acceptance region (cm/s)		(18.3,236.7)†	(61.8,198.5)††	
number of spikes identified		258	634	271

* after removal of spikes

† based on $\bar{u} \pm s_u \sqrt{2 \ln N}$ where $\bar{u}$ and s_u are evaluated from the original data ($N = 33995$)

†† based on $u_{50} \pm 5\text{MAD}_u$ where u_{50} and MAD_u are evaluated from the original data

Figure 6.3.5 a) Segment of the original record of contaminated ADV data time series of Goring and Nikora (2002) showing spikes (outliers), b) corresponding segment after removal of outliers detected with the simple max/mean scheme of Cea *et al.* (2007) based on the sample mean and standard deviation for the entire original record (and also with the K-means clustering algorithm), and c) corresponding segment after removal of outliers detected with the simple robust scheme (Hampel's rule) based on sample median and median absolute deviation for the entire original record.

choice of $N = 33995$ yielding a value of 4.57), resulting in the acceptance region shown in the table embedded in Figure 6.3.5.

The second scheme was similar but based on the corresponding robust estimators, i.e., $\zeta = (u - u_{50})/\text{MAD}_u$, where the critical value was chosen as 5 (guided by the relation that $1.48 \text{ MAD} = \sigma$, so that $|\zeta| > 5$ should be equivalent $|z| > 3.4$). This is sometimes termed Hampel's rule (ASTM, 2016), and is the same as the Westerweel and Scarano (2005) scheme in PIV applications, though applied globally rather than locally. The third scheme follows a non-traditional K-means clustering algorithm (see the discussion of clustering in Section 6.3.2, Volume I), in which two clusters (acceptable values and outliers) are pre-specified. The strength and the weakness of the K-means scheme is that neither a critical value nor an acceptance region needs to be specified.

The first and third scheme detected the same spikes for the signal segment shown, though the max/min scheme actually detected *fewer* spikes than the K-means scheme for the entire signal. As is shown in Figure 6.3.5b (note the change in the scale of the ordinate axis), the largest spikes were successfully identified by these two schemes, but the despiked signal segment still exhibits a number of suspect extreme values, such as those near $t = 24$ s. The Hampel's rule scheme, due to its noticeably narrower acceptance region, flagged considerably more points as spikes, and the signal segment shown in Figure 6.3.5c exhibits no obvious outliers. For comparison, Goring and Nikora (2002) reported 194 spikes and $s_u = 13.78$ cm/s (because the Goring-Nikora technique is iterative, it is not clear whether this reported number of spikes refers to the initial or final iteration). Wahl (2003) reported 2149 spikes and $s_u = 13.19$ cm/s.

Data removal and replacement. Once outliers have been detected, the question arises whether they should be removed, and if so, whether they should be replaced. If the data point can be definitely attributed to instrument malfunction or operator error or is deemed to be unphysical, then its removal can clearly be justified. The case is especially convincing when the instrument, such as the ADV, is known to be prone to spikes or spurious measurements. In other contexts, options include the use of robust statistics and the acquisition of more data, possibly with alternative measurement techniques or procedures. In the ADV literature, it is generally acknowledged that, for the evaluation of basic statistics such as the low-order moments, it is not necessary to replace data, but invariably it is stated (Goring & Nikora, 2002; Cea *et al.*, 2007; Doroudian *et al.*, 2010; Islam & Zhu, 2013; see also Section 3.2.4.3, Volume II) that it is necessary for other statistical estimates such as spectra and autocorrelations. This is misleading as the basic definitions of these statistics do not depend on how the data were sampled.

A large literature may be found on statistical estimation of gapped or irregularly spaced data (for a review specifically of spectral analysis, see Babu & Stoica, 2010, and for applications in laser Doppler velocimetry, see Benedict *et al.*, 2000). As will be discussed in Section 6.12.5, Volume I, power spectral densities (and hence related statistics) can be readily determined from gapped ADV data. The techniques for dealing irregularly spaced data are generally more computationally demanding because standard algorithms, such as those based on the fast Fourier transform (FFT), which assumed regularly spaced data, are not applicable. Data replacement should then be considered as a computational convenience rather than a necessity. Where large numbers of long time series need to be evaluated, data replacement may be the practical option. For ADV applications, the common data replacement techniques were considered by Goring and Nikora (2002), who recommended a cubic interpolation, but Cea *et al.* (2007) and Islam and Zhu (2013) reported good results with simple linear interpolation and Parsheh *et al.* (2010) even suggested that the simpler sample-and-hold technique was sufficient.

6.4 HYPOTHESES, STATISTICAL SIGNIFICANCE, AND INTERVAL ESTIMATES

Common questions arising in statistical inference from experimental data include whether an observed value agrees with a theoretical or literature value, and whether an imposed change in condition has resulted in any significant observed difference. In the preceding discussion, the question was posed as to whether one or more observations in a given sample belong to the same population. Such questions can be formalized into the testing of hypotheses, *e.g.*, the so-called 'null' hypothesis, $\mu = \mu_0$, where μ denotes a quantity of interest, and μ_0 its theoretical or literature value. The difficulty in answering such questions is that the point estimate of a sample statistic, such as a sample mean, $\bar{x}$, evaluated from a random sample, is only one possible outcome of the random variable, $\bar{X}$, associated with a (sampling) distribution, which may be quite different from the original density distribution for the variable X (see Figure 6.4.1). By itself, $\bar{x} - \mu_0 \neq 0$ is insufficient to justify a statistical conclusion because $\bar{X}$ may take on a range of values with varying probabilities, *i.e.*, if different samples were taken, then different values of $\bar{X}$ would be found.

A difference is statistically significant only if the probability of such a difference occurring by chance is sufficiently small. The *p*-value approach evaluates the

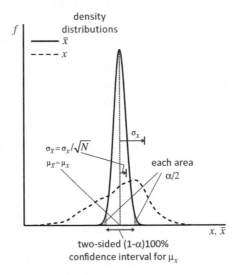

Figure 6.4.1 Original distribution for a random variable, X, sampling distribution for the mean, $\bar{X}$, and two-sided $(1-\alpha)100\%$ confidence interval for the mean, μ_x.

probability or p-value of observing at least the value of the sample statistic, assuming the null hypothesis to be true, and the analyst must subsequently decide whether the p-value is sufficiently small to accept or reject the null hypothesis. The traditional approach chooses a standard 'significance level', $(1 - \alpha)100\%$, e.g., 95%, i.e., $\alpha = 0.05$, and determines a range of values or confidence interval such the probability of observing the value of the statistic within that range is $1 - \alpha$ (see Figure 6.4.1). The acceptance region in the discussion of outliers may be viewed as a type of confidence interval. In both approaches, an assumption regarding the underlying probability distribution is required in order to evaluate the probabilities. Further, both approaches are related in that, if $p < \alpha$, then the traditional approach would conclude that the difference is statistically significant. Needless to say, statistical significance does not imply any practical engineering or even physical importance.

Confidence intervals are constructed by examining the mean and variance of the statistic of interest, under a distribution assumption (often a normal distribution), then replacing unknown population parameters by sample statistics where necessary. In the estimation of the sample mean, $\overline{X}$, the individual observation is assumed to consist of the constant true mean, μ_x, and a random error term, ε_i, assumed $\mathrm{nid}(0, \sigma_x^2)$, i.e., $x_i = \mu_x + \varepsilon_i$. With $\overline{X} = \sum_{i=1}^{N}(\mu_x + \varepsilon_i)/N$, the confidence interval will be based on the mean ($\mu_{\overline{x}}$) and variance ($\sigma_{\overline{x}}^2$) of $\overline{X}$. Because $E(\varepsilon_i) = 0$, $\mu_{\overline{x}} = \mu_x$ and $\mathrm{var}(\varepsilon) = \sigma_x^2$, thus leading to $\sigma_{\overline{x}}^2 = \sigma_x^2/N$, and thus if N is large, $\sigma_{\overline{x}} \to 0$ (see Figure 6.4.1). The unknown population parameters, μ_x and σ_x^2, are replaced by the sample mean, $\overline{x}$, and the sample variance, s_x^2, and so the $(1 - \alpha)100\%$ two-sided confidence interval for the mean μ_x is expressed as

$$(\overline{x}_L, \overline{x}_U) = \left(\overline{x} - t_{n_{\mathrm{df}},\alpha/2}\sigma_{\overline{x}},\ \overline{x} + t_{n_{\mathrm{df}},\alpha/2}\sigma_{\overline{x}}\right) = \left(\overline{x} - t_{n_{\mathrm{df}},\alpha/2}s_x/\sqrt{N},\ \overline{x} + t_{n_{\mathrm{df}},\alpha/2}s_x/\sqrt{N}\right)$$

(6.4.1)

where $t_{n_{\mathrm{df}},\alpha/2}$ is the value of the t distribution for $n_{\mathrm{df}} = N - 1$ degrees of freedom for which the area under the curve is $\alpha/2$, i.e., $P(T > t_{n_{\mathrm{df}},\alpha/2}) = \alpha/2$ (see Figure 6.4.1). For large N (e.g., $N > 30$), $t_{n_{\mathrm{df}},\alpha/2}$ is well approximated by the corresponding normal-distribution value, $z_{\alpha/2}$. As a comparison, for $\alpha/2 = 0.025$, and $N = 30$ (so $n_{\mathrm{df}} = 29$), $t_{29,0.025} = 2.05$ and $z_{0.025} = 1.96$.

The quantity, $\sigma_{\overline{x}} = \sigma_x/\sqrt{N}$, is an example of a standard error, often reported by software packages as it does not depend on a user-specified significance level, and abbreviated as $s.e.$ Two-sided $(1 - \alpha)100\%$ confidence intervals for mean-like statistics are then expressed with an additive uncertainty term $\left(t_{n_{\mathrm{df}},\alpha/2}(s.e.)_{\overline{x}}\right)$ as $[\overline{x} - t_{n_{\mathrm{df}},\alpha/2}(s.e.)_{\overline{x}},\ \overline{x} + t_{n_{\mathrm{df}},\alpha/2}(s.e.)_{\overline{x}}]$, symmetric about $\overline{x}$ (or the relevant mean-like sample value) due to the sampling distribution of means being either the t-distribution or the normal distribution both of which are symmetric. The relevance of this confidence interval for the procedures for conducting the uncertainty analysis discussed in Chapter 7, Volume I is evident.

The frequentist interpretation of the confidence interval about the population mean (or other population parameter) can be rather subtle. Although it can be loosely thought of as the probability that the population mean lies within the confidence interval, the logical problem arises that the frequentist population mean is not a random variable, and such a probabilistic interpretation is inconsistent. Rather the confidence interval itself is the random variable, so that if samples were repeatedly taken and a different confidence interval estimated for each sample, then the confidence interval represents the probability that the confidence interval contains the population parameter.

If the statistic of interest is related to the variance, as is the case for power spectra discussed in Section 6.12.4, Volume I, the appropriate sampling distribution is the chi-squared $\left(\chi^2_{n_{df}}\right)$ distribution, again depending on the number of degrees of freedom, n_{df}, provided that the underlying distribution from which the sample is taken is normal. The confidence interval for σ_x^2 in this case is $[(N-1)s_x^2/\chi^2_{n_{df},\alpha/2}, (N-1)s_x^2/\chi^2_{n_{df},1-(\alpha/2)}]$ that includes a *multiplicative* uncertainty term that is *not* 'symmetric' about the sample estimate, s_x^2, because $\chi^2_{n_{df},\alpha/2} \neq \chi^2_{n_{df},1-(\alpha/2)}$.

Example 6.4.1:   Consider a study of the Darcy-Weisbach pipe friction factor in which a sample mean of $\bar{f}_{DW}$ and a sample variance of $s_{f_{DW}}^2$ were estimated from a random sample of N measurements. A literature value of $f_{DW,0}$ was expected for the given conditions. If $\bar{f}_{DW} > f_{DW,0}$, does that mean that the observed mean friction factor differs from the literature value? To answer this question, a p-value approach would evaluate the probability that values greater than $\bar{f}_{DW}$ would be observed, namely

$$P\left(T > \left|\frac{\bar{f}_{DW} - f_{DW,0}}{s_{f_{DW}}/\sqrt{N}}\right|\right) = p \ \text{ or } \ P\left[-\left(\frac{\bar{f}_{DW} - f_{DW,0}}{s_{f_{DW}}/\sqrt{N}}\right) < T < \left(\frac{\bar{f}_{DW} - f_{DW,0}}{s_{f_{DW}}/\sqrt{N}}\right)\right] = 1 - p$$

where a t-distribution is used with $n_{df} = N - 1$ (for large N, a normal distribution could be justified). If p were sufficiently small, *e.g.*, 0.05, then a statistically significant difference would be inferred. Alternatively, the traditional approach would determine a confidence interval corresponding to a significance level of α,

$$\left[\left(\bar{f}_{DW}\right)_L, \left(\bar{f}_{DW}\right)_U\right] = \left[\bar{f}_{DW} - t_{n_{df},\alpha/2}(s.e)_{\bar{f}_{DW}}, \bar{f}_{DW} + t_{n_{df},\alpha/2}(s.e.)_{\bar{f}_{DW}}\right],$$

where $(s.e.)_{\bar{f}_{DW}} = s_{f_{DW}}/\sqrt{N}$, $t_{n_{df},\alpha/2}$ is the critical value of the t-distribution for $\alpha/2$ and $n_{df} = N - 1$. If the literature value is located within the interval, $\bar{f}_{DW} - t_{n_{df},\alpha/2}(s.e)_{\bar{f}_{DW}} < f_{DW,0} < \bar{f}_{DW} + t_{n_{df},\alpha/2}(s.e.)_{\bar{f}_{DW}}$, then $\bar{f}_{DW}$ would be judged not statistically significant from $f_{DW,0}$. The same inference would be arrived at using the criterion that $p < \alpha$ for statistical significance.

Example 6.4.2:   The sample size, N, may be chosen to obtain a desired uncertainty level (or half width of the confidence interval, Δx_0) by applying the above results. Thus, with $\Delta x_0 = t_{n_{df},\alpha/2}\left(s_x/\sqrt{N}\right)$, then N can be chosen as $\left(t_{n_{df},\alpha/2}s_x/\Delta x_0\right)^2$, where rough approximations for both $t_{n_{df},\alpha/2}$ and s_x are made for a usable criterion. Such an estimate also provides the theoretical basis for the sample time duration for a time series in Section 5.2.4, Volume I, where however the problem of serial correlation (see the above discussion in Example 6.2.2) complicates the analysis. Similar arguments can be made for variances.

The commonly used confidence interval about a mean says nothing about the probability of any *single* or *larger* number of future observations. For this reason, the utility of the population mean may be limited within the context of a conservative design. As an example, in the problem of pier scour, observations may yield a range of (maximum) scour depths for any given flow and sediment conditions. A conservative design would *not* be based on the confidence interval about the mean scour depth because i) the mean scour depth will by definition be exceeded a rather large fraction of time, and ii) a single (rather than a sample of) event may have disastrous consequences. A probabilistic design criterion might be better based on a prediction or tolerance interval (Walpole *et al.*, 2007). For $N \to \infty$, the prediction interval is given by $\bar{x} \pm z_{1-\alpha/2}s_x$ which is much wider than the confidence interval ($\bar{x} \pm z_{1-\alpha/2}s_x/\sqrt{N}$ or equivalent).

6.5 BOOTSTRAPPING

Traditional statistical inference typically proceeds through an analytical argument, sometimes starting from strong assumptions, such as sampling from a specific distribution (*e.g.*, a normal distribution), and leading to simple closed-form results for the statistic of interest, such as a confidence interval discussed in the preceding section. For some practical problems, such analysis might be extremely complicated and/or laborious, and bootstrapping, a resampling method often used in conjunction with Monte-Carlo computer simulations (see also the discussion in Chapter 7) is an attractive alternative.

The basic premise is straightforward for simple problems; the sample observations, x_i, are used as a surrogate population from which random samples, *with* replacement allowed, are taken. Each replicate would consist of N independent observations, $\tilde{x}_i$, that would generally be different from the original observations because replacement is allowed, and a sample mean, $\bar{x}_k$, can be computed in the usual manner. Selection of samples would be repeated N_R times, to obtain the bootstrapped empirical sampling distribution of the sample mean random variable, $\overline{X}$, which could then be used to estimate any statistic related to $\overline{X}$ including the confidence intervals. No assumption about the population distribution need to be made (though it could be). A number of variations on the basic technique appropriate for different problems as well as limitations of the technique are discussed in Davison and Hinkley (1997). An earlier technique, termed the jackknife, applies the simple resampling strategy of repeatedly leaving one observation out in a sequence of estimates of a statistic, thus providing the requisite resampled sample. In routine applications, particularly for large samples, traditional and bootstrapped approaches lead to similar estimates. Maindonald and Braun (2013) suggest that bootstrapping is not likely to be useful for statistics related maxima or extremes, but this is not restricted to bootstrapping as the statistics of extremes has the inherent problem of small samples, yet may be of particular relevance to design. Bootstrapping is subsequently applied to problems in regression (Section 6.6.5, Volume I) and in time series analysis(Section 6.12.2, Volume I) with specific comments provided in those sections.

6.6 REGRESSION

Hydraulic experiments are often performed in order to characterize the variation of a flow or transport variable in space or time or as other variables change. The variation, also termed the response, is generally denoted as f, with other (control) variables, termed the regressors or predictors denoted generally as x_k (or more compactly as a vector, $\mathbf{x}$). Thus, measurements of the centerline time-averaged point velocity or suspended sediment concentration may be taken as f, to be related to the normal distance to a fixed (or erodible) plane bed, which would be chosen as the regressor, x. In this case, basic models such a log-law velocity profile or a Rouse-Ippen sediment profile may be available for testing. On the other hand, if a model for predicting the scour depth at a cylindrical bridge pier is to be developed using data acquired for various flow and pier characteristics, then even the functional form of the model needs to be formulated. The estimation of model parameters and their associated statistics relating the response to regressor variables is generally termed regression. Linear regression is covered in elementary texts such as Walpole *et al.* (2007), but more advanced treatments can be found in Montgomery *et al.* (2006), and in general texts (*e.g.*, Draper & Smith, 1998; Ryan, 2009). Uncertainty analysis related to regression is discussed in Coleman and Steele (2009).

The variation in $\mathbf{x}$ raises the question whether the 'population' remains the same as $\mathbf{x}$ varies. From the classical regression point of view, the mean variation in f due to the variation in $\mathbf{x}$ is treated as deterministic, with a residual 'error', ε, that is assumed statistically homogeneous with respect to $\mathbf{x}$. Thus, despite the deterministic variation in $\mathbf{x}$, the random sampling is considered as being taken from the same population of ε. Regression seeks to estimate the (population) parameters for the deterministic model. If the model is 'incorrect' over some or all of the parameter range of interest, then the removal of the contribution of the assumed deterministic model will not result in a statistically homogeneous population. Even the form of the model may play a role, as transformed variables may lead to a statistically different and more homogeneous population. Assuming that the population is the same, which population is it? While measurements are obtained under specific physical conditions, they are usually intended to illuminate a much broader range of physical phenomena, and hence are compared with measurements performed under different conditions and scales. As remarked previously, this will usually entail choosing appropriate scales, thus making the regressors dimensionless, *i.e.*, 'scale-free', which may be seen as a type of variable transformation, and as a type of standardization in the statistical sense. As will be seen in an extended example (in Section 6.8.2, Volume I), the definition of a population may however be more explicitly made without necessarily appealing to a dimensionless formulation.

6.6.1 The linear model and interval estimates

Models play an essential role in classical regression where the functional form of the model must be pre-specified. Statistical inferences such as confidence intervals apply strictly speaking *only* to the assumed model. If the assumed model is incorrect, any statistical inference may be questioned. This subsection focuses on the linear model, which can be expressed as

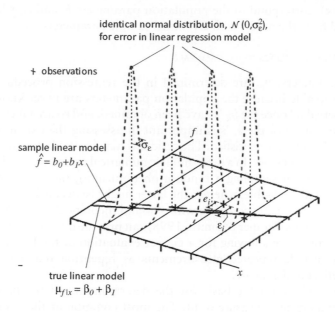

identical normal distribution, $\mathcal{N}(0, \sigma_\varepsilon^2)$, for error in linear regression model

+ observations

sample linear model
$\hat{f} = b_0 + b_1 x$

σ_ε

f

e_i

ε_i

true linear model
$\mu_{f|x} = \beta_0 + \beta_1$

x

Figure 6.6.1 Definition sketch of the linear regression model with a single regressor ($M = 1$), including the assumed normally distributed error.

$$f_i = \beta_0 + \beta_1 x_{1,i} + \beta_2 x_{2,i} + \ldots + \beta_n x_{M,i} + \varepsilon_i, \quad \text{or in vector form,} \quad \mathbf{f} = \mathbf{X}\boldsymbol{\beta} + \boldsymbol{\varepsilon}$$

$$(6.6.1)$$

where f_i (or the vector, $\mathbf{f}$), are the N (continuous) observed responses, $x_{k,i}$, (or $\mathbf{X}$) the $N \times M$ (matrix of) known values of the M regressors giving rise to f_i, and β_k (or $\boldsymbol{\beta}$), the model (population) parameters. Shown in Figure 6.6.1 is a sketch of the simple case for one regressor,($M = 1$), and the random variable, ε_i, representing the deviation of the response from the assumed model. The error term, ε_i, is assumed statistically homogeneous and randomly sampled across the entire range of x_k's, with zero mean and constant variance, σ_ε^2 (either iid(0, σ_ε^2) or more restrictively nid(0, σ_ε^2)) (see Figure 6.6.1). The matrix-vector form is convenient in the analysis and discussion of multi-variate models.

It is emphasized that the linearity refers to $\boldsymbol{\beta}$, and not to $\mathbf{x}$, such that polynomial (*e.g.*, $x_1 = x$, $x_2 = x^2$, ..., $x_M = x^M$) or transcendental-function (*e.g.*, $x_1 = \sin x$, $x_2 = \log x$, etc.) models would still be considered linear regression. In addition, transforma-tion of variables (see also discussion in Section 6.6.4, Volume I) can also be applied to obtain a linear model, *e.g.*, log transforming a power law. Nevertheless, experience in known cases such as the drag curve for a sphere, the Shields curve for incipient sediment motion, or the Nikuradse pipe-resistance curves suggest that simple linear or power-law models will not apply over the entire range of values of $\mathbf{x}$, but will more likely be applicable for quite a restricted range.

The distinction between sample and population variables, discussed above for simple statistics, needs also to be made in regression, where the sample model differs from the population (or true) model (see Figure 6.6.1). The sample parameters, b_i (or $\mathbf{b}$) and

$e_i = f_i - \hat{f}_i$ (or **e**), correspond to the population parameters β_i and ε_i, where $\hat{f}_i$ is the value predicted from the model with estimated sample parameters, *i.e.*,

$$\hat{f}_i = b_0 + b_1 x_{1,i} + b_2 x_{2,i} + \ldots + b_M x_{M,i}. \tag{6.6.2}$$

The sample parameters, b_k, are determined in the regression procedure. Unlike ε_i, which is unobservable because the population parameters are never known, the residual, e_i, is 'observable' once the b_k's have been obtained, and so are taken as surrogate for ε_i. An examination of the e_i's is important in assessing the extent to which the assumptions regarding ε_i are satisfied. The residuals may be attributed to either experimental or to lack-of-fit errors. In a deterministic numerical experiment (which does not quite fit the concept of sampling a random phenomenon), the former error can be generally considered negligible, whereas the latter might be deemed as negligible if the fitted model were reasonably complete with a sound theoretical basis. Only genuinely repeated measurements (Draper & Smith, 1998) can provide information regarding the experimental error hence allowing for a clearer evaluation of the lack of fit. Typical datasets rarely include repeated measurements as repetition requires considerable experimental effort and expense.

The residuals, e_i's, form the basis for the determination of the b_k's, which are obtained by minimizing a measure of fit. The most common method is the ordinary least squares (OLS) which minimizes the sum of the squared residuals, $\sum_i e_i^2$. Under certain standard assumptions, OLS has desirable statistical properties such as the so-called best linear unbiased estimate or BLUE. Ryan (2009) points out that the BLUE characteristic will not apply for OLS applied to an incorrect model. As will be seen in the discussion of Bayesian linear regression (Section 6.7.1, Volume I), OLS can also be justified from a maximum likelihood argument. Criteria other than OLS are available (Ryan, 2009), mainly so as to decrease sensitivity to outliers (robust regression) and to address the problem of highly correlated regressors also known as multicollinearity (*e.g.*, through ridge regression). In the machine learning context, ridge regression is viewed as an example of a regularization or shrinkage technique that seeks a smoother fit less sensitive to noisy data through a measure of fit (other than OLS) that penalizes large values of b_k's.

In some problems, one or more of the model parameters, β_k, may possess some physical meaning, or may be theoretically predicted, and so may be of interest in themselves. In other problems, the main concern is the value, $\hat{f}$, for any arbitrary **x**, and the values of b_k's are only of secondary concern (though probably should not be entirely ignored as very large or otherwise 'unrealistic' values may signal an inappropriate model). If the relationship between f and **x** is viewed as a joint (multivariate) distribution, then the mean response can be seen as conditional on **x**, and so can be denoted as $\mu_{f|\mathbf{x}}$. With the usual assumption of normally distributed errors, statistical inferences for the β_k's and f, such as the interval estimates, can be made, in a manner similar to those already discussed in Section 6.4, Volume I. Thus, two-sided interval estimates for the β_k at a significance level of α may be expressed as the $b_k \pm t_{n_{df}, \alpha/2}(s.e.)_{b_k}$, where $(s.e.)_{b_k} = \sqrt{\text{var}(b_k)}$ denotes the standard error for β_k (see Appendix 6.A.2 or any standard statistics text for formulae for $(s.e.)_{b_k}$ for the case of simple regression), and $n_{df} = N - M - 1$ is the number of degrees of freedom for the t-distribution. Similarly, the two-sided interval estimate for the *mean* response, $\mu_{f|\mathbf{x}_0}$, for a significance level of α is $\hat{f}(\mathbf{x}_0) \pm t_{n_{df}, \alpha/2}(s.e.)_{\mu_{f|x_0}}$, while the corresponding two-sided *prediction* interval (similar to the tolerance interval discussed in Section 6.4) for a *single* observation is $\hat{f}(\mathbf{x}_0) \pm t_{n_{df}, \alpha/2}(s.e.)_{\hat{f}, \text{pred}}$. Note that the standard error for the different statistic will differ.

6.6.2 The coefficient of determination and correlation

The coefficient of determination, R^2, (or an adjusted R^2_{adj}, which attempts to incorporate the effect of the number of model fitting parameters) is often reported and interpreted as a measure of how well a model fits the sampled data. It can be analyzed in terms of various sums of squares, *i.e.*, variances, namely the total sum of squares (*SST*), the regression sum of squares (*SSR*), and the error sum of squares (*SSE*), or a covariance, where

$$SST = \sum_{i=1}^{N} \left(f_i - \bar{f} \right)^2, \quad SSR = \sum_{i=1}^{N} \left(\hat{f}_i - \bar{f} \right)^2, \quad SSE = \sum_{i=1}^{N} \left(f_i - \hat{f}_i \right)^2 = \sum_{i=1}^{N} e_i^2,$$

$$(6.6.3)$$

and

$$SST = SSR + SSE \qquad (6.6.4)$$

where $\bar{f}$ is the sample mean of f_i. The partition of *SST* into *SSR* and *SSE* is an example of an analysis of variance frequently performed in statistics, which seeks to identify the source of different components of the total variance. If (genuinely) repeated measurements are available, a somewhat more refined analysis could be done (Draper & Smith, 1998). As defined above, *SST*, *SSR*, and *SSE* can be evaluated for any arbitrary model, whether linear or non-linear, and whether or not the model was derived using the data, but Equation (6.6.4) is true *only* for a linear (OLS) model. Two definitions of R^2 may be found:

$$R^2 = \frac{SSR}{SST} = \frac{\sum_i \left(\hat{f}_i - \bar{f} \right)^2}{\sum_i \left(f_i - \bar{f} \right)^2} \quad \text{or} \quad R^2 = 1 - \frac{SSE}{SST} = 1 - \frac{\sum_i e_i^2}{\sum_i \left(f_i - \bar{f} \right)^2} \qquad (6.6.5)$$

which motivates the interpretation of R^2 as the fraction (hence with $0 \leq R^2 \leq 1$) of the total variance (*SST*) 'explained' by the regressed model (*SSR*). The two definitions are equivalent but the interpretation as a fraction is valid *only* for a linear model based on the data, and R^2 so defined is limited by these constraints. The choice of *SST* for scaling (or non-dimensionalizing) of R^2 merits comments. *SST* may be viewed as the variance associated with a no-trend model, *i.e.*, with $\beta_k = 0$ $k \geq 1$ and so $\beta_0 = \bar{f}$), and this reference model (*i.e.*, a null hypothesis of $\beta_k = 0$, $k \geq 1$) is for some problems the most meaningful for comparison. In other problems, however, this reference model is not necessarily relevant, and R^2 may be misleading if interpreted with this conventional reference. In curious cases, such as when the second equation of Equation (6.6.5) is evaluated for a model which was not fitted to the data, a *negative R^2* may result.

A related measure of model fit is the sample correlation coefficient (*cf.* Equation (6.3.3)), denoted as r_{fx}, between observations of two variables, f and x, and evaluated as

$$r_{fx}^2 = \frac{\left[\sum_{i=1}^{N}\left(f_i - \bar{f}\right)(x_i - \bar{x})\right]^2}{\left[\sum_{i=1}^{N}\left(f_i - \bar{f}\right)^2\right]\left[\sum_{i=1}^{N}(x_i - \bar{x})^2\right]} \tag{6.6.6}$$

Unlike R^2, no regression between f and x is assumed for r_{fx}^2, and the inequality, $0 \le r_{fx}^2 \le 1$, is always satisfied. It is also interpreted as measure of the linear association between f and x, and used as measure of model fit by choosing f_i and x_i as model prediction and observation respectively. For a simple linear regression, r_{fx}^2 is then equivalent to R^2.

6.6.3 Miscellaneous topics in linear regression

A range of issues can arise in practical regression that often receives little attention in elementary discussions. Regression through the origin or the zero-intercept constraint sometimes seems attractive from a physical viewpoint, but should be viewed with some caution (Ryan, 2009). Data near the origin should be available, and a preliminary analysis with a non-zero intercept should be performed to test whether the intercept would be statistically different from zero. Conventional statistical measures such as R^2 and their interpretation may also require modification (Montgomery *et al.*, 2006).

The standard linear model also assumes that the regressors are fixed (*i.e.*, known with negligible uncertainty) so that the residual used in the OLS algorithm is the (vertical) distance between the observed and predicted values at a *fixed* value of the regressor. In many cases, particularly under laboratory conditions, the uncertainty in the regressors is much smaller than the uncertainty in the response, justifying this standard assumption. In other cases, the 'true' values of the regressors are unknown, and the observed values may be viewed as also being randomly sampled, giving rise to the problem of so-called errors-in-variables. For the simple case of a single regressor with error, the conventional approach can be shown to give an estimate for the slope that is biased (smaller than the true value), with the bias depending on the variance of the regressor (Draper & Smith, 1998).

Another somewhat related question arises in calibration when the value(s) of the regressor(s) for given value(s) of the response variable is desired. The common approach inverts the already determined model for the response variable, and it is even possible to obtain confidence intervals (Draper & Smith, 1998). A more controversial alternative, termed inverse regression (Osborne, 1991; Brown, 1993), reverses the roles of the regressor(s) and the response variable, which may lead to an estimate with reduced variance though may be biased.

6.6.4 Model transformations and nonlinear regression

Before a discussion of non-linear regression models, the common practice of transforming a (physically) nonlinear model into a statistically linear model is considered. The most common example is a power-law model, such as $f = \exp(\beta_0)x_1^{\beta_1}x_2^{\beta_2}...x_k^{\beta_k}$, which is often analyzed by log-transforming into $\ln f = \beta_0 + \beta_1\ln x_1 + \beta_2\ln x_2 + ... + \beta_k\ln x_k$. This apparently *statistically* linear problem has the advantage that results on linear models can be used, but for statistical inference, requires that the model error term, ε, be *additive* (as in the standard linear model), *i.e.*,

$$\ln f_i = \beta_0 + \beta_1\ln x_{1i} + \beta_2\ln x_{2i} + ... + \beta_k\ln x_{ki} + \varepsilon_i = \varepsilon_i + \beta_0 + \sum_{j=1}^{k}\beta_j\ln x_{ji}$$

$$(6.6.7)$$

implying that in the original form the error is *multiplicative*. Further if the error in the transformed form is assumed normally distributed, the error must be *lognormally* distributed in its original form. A practical criterion for identifying a multiplicative error structure is a variation in the response variable (and possibly also the regressors) over more than one order of magnitude, *i.e.*, whenever a log-log plot might be considered appropriate. In such a case, an additive error of constant variance would be either completely negligible or completely dominant compared to the signal over a large part of the range of the response variable. A multiplicative error would however allow the error to be commensurate with the value of the response variable, *e.g.*, a constant fraction, ensuring that the error does not become negligible (or completely dominant) over the range of interest for the variable. For example, in the case of the suspended sediment concentration profile in a two-dimensional uniform flow over a plane bed, a multiplicative original error is plausible.

Despite the convenience in exploiting results on linear models, the transformation is accompanied by the issue of bias when it is desired to predict the response variable. The example of log transformation illustrates the issue, which arises because while the additive error is assumed to have zero mean in log coordinates, its mean in the original coordinates is not unity. With ε in the log coordinates assumed to be nid$(0, \sigma^2)$, then the mean, $E(f)$, in the original coordinates (so not $E(\ln f)$), will involve $E[\exp(\varepsilon)] = \exp(\sigma_\varepsilon^2/2)$, for $\exp(\varepsilon)$ would be lognormally distributed with mean $\exp(\sigma_\varepsilon^2/2)$ (see the discussion of the lognormal distribution in Appendix 6.A.1). This argument assumes that the population parameters, β_k, are known, leading to an approximate estimate for the mean response, $\hat{f}$, for given x_k:

$$\hat{f} = \exp\left[\beta_0 + \sum_{j=1}^{k}\beta_j\ln x_j\right]\exp[s_\varepsilon^2/2], \quad s_\varepsilon^2 = \frac{\sum_{i=1}^{N}\left[(\ln f)_i - \beta_0 - \sum_{j=1}^{k}\beta_j\ln x_{ji}\right]^2}{N-2}$$

$$(6.6.8)$$

where the bias-correction term, $\exp(\sigma_\varepsilon^2/2)$, becomes negligible only when σ_ε^2 (or s_ε^2) $\to 0$. This argument and other approximate results are discussed in Helsel and Hirsch (2002). More general results are given in Beauchamp and Olson (1973) where it is also shown that the above approximate result is valid for small s_ε^2.

If it is deemed that the error term is *not* multiplicative but additive (as in the conventional linear model), *i.e.*,

$$f_i = f_{nl}(\beta_0, \beta_1, ...\beta_k; x_{1,i}, x_{2,i}, ..., x_{k,i}) + \varepsilon_i \qquad (6.6.9)$$

where $f_{nl}()$ denotes a non-linear function of the (population) parameters, $\beta_0, \beta_1, ...\beta_k$. then a non-linear regression model might be considered. Nonlinear least squares regression can still be performed, but simple closed-form formulae for the sample estimates $b_0, b_1, ..., b_k$, are no longer available. A computational solution to the least-squares minimization problem must be obtained that may have to deal with nonlinear issues such as multiple local minima and sharp gradients. Further, determining mean and variances of the $b_0, b_1, ..., b_k$ in order to construct confidence intervals become in general more challenging. The standard approach (Bates & Watts, 1988) linearizes the non-linear model in the vicinity of the already obtained least squares estimates, and then constructs confidence intervals for the individual parameters in the same way as for a linear model.

6.6.5 Residuals and bootstrapping regression

The assessment of a regression model should include a study of the residuals (or quantities derived from them) and their variation with the predicted response variable and possibly other predictors along with their deviation from any assumed distribution. For non-linear models, the raw residuals, e_i, are used. For linear models, standardized (or so-called Studentized) residuals are preferred as they facilitate comparisons. The basic residuals diagnostics are graphical and hence mainly qualitative, searching for systematic variations in plots of residuals against either predictions or individual regressors that would indicate gross violations of the inference assumptions of linearity, independence, and homogeneity. A normal-distribution assumption can be examined through quantile-quantile plots (see Section 6.3.1, Volume I) in which the relevant residuals quantiles are plotted against the ideal normal distribution quantiles, so that large deviations from a straight line would be interpreted as non-normal behavior.

If the residual diagnostics provide strong evidence that inference assumptions are invalid, then bootstrapping becomes an alternative to obtain inferences, specifically confidence intervals, for the statistical estimates of interest. For non-linear regression, for which standard inferences are approximate due to linearization even if the standard assumptions on the error model are satisfied, bootstrapping might also be considered. A subtlety in bootstrapping for regression is illustrated in two possible different resampling strategies (Davison & Hinkley, 1997). In the first, termed case resampling, a replicate consists of regressors (the x_{ki}'s) being resampled (with replacement, therefore allowing omission and repeated sampling of some points) and fitting a linear or nonlinear model, to obtain an instance of the statistic of interest, such as the coefficients. In this type of resampling, no assumption is made regarding the distribution or the constant variance of the residuals (the assumption of independence is still retained). Davison and Hinkley (1997) point out that as the regressors are resampled the information content may vary from replicate to replicate. In contrast, the second resampling strategy, termed model or fixed- x resampling, uses all of the regressors in each replicate, but the constant-variance assumption is made and the residuals (or possibly a modified form of the residuals) of the original model rather than the regressors are resampled. For each replicate, the appropriate model is then fitted to the response variable constructed with the resampled residuals, and the desired bootstrap statistics evaluated by repeating this for a specified number of replicates.

6.6.6 Spurious correlations

The problem of spurious (self)-correlations has been often pointed out in the water resources literature (Benson, 1965; Yalin & Kamphuis, 1971; Kenney, 1982) and deserves a broader recognition. The term refers to a correlation occurring when one variable group, $\prod_1$, such as a sum, ratio or product, is correlated with a second variable group, $\prod_2$, primarily through a common variable. Kenney (1982) gives theoretical results for some simple examples, e.g., when $\prod_1 = f_a + f_b$, or $\prod_1 = f_a/f_b$ and $\prod_2 = f_b$ with the common variable, f_b. Even if the correlation between the individual variables, f_a and f_b, were zero, a non-negligible correlation is found between $\prod_1$ and $\prod_2$ due solely to the self correlation of f_b, the common variable. Because empirical models in hydraulics are often (and should be) formulated in terms of dimensionless groups and determined using repeating variables, the analyst should be on guard against potential spurious self-correlations in regression models. This may appear in various forms, most obviously when the response variable as a whole, not just as part of a product or a ratio, appears as part of a regressor variable. Although such a case may seem straightforward to identify, it can be easily obscured by notation.

6.7 BAYESIAN INFERENCE

This chapter has so far mainly taken the traditional frequentist point of view in which the (objective) probability of an event is interpreted as a limiting relative frequency (see Section 6.2.2, Volume I). This section introduces the Bayesian approach in which the probability is quantified in terms of not only the available data but also the prior belief of the analyst. Due to the latter, the term subjective probability is sometimes applied to contrast the Bayesian interpretation of probability to the frequentist probability. Whereas subjective probability can be assigned to non-repeatable events, such as the probability of failure of an existing dam, an objective probability cannot be meaningfully defined. Readers are referred to Bolstad (2007) for an accessible introduction to Bayesian methods, Jaynes (2003) for a discussion of Bayesian philosophy, Gill (2014) for applied Bayesian inference, and Barber (2012) for connections between Bayesian inference and machine learning. Hydraulic applications of Bayesian techniques have so far been limited to model calibration and sensitivity and uncertainty analyses (e.g., Ruark et al., 2011; Le Coz et al., 2014; Mansanarez et al., 2016), but Schmelter et al. (2011) proposed a Bayesian sediment transport model.

In the Bayesian approach, the analyst's prior belief about a random event, A, is updated based on the observed data, D, through Bayes' rule:

$$P(A|D) = \frac{P(D|A)P(A)}{P(D)}. \tag{6.7.1}$$

The prior belief about A *before* observing the data is captured in the *prior* probability, $P(A)$, while the evidence provided by the data D is captured in the conditional probability $P(D|A)$, called the likelihood function. The denominator $P(D)$ can be simply-treated as a normalizing constant. In the *posterior* probability, $P(A|D)$, Bayes' rule (Equation (6.7.1)) gives the updated uncertainty in A *after* taking the data into account. This general approach can be applied to any statistical inference problem including estimating parameters, testing hypothesis, identifying best model among candidate models, and determining relevant variables.

6.7.1 Bayesian linear regression

The Bayesian approach is illustrated by its application to the problem of linear regression. For the linear regression model of Section 6.6.1, Volume I, the unknown parameters, expressed as a vector, θ, consists of β, and the variance of the error vector σ_ε^2, *i.e.*, $\theta = [\beta, \ \sigma_\varepsilon^2]^T$, (the superscript, T, indicates a transpose), while the observed data, $D = \{X, f\}$. An estimate of θ, and possibly of the response f_0 and associated uncertainty for a new value of the regressor x_0, are desired. Application of Bayesian approach to the linear regression problem involves at least three steps: (i) prescribing a likelihood function, $\lambda_{like}(D|\theta)$, (ii) encoding prior belief about θ in a *prior* distribution, $\lambda_{prior}(\theta)$, and (iii) obtaining the *posterior* distribution, $\lambda_{post}(\theta|D)$, through Equation (6.7.1). With $\lambda_{post}(\theta|D)$ obtained, any other desired derived distribution can be subsequently evaluated, such as for example the distribution, λ_{pred}, of f_0 for any given x_0.

Step 1. The prescription of a likelihood function, $\lambda_{like}(D|\theta)$, in Bayesian practice follows that in the classical *maximum likelihood* (ML) approach (see Section 6.2.4, Volume I), and depends on the assumed distribution of the model error, $\varepsilon = (f - X\beta)$. If as in classical regression, the latter is assumed nid$(0, \sigma_\varepsilon^2)$ or $\mathcal{N}(0, \sigma_\varepsilon^2) = \mathcal{N}(f|X\beta, \sigma_\varepsilon^2 I))$, then $\lambda_{like}(D|\theta)$ takes the form, similar to the classical ML result,

$$\lambda_{like}(D|\theta) = \lambda_{like}(f|X; \beta, \sigma_\varepsilon^2) = \mathcal{N}(f|X\beta, \sigma_\varepsilon^2 I) = \frac{1}{\left(2\pi\sigma_\varepsilon^2\right)^{M/2}} \exp\left[-\frac{1}{2\sigma_\varepsilon^2}(f - X\beta)^T(f - X\beta)\right]$$

(6.7.2)

where I denotes the identity matrix. Maximizing $\lambda_{like}(D|\theta)$ with respect to θ is seen to be equivalent to minimizing the sum of squares of the errors, *i.e.*, $(f - X\beta)^T(f - X\beta)$.

Step 2. Whereas the likelihood function of the previous step also figures prominently in the frequentist interpretation, the distinctive feature of the Bayesian approach is the role of the prior distribution, $\lambda_{prior}(\theta)$, which because of its basis in subjective belief has been the subject of much controversy. Gelman (2008) compiles the main criticisms directed at the Bayesian approach. Prior distributions can be grouped into two classes: i) uninformative priors to be used when no prior knowledge exists, such as uniform priors, Jeffreys prior, and other 'diffuse' priors, and ii) informative priors exploiting prior typically context-specific knowledge of θ (the subjective element). If both prior and posterior distributions are chosen from the same family of distributions, the prior is termed conjugate, and with appropriate choice of parameters, can be informative or uninformative. Conjugate priors have traditionally been popular because they permit an analytical solution for $\lambda_{post}(\theta|D)$. For the linear regression model, the conjugate prior is a Gaussian-inverse gamma (IG) prior of the form

$$\lambda_{prior}(\theta) = \lambda_{prior}(\beta, \sigma_\varepsilon^2) = N(\beta|\beta_0, \sigma_\varepsilon^2 I)IG(\sigma_\varepsilon^2|a_0, b_0)$$

$$\propto \left\{\left(\frac{1}{\sigma_\varepsilon^2}\right)^{M/2} \exp\left[-\frac{1}{2\sigma_\varepsilon^2}(\beta - \beta_0)^T(\beta - \beta_0)\right]\right\}\left\{\left(\frac{1}{\sigma_\varepsilon^2}\right)^{a_0+1} \exp\left(-\frac{b_0}{\sigma_\varepsilon^2}\right)\right\} \quad (6.7.3)$$

i.e., $\boldsymbol{\beta}$ is assumed to follow a normal distribution with mean, $\boldsymbol{\beta}_0$, and constant variance, σ_ε^2, assumed to follow an inverse gamma distribution with the two standard parameters, the shape, a_0, and the (possibly dimensional) rate, b_0 (or the precision, $1/\sigma_\varepsilon^2$), follows a gamma distribution. The so-called hyperparameters associated with the prior distribution, $\boldsymbol{\beta}_0$, a_0, and b_0 to be distinguished from the model parameters, $\boldsymbol{\theta}$, can be adjusted to reflect prior knowledge about $\boldsymbol{\theta}$. If the hyperparameters are also uncertain, they too can be estimated using Bayes' rule by first assigning them each a prior distribution. Such multilevel priors are called hierarchical priors. Small values for a_0 and (a suitably non-dimensionalized) b_0 (say 0.001) imply a large variance (see the Appendix 6.A.1 on the gamma and inverse gamma distributions), and will result in a diffuse prior, reflecting ignorance or large uncertainty in σ_ε^2 before observing the data. If information is available on the sign of $\boldsymbol{\beta}$, then $\boldsymbol{\beta}_0$ can be suitably chosen. A zero value of $\boldsymbol{\beta}_0$ suggests no prior preference for positive or negative weights but rather a preference for smaller weights resulting in a model closely resembling the regularized least squares model. Proper selection of $\lambda_{\text{prior}}(\boldsymbol{\theta})$ is key to effective Bayesian inference – a carefully chosen prior encoding existing knowledge of the physics of the problem can yield significant results with limited data, whereas a poorly designed prior can give poor results with high confidence even with large reliable datasets. Even if the prior information is available, how this can be encoded in a prior distribution is however not a trivial task.

Step 3. Estimating the posterior distribution, $\lambda_{\text{post}}(\boldsymbol{\theta}|D)$, once λ_{like} and λ_{prior} have been defined, is conceptually a straightforward application of Bayes' rule Equation (6.7.1). For most non-trivial statistical inference problems, a closed-form analytical solution for $\lambda_{\text{post}}(\boldsymbol{\theta}|D)$ is not available because the integral denominator ($P(D)$ in Equation (6.7.1)) cannot be evaluated in closed form. Methods for evaluating $\lambda_{\text{post}}(\boldsymbol{\theta}|D)$ can be either deterministic or stochastic. The former obtains analytical approximations to $\lambda_{\text{post}}(\boldsymbol{\theta}|D)$ by making additional assumptions. As examples, the variational Bayes method assumes that $\lambda_{\text{post}}(\boldsymbol{\theta}|D)$ can be factored in a specific form, while the Laplace method assumes a Gaussian $\lambda_{\text{post}}(\boldsymbol{\theta}|D)$. The deterministic approach can be effective for large-scale problems but can never produce 'exact' results. Stochastic schemes numerically approximate $\lambda_{\text{post}}(\boldsymbol{\theta}|D)$ using sampling algorithms, and can produce 'exact' results with sufficiently large number of samples but become computationally demanding for large-scale problems. Markov-chain Monte-Carlo (MCMC) methods, developed in the physics literature, are the most popular sampling method in Bayesian inference because they are effective in sampling high-dimensional (large number of unknown parameters) spaces, easy to design, and applicable to a large class of distributions. These methods sample by constructing a Markov chain whose stationary (equilibrium) distribution is the desired posterior distribution. Two MCMC algorithms, Metropolis–Hastings (Metropolis *et al.*, 1953) and Gibbs sampling (Geman & Geman, 1984), are widely used as they are amenable to problem-independent programming. The wide availability of free software such as BUGS and JAGS implementing these sampling algorithms for a user-defined statistical model of arbitrary complexity has contributed greatly to the increased interest in Bayesian techniques (Lunn *et al.*, 2009). Details of MCMC methods may be found in Gill (2014).

Step 4. At the end of Step 3, the basic estimation problem is solved, and any statistic of interest related to $\boldsymbol{\theta}$ can be evaluated from the posterior distribution, $\lambda_{\text{post}}(\boldsymbol{\theta}|D)$. In many linear regression problems, interest is not restricted solely or even primarily to θ, but rather a prediction of the response, f_0, for a new point x_0, unconditioned on θ, is desired. This also must be expressed probabilistically in terms of a predictive distribution, $\lambda_{pred}(f_0|x_0, D)$, and is obtained from $\lambda_{post}(\theta|D)$ by integrating over θ to obtained the predicted distribution of f_0:

$$\lambda_{pred}(f_0|x_0, D) = \int \lambda_{like}(f_0|x_0, \theta)\lambda_{post}(\theta|D)\, d\theta \qquad (6.7.4)$$

where $\lambda_{like}(f_0|x_0, \theta)$ is the likelihood function of Step 1, with f_0 and x_0 replacing f and X. $\lambda_{pred}(f_0|x_0, D)$ combines the model error (in $\lambda_{like}(f_0|x_0, \theta)$) and the parameter uncertainty (in $\lambda_{post}(\theta|D)$) to determine the prediction uncertainty. If $\lambda_{post}(\theta|D)$ can be found in closed form, $\lambda_{pred}(f_0|x_0, D)$ can be determined analytically; otherwise numerical integration using the samples generated from $\lambda_{post}(\theta|D)$ can be performed.

With the entire (posterior) distribution, $\lambda_{post}(\theta|D)$, available, interval estimates for θ, if desired, can be obtained. The credible interval, (θ_L, θ_U), is obtained from $P(\theta_L < \theta < \theta_U) = 1 - \alpha$ such that the interval is either the narrowest interval or symmetric. The subtle difference from the traditional frequentist interpretation of confidence interval may be noted. The frequentist considers θ to be *fixed*, i.e., the 'true' value, and the interval to be random, depending on the sample on which the interval estimate is based. The Bayesian considers only one dataset, namely the one observed, and views θ as a random variable with an uncertainty expressed in $\lambda_{post}(\theta|D)$. In this respect, the Bayesian interpretation arguably corresponds more closely to the intuitive (confidence) interval, though it raises controversial philosophical questions as to the interpretation of θ, e.g., if θ is a random variable, are there 'true' values of θ?

6.7.2 Comments on Bayesian applications in hydraulics

The Bayesian approach, especially with the advent of numerical evaluation of the posterior distribution using MCMC methods, offers great flexibility in statistical analysis, e.g., regarding the distributional assumptions as well as in dealing with other issues. For example, when the response variable is transformed by a logarithmic function before being used in regression models (as discussed in Section 6.6.4, Volume I), the re-transformation of the mean (or other higher order statistics) of the dependent variable estimated in the transformed space to the original space induces bias. In the Bayesian construct, the re-transformation bias can be avoided in a straightforward way by re-transforming MCMC samples of the predictive distribution from the transformed-space to the original-space before estimating the statistics required for statistical inference (Stow et al., 2006). A Bayesian approach does require greater statistical sophistication in its use, but it also forces the analyst to make explicit the assumptions made in the analysis and its results are inherently probabilistic. Familiarity with Bayesian ideas is also useful as innovations in data analysis are often formulated from that perspective. In cases where reliable prior information is unavailable and so an uninformative or diffuse prior must be assumed, and the distributional assumptions of the traditional approach are reasonable, as may be the case in many hydraulic experiments, results of Bayesian analysis are unlikely to differ substantially from those of the traditional approach. A Bayesian model is likely to be most advantageous when strong prior information can be captured in a prior distribution assumption and the sample size is small. An operational context in which new data is continually being received would also render Bayesian techniques more attractive, as updating of probabilities can be carried out in a more natural setting.

6.8 EXTENDED EXAMPLES IN REGRESSION

This section illustrates through examples several of the points discussed above, emphasizing details related to the choice of populations, differences between linear, log-transformed linear, and nonlinear regression models, Bayesian and bootstrapping approaches, and spurious correlations. The analyses are not intended to be definitive; rather issues are raised that might be easily overlooked. The R software environment was used throughout. Bootstrapping for regression was restricted to case resampling as discussed in Section 6.6.5, Volume I, and Bayesian priors were diffuse (normal distributions for the means and the standard gamma distribution with small values for the precisions as discussed in Section 6.7, Volume I).

6.8.1   The velocity profile in a uniform open-channel suspension flow

The effect of suspended sediment on the time-averaged centerline velocity profile, $u(y)$, in a uniform open-channel flow (y denotes the normal distance from the bed) of depth h has been studied since Vanoni (1946), with more recent measurements acquired with various experimental techniques by Coleman (1981), Lyn (1987, 1988), Muste and Patel (1997), and Bennett $et\ al.$ (1998), among others. The main question addressed here is whether the presence of suspended sediment changes measurably the vertical velocity profile. More specifically, the focus is on the validity of the classic log-law (with a von Karman constant, $\kappa_s = \kappa_0 \approx 0.41$ as discussed in Section 2.2, Volume I) for describing the profile in the sediment-laden case. This question would be the basis for the null hypothesis (the subscript s refers to the value in a sediment-laden flow, while the zero subscript refers to the accepted value for the sediment-free flow). For illustration purposes, the data of Bennett $et\ al.$ (1998), obtained with phase-Doppler anemometry, in a fine-sand suspension flow (arithmetic mean sediment diameter of 0.26 mm) over an equilibrium bed is used.

Because only a single set of flow and sediment conditions is being considered, the question of appropriate scaling can be avoided (at the cost of rather limited generality), and the regression model is applied to mean-velocity data expressed in dimensional form as

$$u = \beta_{u0} + \beta_{u1} \ln y. \tag{6.8.1}$$

Equation (6.8.1) has a form permitting simple linear regression, with model parameters β_{u0} and β_{u1}. For simplicity, the model error term has been omitted in Equation (6.8.1) and in other similar model equations of this section. Attention is focused on β_{u1} which for a sediment-free flow would be $\beta_{u1} = u_*/\kappa_0$, with u_* being the shear velocity. Although the issue of scaling has been avoided, the question of a homogeneous population still arises in another guise, namely in the question of the appropriate region over which the model should be applied. The traditional approach (*e.g.*, Vanoni, 1946), fitted the log-law profile over the entire depth, h, but it is now widely accepted that this profile should only be applied to the near-bed region. Whereas Bennett *et al.* (1998) considered $y/h \leq 0.2$ ($h = 0.11$ m) as a fitting region, the more stringent choice of $y/h \leq 0.12$, to be referred to as the limited-population velocity fit, is made here. Because the value of κ_s is in question, an independent estimate of u_* is required, and is obtained from Reynolds shear-stress profile measurements. This entails another regression model, applied in this case over the entire flow depth for the shear stress profile, which for a uniform flow, should be linear, and zero at the free surface:

$$\tau = \beta_{\tau 1}(1 - y/h), \tag{6.8.2}$$

where the model parameter, $\beta_{\tau 1} = \rho u_*^2$ (ρ is the fluid density). Note that this is a regression through the origin (in the transformed variable, $1 - y/h$), with a zero-intercept constraint.

The results of linear regression analyses are summarized in Table 6.8.1 and plotted in Figure 6.8.1. Several questions may be posed. Are the two populations, that defined by $y/h \leq 1$ and that defined by $y/h \leq 0.12$, statistically different as far as u and specifically β_{u1} are concerned? From Table 6.8.1, a comparison of the confidence intervals for the two populations indicates that the difference in b_{u1} is *not* statistically significant, *i.e.*, the data do not strongly support distinguishing the two populations. Because R^2 is larger in the case of fitting over the entire depth, it is tempting to infer that the log-law model should be applied over the entire depth. This illustrates the limitations, if based on limited data, of regression (and statistics in general) in evaluating physics; statistical results should be evaluated in the wider context of theoretical and empirical results regarding general wall-bounded flows.

The main question asked whether κ_s differs from the clear-water value of κ_0 leads to the null hypothesis, $\kappa_s = \kappa_0$. This requires a confidence interval for the ratio, $\left(\sqrt{b_{\tau 1}/\rho}\right)/b_{u1}$; for the limited-population fit, a two-sided 95% interval is found as $(0.20, 0.36)$, from a standard error of 0.036 and $N_u + N_\tau - 2 = 26$ degrees of freedom for the critical t-value (see Appendix 6.A.3). If κ_0 is taken to be 0.41, then as this

Table 6.8.1 Regression results for the time-averaged velocity and shear stress profiles in a sediment-laden uniform open-channel flow

Model	population restriction	no. of points, N	R^2	b_1 (m/s or Pa)	$(s.e.)_{b_1}$ (m/s or Pa)	95% conf. interval for β_1 (m/s or Pa)
$u = \beta_{u0} + \beta_{u1}\ln y$	$y/h \leq 1$	21	0.99	0.138	0.003, 0.003†	(0.13,0.15), (0.13,0.14)†
	$y/h \leq 0.12$	7	0.91	0.164	0.023, 0.032	(0.10,0.22), (0.13,0.22)
$u = \beta_{u0} + \beta_{u1}\ln y$ (Bayesian)	$y/h \leq 1$	21		0.138		(0.13,0.15)
	$y/h \leq 0.12$	7		0.166		(0.066,0.27)
$\tau = \beta_{\tau 1}\left(1 - \frac{y}{h}\right)$	$y/h \leq 1$	21	0.99	2.03	0.048, 0.053	(1.93,2.13), (1.92,2.13)

† The first entries for $(s.e.)_{b_1}$ and the 95% confidence interval for β_1 are obtained using classical results, while the second entries rely on case-resampling bootstrapping.

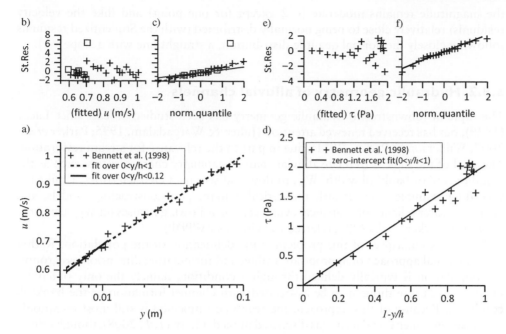

Figure 6.8.1 Regression analysis of data of Bennett *et al.* (1998); a) mean-velocity data and two fits, one based on measurements over the entire depth and the other restricted to measurements satisfying, $y/h \leq 0.12$, b), c) Studentized residuals for the mean-velocity data (squares from the more limited data) as a function of the fitted variable and the normal distribution quantile, d) mean shear stress data and fit through the origin, e, f) Studentized residuals for the mean-shear-stress data as a function of the fitted variable and the normal distribution quantile.

confidence interval does not include κ_0, the null hypothesis is rejected at the 95% significance level. This conclusion would not change if the velocity fit were taken over the entire depth or over $y/h \leq 0.2$ (as was done by Bennett *et al.*, 1998). On the other hand, one could also imagine a situation in which measurements in a 'reference' sediment-free flow were also made, such that a confidence interval for κ_0 could also be determined, then a comparison of the two confidence intervals might lead to a different conclusion. The inference does not make a statement about the physics, but rather about model parameters. The model encapsulates the physics, but the model is *assumed* rather than analytically derived. The inferences assume that the model (the log velocity profile) is correct; if the model is incorrect, then the inferences may be of little value. Thus, even though the difference between the observed κ_s and κ_0 is found to be statistically significant, the statistical argument given does *not* imply that a log-law with a reduced value of κ_s is a physically sound model.

The (externally) Studentized residuals for each fit are also plotted as a function of the fitted variable and of the normal distribution quantile in Figure 6.8.1. Except for one point for the limited-population velocity fit with an unusually large value (> 6) suggestive of an outlier, and perhaps a hint of serial correlation, the velocity-residuals are largely unremarkable. The shear-stress residuals are notable in the clear increase, but the magnitude remains moderate (< 2 except for one point) and (like the velocity residuals) relatively close to being normally distributed (with the Studentized residuals following closely the normal quantile distribution, a straight line with a slope ≈ 1).

6.8.2 Hydraulic geometry of alluvial channels

The topic of (downstream) hydraulic geometry has been studied at least since Lacey (1929), but has received renewed attention (Julien & Wargadalam, 1995; Parker *et al.*, 2007; Wilkerson & Parker, 2011), due in part to the relevance for stream restoration. The goal of the topic is to relate alluvial channel geometry characteristics, such as the 'equilibrium' or bankfull width, W_{bf}, to flow and sediment characteristics. Unlike the preceding example, there is little theoretical guidance for constructing models, and hence a greater reliance on statistical evidence. The following focuses on W_{bf}, using the dataset of Parker *et al.* (2007, referred to below as PWPDP).

An interesting aspect of this problem is the delineation of the population. Rather than the usual approach of appropriate scaling and formulating dimensionless groups, the population is typically defined through a condition, namely the only outcomes included are those thought to be associated with channel formation or the bankfull condition. Because of this approach, the regression model(s) is still most commonly expressed in dimensional terms (as discussed in Biedenharn *et al.*, 2008), though efforts at dimensionless relations such as PWPDP have appeared. The channel-forming condition by itself may not however always ensure a homogeneous population; Wilkerson and Parker (2011) have given a statistical argument for a distinction between sand-bed (or suspended-load-dominated) and gravel-bed (or bedload-dominated) cases. This example will therefore follow PWPDP in considering only the gravel-bed case, with the additional restriction, $d_{50} > 25$ mm, where d_{50} is the median sediment diameter.

Traditional models have related only the channel-forming or bankfull discharge, Q_{bf}, to W_{bf} via a power law, but more recent models have included additional variables, such as d_{50} (see Biedenharn *et al.*, 2008 for a list of proposed models). In the following, the basic model to be examined is of the dimensional power-law form, $W_{bf} = \beta_0 Q_{bf}^{\beta_1} d_{50}^{\beta_2}$ (this includes the traditional model if $\beta_2 = 0$). PWPDP preferred a dimensionless form, namely, $W_{bf}/[Q_{bf}^2/g]^{1/5} = \beta_0 \left(Q_{bf}/\sqrt{g d_{50}^5} \right)^{\beta_1}$, where g is the acceleration due to gravity. While this dimensionless form does also involve d_{50}, the difference from the dimensional form should be emphasized; in the latter, the variation in d_{50} is independent of the variation of Q_{bf}, whereas in the former, it is not (note the one fewer number of fitting parameters resulting from the dimensional analysis).

Results of the regression analyses of various models are summarized in Table 6.8.2 and plotted in Figure 6.8.2a. Because the various models are in the form of a power law, the issue of whether a log-transformation should be applied also arises. In this particular case, the results of a nonlinear (model I) and a linear log-transformed (model II) regression are different but the difference is not statistically significant. This will not be so in general. Because of large variation in values of observed W_{bf} (5 m to 280 m), a multiplicative error structure is more likely than an additive structure, and a log-transformed model is therefore preferable. Residuals analysis may also provide guidance. The residuals for the simpler nonlinear fit (model IVc) and log-transform fit

Table 6.8.2 Results of regression analyses for various models (metric units, where appropriate)

Model	R^2	b_1	b_2	95% conf. int. for β_1	95% conf. int. for β_2
I. $W_{\mathrm{bf}} = \beta_0 Q_{\mathrm{bf}}^{\beta_1} d_{50}^{\beta_2}$ (nonlinear)		0.486	−0.048	(0.45,0.52)	(−0.17,0.08)
II. $\log W_{\mathrm{bf}} = \beta_0 + \beta_1 \log Q_{\mathrm{bf}} + \beta_2 \log d_{50}$	0.91	0.495	−0.034	(0.46,0.54)	(−0.18,0.11)
III. $\log \dfrac{W_{\mathrm{bf}}}{(Q_{\mathrm{bf}}^2/g)^{1/5}} = \beta_0 + \beta_1 \log \dfrac{Q_{\mathrm{bf}}}{\sqrt{g d_{50}^5}}$	0.24	0.065		(0.038,0.092)	
IVa. $\log W_{\mathrm{bf}} = \beta_0 + \beta_1 \log Q_{\mathrm{bf}}$	0.91	0.500		(0.46,0.54), (0.46,0.53)†	
IVb. $\log W_{\mathrm{bf}} = \beta_0 + \beta_1 \log Q_{\mathrm{bf}}$ (Bayesian)		0.498		(0.46,0.54)	
IVc. $W_{\mathrm{bf}} = \beta_0 Q_{\mathrm{bf}}^{\beta_1}$ (nonlinear)		0.485		(0.45,0.52), (0.43,0.52)†	

† These are bootstrapped estimates (based on case resampling).

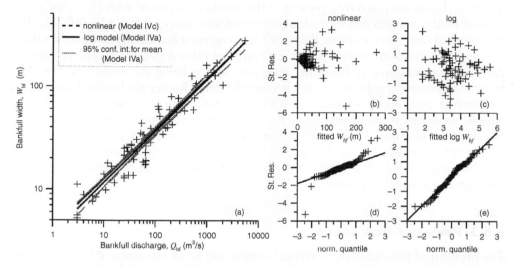

Figure 6.8.2 Results of regression analysis on the data of Parker *et al.* (2007): a) data with fits to model IVa and IVc, b) and d) plots of Studentized residuals for nonlinear fit (model IVc), and c) and e) plots of Studentized residuals for log-transform fit (model IVa) varying with fitted variable and the normal distribution quantiles.

(model IVa), plotted in Figure 6.8.2, lend support to a (statistical) preference for the latter. The nonlinear-fit residuals exhibit a systematic increase in variance over much of the range of the fitted values of W_{bf} and a more marked non-normality.

The results for both models I and II suggest that the dependence on d_{50} is weak at best: the null hypothesis, $\beta_2 = 0$, *cannot* be rejected at the 95% level (the *p*-value is in fact 0.60). Re-expressed in (dimensional) terms, model III, *i.e.*, the PWPDP model, leads to $W_{\mathrm{bf}} \propto Q_{\mathrm{bf}}^{0.465 \pm 0.026} d_{50}^{-0.163 \pm 0.065}$ at the 95% significance level. On the one hand, the values of both exponents are statistically consistent with those obtained with the (dimensional) model II; on the other hand, the confidence interval given suggests an exponent on d_{50} statistically different from zero (at the 95% level). The latter inference is however argued to be questionable in that the confidence interval on β_1 in the PWPDP model is dominated by the variation in Q_{bf}, rather than by any (independent) variation in d_{50}. Although a dimensionless formulation is attractive, the lack of stronger statistical case for a definite dependence on d_{50} suggests some skepticism with regards to the PWPDP model.

The strikingly low value of R^2 for model III deserves comment since the results of models II and III, when expressed in comparable dimensional physical terms, are largely consistent. The 'same' mathematical model may be associated with quite different values of R^2 ! This can be explained as follows: from the results for model II, if $\log W_{bf} - \beta_1 \log Q_{bf} = \log\left(W_{bf}/Q_{bf}^{\beta_1}\right) = \beta_0 + \beta_2 \log d_{50}$ were examined, then the very weak dependence on $\log d_{50}$ by itself would imply a low value of R^2. The scaling of W_{bf} by $\left(Q_{bf}^2/g\right)^{1/5}$ acts in a similar manner in that a large part of the variation in W_{bf} explained by the variation in Q_{bf} has been taken out by the scaling, thereby reducing substantially the value of R^2. The low value does nevertheless reflect the weak dependence on d_{50}, or alternatively a poor fit of the 'residual variation' with $Q_{bf}/\sqrt{gd_{50}^5}$.

Some final comments may be made about the relative uncertainties in W_{bf} and Q_{bf}. In the standard regression problem, the value of the regressor is chosen, and then the value of response variable corresponding to this regressor value is observed, as is the case in the preceding example (y being the regressor, and u being the response). For the hydraulic geometry problem, this is not generally done. Rather the bankfull condition is identified, and the corresponding width W_{bf} and discharge Q_{bf} observed (indeed Q_{bf} may be estimated from a further model equation with uncertain parameters). While W_{bf} itself and to a much lesser extent Q_{bf} can be precisely measured, the uncertainty lies mainly in the identification of the bankfull condition. As a result, the uncertainty in Q_{bf} may be comparable to that in W_{bf}, and the standard assumption of negligible error in the regressor is open to question.

6.8.3 Pressure-flow scour

The problem of pressure-flow or vertical-contraction scour has received less attention than bridge pier or abutment scour, but recent studies include Umbrell et al. (1998), Lyn (2008), Hahn and Lyn (2010). The situation occurs when an approach open-channel flow is forced to contract *vertically* due to the presence of the solid obstacle, such as a bridge deck (see Figure 6.8.3 for a definition sketch). Due to the flow acceleration and the concomitant potential increase in sediment-transport capacity, scour may be induced. As in other local-scour problems, the equilibrium maximum scour depth, y_s, for given flow, obstacle, and sediment characteristics is of interest. Guidance in formulating a predictive model is sparse, and more exploratory analysis may be warranted. Much of the following is taken from Lyn (2008), in which the relatively large-scale laboratory measurements of Arneson (1997) are examined, but data-analysis aspects rather than process-related issues are highlighted.

The directly *linear* model proposed by Arneson (1997) was initially expressed as:

$$\frac{y_s}{y_1} = \beta_0 + \beta_1\left(\frac{y_1}{H_b}\right) + \beta_2\left(\frac{H_b + y_s}{y_1}\right) + \beta_3\left(\frac{V_a}{V_c}\right) \tag{6.8.3}$$

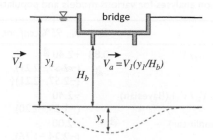

Figure 6.8.3 Definition sketch of pressure-flow scour (Reproduced with permission of ASCE).

where y_1 is the approach flow depth, H_b the initial (prior to scour) vertical distance between the obstacle low chord and the undisturbed bed, V_a the initial (prior to scour) velocity through the opening, and V_c the critical velocity for incipient sediment motion based on y_1 (see Figure 6.8.3). The appearance of y_s/y_1 on both sides of Equation (6.8.3) signals the potential for spurious self-correlation; an alternative model mathematically but not necessarily statistically equivalent to Equation (6.8.3) is

$$\frac{y_s}{y_1} = \beta_0 + \beta_1 \left(\frac{y_1}{H_b} \right) + \beta_2 \left(\frac{H_b}{y_1} \right) + \beta_3 \left(\frac{V_a}{V_c} \right) \tag{6.8.4}$$

where the omission of y_s/y_1 on the right hand side should diminish the importance of spurious self-correlation. Arneson (1997) applied Equation (6.8.3) (model I in Table 6.8.3) over all experiments (total number of points, $N_{tot} = 116$) in which the bridge projected into and hence disturbed the flow, reporting a relatively high value of $R^2 = 0.90$ that might be naively interpreted as indicative of a good fit. These results contrast however with the results over the same population for model II Equation (6.8.4), which gives a sharply reduced value of R^2. Unlike the preceding example, the reduced R^2 is not due to a scaling that changes the response variable, but rather is attributed to the poorer fit when the spurious self-correlation contribution is removed or greatly reduced.

After some exploratory analysis, a simple power-law model was proposed by Lyn (2008) for fitting the Arneson data, namely model III in Table 6.8.4. With the change in model, the issue of population again arises. The Arneson data included points with negative values, *i.e.*, deposition rather than scour. These cannot be predicted by a power-law model, and so were omitted on the physical basis that the absolute values were small and so were treated as noise. Excluding these points may still not ensure a

Table 6.8.3 Comparison of results of regression analyses for Arneson-type models applied to the entire Arneson data set (total number of points, $N_{tot} = 116$)

Model	R^2	b_1, 95% conf. int.	b_2, 95% conf. int.	b_3, 95% conf. int.
I. Eqn. (6.8.3)	0.90	0.23 (0.20,0.27)	0.035 (0.0059,0.0063)	0.82 (0.71,0.92)
II. Eqn. (6.8.4)	0.69	−0.125 (−0.22,−0.032)	−0.455 (−0.76,−0.15)	0.21 (0.18,0.24)

Table 6.8.4 Results of regression analyses for various models and populations

Model	R^2	b_0, 95% conf. int.	b_1, 95% conf. int.
IIIA. $\log(y_s/y_1) = \beta_0 + \beta_1\log(V_a/V_c)$	0.88	−2.40 (−2.59,−2.21), (−2.57,−2.21)†	2.95 (2.54,3.35), (2.41,3.50)†
IIIB. $\log(y_s/y_1) = \beta_0 + \beta_1\log(V_a/V_c)$ (Bayesian)		−2.40 (−2.59,−2.20)	2.94 (2.53,3.36)
IVA. $y_s/y_1 = e^{\beta_0}(V_a/V_c)^{\beta_2}$ (nonlinear)		−2.00 (−2.24,−1.76), (−2.21,−1.82)†	1.81 (1.31,2.31), (1.48,2.23)†
IVB. $y_s/y_1 = e^{\beta_0}(V_a/V_c)^{\beta_2}$ (non-linear, Bayesian)		−2.00 (−2.24,−1.80)	1.81 (1.36,2.28)

† These are bootstrapped estimates (using case-resampling)

sufficiently homogeneous population; Lyn (2008) applied model III to a subpopulation with the further restrictions that $V_1/V_c < 0.9$ (essentially a clear-water-scour restriction) and $V_a/V_c < 2$, the latter roughly based on a tendency to an asymptotic constant behavior (see Figure 6.8.4). The results for the power-law model are summarized in Table 6.8.4 and the fit with the data is shown in Figure 6.8.4.

Unlike in the preceding example, values of y_s/y_1 for model III lie over a relatively limited range, and therefore the usual argument in favor of a log-transformation (*i.e.*, orders of magnitude variation in the response variable) is not strong. Analysis of residuals (not shown) gives somewhat limited, but not decisive, support to a preference

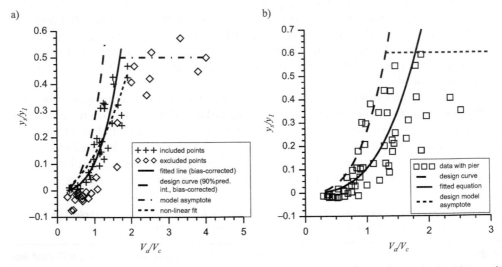

Figure 6.8.4 a) No-pier cases, with points included and excluded in fit, the fitted (power-law and nonlinear) curves, the prediction design curve, and a model asymptote for data satisfying $V_a/V_c > 2$, b) With-pier cases used as for cross-validation purposes compared with the fitted (power-law) curve and the prediction design curve (Arneson, 1997, data reproduced with permission from ASCE).

for a log-transformation. As seen in Table 6.8.4 and Figure 6.8.4a, the nonlinear model IV leads to a (statistically) significantly different value of the exponent.

A few other aspects not dealt with in previous examples should be highlighted. The regression of log-transformed quantities introduces a bias when returning to the original physical scale, and should be corrected. The approximate bias-correction, $i.$ $e.$, a factor of $e^{s^2/2} \approx 1.15$. As mentioned briefly above, the mean response provided by the fitted model is inadequate as a design equation because it will be exceeded about half the time for the given flow, sediment, and obstacle geometry. For the pier-scour problem, the standard design equation was obtained as an envelope curve (presumably estimated visually) over available data. Lyn (2008) proposed a regression-based approach using a prediction interval. A 90% one-sided prediction interval leads to an additional multiplicative factor, ≈ 2.5 over the range being fitted. The predictive 'design' equation is cross-validated (Figure 6.8.4b; see also the discussion of cross-validation in Section 6.10.4, Volume I) using Arneson data with piers (a total of 56 points) which were not used in the original model fitting. This assumes that the presence of piers do not significantly affect pressure scour, given the small pier diameter to channel width ratio. The design curve in Figure 6.8.4b does perform as expected as its values exceed most of the observed values (a more conservative choice of a 95% or even 99% prediction interval would have led to fewer points exceeding the design values). It should be stressed that model III discussed above is not intended as a general pressure-scour model, but rather illustrates data analysis issues.

6.8.4 Comments on Bayesian and bootstrapped estimates

The above discussion was restricted to the classical inference interval estimates. Table 6.8.1 through 6.8.3 also include Bayesian and bootstrapped estimates, which do not differ much from the classical estimates. This confirms that, in the absence of strong prior knowledge of either the coefficient values or the underlying distributions in typical regression problems, and for relatively large number of points in a dataset, the classical estimates are quite reliable, and therefore alternative analyses would lead to similar results. As such, these techniques are more appealing when classical estimates are not readily available, such as in the case of machine learning approaches (see Section 6.10, Volume I) or of the autocorrelation function (see Sections 6.12.2 and 6.12.7, Volume I).

6.9 CLASSIFICATION ANALYSIS: LOGISTIC REGRESSION, LINEAR DISCRIMINATION ANALYSIS, AND TREE CLASSIFICATION

Statistical classification or discriminant analysis (Huberty, 1994; Hastie et al., 2009; James et al., 2013) has been widely applied in the medical and social sciences but rarely in hydraulic engineering. It differs from OLS in that the dependent variable, Y, is discrete rather than continuous, and may even be categorical. In the simplest case, Y is binary, with values limited to 0 or 1 (or success or failure). In hydraulic engineering, classification problems may arise in situations where an instability is involved with a threshold (to be quantified) dividing two qualitatively different behavior. A simple example might be the classification of laminar and turbulent flow states, for which $Y = f(Re)$ might be considered, with Y assuming two possible values (laminar or turbulent), and the Reynolds number, Re, as the continuous independent variable (the predictor). Clustering, discussed in Section 6.3.2, Volume I, may be viewed as a type of classification problem where a quantitative specification of the cluster boundaries is not desired. The standard assumptions of OLS are violated and hence other approaches must be considered. Various techniques are available for this purpose. Details on three techniques (also discussed in Hastie *et al.*, 2009) are given, two traditional (*i.e.*, logistic regression and linear discriminant analysis) and one more modern (*i.e.*, tree-based classification).

In logistic regression, the probability of an successful outcome, here represented numerically as 1, *i.e.*, $P(Y = 1)$, is modeled using the logistic or sigmoid function. For the simple case of two predictors, x_1 and x_2, and a binary outcome:

$$P(Y = 1) = \frac{1}{1 + e^{-(\beta_0 + \beta_1 x_1 + \beta_2 x_2)}}, \quad P(Y = 0) = 1 - P(Y = 1), \tag{6.9.1}$$

and

$$\ln\left[\frac{P(Y = 1)}{P(Y = 0)}\right] = \ln(\text{odds}) = \beta_0 + \beta_1 x_1 + \beta_2 x_2 \tag{6.9.2}$$

The logarithm of the ratio of probabilities (also termed the odds) is *linearly* related to x_1 and x_2, where the model parameters, β_0, β_1, and β_2 are estimated from the data (see Equation (6.9.2)). The OLS algorithm cannot be applied because the values of the probabilities are unknown. Instead the method of maximum likelihood (see Section 6.2.4 in Volume I) is applied, involving numerical optimization of the relevant likelihood function. With estimates, b_0, b_1, and b_2 obtained and given x_1 and x_2, then $P(Y = 1)$ can be evaluated. Further, a class boundary separating the two outcomes (or groups of data) can be defined as that straight line (for two predictors) where $P(Y = 1)/P(Y = 0) = 1$ or $b_0 + b_1 x_1 + b_2 x_2 = 0$. As in the case of OLS, linearity here refers to the β's; the predictors (the x's) may be known or specified nonlinear functions. Confidence intervals for the β's and the class boundary can be obtained in the usual manner, assuming an asymptotic normal distribution for large samples. The confidence interval for a predictor coefficient can be used to assess whether a model predictor adds significantly to the model. There is however no equivalent to R^2 as an overall quality measure. A misclassification error rate expressed as the fraction of the testing sample that is misclassified is often cited as a basic performance measure.

Because of the linear model for ln(odds), logistic regression may be viewed as a generalized linear model in which the dependent variable is no longer limited to y itself but can be a function (termed the link function) of y. The distribution of y is no longer limited to be normal, and could be another distribution, such as binomial in the case of a binary-outcome problem. More general logistic regression permits more than two (*i.e.*, multinomial or polytomous) outcomes and even nonlinear relationships.

Another traditional linear approach, termed linear discriminant analysis (LDA), shares similarities with logistic regression in the model equation, but makes stronger assumptions regarding the probability distributions characterizing each group to be classified. The stronger assumptions allow a closed-form solution, and whereas logistic regression can become unstable in problems with perfectly linearly separable groups, the LDA approach will remain stable. LDA also has an appealing geometric interpretation in terms of the centroids of each group and the distances of a point from the centroids. Hastie *et al.* (2009) compares the two methods and their advantages and weaknesses. Both logistic and LDA are frequently formulated explicitly in terms of Bayes' theorem, and so may be considered Bayesian in approach.

Tree-based classification (Breiman *et al.*, 1984; Hastie *et al.*, 2009) is an example of a machine- or statistical learning (or data-driven) approach. The *tree* refers to a flow-chart-like structure that is followed to predict an outcome as illustrated in Figure 6.9.1b. Tree-based regression, similar in concept to classification, was used by Bhattacharya *et al.* (2007) for sediment-transport modeling. Unlike logistic regression or LDA, it makes few model assumptions, *e.g.*, it does not assume a linear class boundary or any type of distribution. Rather it partitions the entire space of predictors into smaller regions, each containing as many of the same type of points as possible. In a simple form, for a two-predictor problem, the regions would be rectangular, and any point falling in a particular region would be predicted as belonging to the majority in that region. Although rectangular regions may seem quite restrictive, an analogy may be made with numerical integration of a complicated function over a specified interval. Division of the interval (possibly in an adaptive manner) into subintervals is performed and the integrand is approximated as a simple function such as a constant value within each subinterval, allowing simple evaluation of the integral. The main difficulty, common to machine-learning approaches (and to a lesser extent to OLS) arises due to noisy data, and the possibility that the partitioning would proceed to such an extent that the algorithm begins to fit the noise specific to the calibration data rather than the signal, leading to what is termed 'over-fitting'. As such, a data-driven stopping criterion becomes an important part of the method. The details of the algorithm for partitioning or "growing", and its inverse "pruning", a tree to avoid overfitting are beyond the scope of this discussion, but typically involves recursively choosing a split variable and a split value and the tree structure to minimize an error measure, and stopping based on a measure of model complexity and any gain attributed to additional complexity. Hastie *et al.* (2009) cites the ease of interpretation of tree results as being one of its advantages (presumably relative to other machine-learning approaches), but the detailed results, such as the split values, may have little physical basis. Similarly, like many machine-learning approaches, its almost exclusive focus on prediction for specific given values of the predictors may be useful for many practical purposes, but may be less useful for others. If the analogy with numerical integration is pursued, then numerical integration is useful for many problems, but analytical integration is required for others.

The literature on machine-learning techniques is rapidly growing, and a brief broad overview is given in Section 6.10, Volume I; the above is intended to introduce the topic within a specific application context. More sophisticated approaches are available. In a sediment-transport application (similar to the extended example considered below), Dogan *et al.* (2009) applied a relevance-vector-machine (RVM) approach including a logistic model for classification. The RVM approach (Tipping, 2001) is discussed further in Section 6.10.2, Volume I.

6.9.1 Extended example: Classification of bedform channel regimes

Bedform regimes (discussed in Section 2.5.1.3.4 in Volume I) are triggered by bed instabilities that eventually lead to the development of bedforms, such as ripples, dunes, antidunes, and these can dramatically alter flow resistance relationships. For most practical purposes, it suffices to classify the flow as being either in the upper regime (comprising plane-beds with transport, anti-dunes, chutes and pools) or a lower regime (compromising ripples and dunes). Various classifications have been proposed in the literature with regime boundaries typically drawn by eye (Brownlie, 1983; van Rijn, 1984). In this example, quantitative boundaries are obtained based on the data set compiled by Dogan *et al.* (2009) who extended the dataset of Brownlie (1981). As in the preceding examples, the population and the data to be used should receive careful consideration. In the following, an explicit distinction will also be made between training (calibration) data and testing data, where the model parameters are estimated from *only* the training data. In addition to the data restrictions imposed by Dogan *et al.* (2009) [*i.e.*, on the channel width to depth ratio $(B_0/h > 4)$, sand sizes (a median diameter, d_{50}, ranging between 0.062 mm and 2 mm), minimum flow depth $(h/d_{50} > 100)$], a number of outliers were omitted and a constraint on bed slope, $S_0 < 0.006$, was added in order to facilitate a more direct comparison with the Brownlie model. The training set contains 952 laboratory-data points (196 in the upper regime and 756 in the lower regime, as no points were identified in a transition regime) while the testing set had 354 points (38 and 316 in the upper and lower regimes respectively) obtained from field measurements. The present analysis assumes that field and laboratory data are drawn from the same population. This might be a point of debate (see Dogan *et al.*, 2009) but still can be used for purposes of demonstrating the analysis.

A simplified Brownlie model is examined, namely that the upper regime is defined by the condition, $F_{gBr} > 1.74 S_0^{-1/3}$, for $S_0 < 0.006$, where $F_{gBr} = U/\sqrt{g(s-1)d_{50}}$, U being the average velocity, g the acceleration due to gravity, and s the sediment specific gravity. Like the Brownlie model, the three classification approaches consider for simplicity only two predictors, either F_{gBr} and S_0 (*i.e.*, the Brownlie variables), or the alternate pair, $q_* = qS_0/\sqrt{g(s-1)d_{50}^3}$ and $\tau_* = R'_h S_0/[(s-1)d_{50}]$, where q is the discharge per unit width, and R'_h is the hydraulic radius associated with grain friction. The hydraulic radius is estimated using the Einstein fixed rough-bed model as $R'_h = [Ud_{65}^{1/6}/(7.66\sqrt{gS_0})]^{3/2}$, d_{65} being the diameter for which 65% (65th percentile) of the sediment are finer by weight. In all models, physical variables are used in their log-transformed form, such that in the linear models the classification boundary can be expressed in power-law form.

The results and performance of the various models are compiled in Table 6.9.1, while the training (laboratory) data are plotted in Figure 6.9.1a in Brownlie coordinates with the various classification boundaries shown. The error rate or the fraction of points that are misclassified is given separately for upper and lower regimes, and for training and testing data. The results of the logistic model with Brownlie variables are quite close to the Brownlie model (also based on both flume and field data). The performance in terms of error rates is therefore very similar for both. Consideration of the confidence

Table 6.9.1 Results for bedform regime classification for various models

Model (variables)	(Mean) upper regime criterion	error rate (flume, i.e., training, data)	error rate (river, i.e., testing, data)
Brownlie (1983)	$F_{gBr} > e^{0.55} S_0^{-1/3}$	0.17,0.017	0.42,0
logistic(F_{gBr}, S_0)	$F_{gBr} > e^{0.52\pm0.33} S_0^{-0.33\pm0.054}$	0.15,0.034	0.37,0.0064
LDA(F_{gBr}, S_0)	$F_{gBr} > e^{1.45} S_0^{-0.14}$	0.015,0.10	0.16,0.51
tree (F_{gBr}, S_0)	n/a	0.021,0.061	0.26,0.39
logistic (q_*, τ'_*)	$q_* < e^{5.24\pm0.21} \tau'^{2.85\pm0.40}_*$	0.077,0.026	0.58,0.0032
LDA(q_*, τ'_*)	$q_* < e^{5.58} \tau'^{2.66}_*$	0.0051,0.062	0.24,0.039
tree(q_*, τ'_*)	n/a	0.021,0.044	0.21,0.34

Notes:
(i) the logistic-, the linear discriminant (LDA)-, and the tree-model analyses were performed with the "glm", the "lda", and the "rpart" functions respectively from the R statistical software system.
(ii) the two numbers in each error rate column refer to the error rate for lower and upper regime, respectively.

intervals (95% intervals are given in Table 6.9.1) for each coefficient (exponents) in the logistic model shows that the difference between the Brownlie and the logistic models is not statistically significant. On the other hand, the linear discriminant approach (LDA) yields quite different results, being noticeably superior in predicting upper regime, but weaker in predicting lower regime (especially for the testing dataset).

The results of the tree-based classification are shown in Figure 6.9.1b. Shown in the figure are both the split conditions as well as a pair of values indicating the number of points in each regime after each split (*e.g.*, starting with 756 lower regime and 196 upper regime points at the top, the first split to the left satisfies $\log(F_{gBr}) < 2.35$ and has 684 lower regime and 11 upper regime points, and so on). The terminal nodes, those at the bottom of each branch should ideally have a zero in the pair of values, but this is the

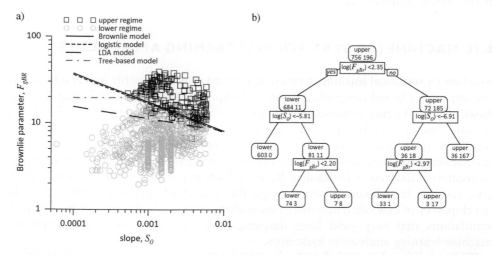

Figure 6.9.1 a) training (laboratory) data points in the $S_0 - F_{gBr}$ planes together with the classification boundary between upper and lower regimes according to the various models, b) detailed results from the tree-based classification, including split values and class distribution at each node.

case only for one terminal node (the leftmost in the figure), all the others nodes involving some degree of misclassification. The tree in Figure 6.9.1b partitions the S_0 − F_{gBr} plane into rectangles, which for this particular case is equivalent to the stairstep boundary shown in Figure 6.9.1b for the tree model. This implies that the boundary can be expressed as a piecewise constant function as the stairstep function could indeed be considered as an approximation to the linear function. While the performance of the tree-based model for the training data is arguably the best (among models based on the Brownlie variables) in terms of balance between upper and lower regime prediction, its performance on the testing data is quite weak with regards to lower-regime prediction. The effect of choice of variables is examined with the alternate coordinates, τ'_* and is discussed next. On the training data, the performance of all models is improved, suggesting that regime separation is clearer in the alternate variables. This does not necessarily imply better performance on the testing data, as the performance of the logistic model has deteriorated noticeably in its upper-regime prediction. Despite its stronger assumptions, not necessarily satisfied by the data, the LDA model arguably performs the best overall on both training and testing data. Interestingly, the difference in the boundaries of the logistic and the LDA models is significant only in the coefficient but not in the exponent, while the performance for upper-regime prediction seems quite different. The small number of upper-regime points may have exaggerated the differences.

The discussion is not intended to provide physical insights in the bedform regime discussion but rather to illustrate results that can be obtained with statistical classification and to reveal method implementation issues. Specifically, the selection of predictor variables, which form the coordinate system, allowing a clearer regime separation and hence facilitating classification, may be even more important than the statistical technique used. Overfitting can be a serious problem for machine-learning (and to a lesser extent traditional) techniques, such that performance on training data is not reflected in performance on testing data.

6.10 MACHINE (OR STATISTICAL) LEARNING APPROACHES

Machine or statistical learning (terms that are used interchangeably in this subsection) encompasses a broad range of disparate techniques or algorithms, including the tree-based schemeal ready discussed above. Such techniques have been applied in diverse domains such as marketing or machine vision or artificial intelligence that are only weakly if at all constrained by principles of mechanics. The applications generally involve very large data sets with many variables (features), which may not characterize conventional datasets in hydraulics. Recent efforts in compiling large hydraulic datasets, such as in sediment transport, e.g., the dataset of Brownlie (1981), and the development of new often automated measurement technologies as well as numerical simulations that may yield large datasets, have stimulated interest in exploring machine-learning analyses in hydraulics.

While traditional statistical methods as applied in engineering experimental hydraulics place emphasis on both explanation and understanding of the physical phenomenon as well as model prediction, machine learning methods focus almost exclusively on prediction. The issue of interpretability (e.g., Hastie et al., 2009; James et al., 2013) has received attention in discussions of machine learning, but this should not be confused with illuminating a physical phenomenon. As noted above, interpretability is considered one of the major advantages of tree-based schemes, but as the results in Figure 6.9.1b illustrate, the tree classification while readily interpretable in a narrow sense may yield little insight into physical mechanisms. General references on machine learning include Bishop (2006), Murphy (2012) and Abu-Mostafa *et al.* (2012), and comparison of machine learning with traditional statistical models can be found in Hastie *et al.* (2009), Brieman (2001), and Shmueli (2010). These techniques can be applied to classification or regression problems, though in hydraulics, regression applications have predominated. Notable studies include those developed for predicting sediment transport (Nagy *et al.*, 2002; Bhattacharya *et al.*, 2007; Dogan *et al.*, 2009) and scour (Liriano & Day, 2001; Azmathullah *et al.*, 2005; Goel & Pal, 2009), making use of available larger datasets. In this section, only two machine-learning techniques are described: artificial neural networks and relevance-vector machines. They are subsequently applied to the regression problem of predicting sand transport in channels.

6.10.1 Artificial neural networks

Artificial neural networks (ANN) refer to a widely applied non-linear statistical modeling technique, in which input variables (the regressors) are linearly combined, then non-linearly transformed to produce (for regression problems) output (or response) variables. Hastie *et al.* (2009) introduces the basic concepts of ANNs. In contrast to the traditional equation-based model, the ANN model is most conveniently expressed in a network diagram, made up of nodes and links, as illustrated in Figure 6.10.1. The network consists of an input layer, with each node defining an input variable, an output layer again with each node defining an output variable, and one or more so-called hidden layers, with nodes and links graphically representing how the inputs are transformed to outputs. Each layer (other than the output layer) consists of a constant or bias term (shown as the top node in Figure 6.10.1, represented as a 1 inside a circle) plus a number of other nodes. The bias, often omitted in network diagrams, plays the role of the intercept in linear regression. Most commonly, the bias and each unit in a given layer is 'connected' to all of the units of the subsequent layer, but there are no connections within the same layer, leading to the term feedforward (lines between units in Figure 6.10.1 represent the connections), Each connection is associated with a numerical value termed the weight, denoted by subscripted w or v in Figure 6.10.1, which plays the role of a fitting or learning parameter.

The network shown in Figure 6.10.1a has two hidden layers, nine units (excluding the bias terms), and 59 connections (lines) and weights (including the bias weights) that are used for fitting. In this figure, the number of units at layer i is denoted as M_i (with M_0 being the number of input variables and M_{K+1} the number of output variables with K being the number of hidden layers) hence the total number of connections (and weights) is $\sum_{i=0}^{K}(M_i + 1)M_{i+1}$. The non-linear aspect of the model is effected in the so-called activation function, $A(t)$, that performs a non-linear transformation on the summed inputs. This function is typically nonlinear (but could be chosen as linear)

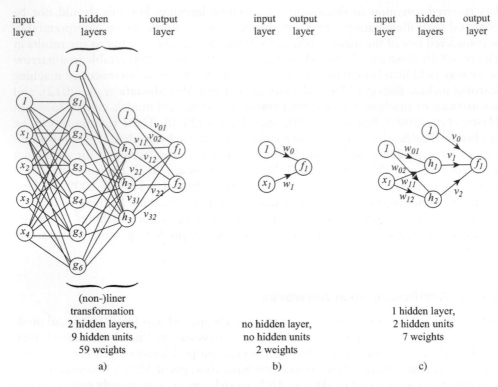

input layer | hidden layers | output layer   input layer | output layer   input layer | hidden layers | output layer

(non-)liner transformation
2 hidden layers,
9 hidden units
59 weights

a)

no hidden layer,
no hidden units
2 weights

b)

1 hidden layer,
2 hidden units
7 weights

c)

Figure 6.10.1 Schematic of various ANN network diagrams, a) ANN with two hidden layers b) ANN with no hidden layer, and c) ANN with one hidden layer; weights in a) shown only for the output layer and the preceding hidden layer.

for each unit in each hidden layer. Of the many activation functions available in the literature, the hyperbolic tangent, $A(t) = \tanh(t)$, is used herein. Although in principle the output layer could also be associated with a non-linear activation function, it is most commonly associated with a pass-through linear identity function, $A(t) = t$, in which case it can be omitted.

For the two hidden layers displayed in Figure 6.10.1a, the ANN can be mathematically expressed as

$$g_i = A\left(\sum_{m=0}^{M_0} w_{mi}x_m\right), \quad h_j = A\left(\sum_{m=0}^{M_1} u_{mj}g_m\right), \quad \text{and} \quad f_k = \sum_{m=0}^{M_2} v_{mk}h_m \quad (6.10.1)$$

where w_{mi}, u_{mj}, and v_{mk} are the connection weights between layers, $M_0 = 4$, $M_1 = 6$, $M_2 = 3$, $M_3 = 2$, $k = 1, ..., M_3$, and $x_0 = g_0 = h_0 = 1$ the bias. This could be a model for the $f_1 =$ depth, and $f_2 =$ sediment concentration, with input variables, $x_1 =$ discharge per unit width, $x_2 =$ median sediment diameter, $x_3 =$ slope, and $x_4 =$ grain shear stress.

The design of the basic ANN model under discussion involves mainly the specification of the hidden layers: i) the number of hidden layers, ii) the number of nodes in each hidden layer, and iii) the choice of an activation function. For the limiting simple case where no hidden layer is chosen, and only a single input unit ($M_0 = 1$) and single output unit, f_1, as depicted in Figure 6.10.1b, the model reduces to simple linear regression with two fitting parameters (the intercept and the slope). From this perspective, ANN can be viewed as non-linear extension of linear regression models. For the other simple case depicted in Figure 6.10.1c, with $M_0 = 1$, a single hidden layer with $M_1 = 2$, and still only a single output, and the hyperbolic tangent as activation function, the response, f_1, can be expressed explicitly as

$$f_1 = v_0 h_0 + v_1 \tanh(h_1) + v_2 \tanh(h_2) = v_0 + v_1 \tanh(w_{01} + w_{11}x_1)$$
$$+ v_2 \tanh(w_{02} + w_{12}x_1) \qquad (6.10.2)$$

which is a model nonlinear equation with a total of $(M_0 + 1)M_1 + (M_1 + 1)1 = (1 + 1)2 + (2 + 1)1 = 7$ weights (v_0, v_1, v_3 and $w_{01}, w_{11}, w_{02}, w_{12}$) to be determined by fitting to the data. The hidden layer is seen to allow a possibly non-linear nesting of linear combinations of the nodes or variables in the preceding layer.

For ANN models with a large number of nodes and/or multiple hidden layers, the model equation becomes extremely complicated and opaque, and the network representation of Figure 6.10.1 is more useful than the corresponding equation. In the ANN model of Nagy et al. (2002) for predicting sediment transport in rivers, the input layer consisted of eight ($M_0 = 8$) dimensionless variables, a single hidden layer, and the sediment concentration as the single output variable. The number of units in the hidden layer was selected by calibration with test data (81 points from field data) and found to be $M_1 = 12$ (yielding therefore a total of 121 fitting parameters).

Apart from the important issues of selecting the input variables (dimensional or dimensionless, log transformed or not, scaling) and the population, the choice of network architecture (number of hidden layers and number of units in each hidden layer) and dealing with the problem of overfitting (already discussed in connection with the other machine-learning technique of tree classification), must be addressed. The obtained solutions may be problem-specific and general recommendations may not be available for the variety of cases (see, however, Hastie et al., 2009). The simple ANN described above does not yield interval estimates, which limits its usefulness. Bootstrapping and Bayesian neural nets are options to obtaining interval estimates (Tibshirani, 1996). Hastie et al. (2009) report on a classification competition where various methods were tested on benchmark classification problems. A procedure based on Bayesian neural nets was found to have perform the best. Hastie et al. (2009) also comment that ANN techniques "are especially effective in problems with a high signal-to-noise ratio and settings where prediction without interpretation is the goal". More sophisticated ANN techniques involve deep learning (see the review of Arel et al., 2010), which aims at a better data representation through in part a relatively high number of hidden layers, and is reported to be capable of achieving significantly better performance than traditional ANNs.

6.10.2 Relevance vector machines

The relevance vector machine (RVM) is an example of a kernel-based technique in which the function to be predicted is expressed in terms of a scalar kernel basis function, $K_{RVM}(\mathbf{x}, \mathbf{x}')$, that measures the 'similarity' between two points, $\mathbf{x}$ and $\mathbf{x}_i$:

$$f(\mathbf{x}) = w_0 + \sum_{i=1}^{N} w_i K_{RVM}(\mathbf{x}, \mathbf{x}_i) + \varepsilon \qquad (6.10.3)$$

where $\mathbf{x}_i$ $i = 1, ..., N$, are the values of the input variables (regressors) for N observations, w_i, $i = 0, ..., N$ are the fitting (model) parameters, and ε is the usual error term. The widely known support vector machine (SVM) is another kernel-based method. For an introductory discussion of SVMs, see Hastie *et al.* (2009) and James *et al.* (2013). Goel and Pal (2009) applied SVM to a scour problem. Various standard kernel functions have been used including the linear kernel (used herein), the Gaussian kernel (Dogan *et al.*, 2009), as well as the polynomial kernel (Goel & Pal, 2009). The number of weights, w_i, (or $\mathbf{w}$), to be determined in such kernel methods is $N + 1$, where N is the number of observations. Although this may seem large from the traditional perspective (though not from an ANN point of view), methods like SVM and RVM seek a sparse solution in which only a small subset of weights are non-zero.

The distinguishing feature of the RVM is its Bayesian approach, which involves (see, also Section 6.7, Volume I): i) defining the likelihood, ii) choosing the prior probability distributions, and iii) obtaining the posterior and prediction probability distributions. In RVMs, the standard likelihood function assumes a joint normal distribution similar to that described by Equation (6.2.4). The priors on $\mathbf{w}$ are chosen similar to those in Section 6.7.1 (joint normal on $\mathbf{w}$ and gamma on the precision, $1/\sigma_\varepsilon^2$) with two important differences. The first is that the prior on w_i has a zero mean, expressing a preference (prior knowledge) for smaller values of w_i, resulting in smoother, less erratic functions. The other major difference is that each weight can have its own variance (which requires therefore its own prior or hierarchical prior, which is modeled like σ_ε^2), which is responsible for the sparsity of the solution (for details, see Tipping, 2001). In the sparse solution, the vectors associated with the non-zero weights (other than the bias weights) were termed relevance vectors by Tipping (2001) to distinguish them from support vectors in the SVM technique. This terminology was inspired by the concept of automatic relevance determination, first proposed by Mackay (1994) and Neal (1996) in the context of Bayesian neural networks. The posterior distribution for $\mathbf{w}$ and a prediction distribution for a new point can be found in the usual manner, though Tipping (2001) makes an additional assumption regarding hyper-parameters in order to avoid the use of a computational sampling solution. Further details of the RVM can be found in Tipping (2001), Tipping (2004) and Bishop (2006).

Whereas the main step in an ANN application is the specification of the hidden layer, the main step in developing an RVM is the selection of a kernel function (K_{RVM}) that maps the input data into a generally non-linear feature space such that the original non-linear relationship between regressors and response variables transforms into a relationship that is linear in the weights. If the choice of the kernel function in the RVM is considered akin to the choice of the activation function in the ANN, then the ANN must also specify the number of units in the hidden layer. Further, ANNs must deal with the problem with overfitting, while RVMs incorporate a built-in sparsity mechanism that automatically handles this problem.

6.10.3 Other general issues in machine learning techniques

Advances in machine learning in fields other than hydraulics have led to increased access to readily available software and tools supporting their application. While this makes the methods appealing for the analysis of hydraulic data, caution is required to avoid over-fitting and, more importantly, the interpretability of the results for understanding the physical phenomena. Physical constraints such as dimensional homogeneity and a physically reasonable asymptotic behavior may be difficult to ensure in machine learning techniques unless these have been already incorporated in the selection and formulation of input variables (which in return restricts the generality of the learning process and may compromise predictive performance). Keijzer and Babovic (1999) attempted to overcome these limits with genetic programming. For well understood hydraulic problems, where the correct relevant dimensionless groups are well known, models for which predictive ability is the sole performance measure, machine-learning methods may have particular attraction, especially in computer models. An example is the use of high-order polynomials in computer models of culvert inlet control conditions (as in the standard HY-8 culvert hydraulics software). Like many machine learning techniques, the coefficients of such polynomials have no physical meaning, but their predictive performance is likely superior to the more traditional models that are more readily interpretable in terms of analogies to weir or gate behavior.

Unlike the traditional statistical modeling approach, selection of models in an appropriate machine learning approach is non-trivial. Many different models have been developed including ANNs, SVMs, trees, and probabilistic graphical models, and it is not clear that any single model will be superior to all others for hydraulic problems. In practice, a few models are tested and that yielding the best performance is selected. For problems where predictive uncertainty is desired, Bayesian variants of machine learning techniques will have a marked advantage. Machine learning models may also be subject to problems in computer transferability (*i.e.*, the ease of using already developed models on other datasets as discussed in Boulesteix & Schmid, 2014). The number of parameters, such as weights, in techniques such as ANN, SVM, RVM, or trees often grows with number of training samples, and complete specification of these models may not be practically feasible, making them less transferable and transparent. This is in contrast with the conventional statistical models like multiple regression or logistic regression, where a much more limited number of weights is typical.

6.10.4 Cross-validation

Cross-validation refers to a computational resampling technique (similar to bootstrapping discussed in Section 6.5, Volume I) for quantifying the predictive performance of a statistical model. Although it is discussed in this section on machine learning, the technique can also be used with traditional models for any supervised learning problem. It is of particular importance in the machine-learning context because of: i) their almost exclusive focus on response prediction, and ii) their often purely algorithmic nature precludes any non-computational approach, such as those used for conventional regression models. Browne (2000) and Arlot and Celisse (2010) review cross-validation methods and provide guidelines for selecting best cross-validation procedure according to the features of the data and model (see also Hastie *et al.*, 2009 and James *et al.*, 2013 for introductory overviews).

Only one technique of cross-validation is dealt with in detail herein, namely N_{cr}-fold cross-validation, in which the sample of N observations is first randomly partitioned into $N_{cr} \leq N$ non-overlapping subsets. The model is trained on all but one of these subsets, and one realization of its prediction, *i.e.*, generalization, performance is determined by testing on, *i.e.*, applying the model to, the subset that was not used in the training. This procedure is repeated N_{cr} times, each time omitting a different subset for training, and using the omitted subset for testing. Averaging results over the N_{cr} trials, provides an estimate of the uncertainty in the response prediction. The case of $N_{cr} = N$, also termed leave-one-out cross-validation and resembling the jackknife procedure, can be computationally costly for large N and complex models, and have high variance (Hastie *et al.*, 2009). The choice, $N_{cr} = 5$ or 10, is usually recommended (Hastie *et al.*, 2009) as a compromise between bias and variance.

In the traditional hydraulics literature, formal cross-validation has been rarely applied, even where results have been intended to be predictive. This is because for almost any process there is some theoretical basis for the proposed model, and the data analysis serves mostly to confirm the theory. For problems where empiricism and hence more purely statistical models play a greater role, cross-validation should become a standard practice. Cross-validation by itself will *not* necessarily lead to reliable models. A flawed statistical model, *e.g.*, one in which spurious correlations are present, may not necessarily be detected through cross-validation analysis. Data snooping, which refers to a situation where a dataset used in model development (parameter estimation or model selection) is also used in cross-validation can compromise the process. Even a seemingly innocuous exposure of data during model development, such as during variable selection or transformation, may result in overoptimistic, but erroneous, estimation of predictive power.

6.10.5 Extended example: prediction of sand transport in channels using machine learning

Due to the availability of large datasets (such as that originally compiled by Brownlie, 1981), the prediction of sediment transport in channels has attracted the attention of researchers using machine learning techniques (Nagy *et al.*, 2002, using ANN; Bhattacharya *et al.*, 2007, using ANN and tree regression; Dogan *et al.*, 2009, using RVM). In this section, ANN and RVM are applied to the same dataset used by Dogan *et al.* (2009; similar to that used in the extended example of classification in Section 6.9.1, Volume I, except that the restriction to $S < 0.006$ was removed). The data are divided into those obtained in the laboratory (1210 points) and those obtained in the field (2911 points), as potentially taken from different populations. For both ANN and RVM approaches, the same input variables were chosen, namely $\mathbf{x} = [\log(q_*), \log(\tau_*), \log(\tau'_*), \log(\tau_{*c})]^T$ where $\tau_* = R_h S / [(s-1)d_{50}]$ is the Shields stress, R_h the hydraulic radius, τ_{*c} the critical Shields stress as defined by Yalin and Karahan (1979), and all other variables are as previously defined in Section 6.9.1, Volume I. The output variable is $\log C$, where C is the sediment concentration (ppm by weight).

For the ANN model, the MATLAB Neural Networks toolbox was used. The number of input variables are therefore $M_0 = 4$ and the number of outputs is 1. The architecture was decided by a trial and error procedure, in which the number of hidden layers was chosen as one, the activation function as the hyperbolic tanh, and the number of nodes, M_1, in the hidden layer was varied. The most parsimonious architecture that gave the least test error was selected, resulting in an ANN with $M_1 = 3$, and a total of 19 weights (including bias weights) to be determined. The Levenberg-Marquardt nonlinear optimization algorithm was used for finding the weights by minimizing the sum of mean squared errors. The available data set (whether from solely laboratory or solely field observations) was randomly divided into three parts: (i) a training data set (75%) used to determine the weights, (ii) a validation data set (15%) used to measure the generalization performance of the model, and stop training when the generalization performance degrades, and (iii) a testing data set (15%) which was used to get an independent measure of the predictive performance of the developed ANN.

While the RVM, like the ANN, is most often used for nonlinear models through the choice of nonlinear kernel functions, for the present case, the frequently overlooked linear kernel function, $K_{RVM}(\mathbf{x}, \mathbf{x}_i) = \mathbf{x}^T\mathbf{x}_i = \sum_{k=1}^{M} x_k x_{ki}$ was chosen (as was also done by Dogan et al., 2009). A constant term could also be added, but for the present model is not necessary. As will be seen, this simple choice is effective, does not require any determination of additional kernel parameters, and leads to a readily interpretable closed-form solution similar to, and hence allowing comparisons with, the traditional solution. With this choice, the model equation can be expressed as

$$f = w_0 + \sum_{i=1}^{N} w_i \left(\sum_{k=1}^{M_0} x_k x_{ki} \right) + \varepsilon = w_0 + \sum_{k=1}^{M_0} x_k \left[\sum_{i=1}^{N} w_i x_{ki} \right] + \varepsilon = w_0^* + \sum_{k=1}^{M_0} x_k w_k^* + \varepsilon$$

(6.10.4)

where $w_0 = w_0^*$ and $w_k^* = \sum_{i=1}^{N} w_i x_{ki}$. Thus, the original RVM formulation in terms of a sum over $N + 1$ weights is, due to the linear kernel, equivalent to the traditional model where the sum is taken over only the M_0 input variables (and may be compared to Equation (6.6.1)). The choice of a linear kernel for an RVM has therefore resulted in a Bayesian linear regression problem, and could have been solved as such. Although RVM routines are now available (in R or in MATLAB), the results given below were obtained with custom-written software based on the RVM algorithm outlined in Tipping (2001) and used in Dogan et al. (2009).

The ability of the specified ANN and RVM models to predict new data was evaluated by training each of the models twice, once exclusively on the laboratory data set (these models were labeled as ANN_L and RVM_L) and again exclusively on the field data set (labeled as ANN_R and RVM_R). The model performances were compared based on a discrepancy ratio, $DR = C_{pred}/C_{obs}$, where C_{pred} and C_{obs} are predicted and observed values of the sediment concentration. The fraction of the total number of data points satisfying $0.5 \leq DR \leq 2$, denoted as DR_2, was calculated. Following Brownlie (1981), DR was assumed to be lognormally distributed, and so its geometric mean (GM) and geometric standard deviation (Gstd) were estimated. In addition, the mean squared error (MSE) and the coefficient of determination (R^2) were also estimated.

Table 6.10.1 compares the performance of laboratory-data trained models (ANN_L and RVM_L) and field-data trained models (ANN_R and RVM_R) on laboratory and field data respectively. As expected, the models typically perform better on the data on which they were developed, *i.e.*, the laboratory-trained models, ANN_L and RVM_L, perform better on laboratory data than the field-trained models, ANN_R and RVM_R, while ANN_R and RVM_R perform better on field data than ANN_L and RVM_L. More interesting, the nonlinear ANN models perform better than the linear RVM models for the data sets on which they were trained, but their performance deteriorates substantially for the data set not used in their training. If it is assumed that the laboratory and the field data were taken from the same population, then this example shows the nonlinear ANNs as more capable of fitting the (training) data, yet quite weak in generalization, *i.e.*, in predicting new data. The more balanced and generally competitive performance of the linear RVM models for both laboratory and field data, irrespective of the dataset on which the RVM was trained, lends some support to the hypothesis that the laboratory and field data population were taken from the same population (as argued in Dogan *et al.* (2009)). On the other hand, the ANN results by themselves might have been interpreted as supporting the claim that laboratory data and field data stem from statistically different populations.

The comparison does not necessarily support a superiority in terms of generalization of an RVM in general with an ANN in general, as much of the generalization performance seen in Table 6.10.1 can be attributed to the linear kernel of the RVM. A degenerate ANN designed to be linear, *e.g.*, with a single hidden layer and a single unit and an identity activation function, $A(t) = t$ would likely have yielded comparable generalization performance. The main advantages of the RVM would be i) its inherent probabilistic interpretation due to its Bayesian construct, which permits interval

Table 6.10.1 Performance of sediment transport prediction models ANN_L and RVM_L trained on laboratory data, and ANN_R and RVM_R trained on field data

Data sets	Performance Measures	models			
		ANN_L	ANN_R	RVM_L	RVM_R
Laboratory (N=1210)	DR_2	0.78	0.50	0.69	0.71
	GM	0.98	0.98	1.00	1.08
	Gstd	1.83	2.51	2.08	2.14
	MSE (log-space)	9.16	13.90	11.06	11.54
	R^2(log-space)	0.90	0.77	0.86	0.84
field (N =2911)	DR_2	0.41	0.66	0.59	0.62
	GM	1.47	1.01	1.03	1.00
	Gstd	3.64	2.21	2.55	2.44
	MSE (log-space)	31.61	18.55	21.90	20.89
	R^2(log-space)	0.64	0.88	0.83	0.84

DR_2: fraction of points with DR<2, where DR is the discrepancy ratio, *i.e.*, the ratio of predicted to observed sediment concentration
GM and Gstd: geometric mean (GM) and geometric standard deviation (GStd) of DR
MSE: mean squared error (in log-transformed coordinates)
R^2: coefficient of determination between predicted and observed (in log-transformed coordinates)

estimates and much else, and ii) its built-in sparsity-inducing mechanism, which automatically deals with the issue of overfitting.

In contrast to the simple non-linear ANN model described above with a total of 19 non-zero weights, the RVM algorithm with the linear kernel started with a total of N (= 1210 for the laboratory dataset, and = 2911 for the field dataset) weights, but eventually reduced this to a total of $M_0 + 1 = 5$ (or possibly fewer) non-zero weights through its sparsity-inducing mechanism. The linear-kernel results for the mean weights, e.g., $(\mu_{w*})_L = (9.28, -2.90, 2.61, 3.75, 0.54)^T$ for the laboratory-trained model, are the same as the traditional linear regression result, and so the result for the prediction equation can be similarly expressed simply as:

$$C_{\text{pred}} = 10^{9.28} q_*^{-2.90} \tau_*^{2.61} \left(\tau_*^{\prime 3.75} \right) \tau_{*c}^{0.54} \qquad (6.10.5)$$

The large value of the constant multiplier $10^{9.28}$ is due to some extent to the parts per million (ppm) units of C_{pred} but may also indicate the effect of other (unknown) variables not explicitly included in the model. The Bayesian advantage of inherently obtaining distributions rather than simply point estimates can be exploited in hypothesis testing. For example, Dogan et al. (2009) examined whether the predictions of the RVM_L and RVM_R models were statistically significant.

6.11 DATA CONDITIONING: TIME SERIES AND FILTERING

In general, filtering may be viewed as an operation in which a subset or subrange of the data is preferentially selected, and so may be considered as a type of restriction to a subpopulation of interest, similar to conditional sampling (recall the discussion of Section 6.2.3 in Volume I). More narrowly and most commonly, filtering is applied to a continuous analog signal or discrete time series, x_k, $k = 1, ..., N$, reflecting a time-continuous process (or a process involving time-like variable). In flow studies, e.g., when dealing with turbulent flows, observations are made of a randomly fluctuating quantity, $u(t)$, over a sampling duration, T_s. The quantity may be a vector (such as velocity) or a scalar (such as concentration or temperature), and the sampling is usually performed at a constant rate or frequency, $f_s = 1/\Delta t_s$, where Δt_s is the time between samples. If it is desired to restrict attention to only a range of the variable frequencies, say $f_L < f < f_U$, then a filtering of the acquired data can be applied where the desired range is termed the passband, defined by the cut-off frequencies f_L and f_U as illustrated in Figure 4.5.7, Volume I. Frequencies outside of the passband will be simply termed here the stopband, though filter-design references (e.g., Oppenheim et al., 1999; Stearns, 2003) also include a transition band between the passband and stopband. The angular frequency, $\omega = 2\pi f$, will often be used in the basic theory. After the signal is subjected to an ideal low-pass (or respectively high-pass) filter with an upper cut-off frequency, f_U, (or respectively, f_L), the resulting output signal should consist of only frequencies, $f < f_U$ (or respectively, $f > f_L$). An example of low-pass filtering is shown in Figure 6.11.1a in which two different low-pass filters have been applied to a (raw) digital time series sampled at 100 Hz (pressure on the side of a square cylinder in a vortex-shedding flow). For the present purposes, it could be any signal, e.g., velocity or

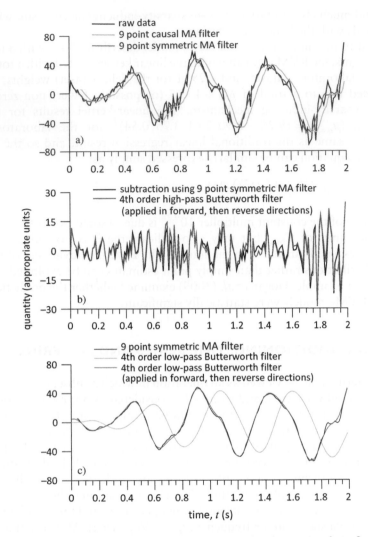

Figure 6.11.1 Comparison of filtering effects: a) the original signal with outputs of two 9-point moving-average (MA low-pass) filters applied to an original signal, one symmetric and non-causal, and the other causal, b) outputs of two high-passed filtering operation on the same input signal as in a), one subtracting the output of the 9-point symmetric MA filter from the original signal, and the other directly applying a 4th-order Butterworth filter in two passes to achieve a zero-phase-lag output, and c) three low-pass filter outputs, one obtained with the 9-point symmetric MA filter as in a), one obtained with a 4th-order Butterworth filter in a single (forward) pass, and one obtained with a 4th-order Butterworth filter in a two-pass (forward and reverse) operation for a zero-phase lag output (Butterworth filter coefficients obtained with the routine "butter", and the Butterworth filter outputs obtained with routines "filter" and "filtfilt", all in the R package signal, MA filter output obtained with the routine "filter" in the R package stats).

temperature, and so the values and the units are not important, but the strong periodic feature is clear and could motivate filtering.

6.11.1 Linear time-invariant filters

The filtering may be performed by analog means through an electronic circuit (see Table 4.2.1 in Volume I). If the analog signal is to be digitized and if aliasing is of concern then analog filtering will be *necessary* (see discussion of sampling and aliasing in Section 4.5, Volume I). A sensor or measuring instrument system will generally perform an implicit filtering, even if the measuring system outputs a digital time series due to finite time for sampling or spatial resolution or intentional averaging (see the discussion of frequency response and spatio-temporal resolution of the instrument in Sections 4.2.4 and 4.4.4.1, Volume I). Garcia *et al.* (2005) examined ADV measurements undergoing a form of low-pass filtering when the recording frequency is lower than the sensor sampling frequency.

This section deals mainly with digital filtering, assuming a digital series of discrete data, sampled at a fixed constant sampling interval, Δt_s. Much of the terminology and theory are however applicable to analog filters also. Further, the discussion is restricted to linear time-invariant filters, where the filter output denoted as y_k, for any specific time instant t_k (so indexed by the integer k) is linearly related to the past, present, and future inputs, x_{k-m}, $m = -M_f, ..., M_p$, as well as past outputs, y_{k-r}, $r = 1, ..., R_p$, where $R_p < M_p$:

$$y_k = \sum_{m=-M_f}^{M_p} b_m x_{k-m} - \sum_{r=1}^{R_p} a_r y_{k-r}, \qquad (6.11.1)$$

with b_m and a_r being constant filter coefficients or weights. If $a_r = 0$ for all r, the filter is termed non-recursive or finite-impulse-response (FIR). Non-zero values of a_r provide feedback to the filter, which is then termed recursive or infinite-impulse-response (IIR). If $M_f = 0$, such that no 'future' (relative to the index i) inputs is required, the filter is said to be causal (and hence can potentially be operated in real-time or on-line mode). If all of the data have already been acquired and available, then a causal filter is not necessary as 'future' data are available for use.

The moving-average (MA, also termed running-mean by Thomson & Emery, 2014) scheme, in which the filter output is a uniformly weighted average of adjacent inputs is one of the simplest low-pass filter. It can be mathematically expressed as:

$$y_k = \sum_{m=-M_f}^{M_p} b_m x_{k-m}, \quad b_m = \frac{1}{M_f + M_p + 1}. \qquad (6.11.2)$$

The MA scheme is therefore non-recursive, and if $M_f = 0$ then it is also causal. If $M_p = M_f > 0$, then the MA filter is symmetric about the kth point. The results of two different MA filters, a symmetric non-causal 9-point filter with $M_p = M_f = 4$, and a causal 9-point filter with $M_p = 8$, $M_f = 0$ are illustrated in Figure 6.11.1a. An implicit non-causal scheme ($M_p = 0$, $M_f > 0$) was applied to ADV datasets by Garcia *et al.* (2005).

6.11.2 Filter performance: power-gain and phase frequency response

The performance of a filter is characterized by its complex frequency response, $H(\omega)$, with real and imaginary parts, a (transfer) function with an amplitude, $|H(\omega)|$, and a phase, $\varphi(\omega)$, both of which may vary with ω. A power gain, $G_p = |H|^2$, often measured in decibels, i.e., $G_{p,db} = 10\log G_p = 20\log|H|$, is also used as a performance measure instead of $|H|$. Similarly, rather than φ, the group delay, $d\varphi/d\omega$, might also be used to characterize the phase performance. For the 'ideal' low-pass filter, $|H| = 1$ or $G_{p,db} = 0$ in the passband, $\omega/\omega_U = f/f_U < 1$, while $|H| = 0$ or $G_{p,db} = -\infty$ in the stopband, $\omega/\omega_U = f/f_U > 1$, with f_U being the cut-off frequency. For a real filter, the cut-off frequency is defined as that where $G_p = 1/2$ or $G_{p,dB} = -3$ dB. The filter roll-off refers to the asymptotic rate of decrease (in the case of a low-pass filter) in $G_{p,dB}$ as ω is increased in the stopband, typically reflected in the steepness of the $G_{p,dB}(\omega)$ curve. The transfer function $H(\omega)$ can be related to the filter weights through the discrete Fourier transform, leading to

$$H(\omega) = \frac{\displaystyle\sum_{m=-M_f}^{M_p} b_m e^{-im\omega\Delta t_s}}{1 + \displaystyle\sum_{r=0}^{R_p} a_r e^{-im\omega\Delta t_s}}, \quad i = \sqrt{-1}. \tag{6.11.3}$$

In the simple case of the MA filter, this yields with $a_r = 0$ (Oppenheim et al., 1999):

$$H(\omega) = \frac{\displaystyle\sum_{m=-M_f}^{M_p} e^{-im\omega\Delta t_s}}{M_f + M_p + 1} = \underbrace{\frac{\left\{\sin\left[\omega\left(M_f + M_p + 1\right)\Delta t/2\right]/\sin\left(\omega\Delta t_s/2\right)\right\}}{M_f + M_p + 1}}_{\text{amplitude, }\pm|H|} e^{\overset{-i\Delta t_s\left[\omega\left(M_p - M_f\right)/2\right]}{\underset{\text{phase, }\varphi}{}}}$$

$$= \frac{\left\{\sin\left[\left(M_f + M_p + 1\right)(\pi/2)(\omega/\omega_N)\right]/\sin\left[(\pi/2)(\omega/\omega_N)\right]\right\}}{M_f + M_p + 1} e^{-i(M_p - M_f)(\pi/2)(\omega/\omega_N)}$$

$$\tag{6.11.4}$$

where for convenience Δt_s has been eliminated in favor of the Nyquist frequency (discussed in Chapter 4, Volume I), $\omega_N = 2\pi f_N = 2\pi/(2\Delta t_s)$. The phase $\varphi = -(M_p - M_f)(\pi/2)(\omega/\omega_N)$ for the MA filter varies linearly with ω and so the MA filter is termed a linear-phase filter, which is important for applications where the shape of the input waveform needs to be preserved. Moreover, for a symmetric MA filter ($M_p = M_f$), then $\varphi = 0$, so that the output phase is exactly the same as the input, which might be considered as ideal. In Figure 6.11.1a, the results of two different MA filters are shown, each involving $M_f + M_p + 1 = 9$ points, but one is symmetric ($M_f = M_p = 4$), while the other is causal $(M_f = 0)$. As expected from Equation (6.11.4), the output of the symmetric filter is exactly in phase with the input signal, while the causal filter has a phase shift but the waveform shape is preserved. The variation of the power gain, $G_{p,dB}$, and the phase, φ, with the normalized frequency, ω/ω_N, for both MA filters is shown in

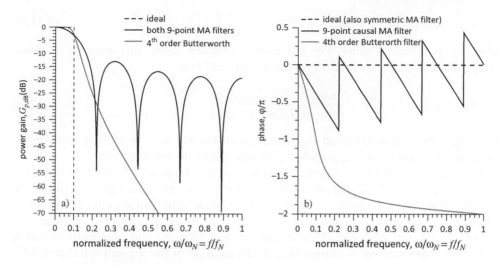

Figure 6.11.2 Variation of a) power gain, $G_{p,dB}$, and b) phase, φ/π, with normalized frequency, $\omega/\omega_N = f/f_N$, for 9-point MA filters and a 4th-order Butterworth low-pass filter (determined using the routine "freqz" in the R software package).

Figure 6.11.2. Because $G_{p,dB}$ depends only on $M_f + M_p + 1$, which is the same for both MA filters, $G_{p,dB}$ is the same for both MA filters. Their cut-off frequency is $\omega_U/\omega_N = f_U/f_N \approx 0.1$, implying $f_U = 5$ Hz for $\Delta t_s = 0.01$ s (a sampling frequency of 100 Hz, and so a Nyquist frequency, $f_N = 50$ Hz).

The power gain for both 9-point MA filters is notable for the relatively large 'lobes' in Figure 6.11.2a in the stopband ($f/f_N > 0.1$), implying that the attenuation of high-frequencies content by the MA filters is relatively poor. For example, at $f/f_N \approx 0.8$ (or $f \approx 40$ Hz), Figure 6.11.2a indicates $G_{p,dB} > -20$ dB or $|H(f = 40\ \text{Hz})| > 0.1$, so that the frequency content fairly close to 50 Hz is only reduced in amplitude by less than a factor of 10. The $G_{p,dB}$ -plot for the MA filters is somewhat misleading in the apparent local minima, which in fact corresponds to zeroes of $H(\omega)$ (and so $G_{p,dB} = -\infty$) occurring at $\omega/\omega_N = f/f_N = 2k/(M_p + M_f + 1)$, k a positive integer. As noted already, $\varphi/\pi = 0$ for the symmetric MA filter, while φ/π varies linearly with f/f_N (slope of $-(M_p - M_f)/2$) otherwise. The apparent discontinuities in $\varphi/\pi = 0$ occurring at the zeroes of $H(\omega)$ should be interpreted carefully, since it merely reflects a change in sign of $H(\omega)$.

While the simplicity and the linear (or zero)-phase response of the MA filters are appealing, their poor power-gain performance in the stopband motivates the search for more effective filters. A wide range of filter options is available, such as Chebyshev and elliptic filters (*e.g.*, Oppenheim *et al.*, 1999; Stearns, 2003). Here only the popular Butterworth filter is described. This filter is causal and IIR (so $M_f = 0$ and at least one of the a_r coefficients in Equation (6.11.1) is non-zero), and has long been implemented in analog filters. The (ideal) low-pass Butterworth filter for continuous (analog) signals is defined by its power gain (*e.g.*, Stearns, 2003):

$$G_p = |H(\omega)|^2 = [1 + (\omega/\omega_U)]^{-2q_f} \tag{6.11.5}$$

where q_f is terms its order. For a discretely sampled series, this must be modified (Oppenheimer et al., 1999; Thomson & Emery, 2014) to

$$G_p = |H(\omega)|^2 = \left\{1 + \left[\frac{\tan\{(\pi/2)(\omega/\omega_N)\}}{\tan\{(\pi/2)(\omega_U/\omega_N)\}}\right]\right\}^{-2q_f}. \tag{6.11.6}$$

For large ω/ω_U, it is expected that the slope of the G_p vs ω in log-log coordinates asymptotes to a slope of $-2q_f$ and so q_f characterizes the filter roll-off.

A discrete 5th order Butterworth filter was tested by Goring and Nikora (2002) as a despiking tool for ADV signals. Whereas in the case of the simple MA filter the number of points are chosen (thus defining the filter weights but the cut-off frequency and roll-off characteristics are not known), in the case of the Butterworth filter, the cut-off frequency and the desired roll-off are chosen, and the corresponding weights are determined. A comparison of the outputs of the 9-point symmetric MA filter and the 4th-order Butterworth filter (with $M_p = 4$ and $R_p = 4$, and so the same computational effort as the 9-point MA filters) of approximately the same cut-off frequency is shown in Figure 6.11.1c for the same input signal as in Figure 6.11.1a. The 4th-order Butterworth filter output is noticeably smoother than the MA filter output, but like the causal MA filter, the causal Butterworth filter also exhibits a clear phase shift.

The frequency responses, in terms of both $G_{p,\mathrm{dB}}$ and φ/π, of the 4th-order Butterworth filter and the 9-point MA filters are compared in Figure 6.11.2. The Butterworth filter has a flat response in the passband, and a monotonic decrease in the stopband. For comparison, for $\omega/\omega_N = 0.8$ and $\omega_U/\omega_N \approx 0.1$, the 4th-order Butterworth filter reduces the power by a factor of 10^5, i.e., $G_{p,\mathrm{dB}} < -100$ dB, while as noted above, the 9-point MA filter reduces the power by a factor of only 10 (or $G_{p,\mathrm{dB}} > -20$ dB, see Figure 6.11.2a). The relative absence of high-frequency features in Figure 6.11.1c in the Butterworth filter output compared to the MA filter output reflects its much better power-gain performance. The phase response of the Butterworth filter (Figure 6.11.2b) is overall non-linear, but in much of the passband, it is nearly linear. Thus, in Figure 6.11.1c, the shape of the input waveform is reasonably well preserved in the Butterworth filter output, but the phase shift is evident.

In some situations, it is desired to have a zero-phase shift, as can be obtained by a symmetric (non-causal) MA filter. While this cannot be achieved by a simple single application of the *causal* Butterworth filter, it can be achieved by an effectively non-causal application in which the Butterworth filter (or any other IIR filter) is applied in two passes, once in the usual forward direction, followed by subsequent application in the reverse direction (Stearns, 2003; Thomson & Emery, 2014). This commonly-used operation is available in many software associated with data acquisition systems (the routine is named "filtfilt" in both R and MATLAB). Any phase shift incurred at any frequency in the forward pass is exactly cancelled out in the reverse pass, so that at the end of the two-pass operation, an exactly zero-phase shift is achieved. Because the filter is applied twice, the order of the two-pass filter is also doubled that of the single-pass filter. The output of such a two-pass Butterworth filter is compared with those of the one-pass Butterworth filter and the symmetric MA filter in Figure 6.11.1c, where a zero-phase shift is seen to have been achieved with the two-pass procedure.

6.11.3 Ringing and end effects, and other aspects

For the same type of filter, higher roll-off (steeper power-gain decay curves) requires longer filter lengths (more points) and hence more computational effort, but more importantly potential disadvantages in signal distortion may also result. Sharp time transients such as a step input induces a 'ringing' behavior characterized by 'unrealistic' overshoots and undershoots, also termed the Gibbs phenomenon), which are more pronounced the higher the filter order. This is illustrated in Figure 6.11.3, where a square-wave input (starting at a value of 1) is subjected to various Butterworth filter operations. The single-pass filters yield an asymmetric phase-shifted output, with the lower (4th)-order filter yielding a smaller phase shift and a less extensive and lower magnitude ringing effect in the vicinity of the step increase and step decrease than the higher (16th)-order filter. Artifacts are also noted at the beginning of the data series but *not* at the end.

The two-pass (forward and reverse) filters yield symmetric (zero-phase lag as expected) outputs, but ringing behavior is still clearly evident in the regions near the step rise and drop, though somewhat less marked effect than in the single-pass filter outputs. In contrast to the single-pass outputs, artifacts appear at both the beginning and the end of the data record due to the reverse pass. Thomson and Emery (2014) recommend limiting filter order to less than 10 to reduce the impacts of ringing effects.

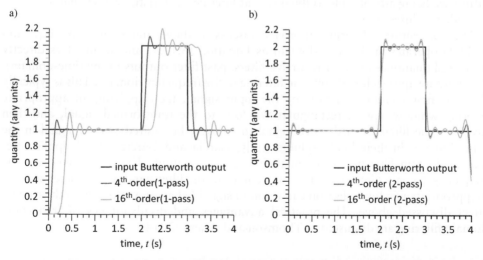

Figure 6.11.3 'Ringing' and end effects caused by various Butterworth filters with a square-wave input (input signal sampled at 100 Hz, cut-off frequency of 5 Hz) starting at a value of 1 with step rise at $t = 2$ s and step drop at $t = 3$ s: a) single-pass results, b) two-pass results, with 4th-order and 16th-order filters (Butterworth filter coefficients obtained with the routine butter, and the Butterworth filter outputs obtained with the routines "filter" and "filtfilt" in the R package).

Increasing filter lengths also causes problems with endpoints (at the beginning, but in non-causal filters also at the end, of the dataset) as the chosen filter may involve data points that are not available. Thus, for the 9-point symmetric MA filter, the outputs for y_k, $k \leq 4$ or $k \geq N - 3$, are not defined because the required data for their evaluation using Equation (6.11.2) are not available. Thomson and Emery (2014) discuss various common practices such as zero padding (often the default in software as reflected in Figure 6.11.3). Zero padding before the start or after the end of the data series may potentially be treated as a step change, resulting in ringing-like effects. To avoid these effects, the minimum filter order that produces an acceptable output might be generally recommended.

While the above discussion has focused on low-pass filters, high-pass filtered output can be obtained by subtracting low-pass filtered output from the original data or by directly applying a high-pass filter (usually constructed from low-pass filters) to the original data. The results of filtering the original data of Figure 6.11.1a according to these approaches are shown in Figure 6.11.1b. The clear large-scale periodic feature has been filtered out, leaving a more 'random' residual, which might be of greater interest. Durão *et al.* (1988) in a study of the turbulent vortex-shedding flow around a square cylinder applied a 7th-order (Chebyshev) filter to remove frequencies within a range centered about the major vortex-shedding frequency. Also, Garcia and Garcia (2006) high-pass filtered the velocity signal in a study of bubble plumes, to separate out the large-scale motion of the plume from the more 'relevant' turbulent characteristics. Such schemes requires a distinct separation of scales, such that the resulting high-passed (or low-passed depending on the context) output does not depend sensitively on the chosen cut-off frequencies or filter. The standard deviations of the two high-passed signals in Figure 6.11.1b are 8.8 (MA filter) and 9.2 (Butterworth filter), the difference being attributable to the different filter characteristics rather than to physical-induced differences.

The conventional filtering approach discussed above and defined by Equation (6.11.1) operates in the time-domain as Equation (6.11.1) and so involves directly measured quantities (and for recursive filters, past filter outputs in the time-domain). An alternate approach performs the filtering in the frequency domain. This is typically done by first Fourier transforming the input signal, then applying an appropriate filtering window, *e.g.*, a rectangular window, to the transformed signal, and then inverting the filtered-transformed signal to recover the desired filtered signal in the time domain. In their bubble-plume study, Garcia and Garcia (2006) applied frequency-domain filtering to obtain their high-pass output. The performance of the frequency-domain filter depends on the choice of the filter window characteristics, as inappropriate window selection can result in signal behavior analogous to the ringing effect discussed above. The practical advantages and disadvantages of frequency-domain filtering are discussed in Thomson and Emery (2014).

6.12 TIME SERIES AND SPECTRAL ANALYSIS

For a time series, the discrete sampling is often regular, *i.e.*, with a constant sampling interval, Δt_s, and so with $N = T_s / \Delta t_s$ data points (with T_s the sampling duration), but may be irregular, due to missing data even under regular sampling or instrument peculiarities, such as in single-sample laser Doppler velocimetry. If Δt_s were sufficiently long such that data points were statistically independent of each other, then the techniques and results previously discussed could be applied without modifications. In practice, Δt_s is often chosen such that the data points in the time series are (serially) correlated (*i.e.*, observations are *not* randomly sampled), and the results so far discussed do not necessarily apply without additional assumptions. Time-averaging of u, defined as

$$\bar{u}_T = \frac{1}{T_s} \int_0^{T_s} u \, dt \approx \frac{1}{N\Delta t_s} \sum_{k=1}^{N} u_k \, \Delta t_s = \frac{1}{N} \sum_{k=1}^{N} u_k, \qquad (6.12.1)$$

is conceptually distinct from the (probability) averaging previously discussed in terms of expectations (Section 6.2, Volume I and Equation (6.2.3)), and the two should not necessarily be equated (see also the discussion of various types of averaging in Chapter 2, Volume I). The ergodic hypothesis discussed in the same chapter refers to the assumption that for T_s (and N) $\rightarrow \infty$, the two averaging procedures are equivalent. While the sample central moments, such as the mean $(\mu_u = \bar{u}_T)$ and variance (s_u^2) are estimated in the usual manner, inferences on confidence intervals cannot be based simply on N due to the presence of serial correlation (see Duchon & Hale, 2011, for results on the estimation of central moments and their confidence intervals for time series data).

As in the case of regression (where the regressors vary), in a time series, the time variable, t, varies and questions of population and homogeneity as t varies may similarly be raised. For a time series, the condition of stationarity may be considered analogous to homogeneity, as it requires that statistics based on any sample of the time series to not differ significantly. Different levels of stationarity may be defined (see Priestley, 1981); here the pragmatic view is taken that whatever level of stationarity is required is assumed to exist. The simplest case, in which all statistics are constant, is the focus of the present discussion. In more complicated cases, the mean value may vary with t, assumed in some non-random fashion (*i.e.*, following a trend), but it is believed or assumed that the removal of the mean variation through detrending or similar technique leaves a residual error that is stationary (again similar to the error assumption in classical regression as discussion in Section 6.6.1, Volume I). In the most complex case, all statistics of interest may vary with t, and the statistical analysis of a non-stationary time series becomes in general much less tractable and quite tentative. Such cases were already briefly mentioned in the discussion of conditional sampling (Section 6.2.3, Volume I) and data conditioning (Section 6.3.3, Volume I). In the laboratory, achieving a reasonable level of stationarity should be straightforward, but in the field this can be much more challenging. A distinct separation of scales is required (Lumley & Panofsky, 1964), such that the sampling duration, T_s, covers the range of smaller (say on the order of minutes) scales of interest in a study, but is much smaller than the time scale of the large-scale often deterministic features (say due to highly-variable flow event, diurnal, or seasonal variation). As will be seen in the extended example below, a time series with non-stationary features may still be analyzed in the conventional manner to yield useful results, but the interpretation must be more cautious.

The analysis of a time series of u differs from the typical regression problem or the statistics of a random sample as the extent to which the value of u at one time instant is related to its value at another time instant (*i.e.*, its autocorrelation) is of particular interest. The information is available in the measurement if Δt_s is sufficiently small, but is integrated out in the evaluation of moments. Further statistical analysis, specifically examining statistics such as autocorrelation and the related power spectra, may yield additional insight into flow dynamics. Many experimental hydraulics studies have used power spectra and/or autocorrelations, including Venditti and Bennett (2000) and Best and Kostaschuk (2002) for examining dune flows, Luketina and Imberger (2001) and Chen *et al.* (2002) for fitting the Batchelor spectrum (with implications for mixing coefficients in lakes and oceans), Lyn *et al.* (1995) and Choi *et al.* (2012) for identifying dominant frequencies in a vortex-shedding flow, and flows in the Great Lakes, respectively. The basic theory for the univariate real-valued case may be found in standard references such as Priestley (1981), Percival and Walden (1993), and Bendat and Piersol (2000), on which the following relies heavily. For a discussion more oriented to flow problems in meteorology and oceanography, Duchon and Hale (2011) and Thomson and Emery (2014) may be mentioned.

Because the interest lies in the relationship between values of u at different instants in time, an important statistic is the autocovariance defined in terms of expectation values as

$$C_{uu}(t, \tau) = cov[u(t), u(t + \tau)] = E[\{u(t) - \mu_u\}\{u(t + \tau) - \mu_u\}] \qquad (6.12.2)$$

where the difference in time instants is τ, termed the lag. The autocovariance is the covariance with the two random variables being the *same* random function at *different* instants in time (see Equation (6.2.7)). Similar to traditional theory, the following discussion will generally be restricted to second-order stationary processes in which all relevant first- and second-order statistics are independent of time. In particular, the mean and variance are constant, so $\mu_{u,t} = \mu_{u,t+\tau} = \mu_0$ and $\sigma_{u,t}^2 = \sigma_{u,t+\tau}^2 = \sigma_0^2$, and $C_{uu}(t, \tau) = C_{uu}(\tau)$ varies only with τ. The additional assumption of a Gaussian process in which any sample follows a multivariate joint normal distribution is also frequently made to simplify analysis and obtain stronger results.

6.12.1 Tests of stationarity

Various statistical tests for stationarity have been proposed (*e.g.*, Bendat & Piersol, 2000; Tavoularis, 2005). They include the runs test (also known as the Wald-Wolfowitz test) and the reverse-arrangement test (also known as the Mann test for trend). Gibbons and Chakraborti (2003) discuss the statistical basis of both tests. Directly applying these two tests to the time series is problematic as both assume independent observations, *i.e.*, random sampling. A more justified approach divides the time series into segments or blocks that can be considered reasonably independent, evaluates the mean or variance (or other statistics) for each segment, and performs the test on the derived-series statistics (Bendat & Piersol, 2000; Tavoularis, 2005). The runs test counts the number of runs, n_{runs}, in a sample, a run being defined as a sequence of consecutive observations that is larger or smaller than a cut-off value (usually the median or the mean) in the entire sample. The reverse arrangement test counts the number of subsequent values that are larger (or smaller) than the current value, and sums these for each value in the sample. Test statistics formed with these counts are compared with what would be expected if the sample were random (*i.e.*, the null hypothesis of no monotone trend), and an appropriate *p*-value is computed. For large samples, the test statistic is approximately normal, simplifying the *p*-value evaluation. The formulae for the runs and the reverse arrangement tests are summarized in Table 6.12.1.

Neither the runs nor the reverse arrangement test deals specifically with frequency aspects of a time series, and so may not be sensitive to certain types of non-stationarity. A more direct means of testing for non-stationarity would segment an entire time series, as is also done for the previous tests, but investigate instead the autocovariance (or power spectrum) of each segment for significant differences between segments. This is the approach taken in the Priestley-Subba-Rao (1969) or PSR test. Rao *et al.* (2003) reviews various stationarity tests, both in the time and the frequency domain.

6.12.2 Autocorrelations and integral time scales

If the ergodic assumption is made, then, for a stationary series, the dimensional autocovariance can be expressed as

$$C_{uu}(\tau) = \lim_{T_s \to \infty} \frac{1}{T_s} \int_0^{T_s} [u(t) - \mu_0][u(t+\tau) - \mu_0] dt = \lim_{T_s \to \infty} \frac{1}{T_s} \int_0^{T_s} u'(t)u'(t+\tau) dt,$$

(6.12.3)

where u' denotes a deviation from the mean value, μ_0. The dimensionless autocorrelation, denoted as $R_{uu}(\tau)$, is equal to $C_{uu}(\tau)$ normalized by its value at $\tau = 0$, namely, $C_{uu}(0) = \text{var}(u) = \sigma_0^2$, so that

$$R_{uu}(\tau) = \lim_{T \to \infty} \frac{1}{T_s} \int_0^{T_s} \frac{u'(t)u'(t+\tau)}{\sigma_0^2} dt, \qquad R_{uu}(\tau = 0) = 1.$$

(6.12.4)

In the statistical literature, $u(t)$ is often assumed to have zero mean and unit variance, thus simplifying the notation (primes and σ_0^2 could then be omitted). $R_{uu}(\tau)$ may be viewed as a correlation coefficient (with magnitude less than or equal to 1) between values of $u(t)$ separated by the lag, τ (*cf.* Equations(6.3.3) and (6.6.6)). For real-valued stationary series considered here, both $C_{uu}(\tau)$ and $R_{uu}(\tau)$ are symmetric about $\tau = 0$.

Two special cases may be noted: i) for a white-noise (*i.e.*, purely random) process, in which the values of u at any arbitrary lag are uncorrelated, $R_{uu}(\tau > 0) = 0$, and ii) for a periodic process with a fixed frequency, $R_{uu}(\tau)$ will exhibit the same frequency with constant (non-decaying) amplitude. For many practical problems, such as in turbulent flows, $R_{uu}(\tau)$ decays gradually to zero, as $u(t)$ becomes decorrelated with itself as τ increases. This can be interpreted as the 'memory' of the process variable. In other words, $R_{uu}(\tau)$ is close to 1 at small τ, when $u(t+\tau)$ is strongly correlated with $u(t)$ and 'remembers' earlier values, while $R_{uu}(\tau)$ is close to zero at large τ, when $u(t+\tau)$ is no longer correlated with $u(t)$, and so has 'forgotten' earlier values. This motivates the definition of an integral time scale,

Table 6.12.1 Formulae for runs and reverse arrangements tests for stationarity

Test	Test parameter	null-hypothesis statistics	(one sided) large-sample p-value
Runs	$n_{runs} = 1 + \sum_{k=1}^{N-1} I(X_{runs}), \; X_{runs} =$ $\lvert sgn(x_{k+1} - x_{co}) - sgn(x_k - x_{co})\rvert$	$\mu_{runs} = \dfrac{2n^+ n^-}{n^+ + n^-} + 1,$ $\sigma^2_{runs} = \dfrac{(\mu_{runs} - 1)(\mu_{runs} - 2)}{(n^+ + n^-) - 1}$	$P(Z < z_{test}),$ $z_{test} = \dfrac{n_{runs} + 0.5 - \mu_{runs}}{\sigma_{runs}}$
reverse arrangements	$n_{ra} = \sum_{k=1}^{N-1}\left[\sum_{j=k+1}^{N} I(X_{ra,jk})\right],$ $X_{ra,jk} = x_j - x_k$	$\mu_{ra} = N(N-1)/4,$ $\sigma^2_{ra} = N(2N+5)(N-1)/72$	$P(Z < -z_{test}),$ $z_{test} = \left\lvert\dfrac{n_{ra} - \mu_{ra}}{\sigma_{ra}}\right\rvert$

Notes:

i) $I(s)$ is an indicator function, defined with value 1 if $s > 0$, and zero otherwise, while $sgn(s)$ is the sign (or signum) function, with value 1 if $s > 0$, 0 if $s = 0$, and -1 if $s < 0$.

ii) For both tests, N is the total number of points in the series.

iii) For the runs test, x_{co} is the cut-off value (usually mean or median), and n^+ and n^- denote the total number of points larger, respectively smaller, than the cut-off value; if x_{co} is chosen as the median value, then this median value is omitted from the series, and N is decreased by 1.

iv) The large-sample p-value based on the normal distribution is a good approximation for $N \geq 10$ for the reverse-arrangement test (Mann, 1945). For the runs test, $p_{exact}/p_{approx} \leq 1.04$ for $n^+ \approx n^- \geq 8$ (for smaller values of , the 0.5- term continuity correction is needed) but p_{exact}/p_{approx} deviates substantially from 1 for small p_{exact} (< 0.05); it does so however in a direction such that if the null hypothesis is rejected using the large-sample p_{approx}, it would also be rejected using p_{exact}. Gibbons and Chakraborti (2003) provide the results to compute p_{exact}.

$$\mathcal{T} = \int_0^\infty R_{uu}(\tau)\, d\tau \qquad (6.12.5)$$

which may be interpreted as a measure of the 'memory time' of the process. For a white-noise process, $\mathcal{T} = 0$.

Regular sampling of $u(t)$ results in a series, $u_j = u(t = t_j), j = 1, 2, ..., N$, where $T_s = N\Delta t_s$ and an estimate, $\hat{R}_{uu}(\tau)$, of $R_{uu}(\tau)$ must be based on this finite data series. This can be done by discretizing its definition, replacing the integral by a summation and population parameters by sample parameters:

$$\hat{R}_{uu}(\tau_k) = \hat{R}_{uu}(k\Delta t_s) = \frac{1}{N} \sum_{j=1}^{N-k} \frac{\left(u_j - \overline{u}_{T_s}\right)\left(u_{j+k} - \overline{u}_{T_s}\right)}{u'^2_{T_s}}. \qquad (6.12.6)$$

This is often referred to as the biased estimate due to the choice of N rather than $N - k$ as a divisor, which is however preferred due to its smaller variance (another example of the bias-variance tradeoff in the choice of estimators). Duchon and Hale (2011) discuss in more detail different estimates of $\hat{R}_{uu}(\tau)$. Actual computations of $\hat{R}_{uu}(\tau_k)$ often use the fast Fourier transform (FFT) rather than the direct summation in the above equation.

The sampling distribution of $\hat{R}_{uu}(\tau_k)$ is complicated and simple general results, such as for confidence intervals, are not available except for the trivial case of a white-noise process (Priestley, 1981). Qualitatively, estimates at small lags are expected to be associated with smaller uncertainty due to the larger number of independent samples, but estimates become increasingly unreliable as τ increases and the number of independent samples becomes small (a main reason for preferring N rather than $N - k$ as a divisor) and $\hat{R}_{uu}(\tau_k) \to 0$. With an argument similar to that for the sampling duration for a specified maximum uncertainty level, ε_{max}, Lumley and Panofsky (1964) arrived at an estimate of a maximum lag, $k_{max}/N \approx \varepsilon_{max}^2 / \left(4\sqrt{2}\right)$ (for $\varepsilon_{max} = 0.2$, this yields $k_{max}/N \approx 0.007$). An empirical rule of thumb (McCuen, 2003; see also Thomson & Emery, 2014) suggests a maximum lag, $k_{max}/N \approx 0.1$ for 'reliable' estimates. Uncertainty in $\hat{R}_{uu}(\tau)$ will propagate into estimates of the integral time scale, $\mathcal{T}$. Due to the large uncertainties in the tails of $\hat{R}_{uu}(\tau)$, the integration of $\hat{R}_{uu}(\tau)$ is not necessarily performed over the entire range of τ. The range is commonly restricted to the first zero crossing (Garcia et al., 2006; Thomson & Emery, 2014); other choices may be found in the literature (O'Neill et al., 2004).

The difficulty in the conventional statistical analysis of $\hat{R}_{uu}(\tau_k)$ makes bootstrapping an attractive option for developing confidence intervals. The main issue in time series that poses difficulty for conventional analysis, namely serial correlation, must also be addressed, such that naive resampling would not be independent. The most common approach is based, like the stationarity tests, on defining blocks (usually overlapping) within the original time series, and subsequently resampling the blocks rather than the individual observations. The size (or length) of a block may be constant (Kunsch, 1989), but this choice leads to a resampled series that is no longer stationary. The stationary bootstrap (Politis & Romano, 1994) seeks to overcome this by using blocks of randomly varying lengths. Garcia et al. (2006a) applied a fixed-block-size bootstrapping procedure to obtain confidence intervals for various turbulent-flow statistics, including $R_{uu}(\tau)$ and $\mathcal{T}$, with $u(t)$ being the streamwise turbulent velocity. The utility of bootstrapping is also seen in obtaining a confidence interval for $\mathcal{T}$, which of course depends on the statistics of $\hat{R}_{uu}(\tau_k)$.

6.12.3 Spectral analysis and the periodogram

The (two-sided) power spectral density, $S_{uu}(f)$, defined for positive as well as negative frequencies, $-\infty < f < \infty$, can be directly related to $C_{uu}(\tau)$ (or $R_{uu}(\tau)$) as they are Fourier transform pairs, namely:

$$S_{uu}(f) = \int_{-\infty}^{\infty} C_{uu}(\tau)e^{-i2\pi f\tau}d\tau, \quad \text{and} \quad C_{uu}(\tau) = \int_{-\infty}^{\infty} S_{uu}(f)e^{i2\pi f\tau}df \quad (6.12.7)$$

with the cycle frequency, $f = \omega/2\pi$, ω being the angular frequency (if ω is used in the definition rather than $2\pi f$, then a factor of $1/(2\pi)$ appears preceding the integral). From Equation (6.12.7), for $\tau = 0$,

$$C_{uu}(0) = \sigma_0^2 = \overline{u'^2} = \int_{-\infty}^{\infty} S_{uu}(f)df \quad (6.12.8)$$

and $S_{uu}(f)df$ is interpreted as the non-negative contribution to σ_0^2 (the 'energy') within an interval, df, centered at f. From a statistical perspective, Equation (6.12.8) may be viewed as the result of an analysis of variance (ANOVA), akin to that in regression analysis, in which the total variance (σ_0^2) is attributed or allocated to various sources (here the frequency bands). Equation (6.12.8) also provides a useful check on different normalizations encountered in software for estimating power spectra (discussed further below).

Special cases illustrate the behavior of $S_{uu}(f)$ compared to $C_{uu}(\tau)$: i) for a white-noise process, $C_{uu}(\tau)$ is a delta function at $\tau = 0$ while $S_{uu}(f)$ is a constant, and ii) for a periodic process with a fixed frequency, $C_{uu}(\tau)$ is a periodic function with the same frequency, whereas $S_{uu}(f)$ is a delta function at the frequency of the periodic process. As $S_{uu}(f)$ and $C_{uu}(\tau)$ are transform pairs, they contain the same information, but specific aspects of a time series may be more readily apparent in either the time (τ) or the frequency (f) domain, analogous to different coordinate systems. An important additional consideration is that the asymptotic statistics of spectral estimates is simpler than those of autocovariance estimates.

An estimate of $S_{uu}(f)$ is based on converting the integral in Equation (6.12.7) into a finite sum:

$$\hat{S}_{uu}(f) = \Delta t_s \sum_{k=-(N-1)}^{N-1} \left[\frac{1}{N_s} \sum_{j=1}^{N-|k|} (u_j - \overline{u}_{T_s})(u_{j+|k|} - \overline{u}_{T_s}) \right] e^{-i2\pi f(k\Delta t_s)}$$

$$= \frac{\Delta t_s}{N_s} \sum_{k=1}^{N} \sum_{j=1}^{N} (u_j - \overline{u}_{T_s})(u_k - \overline{u}_{T_s})e^{-i2\pi f(k-j)\Delta t_s} = \frac{\Delta t_s}{N} \left| \sum_{j=1}^{N} (u_j - \overline{u}_{T_s})e^{-i2\pi f(k\Delta t_s)} \right|^2$$

$$= \frac{\Delta t_s}{N} \left| \sum_{j=1}^{N} u'_j e^{-i2\pi f(k\Delta t_s)} \right|^2 \quad (6.12.9)$$

which is termed the periodogram, where regular sampling is assumed (note that as in Section 6.11.2, Volume I, $i = \sqrt{-1}$ is the complex or imaginary unit). The last relationship also points to the interpretation of $S_{uu}(f)$ as a Fourier series expansion of the signal. The periodogram estimate is asymptotically unbiased for large N, but bias may be substantial for practical N. It also suffers from high variability: its variance is proportional to its value, and does not decrease as N increases (for this reason, the periodogram is termed an inconsistent estimator). For a Gaussian process, the ratio of the estimate to the true value for any f, i.e., $\hat{S}_{uu}(f)/S_{uu}(f)$, follows a chi-square ($\chi^2$) distribution with $(n_{df})_{per} = 2$ degrees of freedom, on which a confidence interval estimate can be based (discussed further below). Also, $S_{uu}(f)$ at adjacent frequencies becomes increasingly uncorrelated as N increases, which simplifies the statistics, but leads to erratic non-smooth behavior.

6.12.4 Smoothing periodogram estimates, confidence intervals and spectral resolution

The issues discussed in this section can be traced mainly to the finite duration of the time series and to the uncertainties previously pointed out in the estimation of $C_{uu}(\tau)$. Dealing with these issues involve data tapering and averaging (applied separately or often together) periodogram estimates. Data tapering aims to reduce bias for finite samples, whereas averaging over independent quantities aims to reduce variability in the estimates. The finite-duration sample, $u(t)$, defined for $0 \leq t \leq T_s$, can be viewed as the product of $u_\infty(t)$, defined for $0 \leq t < \infty$, and a duration function, $D(t) = 1, 0 \leq t \leq T_s$, and zero otherwise. The discontinuity of $D(t)$ at $t = 0$ and $t = T_s$ causes problems in the Fourier transform, which can be mitigated by applying a data taper, $D^*(t)$, to the beginning and end of the original series, such that the product, $D^*(t)u_\infty(t)$, goes more smoothly to zero than $D(t)u_\infty(t)$ as $t \to T_s$ (and $t \to 0$).

Averaging can be performed over adjacent Fourier frequencies, separated by the basic frequency, $\Delta f_0 = 1/(N\Delta t_s)$, in which case it is associated with a window characterized by a half-window bandwidth, $N_{Wf}\Delta f_0$ (in the frequency domain, not to be confused with a similar window in the time domain), or over the entire spectra using segments or orthogonal tapers. A large number of window types for averaging of spectra have been proposed, associated with names such as Hanning, Parzen, and Kaiser. A simple example of the first type of averaging is the Daniell window, where the smoothed estimate, $\hat{S}^*$, at a frequency, f_k, is obtained by averaging estimates from $f_k - N_{Wf}\Delta f_0$ to $f_k + N_{Wf}\Delta f_0$, i.e., over a frequency band of $2N_{Wf}\Delta f_0$ of the 'raw' periodogram, as

$$\hat{S}^*_{uu}(f_k) = \frac{1}{2N_{Wf}\Delta f_0} \int_{f_k - N_{Wf}\Delta f_0}^{f_k + N_{Wf}\Delta f_0} \hat{S}_{uu}(f)df \approx \frac{1}{2N_{Wf}} \sum_{j=k-N_{Wf}}^{k+N_{Wf}} \hat{S}_{uu}(f_j). \qquad (6.12.10)$$

The reduction in variability of the estimates can be measured by the effective degrees of freedom, $(n_{df})_{eff}$, of the corresponding χ^2 distribution. For the Daniell window, $(n_{df})_{eff}/(n_{df})_{per} = 2N_{Wf}$; for the raw periodogram, as there is no averaging, N_{Wf} may be considered as equal to 1/2. Overly aggressive smoothing may substantially increase the bias in the estimates, a good example of the bias-variance trade-off. Another trade-off is that between smoothness and spectral resolution which is actually measured in terms of the bandwidth, b_w. While more precise quantitative definitions of b_w are available in standard references, for the present purposes, b_w will refer to the frequency band within which two nearby spectral peaks are not distinguishable. The smoothness-bandwidth tradeoff may be seen in the approximate relation, $[(n_{df})_{eff}/(n_{df})_{per}]/(b_w/\Delta f_0) \approx 1$, so that large $(n_{df})_{eff}/(n_{df})_{per}$ (implying reduced variability), is accompanied by large $(b_w/\Delta f_0)$ (implying reduced resolution). For the Daniell window, b_w may be taken as $2N_{Wf}\Delta f_0$ (compared to simply Δf_0 for the raw periodogram) which is consistent with the approximate relation. A corresponding window, $N_{Wt}\Delta t_s$, in the time domain has an inverse relationship to $N_{Wf}\Delta f_0$ in that a large $N_{Wf}\Delta f_0$ (i.e., more smoothing), implies a small $N_{Wt}\Delta t_s$. Smoothing in the frequency domain can then be viewed as tapering or limiting the periodogram computation to the region where estimates of $C_{uu}(\tau)$ are reasonably reliable.

The other approach to smoothing the periodogram, which might be termed block averaging, is based on obtaining independent estimates of the entire spectra over all frequencies, and then averaging the estimates (similar to the procedure used in the stationarity analysis and in bootstrapping analysis of the autocorrelation). Welch's overlapped segment averaging (WOSA) due to Welch (1967) divides the original series into N_{seg} overlapping tapered segments, computes the periodogram for each segment, and then averages these over all segments to obtain the final estimates. For the WOSA technique, the precise value of $(n_{df})_{eff}/(n_{df})_{per}$ will depend on the details of the data taper, if any, and the extent of overlap, but can be roughly estimated as $(n_{df})_{eff}/(n_{df})_{per} = N_{seg}$. Its bandwidth will be determined by the resolution in the individual segment, which will depend on N_{seg} and the overlap extent (and any data taper). A rough estimate based on no overlap and no taper leads to $b_w \approx N_{seg}\Delta f_0$, which would be consistent with $[(n_{df})_{eff}/(n_{df})_{per}]/(b_w/\Delta f_0) \approx 1$.

Another block averaging technique relies on defining independent 'segments' through a sequence of approximately orthogonal tapers, and is known as the multi-taper (MTM) technique (Thomson, 1982). A WOSA application using zero overlap and no taper to the individual segments may be interpreted as a special case of the multitaper approach, in which each segment is selected using different rectangular tapers applied to the original entire series. The multi-taper technique is commonly applied using tapers based on so-called discrete prolate spheroid sequences (dpss), but a sequence of orthogonal sine tapers may also be used. Not surprisingly, the number of tapers, N_{tap}, plays a role similar to that played by N_{seg} in WOSA, and $(n_{df})_{eff}/(n_{df})_{per} = N_{tap}$. If dpss tapers are applied, these are naturally associated with a half bandwidth parameter, $N_{mtm, f}$, (in the multitaper literature, this is often referred to as the time-half-bandwidth product and denoted by NW), analogous to N_{Wf}, so that $b_w = 2N_{mtm, f}\Delta f_0$, but the constraint $N_{tap} < 2N_{mtm, f}$ should be satisfied. As N_{tap} is frequently chosen as $2N_{mtm, f} - 1$ to achieve maximum smoothing, $[(n_{df})_{eff}/(n_{df})_{per}]/(b_w/\Delta f_0) \approx 1$ again. Further details on WOSA and MTM approaches are given in Percival and Walden (1993).

Confidence intervals for the smoothed periodogram estimates can be based on the large-sample result that $(n_{df})_{eff}\hat{S}^*_{uu}(f)/S_{uu}(f)$ follows a χ^2-distribution with $(n_{df})_{eff}$ degrees of freedom. Hence, the two-sided $(1 - \alpha)100\%$ confidence interval may be expressed as

$$[(n_{\rm df})_{\rm eff}\hat{S}^*_{uu}(f)/\chi^2_{1-\alpha/2,(n_{\rm df})_{\rm eff}}, (n_{\rm df})_{\rm eff}\hat{S}^*_{uu}(f)/\chi^2_{\alpha/2,(n_{\rm df})_{\rm eff}}], \tag{6.12.11}$$

which shows a confidence interval varying (asymmetrically about $\hat{S}^*_{uu}(f)$) with f. Expressed however on a logarithmic scale, the width of the interval, $\log(\chi^2_{1-\alpha/2,(n_{\rm df})_{\rm eff}}/\chi^2_{\alpha/2,(n_{\rm df})_{\rm eff}})$, is independent of f, and so is often represented on log-log scales as a single line in an inset.

6.12.5 Treatment of irregularly sampled data, normalization, and zero padding

The estimation of power spectra from irregularly spaced data, *i.e.*, obtained with a non-constant sampling interval, merits a brief discussion as standard references tend to focus exclusively on regularly spaced data. Moreover, a number of references in the hydraulics literature imply that irregular sampling precludes estimating power spectra (see *e.g.*, Parsheh *et al.*, 2010, Doroudian *et al.*, 2010; and Section 3.2.4, Volume II in the context of ADV signals). The basic definition of the power spectra provided in Equation (6.12.7) does not assume regularly spaced data (and does not even assume the frequency at which the spectra are to estimated). Rather, the regular-sampling requirement is a computational restriction only for the use of fast Fourier transform (FFT) algorithms. The most well-known approach to dealing with irregularly spaced data is that formulated by Lomb (1976) and further developed by Scargle (1982), hence commonly termed the Lomb-Scargle method. This method is widely implemented in available software. It generalizes the periodogram to allow for irregularly spaced data, and has the practical advantage that techniques and results developed for the conventional periodogram, such as smoothing and confidence intervals, can be applied. It is however computationally more demanding (Press *et al.*, 1992) than an FFT-based algorithm, but should not pose any substantial burden with current computational resources unless estimates for a large number of long time series are required. Best and Kostaschuk (2002) in their study of a dune-bed flow using laser-Doppler anemometry estimated power spectra with the Lomb-Scargle technique. Other approaches to power spectra estimation of irregularly spaced time series (see the reviews by Benedict *et al.* (2000) and Babu and Stoica (2010)) have been proposed that for some specific application may perform better than the Lomb-Scargle approach.

In applications where the interest is restricted to the shape of the spectrum (*e.g.*, the slope over some limited region in a log-log representation, or identifying the frequency of a spectral peak), the absolute value of spectral estimates and hence the normalization assumed in the estimates are not important. In other applications however, greater care is needed to obtain appropriately normalized estimates. An example is the spectrum-fitting used to estimate the rate of turbulent kinetic energy dissipation (Soga & Rehmann, 2004; Garcia *et al.*, 2006). For real-valued time series sampled regularly at a fixed time interval, Δt_s, the frequency, f, will range from 0 to the Nyquist frequency, $f_N = 1/(2\Delta t_s)$; in some discussions (and software), Δt_s is assumed to be unity, so that $0 \leq f \leq 1/2$. The range of f is relevant because of the constraint (Equation (6.12.8)) that the integral of the power spectra over all frequencies be equal to the variance. Tapering and smoothing may also affect this constraint. The single-sided power spectral esti-mate, $\hat{E}_{uu}(f)$, is given by

$$\hat{E}_{uu}(f_k) = a_k \hat{S}_{uu}^*(f_k), \qquad f_k = k/(N\Delta t_s), \qquad k = 0, 1, ..., N, \tag{6.12.12}$$

with $a_k = 1$ for $k = 0, N$ and $a_k = 2$ otherwise. In the extended examples below, the single-sided estimate will be consistently given for all computed spectra.

The practice of zero padding, i.e., appending zeroes to, and therefore artificially extending the time series to a length, N^* $(> N)$, is primarily employed for efficient computation using the FFT algorithm, which is optimized when the length of the series is highly composite, i.e., when the number of points is an exact product of many factors, e.g., if it is a power of 2. A secondary practically useful effect is that the typical FFT-based software will evaluate periodogram estimates on a finer grid (grid spacing, $\Delta f_0^* = 1/(N^*\Delta t_s)$, rather than Δf_0). The bandwidth is however still dictated by Δf_0, the finer grid being essentially an interpolation of the no-zero-padding estimates.

6.12.6 Other techniques in time series analysis

The above discussion has been restricted to univariate or scalar stationary series. Dealing with complex flows without a single dominant direction, or with turbulent shear flows where Reynolds shear stresses, motivates an examination of multivariate or vector series, but these have so far not received much attention in the hydraulic literature. Priestley (1981) and Bendat and Piersol (2000) deal with multivariate series from a more general statistical or signal-analysis perspective, and Thomson and Emery (2014) discuss vector time series in an oceanographic context. Similarly, techniques for studying non-station-ary aspects of time series have been widely applied in fields outside of hydraulics. One of the most-known methods from the latter category is based on wavelets (Percival & Walden, 2000), which may be viewed as using alternate basis functions that, in contrast to the standard trigonometric functions, are localized in time, thereby making them more suitable for representing time-varying processes. The Hilbert-Huang approach (Huang et al., 1998; Huang & Wu, 2008) to dealing with non-stationary processes is also based on alternate basis functions, but instead of being defined a priori mathematically, the basis functions are obtained from the time series itself through an algorithm termed empirical mode decomposition. The Hilbert transform is then applied to the decomposed signal to produce a spectrogram depicting the variation with time of the important (instantaneous) frequencies found in the signal. For some problems, the Hilbert-Huang approach has been found to give sharper results than those obtained using wavelets (Huang et al., 1998). It should also be mentioned that Bayesian approaches to time series and specifically spectral analysis are available, and that bootstrapping can also be applied to obtain confidence intervals for power spectral estimates.

6.12.7 Extended example: Autocorrelation and power spectral estimates

Two univariate velocity time series are here considered. Series 1 was acquired with an acoustic Doppler velocimeter (Sontek 50 Hz micro ADV) in a straight power canal (Schemper & Admiraal, 2002). The micro-ADV was mounted on a vertical pole fixed

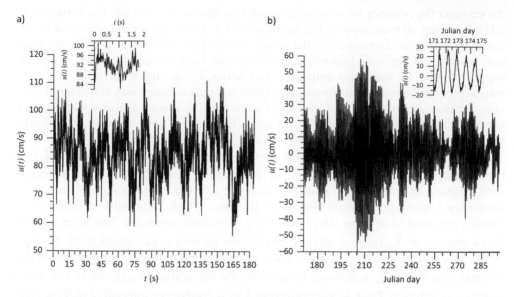

Figure 6.12.1 Velocity time series, a) straight power canal, b) Lake Michigan surface layer.

to a frame placed on the bed, and sampled for $T_s \approx 180$ s ($N = 9203$, no data omitted) at a rate of $f_s = 50$ Hz with the probe mounted 2 m above the local bed at 7 m from the bank. This velocity series provides an example of simple channel flow, with the interest on turbulence characteristics. The entire series is plotted in Figure 6.12.1a. Series 2 was acquired with a single-point vector-averaging current meter (EG&G VACM) in Lake Michigan (Choi *et al.*, 2012), mounted at a depth of 12 m in the middle of the lake using a sampling frequency of $f_s = 96$/day, over a period of $T_s = 125$ days (starting from Julian day 171, so that $N = 12001$). Each measurement was averaged over 15 mins as the main variation of interest was not turbulence but the dynamics of the internal waves occurring due to summer stratification and surface-wave activity. The series is plotted in Figure 6.12.1b. Insets in the two plot of Figure 6.12.1 give details of the flow dynamics revealing weak variation in the first case and a contrasting regular periodic variation for the second case.

The results of stationary tests conducted on both velocity series are summarized in Table 6.12.2. The runs test and the reverse arrangement tests were performed with each series divided into 20 segments for both means and variances. The p-values (obtained using the formulae in Table 6.12.1) for the various tests are given (the lower the values

Table 6.12.2 Large-sample p-values for the null hypothesis that the time series is stationary (alternative hypothesis of a monotone trend)

Series	runs test[†]		reverse arrangement test		PSR test
	means	variances	means	variances	
1	0.59 (0.59)	0.25 (0.24)	0.13	0.39	0.29
2	0.41 (0.41)	0.0058 (0.0045)	0.21	0.037	0

† values in parentheses are exact p-values while those not in parentheses are based on the large-sample normal-distribution approximation, as given in Table 6.12.1.

the stronger the evidence for rejecting the null hypothesis of no-trend). For Series 1, the estimates from all tests support the hypothesis of a stationary series (exact and large-sample p-values for the runs test differ noticeably in value only for the smallest p-value, but rejection of the null hypothesis based on the large-sample estimate would be supported also by the exact estimate). For Series 2, the mean statistics might be stationary but the results on the variance more strongly support non-stationary behavior. A visual examination of Series 2 in Figure 6.12.1b reveals features suggesting a variance changing in time, and it is difficult to believe that over a period of 125 days some physical non-stationarity would not arise in field measurements.

The estimates of autocorrelations for the two series are plotted in Figure 6.12.2. For the turbulent channel flow (Series 1), $\hat{R}_{uu}(\tau)$ decays to zero relatively quickly, with a zero crossing at $\tau \approx 6$ s, and remains small for $\tau > 6$ s. Bootstrapping, using the stationary bootstrap approach (Politis & Romano, 1994, thus, with variably sized blocks) was used to develop the plotted 95% confidence intervals. The latter suggest that estimates of $\hat{R}_{uu}(\tau)$ may already be dubious by $\tau \approx 3$ s, i.e., about $0.02\ T_s$, substantially smaller than the commonly given rule of thumb of $0.1\ T_s$. An estimate of the integral time scale, $\hat{T}$, was made by numerically integrating $\hat{R}_{uu}(\tau)$ up to the zero-crossing, resulting in $\hat{T} \approx 1.47$ s (bootstrapped 95% confidence interval of (0.99 s, 1.96 s)). This may be compared to the scaling estimate, $T \approx \mathcal{L}/\mathcal{U}$, where $\mathcal{L}$ and $\mathcal{U}$ are characteristic integral length and velocity scales (see also Chapter 5.2.4.2 in Volume I). For the local flow depth of $\mathcal{L} = 2$m, and the average velocity of $\mathcal{U} = 84.7$cm/s, this yields $\mathcal{L}/\mathcal{U} = 2.4$ s, which is of the same order of magnitude as $\hat{T}$.

$\hat{R}_{uu}(\tau)$ for Series 2 is quite different, as it exhibits marked periodic features, with amplitudes that do not decay to zero even for $\tau > 25$ days. This reflects the periodic features already pointed out in the inset of Figure 6.12.1b. The peak-to-peak lag time (the number of peaks in Figure 6.12.2b per unit lag time) points to an oscillation period

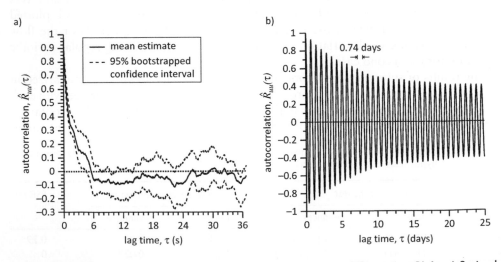

Figure 6.12.2 Estimates of the autocorrelation functions (using the "acf" function in R) for a) Series 1 (using for the stationary bootstrap the "tsboot" function in R), and, b) Series 2.

of ≈ 0.74 day ($= 17.6$ hr), which is consistent with the physically relevant inertial period (estimated as 17.7 hr at the mooring location by Choi *et al.*, 2012). Higher-order frequency components are more difficult to discern directly. Despite the evident second-order non-stationarity in amplitude, $\hat{R}_{uu}(\tau)$ can yield useful information, at least about a dominant frequency. Bootstrapped confidence intervals were also estimated as for Series 1, and similar qualitative results were obtained (large errors for large τ, mitigated however by the larger magnitudes of $\hat{R}_{uu}(\tau)$ at large τ), but were not plotted.

Various estimates of the power spectral density are given in Figure 6.12.3a: one obtained from the raw periodogram, and three using different approaches to smoothing, namely the Daniell window, Welch overlapping segment averaging (WOSA), and multitapering (MTM). The parameters for the smoothing approaches were chosen to obtain approximately the same $(n_{df})_{eff}$ (≈ 22), and comparable resolution of 11 to 12 Δf_0. For the WOSA (with 50% overlap and a Hanning taper), $N_{seg} = 11$, for the MTM, $N_{tap} = 11$ (using discrete prolate spheroid sequences, $N_{MTM,f} = 6$), and for the Daniell window, $N_{Wf} = 5$. The large variability in the raw periodogram estimates contrasts with the much reduced variability of the smoothed estimates (see also the confidence intervals; note that the cross indicates the location of the estimate and not the bandwidth), all three of which are quite similar. For Series 1, which does not contain any strong periodic component excepting a small peak at $f \approx 0.4$ Hz, the resolution of the different approaches is not of any large concern. A line fitted to the WOSA estimates for 0.2 Hz $< f <$ 2 Hz, resulting in a slope estimate of 1.67±0.06 (95% confidence interval, $R^2 = 0.89$), is drawn. This slope agrees well with the Kolmogorov −5/3 law

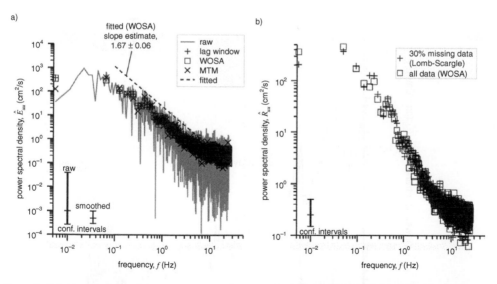

Figure 6.12.3 Power spectral density estimates for series 1: a) the raw periodogram estimate compared with smooth estimates using lag window, WOSA, and MTM methods (all obtained with the R functions "spectrum" or "sdf"), and a line fitted to the WOSA estimate in the region, 0.2 Hz $< f <$ 2 Hz, b) comparison of a WOSA estimate using the entire (≈ 9200 points) time series with a Lomb-Scargle estimate (using the R function, "lsp", followed by smoothing by the R function, "tskernel") based on a series with ≈ 30 % of the points omitted (30 gaps of 100 points).

expected in the inertial subrange (Tennekes & Lumley, 1972; Pope, 2000; see also Section 2.3.3 in Volume I). If the fit were taken with the raw periodogram, the slope would be estimated as 1.61±0.24, with $R^2 = 0.35$, therefore suggesting a close agreement with the theoretical values for the mean estimate but a much larger uncertainty in the fit. It would be expected that the inertial subrange extend well beyond $f = 10$ Hz, but all of the spectra exhibit a tendency to become flatter, which may be more evident in the smoothed estimates. This is mainly attributed to instrument-related noise (Voulgaris & Trowbridge, 1998; Doroudian et al., 2010; see also the discussion in Section 3.2.4, Volume II) but smoothing may also have contributed to this bias at higher frequencies.

The Lomb-Scargle approach to dealing with irregularly sampled data is illustrated by creating 30 gaps of length 100 points (duration 2 s, and so comparable to the integral time scale) at regular intervals in series 1, obtaining Lomb-Scargle spectral estimates, and applying the Daniell smoother. The remaining gapped data consist therefore of only ≈ 67% of the original 9203 points. The results in Figure 6.12.3b show that the Lomb-Scargle estimates based on the gapped data compare very well with the WOSA results plotted in Figure 6.12.3a (an estimate of the slope obtained by fitting the Lomb-Scargle estimates for 0.2 Hz < f < 2 Hz yields 1.70 ± 0.06 with $R^2 = 0.91$).

Whereas Series 1 illustrated features of a typical turbulent-flow power spectra, Series 2 was chosen as an example exhibiting strong periodic wave-like features shown in the inset of Figure 6.12.2b. The power spectrum for Series 2, estimated using the MTM technique with $N_{MTM,f} = 6$ and $N_{tap} = 11$ tapers, is plotted in Figure 6.12.4. It exhibits

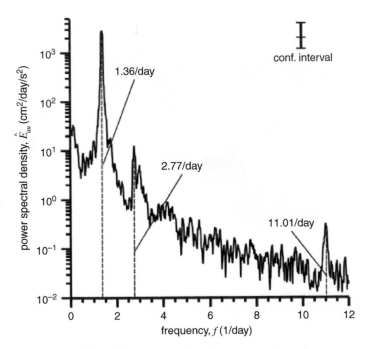

Figure 6.12.4 Smoothed power spectral density estimates for Series 2 (MTM technique with $N_{MTM,f} = 6$ and $N_{tap} = 11$, using the "sdf" function in R).

at least three clear peaks, at $f_{in} = 1.36$/day, 2.77/day, and 11.01/day. The first peak corresponds to that previously identified in the autocorrelation in Figure 6.12.2b, while the latter two are not readily apparent in Figure 6.12.2b. Choi *et al.* (2012) suggested that the peak at 2.77/day is a harmonic of f_{in} ($2f_{in} = 2.72$/day ≈ 2.77/day). The third peak at 11.01/day might also be thought a harmonic ($8f_{in} = 10.88$/day ≈ 11.01/day) but Choi *et al.* (2012) offer an alternative interpretation related to another mechanism previously identified in other studies. The examples above illustrate the advantages of studying the power spectra. Similar to the discussion of the autocovariance, in spite of the non-stationary features of the time series, the frequencies of the spectral peaks have plausible physical interpretations. If this is due to the non-stationarity being mainly in the amplitudes and not in the frequencies, then the absolute values (and of course the confidence interval) of the spectra may not be very meaningful.

6.13 SPATIAL INTERPOLATION, KRIGING, AND SPATIAL DERIVATIVES

As more complicated three-dimensional flows receive increasing attention, and computer-aided systems permit more extensive measurements at multiple points in a plane or a volume, visualization of data acquired in the laboratory or the field becomes an issue. The spatial data are often coarsely sampled at possibly irregular points constrained by access or other experimental conditions. An example is shown in Figure 6.13.1, where streamwise velocities, directly measured using an ADV, *i.e.*, uninterpolated, at the inlet of a 90° laboratory channel bend flow (Cunningham & Lyn, 2016). The channel boundaries are erodible, but at the inlet erosion is negligible, and the initial boundary is an asymmetric trapezoid, shown in Figure 6.13.1 as the thick black line (y' is the vertical distance from the initial bed, while r' is the radial distance with the zero set where the water surface initially meets the inner bank). The flow depth is approximately 12.5 cm, while the cross-sectionally averaged velocity is 22 cm/s. The measurement grid (data points are indicated by crosses in Figure 6.13.1) is regular, 5.1 cm in the lateral (r') direction, and 1.27 cm in the vertical (y'), but with missing points

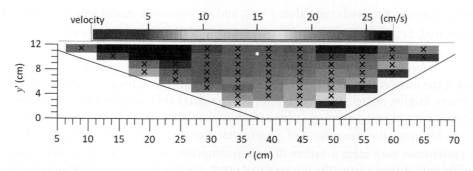

Figure 6.13.1 Original data from time-averaged ADV measurements of streamwise velocity (color-coded are magnitudes) in an asymmetric approximately trapezoidal channel (thick black lines at bottom indicate initial boundaries) at the inlet of a 90° bend (left – inner bank), × indicates the measurement points, y' produced with the "image.plot" function in R.

near the boundaries. Due to the relative coarseness of the measurement grid, the results presented as a color map appear highly pixelated. Interpretation may be facilitated by densely interpolating the data onto a regular grid, allowing ease of manipulation and visualization. In some studies, the interpolated (or fitted) field quantities are subsequently used to estimate other secondary quantities, such as vorticity.

Various options for interpolating (or fitting – here no distinction will be made between the two) irregularly spaced data are available (and implemented in commercial software packages, such as Surfer or Tecplot or ArcGIS, or in more general purpose software such as R or MATLAB), and an overview is given in Eberly *et al.* (2004). Blanckaert and Graf (2001, 2004) applied nonlinear weighted splines to perform a least-squares surface fit of velocity data in a bend flow, and then examined terms involving spatial derivatives of fitted velocities in the relevant balance equations. Jamieson *et al.* (2013a, 2013b) applied kriging to interpolate velocity data, and then estimated vorticities from the interpolated data. The discussion in this section is restricted to the kriging technique (and specifically ordinary kriging), because, unlike other common techniques, its geostatistical basis provides error estimates. The theory of kriging is discussed in standard geostatistics or spatial statistics references such as Isaaks and Srivastava (1989), Cressie (1993), Olea (1999), and Webster and Oliver (2007).

6.13.1 The ordinary kriging model

The ordinary kriging model predicts the interpolated value, $\hat{f}(\mathbf{x}_0)$, at a point $\mathbf{x}_0$, as a *linear* weighted sum of all of the other observed values at $\mathbf{x}_i$, $i = 1, ..., N$, *i.e.*,

$$\hat{f}(\mathbf{x}_0) = \sum_{i=1}^{N} \lambda_i f(\mathbf{x}_i), \quad \sum_{i=1}^{N} \lambda_i = 1, \tag{6.13.1}$$

where λ_i are the weights (Olea, 1999). As a weighted average of 'neighboring' values, such an interpolation is similar to a low-pass filter, and so a smoothing effect may be expected. In this respect, the kriging prediction does not differ from other spatial interpolation techniques such as inverse-distance weighting. The variance of the difference between the predicted value, $\hat{f}(\mathbf{x}_0)$, and an unknown model value, $f_{\mathrm{mod}}(\mathbf{x}_0)$, is formulated, *i.e.*, $\sigma_K^2(\mathbf{x}_0) = \mathrm{var}[\hat{f}(\mathbf{x}_0) - f_{\mathrm{mod}}(\mathbf{x}_0)]$, and λ_i are weights to be determined by minimizing $\sigma_K^2(\mathbf{x}_0)$. In this way, kriging is considered the best linear unbiased predictor. Different models, $f_{\mathrm{mod}}(\mathbf{x}_0)$, may be considered but here only the ordinary kriging model, the most common and usually the default model in software, is discussed. The ordinary kriging model, denoted as $f_{OK}(\mathbf{x}_0)$, is remarkably simple in that it assumes a constant but unknown mean, μ_f, plus a random zero-mean error term, $\varepsilon(\mathbf{x}_0)$: $f_{OK}(\mathbf{x}_0) = \mu_f + \varepsilon$. Although the assumption of a mean that is constant everywhere in the region of measurements may seem a rather drastic assumption, in practice, the weights decay rapidly with distance from the interpolated point, so that the contribution of far points become negligible and the constant mean is effectively a locally constant assumption (Webster & Oliver, 2007). In software implementations, localization can be explicitly imposed by limiting the sum to points near to the interpolation location.

For ordinary kriging, $\sigma_K^2(\mathbf{x}_0)$ can be related to the spatial covariance of f between any two points, or to the variogram (sometimes also termed the semi-variance or semi-variogram), which is defined as

$$\gamma_{ij} = \gamma(\mathbf{x}_i, \mathbf{x}_j) = \frac{1}{2}\mathrm{var}[f(\mathbf{x}_i) - f(\mathbf{x}_j)] = \frac{1}{2}\mathrm{var}[f(\mathbf{x}_i) - f(\mathbf{x}_i + \mathbf{h})] = \gamma(\mathbf{h}), \qquad (6.13.2)$$

where $\mathbf{x}_j$ is expressed more conveniently in terms of the (vector) lag distance, $\mathbf{h}$. In the last equation in Equation (6.13.2), the (second-order) stationarity (or homogeneity) assumption, analogous to that used in time-series analysis (see Section 6.12.1) has been made, such that the covariance or the variogram depends only on the lag distance between the two points and not on the individual locations. A complexity stems from the vectorial nature of $\mathbf{h}$ (compared to the scalar time variable), but if an isotropy assumption is made, as is done in the following, then $\gamma(\mathbf{h}) = \gamma(h)$, and the variogram depends only on the scalar distance and is independent of the direction of $\mathbf{h}$. Similar to the temporal covariance, the variogram measures the extent to which changes in f at a given location is related to changes at a different location. Thus, it would be expected that, beyond a certain lag distance that could be interpreted as an integral length scale, changes at two points would be uncorrelated. In this case, the covariance becomes negligible, but the variogram asymptotes to a constant value.

The linear system of equations determining λ_i by minimizing $\sigma_K^2(\mathbf{x}_0)$ can be expressed as (Webster & Oliver, 2007):

$$\begin{bmatrix} \gamma_{11} & \gamma_{12} & \cdots & \gamma_{1N} & 1 \\ \gamma_{21} & \gamma_{22} & \cdots & \gamma_{2N} & 1 \\ \vdots & \vdots & \vdots & \vdots & \vdots \\ \gamma_{N1} & \gamma_{N2} & \cdots & \gamma_{NN} & 1 \\ 1 & 1 & \cdots & 1 & 0 \end{bmatrix} \begin{bmatrix} \lambda_1 \\ \lambda_2 \\ \vdots \\ \lambda_N \\ \psi_0 \end{bmatrix} = \begin{bmatrix} \gamma_{10} \\ \gamma_{20} \\ \vdots \\ \gamma_{N0} \\ 1 \end{bmatrix} \quad \text{or} \quad \mathbf{A}\lambda = \mathbf{b}, \qquad (6.13.3)$$

where ψ_0 is the Lagrange multiplier used to impose the condition, $\sum_i \lambda_i = 1$ (the last row in Equation (6.13.3)). The matrix $\mathbf{A}$ is evaluated from the variogram for observed values, while the vector, $\mathbf{b}$, depends on the variogram values for the location, $\mathbf{x}_0$, for which the prediction is desired. Moreover, the kriging variance, on which the error estimate is based, is found as

$$\sigma_K^2(\mathbf{x}_0) = \mathbf{b}^T\lambda = \sum_{i=1}^{N} \lambda_i \gamma_{i0} + \psi_0. \qquad (6.13.4)$$

6.13.2 The variogram and its modeling

The variogram, $\gamma(h)$, plays a central role in kriging, and its estimation/modeling becomes the focus of attention. Each pair of observations provides an observed value:

$$\hat{\gamma}_{ij} = \hat{\gamma}(\mathbf{x}_i, \mathbf{x}_j) = \frac{1}{2}[f(\mathbf{x}_i) - f(\mathbf{x}_j)]^2 = \frac{1}{2}[f(\mathbf{x}_i) - f(\mathbf{x}_i + h_{ij})]^2 = \hat{\gamma}_{ij}(h_{ij}), \qquad (6.13.5)$$

where h_{ij} is the (scalar) distance between $\mathbf{x}_i$ and $\mathbf{x}_j$. For the data of Figure 6.13.1, the $N = 67$ data points yield $N(N-1)/2 = 1830$ pairs, and the $\hat{\gamma}$-values for all pairs are plotted in Figure 6.13.2a as a 'cloud', with a maximum lag of ≈ 56 cm. For any given value of h, a wide range of values of $\hat{\gamma}$ is found. As with autocorrelation estimates for time series, estimates of $\hat{\gamma}$ for large h are prone to scatter because of the few samples at large lags. Even when binned, *i.e.*, measurements points grouped together in bins of width 1.5 cm similar to that performed in developing histograms (as in Figure 6.13.2b), a large number of points might be considered outliers for each bin. Such outliers may reflect the limitations of the assumptions (constant-mean and isotropy) of the ordinary kriging model. The bins with the widest interquartile ranges (those centered on $h = 9$ cm, 13.5 cm, and 18 cm) had the smallest number of samples, namely 17, 6, and 6 respectively; all other bins had 38 or more pairs. A rule of thumb (Olea, 1999) suggests limiting binned estimates to bins with more than 30 pairs.

Rather than using noisy estimates of γ directly obtained from observations, a smoothed model for γ is actually used in predictions. Different standard (also termed authorized as they satisfy certain constraints) models may be considered. In Figure 6.13.2b, an exponential model for γ of the form:

$$\gamma(h) = c_{nug} + c_{psill}\left(1 - e^{-h/h_{sc}}\right), \tag{6.13.6}$$

was fitted using a maximum-likelihood approach to the average binned values. Thus, the nugget variance, $c_{nug} = 5.7$ cm^2/s^2, the partial sill, $c_{psill} = 200.8$ cm^2/s^2, (the total sill, $c_{nug} + c_{psill}$, is the value that γ attains for large h), and the range (parameter), $h_{sc} = 43.1$ cm, which can be interpreted as an integral length scale, were found. As pointed out previously, the data in Figure 6.13.2b potentially contains outliers, and omitting such outliers, could change the estimates of smoothed-model parameters of Equation (6.13.6). For the present illustrative purposes, this was not further explored.

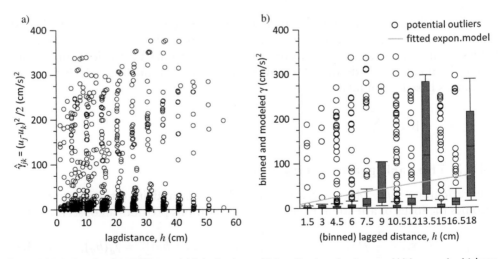

Figure 6.13.2 For the data of Figure 6.13.1, a) values of $\hat{\gamma}$ for all pairs of points, and b) box-and-whiskers plot of binned $\hat{\gamma}$ (bin width, $\Delta h = 1.5$ cm). Variogram cloud and binned variogram obtained with the "variog" function, and an exponential model variogram obtained using the maximum likelihood estimator in the routine "likfit" in package geoR.

6.13.3 Kriging predictions and standard prediction error

With the fitted exponential model for $\gamma(h)$, Equation (6.13.3) was solved for the weights, λ_i, for each predicted point on a dense rectangular grid with mesh size $\Delta r' = 0.5$ cm, $\Delta y' = 0.25$ cm (*cf.* the original measurement grid, $\Delta r' = 5.1$ cm, $\Delta y' = 1.27$ cm grid). The resulting interpolated results are plotted in Figure 6.13.3, where predictions for the entire rectangular region are presented to emphasize that the model can make entirely unrealistic predictions, such as those outside of the physical channel. In practice, such unrealistic region would be masked out as was done in Cunningham and Lyn (2016), who also used (default) ordinary kriging in the commercial Surfer software, with a default linear variogram model. Even within the channel, predictions near the boundaries, *i.e.*, outside of the region of measurements are also incorrect, especially on the left (inner-bank) side, where rather high velocities are predicted near the banks. Velocities in boundary regions are expected to be poorly predicted because they will not be as constrained by adjacent measurements and, especially near solid boundaries, high velocity gradients will cause difficulties. Such problems might be mitigated by imposing theoretical conditions such as a zero-velocity or a zero-shear condition at a boundary as was done by Blanckaert and Graf (2001) in their weighted cubic-spline interpolation. Otherwise, it is recommended that the predictions be made only within the convex region spanned by the measurements and possibly known values at unmeasured points.

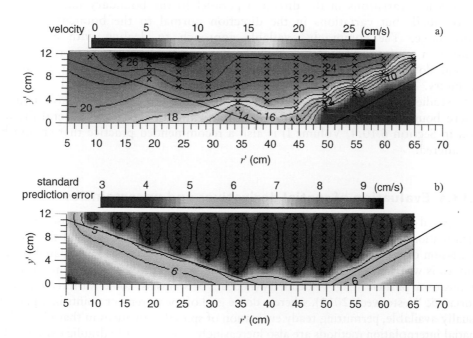

Figure 6.13.3 Interpolated velocity fields for the data in Figure 6.13.1: a) predictions obtained with ordinary kriging on a rectangular grid, and b) standard prediction kriging errors (standard deviations) for the predictions in a). Kriging interpolation and standard errors obtained with "krige.conv" in geoR while the maps and legend were produced with "image.plot" in R.

In the present context of statistical error estimates, the main advantage of the kriging interpolation technique lies in that $\sigma_K(\mathbf{x}_0)$ is directly obtained with Equation (6.13.4) at each prediction point. The corresponding σ_K -field corresponding to the predicted field in Figure 6.13.3a is plotted in Figure 6.13.3b. Consistent with the discussion above, the largest errors are found in the region outside of the measurement region, and the smallest errors in the regions surrounding each measurement point. The values of σ_K are still rather large, 15%–20% of the local velocities (recall that for normally distributed errors, the standard error gives only the 68% confidence interval), even in regions between measurement points. More sophisticated kriging and/or variogram modeling might reduce the values of σ_K, but these values should certainly lead to more caution in interpreting maps based on kriging (or indeed any other type of) interpolation, and especially in using the interpolated results for further analyses.

The above analyses adopted the ordinary kriging isotropic-field model. Although the local near-boundary effect of kriging has been pointed out, high-gradient regions may nevertheless cause difficulties for the constant-mean assumption in ordinary kriging. Universal kriging, in which a deterministic 'regression'-type model of spatial variation replaces the constant-mean assumption, might be considered. This requires however a further model specification, and the linear or polynomial model usually available as standard may be far from adequate. The isotropic assumption is also problematic in high-gradient regions as these often have a pronounced directionality. For example, near solid boundaries, variations in the direction parallel to the boundary might be relatively small, but variations in the direction normal to the boundary may be quite large. The commonly available approach to dealing with anisotropy assumes a uniform anisotropy throughout the entire measurement region, but in real flows the anisotropy may vary due to the highly irregular boundary geometry. These problems with solid boundaries may be exacerbated in laboratory studies due to their smaller scale, and hence the more extensive regions where boundary effects are important. Cross-validation (see the general discussion in Section 6.10.4, Volume I) is also a recommended practice that should be considered.

6.13.4 Evaluation of spatial derivatives and uncertainties

The spatial derivatives discussed in this section are useful for various purposes (*e.g.*, the examination of terms in balance equations or a study of vorticity), and therefore a discussion of techniques for evaluating spatial derivatives and the associated uncertainties is warranted. Not surprisingly, the topic has received much attention in the literature on particle image velocimetry (Foucaut & Stanislas, 2002; Raffel *et al.*, 2007; Adrian & Westerweel, 2011), where a dense grid of measurement points in a plane is usually available, permitting ready evaluation of spatial derivatives in that plane. The spatial interpolation methods are also increasingly relevant for hydraulic experiments as the emerging techniques makes data acquisition more efficient and spatial dense. Examples of measurements and associated evaluation of spatial derivatives in hydraulics include Blanckaert and Graf (2004) and Jamieson *et al.* (2013a, 2013b).

Formulas for estimating spatial derivatives from discretely sampled data on a regular grid (interpolated or not) are finite-difference schemes, similar to those found in computational fluid dynamics, derived from truncation of Taylor series expansions. The central difference scheme is expressed for a regular grid (constant mesh size, Δx) in Cartesian coordinate direction, x, as

$$\frac{\partial f}{\partial x}\bigg|_{x=x_i} = \frac{f_{i+1} - f_{i-1}}{2\Delta x}, \tag{6.13.7}$$

where $f_i = f(x_i)$. From the Taylor-series truncation argument, the error, $\sigma_{df,tr}$, in this approximation is due to the first neglected term, so that $\sigma_{df,tr} \propto (\Delta x)^2 (\partial^3 f / \partial x^3)|_{x=x_i}$, which might be considered a mainly deterministic error. The central-difference scheme is therefore said to be second-order accurate, referring to the exponent of Δx. Lower-order schemes such as the forward- or the backward-difference scheme are only first-order accurate (Raffel $et\ al.$, 2007). This applies only to constant mesh size grids; for irregular grids with non-uniform mesh sizes, even the 'central'-difference scheme is only first-order accurate. The difficulty with such truncation-error estimates is that the partial-derivative term is unknown, but they point to the necessity of resolving any high-gradient regions, where the partial-derivative terms would be large, by reducing Δx, $i.e.$, using a fine grid. It should be stressed that the truncation error can only be reduced by choosing a fine measurement grid and not by a fine interpolated grid.

The other source of error stems from the random uncertainty, $\sigma_{df,\varepsilon}$, in the discrete measurement of f, which has been the focus of this chapter. For a first derivative, $\sigma_{df,\varepsilon} \propto \sigma_f / \Delta x$. The proportionality constants for various schemes, found in Raffel $et\ al.$ (2007), Adrian and Westerweel (2011), are obtained under the assumptions of independent uncertainties in values used, which is unlikely but still reasonable and even conservative. Although $\sigma_{df,\varepsilon}$ will increase with a finer grid (smaller Δx), the relative error, $\sigma_{df,\varepsilon}/(\partial f / \partial x) \sim \sigma_f / f$ should remain acceptable if σ_f / f is acceptable.

6.14 IDENTIFICATION OF COHERENT STRUCTURES

6.14.1 Introduction

Turbulent motions of many scales, from eddies and bulges comparable in size to the width of the river to the smallest scales responsible for the dissipation of turbulent energy to heat, have been observed. Coherent structures are identifiable flow structures that survive for a relatively long time, frequently reappear, and have a typical life cycle (Nezu, 2005; Adrian & Marusic, 2012). Not all turbulence scales equally contribute in determining the statistical properties of the turbulent flows and flows are known to exhibit coherent structures of different types that play an important role in momentum transfer, transport processes and mixing.

Typical coherent structures developing in open-channel flows include low-speed streaks in the inner layer, sweeps and ejections, spanwise vortices, boils or surface-renewal eddies, large coherent shear stress structures (LC3S), alternating high/low speed regions, and long high/low-speed streaks at free surface. These coherent features vary greatly in size, $e.g.$, low-speed streaks in the near-bed region are spaced about 100 y_* apart ($y_* = v/u_*$ is the viscous length scale defined by friction or shear velocity u_* and kinematic viscosity, v), spanwise vortices exhibit a characteristic radius of 15 to 30 y_*, surface boils and LC3S are of the same order of magnitude as the water depth h, and finite velocity-correlation of streamwise structures at the free surface may extend up to 10-20h.

For simplicity, these coherent phenomena are classified into small-scale, large-scale, and super-scale motions. Small-scale motions can be scaled by y_* while the dividing line between large-scale and super-scale motions can be set at 2-3 h. A classification of the coherent structures in open-channel flows according to their length scales is provided in Figure 6.14.1. A schematic of some of the largest coherent structures is provided in Figure 6.14.2, in which is noted the steady closed-loop feedback cycle of the super streamwise vortices and the alignment of the large-scale structures with the downstream movement in open channel flows.

Flow visualization techniques are extensively applied to numerical simulations and experimental results for educing the coherent structures discussed in Chapter 2, Volume II. Caution is needed in the technique implementation as they can be subjective and precise quantitative estimates of structure characteristics can be difficult to obtain. Several approaches have been proposed for isolating and visualizing the coherent structures. A main implementation concern is the differentiation of regions of high vorticity associated with the core of vortices and regions of high vorticity associated

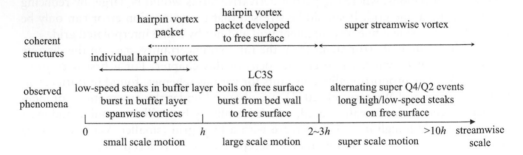

Figure 6.14.1 Major coherent structures identified in open channel flows (Zhong et al., 2015; reproduced with permission).

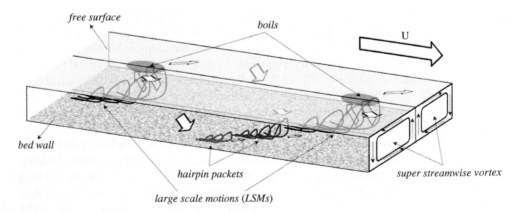

Figure 6.14.2 Closed-loop feedback cycle between super-scale and large-scale structures (Zhong et al., 2015 with permission of Springer).

with attached boundary layers. Examples of eduction techniques based on the single-point measurements include quadrant analysis (Willmarth & Lu, 1972) and variable time-interval averaging (VITA) (Blackwelder & Kaplan, 1976).

The availability of the mean and instantaneous flow-field information as provided by Particle Image Velocimetry (PIV) used in experimental studies and numerical simulations (DNS, LES, hybrid RANS-LES) that resolve at least partially the mean flow unsteadiness and turbulent scales allows for application of feature extraction techniques based on two-point velocity correlations, pressure and various invariants of the velocity-gradient tensor. Most of these methods can be used to identify and visualize coherent structures in both the mean flow and instantaneous flow fields. Other methods such as Proper Orthogonal Decomposition (POD) do not directly visualize coherent structures, but identify the dominant motions which, on average, contain the most energy in the flow, and so are thought to be associated with the unsteady dynamics of the most energetic coherent structures present in the flow.

Several of the most popular methods will be described below and some examples will be included to illustrate the results of their application. For visualizations of three-dimensional (3-D) coherent structures, most of the examples are based on data from 3-D numerical simulations. As the use of 3-D PIV techniques become more popular in hydraulics, these eduction methods can be used in an identical way using data originating from physical experiments rather than 3-D numerical simulations.

6.14.2 Common educing methods

Once the mean or instantaneous velocity fields are available from measurements or simulations they can be used to visualize coherent structures in the mean or instantaneous flow field. Common methods visualize the streamwise velocity or fluctuating streamwise velocity flow fields in planes parallel to wall boundaries (*e.g.*, streaks of high and low streamwise velocity forming in turbulent boundary layers), the pressure distribution in the flow (*e.g.*, the pressure generally decreases in the core of strongly coherent vortices compared to the surrounding flow), or the instantaneous pressure fluctuations. 2-D or 3-D streamlines and out-of-plane vorticity or vorticity magnitude contours can be used to identify vortices in the flow. Though vorticity is commonly used as a quantity to identify vortices, regions of high vorticity magnitude do not always correspond to vortices (*e.g.*, in regions of high shear associated with wall-attached boundary layers). An example where two-dimensional (2-D) streamlines were used to visualize the unsteady laminar horseshoe vortex system developing at the base of a circular cylinder mounted on a flat surface with an incoming Blasius boundary layer is given in Figure 6.14.3a. 2-D velocity measurements were conducted with PIV and laser Doppler velocimetry by Lin *et al.* (2003) in the symmetry plane of the cylinder, on its upstream side. The cylinder Reynolds number was 4,460 in the experiment. The 2-D streamlines in Figure 6.14.3b illustrate the changes in the vortical structure within the horseshoe vortex system during an oscillatory cycle, characterized by the shedding of necklace vortices from the region where the incoming boundary separates and merging events between the middle necklace vortex and the one situated the closest to the upstream face of the cylinder.

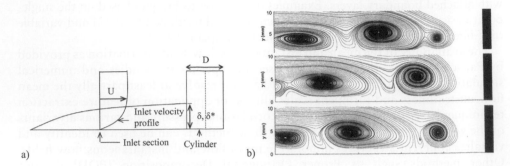

Figure 6.14.3 Visualization of PIV and LDV results in a horseshoe vortex: a) Sketch showing a laminar boundary layer developing on a flat plate starting from a virtual origin. A Blasius velocity profile of thickness δ_{inlet} is imposed at the inlet section, b) Comparison of streamline patterns in the symmetry plane upstream of the cylinder at a time interval of 0.4T, where T is the period of the cycle of the laminar unsteady horseshoe vortex system forming around the upstream base of the circular cylinder. Adapted from Lin *et al.* (2003) with permission from ASCE.

6.14.3 Educing methods based on invariants of the velocity gradient tensor

The most common vortex identification methods are based on a local analysis of the velocity gradient tensor, $\nabla\mathbf{u}$, that satisfies the requirement of Galilean invariance. A detection function related to $\nabla\mathbf{u}$ is evaluated at each point, and is then classified according to the value of the function as being or not part of a vortical structure. The decomposition of $\nabla\mathbf{u}$ into the strain-rate tensor, $\mathbf{S}$, and the rotation or vorticity tensor, Ω, *i.e.*, $\nabla\mathbf{u} = \mathbf{S} + \Omega$, facilitates the interpretation of the widely used detection functions, which are:

a)   the second invariant of $\nabla\mathbf{u}$, conventionally denoted as Q, which can be expressed in terms of $\mathbf{S}$ and Ω, such that the criterion $Q > 0$ (considered by Hunt *et al.*, 1988 along with a condition that the local pressure is lower than the ambient pressure) may be interpreted as rotational motion dominating straining motion,

b)   the second eigenvalue of the symmetric tensor, $\mathbf{S}^2 + \Omega^2$, conventionally denoted as λ_2, is argued by Jeong and Hussain (1995) to be associated with vortical structures for $\lambda_2 < 0$; and

c)   the imaginary part of the complex eigenvalues of $\nabla\mathbf{u}$, conventionally denoted as λ_{ci}, which is also termed the swirling strength, such that regions with high values of λ_{ci} are associated with vortical structures (Zhou *et al.*, 1999).

Chakraborty *et al.* (2005) have shown that the criteria based on Q and λ_2 satisfy the non-local vortex requirement that the orbits of the material points be compact, and that λ_{ci} is an indicator of how fast a fluid is rotating locally. For 3-D turbulent flows, Chakraborty *et al.* (2005) argued that all criteria lead to similarly looking vortices if equivalent thresholds are applied to the detection functions, but Jeong and Hussain (1995) found differences in vortices educed with the criteria based on Q and λ_2. The selection of a threshold is important since the spatial extent of a vortex in a viscous fluid is not clearly defined and depends on the choice of a threshold. One important

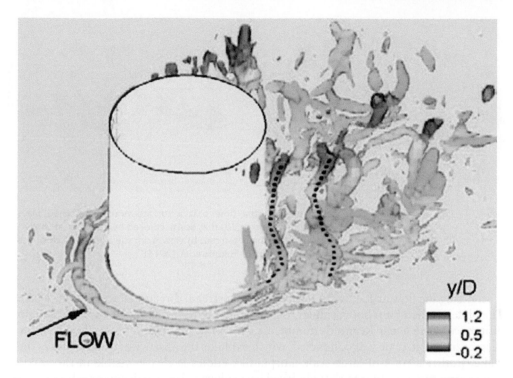

Figure 6.14.4 Large scale coherent structures in an instantaneous flow field around a circular cylinder mounted on a flat bed visualized using the Q criterion. The Q-isosurface is colored with the elevation from the bed, y/D. The dashed lines show the axes of the vortex tubes generated in the separated shear layers. Reproduced from Kirkil & Constantinescu (2015), with the permission of AIP Publishing.

challenge when applying these methods to velocity fields corresponding to high Reynolds number turbulent flows that contain energetic small scales is a separation of large-scale coherent structures from small-scale turbulence.

Example 6.14.1:   An example of using the Q criterion to visualize the vortical structure of the flow past a surface mounted cylinder at a moderate cylinder Reynolds number (Re =16,000) is given in Figure 6.14.4. The instantaneous flow field is taken from a 3-D LES performed by Kirkil and Constantinescu (2015). The Q criterion captures the formation of a large necklace vortex around the upstream base of the circular cylinder. Note the formation of secondary hairpin-like eddies on the core of the main necklace vortex. The Q-criterion also visualizes the deformed vortex tubes extending from the bed to the free surface in the separated shear layers. For this particular case, the same value of Q was used to visualize both types of coherent structures. Depending on the flow type and flow parameters, different values of Q may be needed to educe large-scale coherent structures in different regions of the flow.

Example 6.14.2. This example presents visualizations of coherent structures obtained with the λ_2 criterion applied to flow past around a rectangular plate which

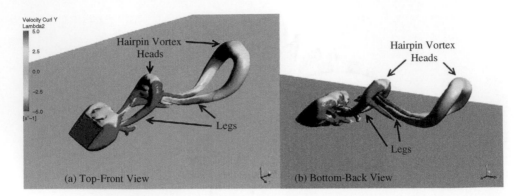

Figure 6.14.5 Large-scale coherent structures in the flow past a surface-mounted inclined plate visualized using a iso-surface of the normalized λ_2 scalar colored by the vertical (y-axis) vorticity component. a) view from the top-front; b) view from the bottom-back of the plate. Reproduced with permission from Fukuda *et al.* (2014).

is set facing the flow at 135° with respect to the upstream direction (See Figure 6.14.5). The plate acts as a bluff body, similar to the cylinder in the previous example. The main difference is the lower Reynolds number at which the simulation of the flow past the surface-mounted plate was conducted, which explains the absence of small scales in the wake. Rather a train of large-scale hairpin-like eddies is clearly visible in the wake, whose structure is similar to that observed past spheres. These hairpins are shed quasi-regularly from the back of the plate. After they have been generated, the hairpins are stretched mostly because the head of the hairpin travels faster than its legs.

Example 6.14.3.   This example visualizes vortical structures ina turbulent flow past evolving bed forms using the swirling-strength criterion (see Figure 6.14.6). Instantaneous velocity fields were generated by 3-D LES (Frias & Abad, 2013) corresponding to the experiments of Fernandez *et al.* (2006). The results show the presence of large-scale hairpin eddies scaling with the size of the bed forms and the change in the flow structure during the amalgamation process. The LES with a fixed bed was conducted for three different representative types of bedforms corresponding to the main stages of the amalgamation process: 1) train of ripples, 2) superimposed bed forms and 3) amalgamated bed forms. In each flow, the same value of $\lambda_{ci} = 25$ was used. The hairpin-like eddies are more defined during the train of ripples stage than during the other two stages where streamwise roll structures are more defined.

6.14.4 The proper orthogonal decomposition (POD) technique

POD aims to extract relevant information for the physical understanding of the investigated flow by separating the effects of appropriately defined modes of the flow from the background (residual random) turbulence. In other words, POD extracts coherent structures in the turbulent flow, irrespective of the definition adopted for the "coherent structure". The method has been applied to extract flow features in

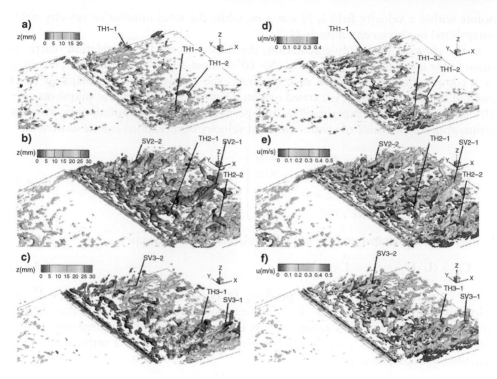

Figure 6.14.6 Coherent structures in channel flow over a train of ripples. a)-c) Swirling strength iso-surfaces with elevation contours (mm) for the train of ripples, superimposed bed forms and amalgamated bed forms respectively; d)-f) Swirling strength iso-surfaces with velocity contours (m/s). Reproduced with permission from Frias and Abad (2013).

different flows of interest in hydraulic engineering such as open channel flows, jets and wakes created behind bluff bodies (see also Section 6.3.2 in Volume I).

The POD technique was first introduced by Lumley (1967); for a more complete description, see Berkooz *et al.* (1993) and Holmes *et al.* (1996). In experimental investigations, POD analysis is usually carried out on a set of spatial-temporal data associated with various flow quantities. For traditional multi-point probe measurements, the data can be velocities from an array of hot-wires (Citriniti & George, 2000) or surface pressure signals from a grid of pressure taps (Kho *et al.*, 2002). Modern 2-D and 3-D PIV techniques make available flow data with a much higher spatial resolution (see also Section 3.7, Volume II). Of particular interest is the capability of POD to identify the motions which on average contain the most energy. This is obtained for 2-D measurements by calculating the average two-point velocity correlation matrix of velocity components, $u(x, y, t)$ and $v(x, y, t)$ measured simultaneously at discrete times t_i with $i = 1, ..., M$. The time intervals between consecutive snapshots should be large enough such that the velocity fields are uncorrelated. High resolution 2-D PIV measurements provide information for the two component velocity fields on a regular grid at m discrete x - locations and n discrete y -locations. The total number of grid points within a velocity field is $N = n \times m$, while the total number of velocity fields (snapshots) in a given ensemble is M.

In the classical POD implementation, the dimension of the correlation matrix is equal to the number of the grid points ($N \approx 10^5$ in most 2-D PIV experiments), resulting in a computationally intensive task to obtain the eigenvalues. The snapshot POD method of Sirovich (1987) is carried out on a time sequence of these snapshots ($M \approx 10^3$ in most 2-D PIV experiments), and permits a more computationally efficient estimation. An ensemble of spatio-temporal velocity observations can be written as:

$$\mathbf{U} = \{\mathbf{U_i}\}_{i=1}^M = \begin{bmatrix} u_{11} & u_{12} & \cdots & u_{1M} \\ u_{21} & u_{22} & \cdots & u_{2M} \\ & \vdots & & \\ u_{N1} & u_{N2} & \cdots & u_{NM} \end{bmatrix} \tag{6.14.1}$$

In the method of snapshots, the $M \times M$ correlation matrix becomes:

$$C_{ij} = \langle \mathbf{U'_i U'_j} \rangle \quad i, j = 1, \ldots, M \tag{6.14.2}$$

where $\langle \, , \, \rangle$ is the Euclidian inner product. In Equation (6.14.2), $\mathbf{U'_i} = \mathbf{U_i} - \overline{\mathbf{U}}$ are velocity fluctuations and $\overline{\mathbf{U}} = \sum_i \mathbf{U}_i / M$ is the mean velocity averaged over the columns for $i = 1, \ldots, M$. Since $\mathbf{C}$ is symmetric, its eigenvalues, λ_i, are non-negative, and its eigenvectors, φ_i, $i = 1, \ldots, M$, form a complete orthogonal set. The orthogonal eigenfunctions are computed as

$$\Phi^{\{k\}} = \sum_{i=1}^M \varphi_i^{\{k\}} \mathbf{U}_i, \quad k = 1, \ldots, M \tag{6.14.3}$$

where $\varphi_i^{\{k\}}$ is the ith component of the kth eigenvector. The index i distinguishes between velocity fields at different instances in time, *not* between different components. The POD analysis is performed on the fluctuating velocity fields, so the ith eigenvalue, λ_i, represents the turbulent kinetic energy contribution of the ith POD mode ϕ_i, and the fractional contribution of ith POD mode to the total turbulent kinetic energy, k_i, can be expressed as $k_i = \lambda_i / k$, where $k = \sum_i \lambda_i$ is the total turbulent kinetic energy of the flow. The snapshot method has been used to study dynamics of energetically dominant coherent structures forming in turbulent jets, wakes and boundary layers (Shinneeb *et al.*, 2006; Singha *et al.*, 2009, among others).

Several recent experimental studies (Liu *et al.*, 1994; Gurka *et al.*, 2006; Roussinova *et al.*, 2010 among others) have employed POD to extract coherent structures from 2-D PIV measurements acquired in the streamwise-wall-normal plane of smooth bed open channel flow. Many environmentally relevant flows occur over surfaces (*e.g.*, containing dunes and ripples) where the effect of the rough boundary is to significantly increase transport of species (heat or mass) and momentum compared to the canonical case of a smooth boundary. Examples of the application of the POD technique to such a flow are discussed next. The focus is on the characterization of the superstreaks forming in a channel on top of the bottom boundary which contains large-scale roughness elements. The problem is relevant to flow in alluvial channels containing bedforms (*e.g.*, dunes).

Example 6.14.4: In this example the turbulent channel flow over a wavy bottom wall is analyzed based on experiments and their analysis carried out by Günther and von Rohr (2003) and Kruse *et al.* (2003). The PIV fields in a horizontal plane situated above the top of the wavy wall were analyzed using a snapshot POD technique to reveal the large energy-containing structures present in the flow and to obtain information on the mean size and spacing of the superstreaks. Günther and von Rohr (2003) acquired 2-D PIV velocity fields in two planes one parallel to the bed at a vertical location $y/\Lambda = 0.26$ (Λ is the mean depth of the channel) and the other inclined to the bed at an angle $\beta = 53°$ at $x/\Lambda = 0.83$ as shown in Figure 6.14.7. The experiment was carried out in an open-channel flow with Reynolds number of 4,500. The snapshot POD decomposition was used to obtain the structure of dominant eigenfunctions of developed turbulent flow on the wavy wall. Figure 6.14.8 shows the eigenfunctions $\Phi_{1,u}$ and $\Phi_{2,u}$ corresponding to the two dominant modes based on their fractional contribution to the turbulent kinetic energy. These modes show the existence of large-scale longitudinal structures with a characteristic spanwise scale of $\Lambda_z = O(1.5)\Lambda$. The superstreaks are visible in the instantaneous streamwise velocity plot in Figure 6.14.9. These structures do not have fixed spanwise locations and meander laterally in time. As more modes are used to reconstruct the velocity field, the spatial extent and size of the superstreaks in the instantaneous flow field visualized in Figure 6.14.9 become clearer. Still, a relatively small number of modes is enough to get this information.

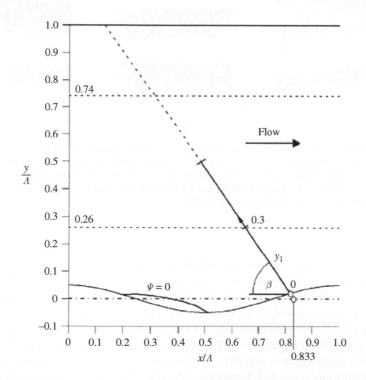

Figure 6.14.7 Sketch of the channel containing the wavy wall at the bottom in which 2-D PIV measurements were conducted in y/Λ = constant planes (Λ is the channel flow depth) and in a (y_1, z)-plane with $\beta = 53°$. From Günther & von Rohr (2003), reproduced with permission.

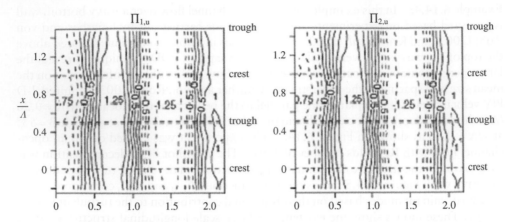

Figure 6.14.8 Eigenfunctions $\Phi 1$, u and $\Phi 2$, u in a horizontal plane situated at $y/\Lambda = 0.26$ from POD analysis. From Günther & von Rohr (2003), reproduced with permission.

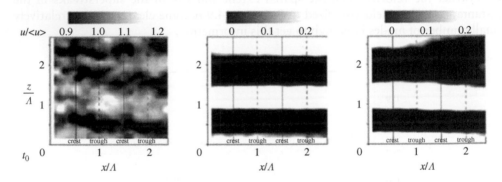

Figure 6.14.9 Instantaneous streamwise velocities in a horizontal plane situated at $y/\Lambda = 0.26$. The left frame shows the raw data. The middle and right frame show the projection of the first two and, respectively, first 6 eigenfunctions on the instantaneous velocity. Solid vertical lines denote the crest of the dunes, while dashed vertical lines the troughs. Reproduced with permission from Kruse et al. (2003).

Example 6.14.5: A second example illustrates the application of the snapshot POD method to 2-D PIV measurements carried out in the near field of an axisymmetric turbulent free jet (Tandalam *et al.*, 2010). The study is focused on extracting the large scale coherent structures by documenting the signatures of vortices for different jet Reynolds numbers. The light sheet was oriented vertically (x, y)-plane and horizontally (x, z)-plane along the jet centerline. PIV measurements were made along the centerline of the jet in the vertical plane as well as in a horizontal plane. The images were recorded using a PIV system (*i.e.*, TSI PowerViewPlus 4 MP 12-bit digital camera with a resolution of 2048 × 2048 pixels operating in dual-capture mode). Two thousand image pairs were acquired at a frequency of 1.04 Hz and a time separation of 200 µs between consecutive frames. The images were analyzed with 16 × 16 pixels interrogation area, which, with a field of view of 44 mm × 44 mm, yielded a spatial resolution of ≈ 0.4 mm.

Figure 6.14.10a shows an instantaneous velocity flow field of the jet with an exit velocity of 1.0 m/s corresponding to a jet Reynolds of 10,000. The fluctuating velocity field, $u' = u(x, y, t) - \bar{u}(x, y)$, is decomposed on an orthogonal basis ($u = \bar{u} + u'$, with u being the instantaneous velocity, $\bar{u}$ the time-averaged velocity, and u' the turbulent fluctuation for the axial component of the jet velocity). In theory, large-scale flow structures are represented by the lower modes, whereas small-scale flow structures are

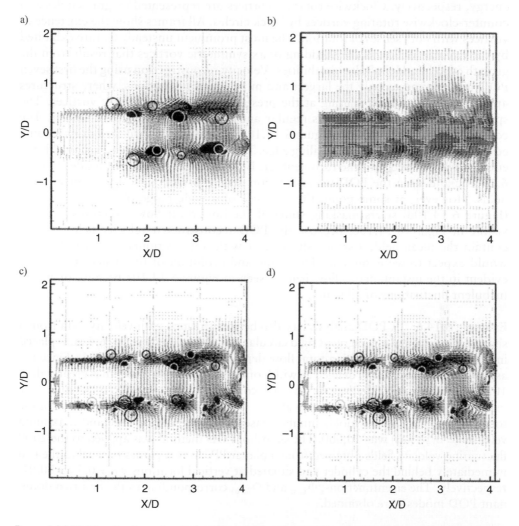

Figure 6.14.10 Visualization of turbulent flow field and vortices using reconstructed velocity fields based on the POD method for a turbulent circular jet. a) Instantaneous velocity; b)-d) POD reconstruction using 25% (b), 50% (c) and 75% (d) of turbulent kinetic energy. Adapted from Tandalam *et al.* (2010), with permission from ASCE.

represented by the higher modes. A beneficial use of POD basis functions is to low-pass filter the velocity realizations in order to study the physics of the large and small scale features of the flow. The question of how many leading POD modes have to be included to reconstruct the velocity so as to capture only the large scale energetic structures is somewhat arbitrary and it does depend on the particular flow investigated. In the results presented in Figure 6.14.10, the number of modes employed in the reconstruction of the velocity fields was chosen to be the minimum number required to capture 25%, 50% and 75% of the surrogate turbulent kinetic energy $\left(\overline{u'^2} + \overline{v'^2} \right)$.

Figure 6.14.10b-d show the same velocity field obtained from POD reconstruction using 25% (7 modes), 50% (24 modes) and 75% (78 modes) of the turbulent kinetic energy, respectively. Clockwise rotating vortices are represented by gray circles and counter-clockwise rotating vortices by black circles. All frames show the existence of vortices with both senses of rotation. The most prominent unsteady feature exhibited by the circular jet flow is the shedding of axisymmetric vortices that result from the growth of Kelvin-Helmholtz instabilities. Velocity reconstruction using the first seven POD modes exposes only the largest and most energetic motions, where structures appear orderly oriented hinting at the presence of the axisymmetric vortices. The spacing between these vortices is regular and correlateswith the jet diameter. The cores of the vortices shown in Figure 6.14.10b induce an outward flow at its leading edge and an inward flow at its trailing edge. This sequence of events is responsible for entrainment of fluid into the turbulent jet. Using a larger number of modes (Figure 6.14.10c and d) in the POD reconstruction adds more detail to the reconstructed fields. However, at some point, the presence of a wide range of scales of motion (Figure 6.14.10d) starts masking some of the important flow structures that are visible in Figure 6.14.10c. Because the POD reconstructed velocity fields do not contain the mean field, a strong shear layer with unidirectional vorticity that one would expect to see is missing. The shape and evolution of the structures are not evident in the instantaneous flow field as seen in Figure 6.14.10a because of strong turbulent fluctuations of all scales.

Example 6.14.6: A POD analysis has also been performed with velocity data from a shallow flow around an emergent rigid circular cylinder (diameter, D = 6.4 mm) placed in an open channel flow with a total flow depth of 70 mm (Heidari, 2016). Since the aspect ratio of the channel defined as width of the channel to flow depth is greater than 10, the flow is termed shallow. It was also confirmed, that the velocity profiles scale logarithmically in a region near the wall and the boundary layer occupies 70% of the total depth. The flow Reynolds number based on the depth of flow and maximum velocity (= 0.16 m/s) was 10,000. The snapshot POD method was applied to the 2000 fluctuating velocity fields obtained from a planar PIV in two planes parallel to the bed immediately behind the cylinder and situated at vertical locations y/d = 0.5 and 0.07, respectively. The eigenfunctions, $\Phi_{1,\,u}$ and $\Phi_{2,\,u}$, corresponding to the first two dominant POD modes were obtained.

At mid-depth plane (y/d = 0.5), modes $\Phi_{1,\,u}$, $\Phi_{2,\,u}$ are symmetric, forming a conjugate pair at $x/D > 1$ with a phase shift of 90°. Any such pair of modes, phase shifted by a quarter cycle, have been termed as regular modes according to Sengupta *et al.* (2010).

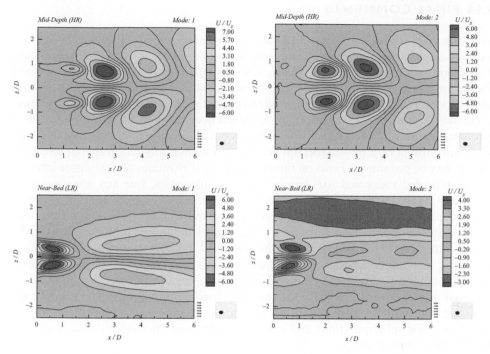

Figure 6.14.11 Eigenfunctions $\Phi_{1,\,u}$ and $\Phi_{2,\,u}$ in two horizontal planes (x-z) situated at (a) y/d = 0.5 and (b) y/d = 0.07 at Re_D = 900. Here, d = 70 mm is the depth of flow and the Reynolds number Re_D is based on the diameter of the cylinder (D = 6.4 mm) and the flow is subcritical. Reproduced with permission from Heidari (2016).

These modes are consistent with the regular von-Karman wake vortices. Very different modes are observed in the near-bed plane at y/d = 0.07. This horizontal plane is located in the layer (25 $< yu_*/v <$ 60, where u_* is the shear velocity and v is the fluid kinematic viscosity) where $\overline{u'^2}$ has a peak. Modes $\Phi_{1,u}$, $\Phi_{2,u}$ shown in Figure 6.14.11 are attached to the cylinder with two symmetric elongated regions. Near the sides of the cylinder, separated shear layer instabilities generate high turbulent kinetic energy. Further downstream, at $x/D>$ 1.5 the imprint of large scale structure similar to that described by Sheng *et al.* (2009) is observed. Sheng *et al.* (2009) conducted digital holographic microscopy experiments in the inner part of a turbulent boundary layer over a smooth wall to examine the link between the quasi-streamwise vortices and generation of extreme wall events. They reported formation of spanwise shear layers (SSLs) in the horizontal plane located in the buffer layer, similar to the one observed in this study. Kirkil *et al.* (2015) reported that the presence of energetic necklace vortices in the vicinity of the bed-mounted cylinder and advection of large scale eddies over the bed surface situated beneath the SSLs are the main reasons for the regions of large bed friction. The energetic necklace vortices interacting with the quasi-streamwise vortices lifting from the bed appear to play a dominant role in the spanwise transport of streamwise momentum in the near bed plane of a cylinder's wake at this subcritical flow condition.

6.15 FINAL COMMENTS

In this chapter, a number of data analysis concepts and techniques, both traditional as well as non-traditional, have been discussed within the specific context of experimental data from hydraulic engineering flows. The chapter is intended as a starting point, illustrating some of the subtleties of statistical data analyses, and the reader is encouraged to delve further in the literature and to become more aware of the statistical bases of the analyses undertaken. A recurring theme has been the uncertainty in any statistical estimate, whether in a simple statistic such as a mean, or a more complex statistic such as a regression fit or a power spectral density or a statistically based spatial interpolation. It may be pedantic to insist on rigorous statistical estimates of confidence intervals for all reported measured (or thereby derived) quantities, but any conclusions or inferences based on measurements should be cognizant of associated uncertainties. Sophisticated statistical analyses however can lead to incorrect inferences if the underlying model assumptions are incorrect. Yet useful if limited physically reasonable results have been obtained even when the statistical model assumptions are questionable, e.g., in identifying important frequencies in a non-stationary time series using conventional power spectra. A broad understanding of the physical phenomena involved must inform the statistical thinking in formulating appropriate models, and avoiding potential problems, such as spurious correlations, and considerations of population (or subpopulation) and sample should guide not only the data analysis but the experimental design.

6.A APPENDIX A

6.A.1 Some density distributions used in statistical inference

The density distribution for the lognormal distribution can be expressed as

$$f\left(x; \mu_{\ln(x)}, \sigma_{\ln(x)}\right) = \frac{1}{\sigma_{\ln(x)} x \sqrt{2\pi}} \exp\left[-\frac{1}{2}\left(\frac{\ln(x) - \mu_{\ln(x)}}{\sigma_{\ln(x)}}\right)^2\right]$$

$$= \frac{1}{\sigma_{\ln(x)} x \sqrt{2\pi}} \exp\left[-\frac{1}{2}\left\{\frac{\ln(x) - \ln(\mu_{gx})}{\ln(\sigma_{gx})}\right\}^2\right], \quad (6.A.1.1)$$

with (possibly dimensional) geometric mean, $\mu_{gx} = \exp(\mu_{\ln(x)})$, and dimensionless geometric standard deviation, $\sigma_{gx} = \exp(\sigma_{\ln(x)})$. The mean is $\mu_x = \exp[\mu_{\ln(x)} + (\sigma^2_{\ln(x)}/2)] = \mu_{gx} \exp\left(\sigma^2_{\ln(x)}/2\right)$, the median, $x_{50} = \mu_{gx}$, and the variance, $\sigma^2_x = \mu^2_{gx}[\exp(\sigma^2_{\ln(x)}) - 1]$.

The standard (Student) t-density distribution can be expressed as

$$f(x; n_{df}) = \frac{\Gamma[(n_{df} + 1)/2]}{\Gamma(n_{df}/2)\sqrt{n_{df}\pi}} \left(1 + \frac{x^2}{n_{df}}\right)^{-\left(\frac{n_{df}+1}{2}\right)} \quad (6.A.1.2)$$

where $\Gamma()$ is the gamma function. For $n_{df} > 1$, the mean and median are $\mu_x = x_{50} = 0$ (the distribution is symmetric about $x = 0$), while $\sigma_x^2 = n_{df}/(n_{df} - 2)$ for $n_{df} > 2$.

The standard two-parameter (c_{sh}, c_{sc}) gamma density distribution may be expressed in so-called scale form as:

$$f(x; c_{sh}, c_{sc}) = \frac{1}{c_{sc}^{c_{sh}}\Gamma(c_{sh})} x^{c_{sh}-1} \exp(-x/c_{sc}), \quad x \geq 0, \ c_{sh} > 0, \ c_{sc} > 0, \quad (6.A.1.3)$$

where c_{sh} is termed the shape, and c_{sc} the (possibly dimensional) scale with $\mu_x = c_{sh}c_{sc}$ and $\sigma_x^2 = c_{sh}c_{sc}^2$. The chi-squared (χ^2) distribution is a special case of the gamma distribution, with $c_{sh} = n_{df}/2$ and $c_{sc} = 2$. Within the context of Bayesian approaches, the popular BUGS and JAGS packages, e.g., used in the R software environment, define the gamma distribution in terms of $1/c_{sc}$, termed the rate. For small values of both c_{sh} and $1/c_{sc}$, the distribution is concentrated at $x = 0$, but the variance is large, which makes it suitable as an uninformative or diffuse model prior probability for the precision, i.e., the reciprocal of the variance, of a normal distribution. Alternatively, the Bayesian variance can be directly modeled by an inverse gamma distribution which can be expressed similarly as

$$f(x; c_{sh}, c_{sc}) = \frac{c_{sc}^{c_{sh}}}{\Gamma(c_{sh})} x^{-(c_{sh}+1)} \exp(-c_{sc}/x), \quad x > 0, \ c_{sh} > 0, \ c_{sc} > 0. \quad (6.A.1.4)$$

A useful compendium of density distributions can be found in an appendix of Gill (2014).

6.A.2 Confidence and prediction intervals in simple linear regression

For the simple linear regression model, $f_i = \beta_0 + \beta_1 x_i + \varepsilon_i$, the two-sided $(1 - \alpha)100\%$ confidence interval for β_0 is $b_0 \pm t_{n_{df},\alpha/2}(s.e.)_{\beta_0}$ where the standard error, $(s.e.)_{\beta_0}$, is

$$(s.e.)_{\beta_0} = s\sqrt{\sum_i \frac{x_i^2}{NS_{xx}}} \quad \text{where} \quad s^2 = \frac{SSE}{N-2} = \sum_i \frac{\left(f_i - \hat{f}_i\right)^2}{N-2}, \quad (6.A.2.1)$$

and $n_{df} = N - 2$, and the other variables are as defined earlier in Section 6.6.2, Volume I. The two-sided $(1 - \alpha)100\%$ confidence interval for β_1 is $b_1 \pm t_{n_{df},\alpha/2}(s.e.)_{\beta_1}$ where $(s.e.)_{\beta_1}$ is

$$(s.e.)_{\beta_1} = s/\sqrt{S_{xx}} \quad (6.A.2.2)$$

The two-sided $(1 - \alpha)100\%$ confidence interval for the _mean_ response, $\mu_{f|x_0}$, for $X = x_0$ is $\hat{f}_0 \pm t_{n_{df},\alpha/2}(s.e.)_{\mu_{f|x_0}}$ where $(s.e.)_{\mu_{f|x_0}}$ is

$$(s.e.)_{\mu_{f|x_0}} = s\sqrt{\frac{1}{N} + \frac{(x_0 - \overline{x})^2}{S_{xx}}}, \quad (6.A.2.3)$$

where $\mu_{f|x_0} = \hat{f}_0 = b_0 + b_1 x_0$. The two-sided $(1 - \alpha)100\%$ prediction interval, *i.e.*, for the *single* response, is $\hat{f}_0 \pm t_{n_{df},\alpha/2}(s.e.)_{\hat{f},\text{pred}}$ where is

$$(s.e.)_{\hat{f},\text{pred}} = s\sqrt{1 + \frac{1}{N} + \frac{(x_0 - \bar{x})^2}{S_{xx}}}. \tag{6.A.2.4}$$

Note that $(s.e.)_{\hat{f},\text{pred}}$ may be substantially larger than $(s.e.)_{\mu_{f|x_0}}$. The corresponding formulae for multiple linear regression can be best expressed in vector form, and can be found in standard texts such as Walpole *et al.* (2007).

6.A.3 Confidence interval of ratios

In a number of problems, confidence intervals are desired for the ratio of the means of two random variables, $\bar{X}_1/\bar{X}_2$. For example, in Section 6.8.1, Volume I, the von Kármán constant, κ, is estimated as the ratio of quantities obtainable from the slopes of two regressed lines, each with a confidence interval that can be estimated by classical formulae, and the confidence interval for κ is desired. The general statistical problem is quite involved, due mainly to the possibility that $\bar{X}_2$ may take on a zero value, leading to a singularity. The classical result for bivariate normally distributed variables is a theorem due to Fieller (1954) which gives the $(1 - \alpha)100\%$ two-sided interval as:

$$(r_L, r_U) = \frac{\bar{r}}{1 - G}\left[\left(1 - G\tilde{s}_{x_1 x_2}\tilde{s}_{x_2}^2\right) \mp t_{n_{df},\alpha/2}\sqrt{\tilde{s}_{x_1}^2 - 2\tilde{s}_{x_1 x_2} + \tilde{s}_{x_2}^2 - G\left(\tilde{s}_{x_1}^2 - \tilde{s}_{x_1 x_2}^2 \tilde{s}_{x_2}^2\right)}\right], \tag{6.A.3.1}$$

where $\bar{r} = \bar{x}_1/\bar{x}_2$, $G = t_{n_{df},\alpha/2}^2 \tilde{s}_{x_2}^2$, the coefficients of variation, $\tilde{s}_{x_1} = s_{x_1}/\bar{x}_1$, $\tilde{s}_{x_2} = s_{x_2}/\bar{x}_2$, the dimensionless covariance, $\tilde{s}_{x_1 x_2} = s_{x_1 x_2}/(\bar{x}_1 \bar{x}_2)$, where $s_{x_1}^2$, $s_{x_2}^2$, and $s_{x_1 x_2}$ are the sample variances and covariance, and the number of degrees of freedom in the t-distribution is $n_{df} = N_1 + N_2 - 2$. This result indicates that if $\tilde{s}$ is not small and hence that G is not negligible, the center of the symmetric Fieller confidence interval is *not*, as might be expected, simply the ratio of the sample means, $\bar{r}$. An approximate result, applicable when $\tilde{s}$ is small and G is negligible, and $\bar{x}_2$ is not near zero, does yield a confidence interval symmetric about $\bar{r}$ with a standard error that can be expressed as

$$(s.e.)_r = |\bar{r}|\sqrt{\frac{s_{x_1}^2}{\bar{x}_1^2} - 2\frac{s_{x_1 x_2}}{\bar{x}_1 \bar{x}_2} + \frac{s_{x_2}^2}{\bar{x}_2^2}}. \tag{6.A.3.2}$$

This result can also be obtained from a Taylor series expansion (also sometimes referred to as the delta rule, see Herson (1975) and Coleman and Steele (2009)) of the variance of $\bar{X}_1/\bar{X}_2$. This simpler result was used in Section 6.8.1, Volume I, with the additional assumption that the covariance, $s_{x_1 x_2}$, is negligible. Bootstrapping or a Bayesian approach are other options for obtaining interval estimates, particularly when the distributional assumptions of Fieller's theorem are not justified though special bootstrapping techniques may be necessary in problematic cases (Hwang, 1995).

REFERENCES

Abu-Mostafa, Y.S., Magdon-Ismail, M., & Lin, H.T. (2012) *Learning from Data: A short course.* 1st edition. Available online at: http://amlbook.com [Accessed 29th November 2016].

Adrian, R.J. & Marusic, I. (2012) Coherent structures in flow over hydraulic engineering surfaces. *J Hydraul Res.* [Online] 5(50), 451–464. Available from: doi:10.1080/00221686.2012.729540 [Accessed 29th November 2016].

Adrian, R.J. & Westerweel, J. (2011) *Particle image velocimetry.* New York, Cambridge University Press.

Antonia, R.A. (1981) Conditional sampling in turbulence measurement. *Annu Rev Fluid Mech.* [Online] 13, 131–156. Available from: doi:10.1146/annurev.fl.13.010181.001023 [Accessed 29th November 2016].

Arel, I., Rose, C.D., & Karnowski, T.P. (2010) Deep machine learning-a new frontier in artificial intelligence research. *IEEE Comput Intell M.* [Online] 5(4), 13–18. Available from: doi:10.1109/MCI.2010.938364 [Accessed 29th November 2016].

Arlot, S. & Celisse, A. (2010) A survey of cross-validation procedures for model selection. *Statistics surveys.* [Online] 4, 40–79. Available from: doi:10.1214/09-SS054 [Accessed 29th November 2016].

Arneson, L.A. (1997) *The effects of pressure-flow on local scour in bridge openings.* Ph.D. thesis, Dept. of Civil Engineering, Fort Collins, CO, Colorado State University.

ASTM (2016) *Standard Practice for Dealing with Outlying Observations.* E178–16, [Online] ASTM International. Available from: doi: 10.1520/E0178-16 [Accessed 29th November 2016].

Azmathullah, H., Deo, M., & Deolalikar, P. (2005) Neural networks for estimation of scour downstream of a ski-jump bucket. *J Hydraul Eng.* [Online] 131(10), 989–908. Available from: doi:10.1061/(ASCE)0733–9429(2005)131:10(898) [Accessed 29th November 2016].

Babu, P. & Stoica, P. (2010) Spectral analysis of nonuniformly sampled data – a review. *Digit Signal Process.* [Online] 20(2), 359–378. Available from: doi:10.1016/j.dsp.2009.06.019 [Accessed 29th November 2016].

Balachandar, R. & Bhuiyan, F. (2007) High-order moments of velocity fluctuations in an open-channel flow with large bottom roughness. *J Hydraul Eng.*, 133(1), 77–87.

Barber, D. (2012) *Bayesian Reasoning and Machine Learning.* 1st edition. New York, Cambridge University Press.

Barnett, V. & Lewis, T. (1988) *Outliers in Statistical Data.* 2nd edition. New York, NY, John Wiley & Sons.

Bates, D.M. & Watts, D.G. (1988) *Nonlinear Regression Analysis and its Applications.* New York, John Wiley & Sons.

Beauchamp, J.J. & Olson, J.S. (1973) Corrections for bias in regression estimates after logarithmic transformation. *Ecology.* 54(6), 1403–1407.

Bendat, J.S. & Piersol, A.G. (2000). *Random Data: Analysis and Measurement Procedures.* New York, Wiley-Interscience.

Benedict, L.H., Nobach, H., & Tropea, C. (2000) Estimation of turbulent velocity spectra from laser Doppler data. *Meas Sci Technol.*, 11(8), 1089–1104.

Bennett, S.J., Bridge, J.S., & Best, J.L. (1998) Fluid and sediment dynamics of upper stage plane beds. *J Geophys Res.*, 103(C1), 1239–1274.

Benson, M.A. (1965) Spurious correlation in hydraulics and hydrology. *ASCE J Hydraul Div.*, 91(4), 35–42.

Berkooz, G., Holmes, P., & Lumley, J.L. (1993) The proper orthogonal decomposition in the analysis of turbulent flows. *Annu Rev Fluid Mech.*, 25, 539–575.

Best, J. & Kostaschuk, R. (2002) An experimental study of turbulent flow over a low-angle dune. *J Geophys Res, Oceans.*, 107(C9), 18-1-18-19.

Bhattacharya, B., Price, R.K., & Solomatine, D.P. (2007) Machine learning approach to modeling sediment transport. *J Hydraul Eng*. [Online] 133(4), 440–450. Available from: doi:10.1061/(ASCE)0733–9429(2007)133:4(440) [Accessed 29th November, 2016].

Biedenharn, D., Watson, C., & Thorne, C. (2008) Fundamentals of fluvial geomorphology. In M. H. Garcia (ed.) *Sedimentation Engineering: Processes, Measurements, Modeling, and Practice*. ASCE Manual of Practice 110(6), 355–386.

Bishop, C.M. (2006) *Pattern Recognition and Machine Learning*. 1st edition. New York, NY, Springer.

Blackwelder, R.F. & Kaplan, R.E. (1976) On the wall structure of the turbulent boundary layer. *J Fluid Mech*., 76(1), 89–112.

Blanckaert, K. & Graf, W. (2001) Mean flow and turbulence in open-channel bend. *J Hydraul Eng*., 127(10), 835–847.

Blanckaert, K. & Graf, W. (2004) Momentum transport in sharp open-channel bends. *J Hydraul Eng*., 130(3), 186–198.

Bolstad, W.M. (2007) *Introduction to Bayesian Statistics*. Hoboken, NJ, John Wiley & Sons.

Boulesteix, A.L. & Schmid, M. (2014) Machine learning versus statistical modeling. *Biometrical Journal*., 56(4), 588–593.

Breiman, L. (2001) Statistical modeling: The two cultures. *Stat Sci*., 16(3), 199–231.

Breiman, L., Friedman, J.H., Olshen, R.A., & Stone, C.J. (1984) *Classification and Regression Trees*. Belmont, CA, Wadsworth.

Brown, P. (1994). *Measurement, regression, and calibration*, Oxford, U. K., Oxford University Press.

Browne, M.W. (2000) Cross-validation methods. *J Math Psychol*., 44(1), 108–132.

Brownlie, W.R. (1981) *Compilation of alluvial channel data: Laboratory and field*. In: W. M. Keck (ed.) Laboratory of Hydraulics and Water Resources, California Institute of Technology, Pasadena, CA. Record number: KH-R-43B.

Brownlie, W.R. (1983) Flow depth in sand-bed channels. *J Hydraul Eng*., 109(7), 959–990.

Cea, L., Puertas, J., & Pena, L. (2007) Velocity measurements on highly turbulent free surface flow using ADV. *Exp Fluids*., 42(3), 333–348.

Chakraborty, P., Balachandar, S., & Adrian, R.J. (2005) On the relationships between local vortex identification schemes. *J Fluid Mech*., 535, 189–214.

Citriniti, J. & George, W. (2000) Reconstruction of the global velocity field in the axisymmetric mixing layer utilizing the proper orthogonal decomposition. *J Fluid Mech*., 418, 137–166.

Chen, H.-L., Hondzo, M., & Rao, A.R. (2002) Segmentation of temperature microstructure. *J Geophys Res*., 107(C12), 3211.

Choi, J., Troy, C.D., Hsieh, T.-C., Hawley, N., & McCormick, M.J. (2012) A year of internal of Poincare waves in southern Lake Michigan. *J Geophys Res*. [Online] 117(C07014). Available from: doi:10.1029/2012JC007984 [Accessed 29th November 2016].

Coleman, N.L. (1981) Velocity profiles with suspended sediment. *J Hydraul Res*., 19(3), 211–229.

Coleman, H.W. & Steele, W.G. (2009) *Experimentation, Validation, and Uncertainty Analysis for Engineers*. 3rd edition. [Online] Hoboken, NJ, John Wiley & Sons. Available from: doi: 10.1002/9780470485682 [Accessed 29th November 2016].

Cressie, N.A.C. (1993) *Statistics for Spatial Data*. Revised edition. New York, Wiley.

Cunningham, R. & Lyn, D. (2016) Laboratory study of bendway weirs as a bank erosion countermeasure. *J Hydraul Eng*., 142(6), doi:10.1061/(ASCE)HY.1943-7900.0001117, [Accessed 6th March 2017].

Davison, A.C. & Hinkley, D.V. (1997) *Bootstrap Methods and their Applications*. Cambridge, Cambridge University Press.

Ding, C. & He, X. (2004) K-means clustering via principal component analysis. In:*Proceedings of the 21st International Conference on machine learning*, 4–8 July 2004, Banff, Canada. New York, ACM. p. 29.

Dogan, E., Tripathi, S., Lyn, D.A., & Govindaraju, R.S. (2009) From flumes to rivers: Can sediment transport in natural alluvial channels be predicted from observations at the laboratory scale? *Water Resour Res.*, 45(W08433), doi:10.1029/2008WR007637, [Accessed 6th March 2017].

Doroudian, B., Bagherimiyab, F., & Lemmin, U. (2010) Improving the accuracy of four-receiver acoustic Doppler velocimeter (ADV) measurements in turbulent boundary layer flows. *Limnol Oceanogr-Meth.*, 8, 575–591.

Draper, N.R. & Smith, H. (1998) *Applied Regression Analysis*. 3rd edition. New Yoirk, NY, Wiley.

Duchon, C. & Hale, R. (2011) *Time Series Analysis in Meteorology and Climatology: An Introduction*. New York, NY, Wiley-Blackwell.

Durão, D.F.G., Heitor, M.V., & Pereira, J.C.F. (1988) Measurements of turbulent and periodic flows around a square cross-section cylinder. *Exp Fluids.*, 6(5), 298–304.

Dwivedi, A., Melville, B., & Shamseldin, A. (2010) Hydrodynamic forces generated on a spherical sediment particle during entrainment. *J Hydraul Eng.*, 136(10), 756–769.

Eberly, S., Swall, J., Holland, D., Cox, B., & Baldridge, E. (2004) *Developing Spatially Interpolated Surfaces and Estimating Uncertainty*. U. S. Environmental Protection Agency, Research Triangle Park, NC. Report number: EPA-454/R-04-004.

Fernandez, R., Best, J., & Lopez, F. (2006) Mean flow, turbulence structure and bedform superimposition across the ripple-dune transition. *Water Resour Res.*, 42(5), 948–963.

Fieller, E.C. (1954) Some problems in interval estimation. *J Roy Stat Soc B.*, 16(2), 175–185.

Foucaut, J.M. & Stanislas, M. (2002) Some considerations on the accuracy and frequency response of some derivative filters applied to particle image velocimetry vector fields. *Meas Sci Technol.*, 13(7), 1058–1071.

Frias, C.E. & Abad, J.D. (2013) Mean and turbulent flow structure during the amalgamation process in fluvial bed forms. *Water Resour Res.*, 49(10), 6548–6560.

Fukuda, K., Balachandar, R. & Barron, R. (2014) *Flow structures in the wake of a bluff body*. University of Windsor, Canada. Technical Report 2014/7/10.

Garcia, M.H. (2008) Sediment transport and morphodynamics. In Garcia, M. H. (ed.) *Sedimentation Engineering: Processes, Measurements, Modeling, and Practice*, Reston, VA, ASCE Press. ASCE Manual of Practice 110, Chapter 2, pp. 21–163

Garcia, C.M., Cantero, M.I., Niño, Y., & Garcia, M.H. (2005) Turbulence measurements with acoustic doppler velocimeters. *J Hydraul Eng.*, 131(12), 1062–1073.

Garcia, C.M. & Garcia, M.H. (2006) Characterization of flow turbulence in large-scale bubble-plume experiments. *Exp Fluids.*, 41(1), 91–101.

Garcia, C.M. Jackson, P.R., & Garcia, M.H. (2006) Confidence intervals in the determination of turbulence parameters. *Exp Fluids.*, 40(4), 514–522.

Gelman, A. (2008) Objections to bayesian statistics. *Bayesian Anal.*, 3(3), 445–449.

Geman, S. & Geman, D. (1984) Stochastic relaxation, Gibbs distributions, and the Bayesian restoration of images. *IEEE T Pattern Anal.*, 6(1), 721–741.

Gibbons, J.D. & Chakraborti, S. (2003) *Nonparametric Statistical Inference*. 4th edition. New York, Marcel Dekker.

Gill, J. (2014) *Bayesian Methods: A Social and Behavioral Sciences Approach*. 3rd edition. Boca Raton, CRC Press.

Goel, A. & Pal, M. (2009) Application of support vector machines in scour prediction on grade-control structures. *Eng Appl Artif Intel.*, 22(2), 216–223.

Goring, D.G. & Nikora, V.I. (2002) Despiking acoustic Doppler velocimeter data. *J Hydraul Eng.*, 121(1), 117–126.

Günther, A. & von Rohr, P.R. (2003) Large-scale structures in a developed flow over a wavy wall. *J Fluid Mech.*, 478, 257–285.

Gurka, R., Liberzon, A., & Hetsroni, G. (2006) POD of vorticity fields: A method for spatial characterization of coherent structures. *Int J Heat Fluid Fl.*, 27(3), 416–423.

Hahn, E. & Lyn, D. (2010) Anomalous Contraction Scour? Vertical-Contraction Case. *ASCE J Hydraul Eng.*, 136(2), 137–141.

Hastie, T., Tibshirani, R., & Friedman, J. (2009) *The Elements of Statistical Learning: Data mining, Inference and Prediction.* 2nd edition. New York, NY, Springer.

Heidari, M. (2016) *Wake characteristics of single and tandem emergent cylinders in shallow open channel flow.* Ph. D. dissertation. Windsor, Ontario, Canada, University of Windsor.

Helsel, D.R. & Hirsch, R.M. (2002) *Statistical methods in water resources.* Techniques of Water-Resources Investigations of the United States Geological Survey Book 4, Hydrologic Analysis and Interpretation Chapter A3.

Holmes, P., Lumley, J.L., & Berkooz, G. (1996) *Turbulence,Coherent Structures,Symmetry and Dynamical Systems.* New York, NY, Cambridge University Press.

Huang, N.E., Shen, Z., Long, S.R., Wu, M.C., Shih, H.H., Zheng, Q., Yen, N.C., Tung, C.C., & Liu, H.H. (1998) The empirical mode decomposition and the hilbert spectrum for nonlinear and nonstationary time series analysis. *P R Soc London A.*, 454(1971): 903–995.

Huang, N.E. & Wu, Z. (2008) A review on Hilbert-Huang transform: Method and its applications to geophysical studies. *Rev Geophys.*[Online] 46(2). Available from: doi:10.1029/2007RG000228 [Accessed 29th November 2016].

Huberty, C.J. (1994) *Applied Discriminant Analysis.* New York, John Wiley & Sons.

Hunt, J.C.R., Wray, A., & Moin, P. (1988) Eddies, streams and convergence zones in turbulent flows. *Proceedings of the Summer Program, 1988*, Center for Turbulence Research. Palo Alto, CA, Stanford University.

Hurther, D., Lemmin, U., & Terray, E.A. (2007) Turbulent transport in the outer region of rough-wall open-channel flows: The contribution of large coherent shear stress structures (LC3S). *JFluid Mech.*, 574, 465–493.

Hwang, J.T. (1995) Fieller's problem and resampling techniques. *Stat Sinica.*, 5(1), 161–171.

Isaaks, E.H. & Srivastava, R.M. (1989) *Applied Geostatistics.* New York, Oxford University Press.

Islam, M.R. & Zhu, D. (2013) Kernel density–based algorithm for despiking ADV data. *J Hydraul Eng.*, 139(7), 785–793.

Jain, A.K. (2010) Data clustering: 50 years beyond k-means. *Pattern Recogn Lett.*, 31(8), 651–666.

Jamieson, E., Rennie, C., & Townsend, R. (2013a) 3D flow and sediment dynamics in a laboratory channel bend with and without stream barbs. *J Hydraul Eng.*, 139(2), 154–166.

Jamieson, E., Rennie, C., & Townsend, R. (2013b) Turbulence and vorticity in a laboratory channel bend at equilibrium clear-water scour with and without stream barbs. *J Hydraul Eng.*, 139(3), 259–268.

James, G., Witten, D., Hastie, T., & Tibshirani, R. (2013) *An Introduction to Statistical Learning, with Applications in R.* New York, NY, Springer.

Jaynes, E.T. (2003) *Probability Theory: The Logic of Science.* Cambridge, Cambridge University Press.

Jeong, J. & Hussain, F. (1995) On the identification of a vortex. *J Fluid Mech.*, 285, 69–94.

Joanes, D.N. & Gill, C.A. (1998) Comparing measures of sample skewness and kurtosis. *J Roy Stat Soc D (The Statistician).*, 47(1), 183–189.

Julien, P.Y. & Wargadalam, J. (1995) Alluvial channel geometry: theory and applications. *J Hydraul Eng.*, 121(4), 312–325.

Keijzer, M. & Babovic, V. (1999) Dimensionally aware genetic programming. In *Proceedings of the Genetic and Evolutionary Computation Conference.* Vol 2, 1069–1076.

Kenney, B.C. (1982) Beware of spurious self-correlation. *Water Resour Res.*, 18(4), 1041–1048.

Khorsandi, B., Mydlarski, L., & Gaskin, S. (2012) Noise in turbulence measurements using acoustic Doppler velocimetry. *J Hydraul Eng.*, 138(10), 829–838.

Kho, S., Baker, C., & Hoxey, R. (2002) POD/ARMA reconstruction of the surface pressure field around a low rise structure. *J Wind Eng Ind Aerod.*, 90(12–15), 1831–1842.

Kirkil, G. & Constantinescu, G. (2015) Effects of cylinder Reynolds number on the turbulent horseshoe vortex system and near wake of a surface-mounted circular cylinder. *Phys Fluids.* [Online] 27, 075102. Available from:doi:10.1063/1.4923063 [Accessed 29th November 2016].

Kottegoda, N.T & Rosso, R. (1997) *Statistics, Probability, and Reliability for Civil and Environmental Engineers*. New York, McGraw-Hill.

Kruse, N., Günther, A., & von Rohr, P.R. (2003) Dynamics of large-scale structures in turbulent flow over a wavy wall. *J Fluid Mech.*, 485, 87–96.

Kunsch, H.R. (1989) The jackknife and the bootstrap for general stationary observations. *Ann Stat.*, 17(3), 1217–1241.

Lacey, G. (1929) Stable channel in alluvium. *Minutes of the P I Civil Eng.*, 229, 259–285.

Le Coz, J., Renard, B., Bonnifait, L., Branger, F., & Le Boursicaud, R. (2014) Combining hydraulic knowledge and uncertain gaugings in the estimation of hydrometric rating curves: a Bayesian approach. *J Hydrol.*, 509, 573–587.

Liang, Y.C., Lee, H.P., Lim, S.P., Lin, W.Z., Lee, K.H., & Wu, C.G. (2002) Proper orthogonal decomposition and its applications – Part 1: Theory. *J Sound Vib.* [Online] 252(3), 527–544. Available from: doi:10.1006/jsvi.2001.4041 [Accessed 29th November 2016].

Lin, C., Lai, W.J., & Chang, K.A. (2003) Simultaneous Particle Image Velocimetry and Laser Doppler Velocimetry measurements of periodical oscillatory horseshoe vortex system near square cylinder-base plate juncture. *J Eng Mech.*, 129(10), 1173–1188.

Liriano, S.L & Day, R.A. (2001) Prediction of scour depth at culvert outlets using neural networks. *J Hydroinform.*, 3, 231–238.

Liu, Z., Adrian, R.J., & Hanratty, T.J. (1994) Reynolds number similarity of orthogonal decomposition of the outer layer of turbulent wall flow. *Phys Fluids.*, 6(8), 2815–2819.

Lomb, N.R. (1976) Least-squares frequency analysis of unequally spaced data. *Astrophys Space Sci.* [Online]39(2), 447–462. Available from: doi:10.1007/BF00648343 [Accessed 29th November 2016].

Lu, S.S & Willmarth, W.W. (1973) Measurements of the structure of the Reynolds stress in a turbulent boundary layer. *J Fluid Mech.*, 60, 481–511.

Luketina, D.A. & Jörg Imberger, J. (2001) Determining turbulent kinetic energy dissipation from Batchelor curve fitting. *J Atmos Ocean Tech.*, 18, 100–113.

Lumley, J.L. (1967) The structure of inhomogenous turbulence. In: Yaglom, A. M. & Tatarski, V. I. (eds.) *Atmospheric Turbulence and Wave Propagation*. Moscow, Nauka. pp. 166–178.

Lumley, J.L & Panofsky, H.A. (1964) *The Structure of Atmospheric Turbulence*. New York, NY, Wiley.

Lunn, D., Spiegelhalter, D., Thomas, A., & Best, N. (2009) The BUGS project: Evolution, critique and future directions. *Stat Med.*, 28(25), 3049–3067.

Lyn, D.A. (1987) *Turbulence and turbulent transport in sediment-laden open-channel flows*. Ph. D. thesis. Pasadena, CA, California Institute of Technology.

Lyn, D.A. (1988) A similarity approach to turbulent sediment-laden flows in open channels. *J Fluid Mech.*, 193, 1–26.

Lyn, D.A., Einav, S., Rodi, W., & Park, J.H. (1995) A laser-Doppler velocimetry study of ensemble-averaged characteristics of the turbulent near wake of a square cylinder. *J Fluid Mech.*, 304, 285–319.

Lyn, D. A. (2008) Pressure-flow scour: A reexamination of the HEC-18 equation. *J Hydraul Eng.*, 134(7), 1015–1020.

MacKay, D.J.C. (1996) Bayesian methods for backpropagation networks. In: Domany, E., van Hemmen, J. L. & Schulten, K. (eds.) *Models of Neural Networks III*. New York, NY, Springer. pp.211–254.

Maindonald, J. & Braun, W.J. (2013) *Data Analysis and Graphics using R*. Cambridge, Cambridge University Press.

Mann, H.B. (1945) Nonparametric tests against trend. *Econometrica*. 13(3), 245–259.

Mansanarez, V., Le Coz, J., Renard, B., Vauchel, P., Pierrefeu, G., & Lang, M. (2016) Bayesian analysis of stage-fall-discharge rating curves and their uncertainties. *Water Resour Res.*, 52, 7424–7443.

McCuen, R.H. (2003) *Modeling Hydrologic Change: Statistical Methods*. Boca Raton, FL, CRC Press.

Mendenhall, W. & Sincich, T. (1995) *Statistics for Engineering and the Sciences*. 4th edition. Englewood Cliffs, NJ, Prentice-Hall.

Metropolis, N., Rosenbluth, A.W., Rosenbluth, M.N., Teller, A.H., & Teller, E. (1953) Equation of state calculations by fast computing machines. *J Chem Phys.*, 21(6), 1087–1092.

Montgomery, D.C., Peck, E.A., & Vining, G.G. (2006) *Introduction to Linear Regression Analysis*. Hoboken, NJ, Wiley.

Murphy, K.P. (2012) *Machine Learning: A Probabilistic Perspective*. 1st edition. Cambridge, MA, MIT Press.

Muste, M. & Patel, V. (1997) Velocity profiles for particles and liquid in open-channel flow with suspended sediment. *J Hydraul Eng.*, 123 (9), 742–751.

Nagy, H.M., Watanabe, K., & Hirano, M. (2002) Prediction of sediment load concentration in rivers using Artificial Neural Network model. *J Hydraul Eng.*, 128(6), 588–595.

Neal, R.M. (1996) *Bayesian Learning for Neural Networks*. New York, NY, Springer.

Nezu, I. (2005) Open-channel flow turbulence and its research prospect in the 21st century. *J Hydraul Eng.*, 131(4), 229–246.

Olea, R.A. (1999) *Geostatistics for Engineers and Earth Scientists*. Boston, MA, Springer.

O'Neill, P.L., Nicolaides, D., Honnery, D., & Soria, J. (2004) Autocorrelation functions and the determination of integral length with reference to experimental and numerical data. [Presentation] *Proceedings of 15th Australasian Fluid Mechanics Conference*, University of Sydney, Sydney, Australia, 13–17 December 2004.

Oppenheim, A.V., Schafer, R.W., & Buck, J.R. (1999) *Discrete-Time Signal Processing*. 2nd edition. Englewood Cliffs, NJ, Prentice-Hall.

Osborne, C. (1993) Statistical calibration: A review. *Int Stat Rev/Revue Internationale de Statistique.*, 59(3)(Dec., 1991), 309–336.

Panik, M.J. (2005) *Advanced Statistics from an Elementary Point of View*. Burlington, MA, Elsevier Academic Press.

Parker, G., Wilcock, P.R., Paola, C., Dietrich, W.E., & Pitlick, J. (2007) Physical basis for quasi-universal relations describing bankfull hydraulic geometry of single-threaded gravel-bed rivers. *Water Resour Res.*, 112(F04005), doi:10.1029/2006JF000549 [Accessed 6th March 2017].

Parsheh, M., Sotiropoulos, F., & Porté-Agel, F. (2010) Estimation of power spectra of acoustic-Doppler velocimetry data contaminated with intermittent spikes. *J Hydraul Eng.*, 136(6), 368–378.

Percival, D.B & Walden, A.T. (1993) *Spectral Analysis for Physical Applications; Multitaper and Conventional Univariate Techniques*. Cambridge, Cambridge University Press.

Politis, D.N & Romano, J.P. (1994) The stationary bootstrap. *J Am Stat Assoc.*, 89(428), 1303–1313.

Pope, S.B. (2000) *Turbulent Flows*. Cambridge, U.K., Cambridge University Press.

Press, W.H., Teukolsky, S.A., Vettering, W.T., & Flannery, B.R. (1992) *Numerical Recipes*. Cambridge, Cambridge University Press.

Priestley, M.B. (1981) *Spectral Analysis and Time Series*. London, Academic Press.

Priestley, M.B. & Subba Rao, T. (1969) A test for non-stationarity of time-series. *J Roy Stat Soc B (Methodological).*, 31(1), 140–149.

Raffel, M., Willert, C.E., Wereley, S.T., & Kompenhans, J. (2007) *Particle Image Velocimetry: A Practical Guide.* 2nd edition. Berlin, Springer-Verlag.

Roussinova, V., Shinneeb, A., & Balachandar, R. (2010) Investigation of Fluid Structures in a Smooth Open-Channel Flow Using Proper Orthogonal Decomposition, *J Hydraul Eng.*, 136 (3), 143–154.

Roweis, S.T. (1998) EM algorithms for PCA and SPCA. In: Jordan, M.I., Kearns, M.J., & Solla, S.A. (eds.) *Advances in Neural Information Processing Systems.* Cambridge, MA, MIT Press. pp. 626–632.

Ruark, M., Niemann, J., Greimann, B., & Arabi, M. (2011) Method for assessing impacts of parameter uncertainty in sediment transport modeling applications. *ASCE J Hydraul Eng.*, 137(6), 623–636.

Ryan, T.P. (2009) *Modern Regression Methods.* Hoboken, NJ, John Wiley & Sons.

Scargle, J.D. (1982) Studies in astrophysical time series analysis, II: Statistical aspects of spectral analysis of unevenly spaced data. *Astrophys J.*, 263, 835–853.

Schemper, T.J & Admiraal, D.M. (2002) An examination of the application of acoustic Doppler current profiler measurements in a wide channel of uniform depth for turbulence calculations. [Online] *Proceedings of ASCE/EWRI Hydraulic Measurements and Experimental Methods (HMEM)*, Estes Park, CO. Available from: doi: 10.1061/40655(2002)75 [Accessed 29th November 2016].

Schmelter, M.L., Hooten, M.B., & Stevens, D.K. (2011) Bayesian sediment transport model for unisize bed load. *Water Resour Res.* [Online] 47(11). Available from: doi:10.1029/2011-WR010754 [Accessed 29th November 016].

Scott, D.W. (2015) *Multivariate Density Estimation: Theory, Practice, and Visualization.* 2nd edition. New York, NY, Wiley.

Sengupta, T.K., Singh, N., & Suman, V.K. (2010) Dynamical system approach to instability of flow past a circular cylinder. *J Fluid Mech.*, 656, 82–115.

Sheng, J., Malkiel, E., & Katz, J. (2009) Buffer layer structures associated with extreme wall stress events in a smooth wall turbulent boundary layer. *J Fluid Mech.*, 633, 17–60.

Shinneeb, A.M. (2006) *Confinement effects in shallow water jets.* Ph. D. dissertation. University of Saskatchewan, Saskatoon, Canada.

Shmueli, G. (2010) To explain or to predict? *Stat Sci.*, 25(3), 289–310.

Singha, A., Shinneeb, A.-M., & Balachandar, R. (2009) PIV-POD investigation of the wake of a sharpedged flat bluff body immersed in a shallow channel flow. *J Fluids Eng.*, 131, 021202-1–021202-12.

Sirovich, L. (1987). Turbulence and the dynamics of coherent structures, Parts I–III, *Quart. J. Appl. Math.* 45, 561–590.

Soga, C. & Rehmann, C. (2004) Dissipation of turbulent kinetic energy near a bubble plume. *J Hydraul Eng.*, 130(5), 441–449.

Stearns, S. D. (2003) *Digital Signal Processing with Examples in MATLAB.* Boca Raton, FL, CRC Press.

Stow, C.A., Reckhow, K.H., & Qian, S.S. (2006) A Bayesian approach to retransformation bias in transformed regression. *Ecology.*, 87(6), 1472–1477.

Tandalam, A., Balachandar, R., & Barron, R. (2010) Reynolds number effects on the near-exit region of turbulent jets. *J Hydraul Eng.*, 136(9), 633–641.

Tavoularis, S. (2005) *Measurements in Fluid Mechanics.* Cambridge, UK, Cambridge University Press.

Tayfur, G., Karimi, Y., & Singh, V.P. (2013) Principle Component Analysis in conjunction with data-driven methods for sediment load prediction. *Water Resour Man.* [Online] 27, 2541–2554. Available from: doi:10.1007/s11269-013-0302-7 [Accessed 29th November 2016].

Tennekes, H., & Lumley, J. (1972) *A First Course in Turbulence*. Cambridge, MA, MIT Press.

Thomson, R.E & Emery, W.J. (2014) Data Analysis Methods in Physical Oceanography. 3rd edition. New York, NY, Elsevier.

Thomson, D.J. (1982) Spectrum estimation and harmonic analysis. *P IEEE.*, 70, 1055–1096.

Tibshirani, R. (1996) A comparison of some error estimates for neural network models. *Neural Comput.*, 8(1), 152–163.

Tipping, M.E. (2001) Sparse Bayesian learning and the relevance vector machine. *J Mach Learn Res.*, 1(3), 211–244.

Tipping, M.E. (2004) Bayesian inference: An introduction to principles and practice in machine learning. In: Bousquet, O., von Luxburg, U., & RÄatsch, G. (eds.) *Lect Notes Artif Int.*, New York, NY, Springer. 41–62.

Tipping, M.E & Bishop, C.M. (1999) Probabilistic principal component analysis. *J Roy Stat Soc B.*, 61(3), 611–622.

Tukey, J.W. (1977) *Exploratory Data Analysis*. 1st edition. Readig, MA, Addison-Wesley..

Umbrell, E.R., Young, G.K., Stein, S.M., & Jones, J.S. (1998) Clear-water contraction scour under bridges in pressure flow. *J Hydraul Eng.*, 124(2), 236–240.

van Rijn, L.C. (1984) Sediment transport: Part III. Bed forms and alluvial roughness. *J Hydraul Eng.*, 110(12), 1733–1754.

Vanoni, V.A. (1946) Transportation of suspended sediment by water. *T Am Soc Civ Eng.*, 111, 67–133.

Venditti, J.G & Bennett, S.J. (2000) Spectral analysis of turbulent flow and suspended sediment transport over fixed dunes. *J Geophys Res-Oceans.*, 105(C9), 22035–22047.

Voulgaris, G. & Trowbridge, J. (1998) Evaluation of the acoustic Doppler velocimeter (ADV) for turbulence measurements. *J Atmos Ocean Tech.*, 15, 272–288.

Wahl, T.L. (2003) Discussion of "Despiking acoustic doppler velocimeter data" by Derek G. Goring and Vladimir I. Nikora. *J Hydraul Eng.*, 129(6), 484–487.

Wallace, J.M., Eckelmann, H., & Brodkey, R.S. (1972) The wall region in turbulent shear flow. *J Fluid Mech.*, 54(1), 39–48.

Walpole, R.E., Myers, R.H., Myers, S.L., & Ye, K. (2007) *Probability and Statistics for Engineers and Scientists*. 8th edition. Englewood Cliffs, NJ, Pearson Prentice Hall.

Webster, R., & Oliver, M.A. (2007) *Geostatistics for Environmental Scientists*. New York, NY, Wiley.

Welch, P.D. (1967) The use of fast fourier transform for the estimation of power spectra: A method based on time averaging over short, modified periodograms. *IEEE T Acoust Speech.* AU-15, 70–73.

Westerweel, J. & Scarano, F. (2005) Universal outlier detection for PIV data. *Exp Fluids.*, 39(6), 1096–1100.

Wilkerson, G.V & Parker, G. (2011) Physical basis for quasi-universal relations describing bankfull hydraulic geometry of sand-bed rivers. *J Hydraul Eng.*, 137(7), 739–753.

Willmarth, W.W & Lu, S.S. (1972) Structure of the Reynolds stress near the wall. *J Fluid Mech.*, 55(1), 65–92.

Yalin, M.S & Kamphuis, J. (1971) Theory of dimensions and spurious correlation. *J Hydraul Res.*, 9(2), 249–265.

Yalin, M.S & Karahan, E. (1979) Inception of sediment transport. *J Hydraul Div, ASCE.*, 105 (11),1433–1443.

Zhong, Q., Li, D., Chen, Q., & Wang, X. (2015) Coherent structures and their interactions in smooth open channel flows. *Environ Fluid Mech.*, 15(3), 653–672.

Zhou, J., Adrian, R.J., Balachandar, S., & Kendall, T.M. (1999) Mechanisms for generating coherent packets of hairpin vortices in channel flow. *J Fluid Mech.*, 387, 353–396.

Chapter 7

Uncertainty Analysis for Hydraulic Measurements

7.1 INTRODUCTION

The results from hydraulic studies involving measurements often require a statement regarding the quality of the measurements produced. Usually the quality of the measurements is expressed quantitatively in terms of the uncertainty associated with them, therefore the terms "data quality" and "uncertainty" are often used interchangeably. Hydraulic measurements are affected by the randomness of the observed physical phenomena and by errors occurring during various phases of the experimental process. Since randomness and experimental error are unavoidable, margins of uncertainty usually surround the true value of a measured result. The best way to define the "true" value of a measurement is to estimate the uncertainty associated with the measurement. Uncertainty analysis (UA) is used for this purpose.

UA is a practical procedure for rigorously addressing the variability and the error sources affecting the measurement process. The primary product of UA is an interval around a determined result, within which the true value is thought to lie; the interval is determined with a certain degree of confidence (Coleman & Steele, 1995). In addition to providing confidence in the reported measurement, UA informs on the efficiency of an experiment, identifying elements of the experimental procedure that have the greatest potential for improvement. UA can be utilized in all phases of an experiment, including initial planning, design, debugging, execution, and post-processing. It is important that there be reasonably consistent approaches for conducting UA across all phases of experimentation as well as across similar experiments so that the quality of experimental findings can be reliably assessed and compared to findings reported from different studies.

In most situations, an experiment involves measuring a number of individual variables that are subsequently used in a functional relationship referred to as the data reduction equation (DRE). A first step in formal UA is estimation of the elemental sources of uncertainty associated with each of the variables in the DRE (whether they were induced from a direct measurement or not). Determined next are the effects of these estimated uncertainties on the final result produced by a DRE. While the methods for estimating elemental sources of uncertainty are similar for a wide range of scientific or engineering fields (*i.e.*, using statistical analysis; reliance on previous experience and expert opinion; or reference to manufacturer specifications), the methods used to determine how those sources of uncertainty are accounted for in the final result may differ widely (*e.g.*, TCHME, 2003).

7.1.1 Standardized methods for uncertainty analysis

For several decades, scientists and engineers have debated the appropriate procedure for conducting UA (*e.g.*, Abernethy & Ringhiser, 1985). These days, opinions regarding UA procedures have considerably converged toward an overall consensus. Early efforts in conceptualizing UA started in the 1950s, when the American Society of Mechanical Engineers (ASME) initiated an effort based on the hallmark paper by Kline & McClintock (1953). After many years of disagreement, ASME's efforts finally achieved consensus in 1986 with the adoption of ASME-PTC 19.1 (ASME, 1986). In 1978, the foremost authority in metrology, the Comité International des Poids et Mesures (CIPM) requested the Bureau International des Poids et Mesures (BIPM) to search for an international consensus on the expression of uncertainty in measurements. For this purpose, BIPM and the International Standard Organization (ISO) assembled a joint group of international experts from at least five other organizations. This multi-organization collaboration has produced the "Guide to Expression of Uncertainty in Measurement" (GUM, 1993). This UA standard was republished with minor modifications in 1995, and is now seen as the first internationally accepted guideline for conducting uncertainty analysis, including for hydraulic experiments. In 1997, the Joint Committee for Guides in Metrology (JCGM) was formed to maintain and, when necessary, update GUM (freely available from the JCGM website).

GUM's framework is based on mathematical statistics that basically address how the propagation of the elemental sources of errors influences final results obtained with a DRE. GUM provides general rules for evaluating uncertainty in measurement, rather than providing detailed and specific instructions linked only to a specific field of measurement. The guide differs from previous UA methodologies in regards to terminology, classification of errors, and procedures (Herschy, 2002). Studies conducted to compare available UA methodologies (*i.e.*, Steele *et al.*, 1994; Coleman & Steele, 1995) conclude that the GUM framework is more robust and mathematically firmer than alternative methods, such as the standard methods proposed by ASME and American Institute of Aeronautics and Astronautics (AIAA) (Taylor & Kuyatt, 1994; NF ENV, 1999). Recently, ASME and AIAA standards have been harmonized with GUM terminology and procedures through new versions [ASME Standard PTC 19.1 (1998, 2005) and AIAA Standard S-071 (1999)]. GUM terminology and procedures have been widely adopted by national and regional metrology and related organizations in Europe and the USA (Taylor & Kuyatt, 1994; ANSI, 1997). Several scientific and engineering areas (*e.g.*, NF ENV, 1999; ISO, 2005, and UKAS, 2007) have also embraced the guidelines in an attempt to use a uniform method of expressing measurement uncertainty.

Hydraulic engineers have continuously sought to improve procedures for assessing measurement errors, but, until recently, have not agreed on an assessment framework. Various hydrometric groups have tested methodologies developed by other communities, such as GUM (1993) or AIAA (1995), and concluded that they can be successfully applied for assessing measurement uncertainty in laboratory and field hydrologic and hydraulic measurements (Bertrand-Krajewski & Bardin, 2002; Muste & Stern,

2000; Muste *et al.*, 2004; Kim *et al.*, 2005; Kim *et al.*, 2007; Gonzalez-Castro & Muste, 2007; UNESCO, 2007). More such initiatives have been launched recently. The Technical Committee on Hydraulic Measurements and Experimentation (TCHME) of the American Society of Civil Engineers' Environmental & Water Resources Institute formed a dedicated Task Committee on uncertainty estimation in hydraulic engineering (Wahlin *et al.*, 2005). A similar effort was launched in 2004 by UNESCO's International Hydrology Program (UNESCO, 2007). These last initiatives indicate that it might be more practical to adopt GUM for conducting UA rather than developing specialized standards for the hydraulic and hydrometry communities. The World Meteorological Organization's Committee on Hydrology has recently recommended the adoption of the GUM methodology for the routine national hydrologic services (Muste & Lee, 2014; HUG, 2007; WMO, 2017).

Selecting and implementing a standard UA method for a professional community is a process that requires long-term commitment and effort. According to Thomas (2002), the following steps are essential in the adoption of a standard: evaluation, prioritization, implementation, planning, accessing standards, getting the standards used, and maintaining the drive. The present chapter aims at providing the reader with an overview of judiciously selected standards and demonstrating how such standards can be successfully applied for assessment of the measurement uncertainty in hydraulic studies.

7.2 CONCEPTS AND TERMINOLOGY

This section introduces basic principles and concepts pertinent to uncertainty analysis. It is assumed herein that the reader has basic knowledge of statistics concepts used in engineering standards, such as probability distribution function, mean, and standard deviation of a sample population. Details on definitions and implementation conditions can be found in Section 6.2, Volume I. In this chapter, we use the terminology and definitions prescribed by GUM, as published in JCGM (2007). A useful complementary resource for terminology is JCGM (2009).

7.2.1 Measurement and uncertainty analysis

The objective of a measurement is to determine the value of the particular quantity being measured (the measurand). The accuracy of a measurement indicates the closeness of agreement between the result of a measurement and the true value of the measurand. Measurements can be regarded as direct (such as the weighing) or, as is often the case, a combination of several measured values used in an analytical relationship to obtain the value of a measurand (*e.g.*, measurement of volume and time to obtain discharge). A measurement has imperfections such that if the quantities were to be measured several times, in the same way and circumstances, a different value would be obtained; this outcome is usually referred to as repeatability uncertainty. The dispersion of the indication values would relate to how well the measurement is made. The dispersion and the number of indication values would provide information relating to the average value as an estimate or approximation of the true (but unknown) value of the measurand. Consequently, this estimate is complete only when accompanied by a statement of its uncertainty.

Uncertainty analysis is a rigorous methodology for determining uncertainties of measurement results using statistical and engineering concepts. The measurement process for a specific measurand does not include just the measurement system used to produce the final result, but also the effects induced by the experimental facility, operator actions, and other measurement environment influences (*e.g.*, change in the room temperature). Collectively, these components form the experimental process, as described in Section 4.1, Volume I. The implementation of GUM protocols requires a measurement model that is associated with the measurement process. The model is referred to as the functional relationship. The items required by a model to define a measurand are labeled as input quantities (independent variables). The output quantity in a measurement model is the measurand (dependent variable). The input and output quantities are treated mathematically as random variables characterized by their probability distributions, mean values, and standard deviations (see also Section 6.2.1, Volume I). The probability distributions are determined from measurements or by using the best available knowledge. Correction terms should be included in the model when the conditions of the measurements are not exactly as stipulated. There will be an uncertainty associated with the estimate of a correction term, even if the estimate is zero, as is often the case. Data about the quantities representing physical constants involved in the functional relationship should also be considered in the model.

It is particularly important to note that in hydraulic applications the measurements vary widely in space and time in comparison with other engineering domains, especially if the measurements are conducted in field conditions. Most notably, the time variation can span a wide range of scales. For example, the flow in a river is subjected to turbulent fluctuations (of the order of seconds), while during a flood event the discharge varies continuously during flood wave propagation (over hours or even days or months). For simplicity, this chapter refer only to measurements of the mean flow characteristics in steady flows. Measurement of the mean characteristics in unsteady flows requires special procedures (*e.g.*, Joannis & Bertrand-Krajewski, 2009). Throughout this chapter is assumed that measurement design, protocols, and data processing were duly implemented. Any deviation from the above rules can introduce uncertainties that are not captured by conventional UA discussed herein.

7.2.2 Errors and uncertainties

Measurement error is defined as the result of a measurement minus the true value of the measurand. Neither the true value of the measurand nor the result of the measurement can ever be known exactly because of the uncertainty arising from various effects. Consequently, distinction should be made between error (conceptual definition) and uncertainty (an estimate of the error). The uncertainty of a measurement reflects the lack of exact knowledge about the value of the measurand. The JCGM (2007) defines uncertainty as the "non-negative parameter characterizing the dispersion of the quantity values being attributed to a measurand".

Uncertainty can be broadly associated with random effects in the measurement process and from imperfect correction of the results for systematic effects; *e.g.*, inadequate calibration of the measuring instrument. In practice, there are many possible sources of uncertainty in a measurement, including: incomplete definition of the measurand, imperfect realization of the definition of the measurand, non-representative sampling, inadequate knowledge of the effects of environmental conditions, imperfections in reading analog instruments, finite instrument resolution or discrimination threshold, inexact values of measurement standards and reference materials, inexact values of constants and other parameters obtained from external sources and used in the functional relationship, approximations and assumptions incorporated in the measurement method and procedure, and variations in repeated observations of the measurand under apparently identical conditions. Formal UA requires that each of these sources be identified and evaluated.

Traditionally, errors were classified as random and systematic as a reflection of their effect on the measurand. Random errors presumably arise from unpredictable or stochastic temporal and spatial variation of the measured variable and other factors that influence the results of the measurement. These errors give rise to variations in repeated observations of the measurand. Although it is not possible to compensate for the random errors of a measurement result, they can usually be reduced by increasing the number of observations. Frequently in engineering practice, a number of measurements are used to establish a conventional true value (see Figure 7.2.1.i).

Systematic error (also termed bias), unlike the random error, cannot be eliminated, but it can be estimated and then corrected through carefully designed experiments (a.k.a. calibrations). If a systematic error arises from a recognized effect of an influence quantity and its effect can be quantified, then a correction can be estimated and applied to the measurement to compensate for the effect. It is assumed that after such a correction has been applied, the expected value of the error arising from the particular effect is zero. The net effect of random and systematic errors on repeated measurements is shown in Figure 7.2.1.ii. Detecting, identifying, estimating, and correcting systematic errors may be extremely difficult in practical situations. Comparison with standards, certified reference material, calibration and verification

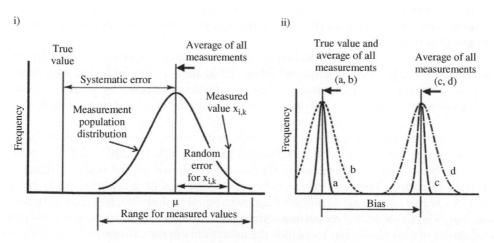

Figure 7.2.1 Errors and their effects: i) errors in the measurement of an input quantity (μ is the mean of the measurement population); and, ii) illustration of the measurement error effect: (a) unbiased, precise, accurate (ideal situation); (b) unbiased, imprecise, inaccurate; (c) biased, precise, inaccurate; (d) biased, imprecise, inaccurate

of instruments, and comparison with alternative measurement methods are some of the means to track systematic errors.

GUM provides a more comprehensive perspective for classifying errors and uncertainties in a measurement in comparison with the classification in traditional UA approaches. Specifically, GUM classifies errors based on the approach used to evaluate them, rather than their effect on the results as is expressed by the traditional engineering, *i.e.*, bias and precision errors. Defining uncertainties as the dispersion of the values that could reasonably be attributed to the measurand, GUM enforces the concept that there is no inherent difference between the uncertainty components arising from random and systematic effects. Both effects are assumed to exist as dispersions around the measured value. Consequently, uncertainties are always estimated using probability density functions or frequency distributions; hence their classification should be based on the method used to estimate their numerical values.

GUM defines Type A uncertainty the quantity that can be evaluated from a sample of actual measurements. Type A uncertainties are evaluated using a set of repeated observations acquired in the measurement environment for which the analysis is conducted. Type B uncertainty should be evaluated by other means. Type B uncertainties are evaluated by assumed probability distributions derived from scientific judgment and consideration of a pool of comparatively reliable information. This pool of information includes previous measurements, calibrations, and experience or general knowledge of the behavior and properties of relevant instruments and measurement procedures.

Another way of interpreting GUM's classification is that it distinguishes between information that comes from sources local to the measurement process (Type A) and information from other sources (Type B). For the final uncertainty, it makes no difference how the components are classified, because GUM treats Type A and Type B evaluations in the same manner. Typically, Type A uncertainty captures the randomness of the measurement process generated by various sources when repeated measurements are made. Type B uncertainty is necessary when a single measurement or no measurements are available. Consequently, previous knowledge or engineering judgment is required.

As an example of Type A evaluation of a systematic uncertainty, consider the situation of a measuring instrument calibrated against a standard. The calibration process normally involves taking a number of readings affected by random effects. The calibration uncertainty is evaluated statistically as Type A using the statistical parameters of the experimental sample. An example of Type B evaluation for a random uncertainty is when a measurement is made with an instrument that displays the reading to just three digits, and is performed just once (UKAS, 2007). This measurement introduces an error defined by the limited resolution of the reading, which is random in nature. The true value of the measurand can lie anywhere in the range of $\pm 0.5 \times$ (value of the least significant digit) with equal probability. It is worth pointing out that, when a calibrated measuring instrument is used in a measurement process, the evaluation of uncertainty has to include the uncertainty in the calibration (estimated as Type A). However, the errors obtained through the calibration will contribute to the new measurements in a systematic manner. The effect of random errors in the calibration process will have become "fossilized" into an effect that is systematic, and they become Type B uncertainty.

7.2.3 Propagation of uncertainties

According to the GUM terminology, determination of the uncertainty of an output variable by means of a functional relationship connecting input and output variables is known as the *law of propagation of uncertainties*. The standard uncertainty of the output quantity is obtained using the functional relationship of the measurement along with the standard uncertainties of the input quantities in a quadrature that includes the so-called sensitivity coefficients. The sensitivity coefficient for an input quantity is obtained as the partial derivative of the functional relationship with respect to that specific quantity. This equation is an approximate for the measurement model and is derived using the Taylor series expansion, neglecting terms higher than first order (AIAA, 1995; GUM, 1993). The sensitivity coefficients associated with each of the input quantities define how the estimate of the output quantity will be influenced by small changes in the estimates of the input quantities. When the input quantities or their respective standard uncertainties contain dependencies, the above uncertainty propagation equation contains covariances (JCGM, 2007) that may decrease or increase the combined total uncertainty of the output variable.

Usually, the result of a measurement is expressed as the best estimate for the measurand and a confidence interval with a specified probability (level of confidence). This interval is expected to contain a large fraction of the distribution of values that can be reasonably attributed to the measurand (see also Section 6.3 in Volume I). Such an interval is defined using a coverage factor that multiplies the combined standard uncertainty of the output quantity. The value of the coverage factor is chosen on the basis of the level of confidence required for the measurement to fall in a given (confidence) interval.

An alternative approach to using a Taylor series expansion to track the propagation of elemental errors to the final results is use of the Monte Carlo Method (MCM). The simulation has as basis the functional relationship for the measurement process. In this approach, the best estimate of each variable in the functional relationship is first input. Then the estimated uncertainty for each variable is set. A uniform random number generator is used to produce distributions (*e.g.*, Gaussian, triangular, log-normal, etc.) of scaled uncertainties for individual running tests. Using the input values for all the variables, the final result is then calculated. This process is repeated for a very large number of iterations (*i.e.*,10,000 to 250,000). This method enables calculation of the mean, the standard deviation, and the coverage interval for a given level of confidence of the distribution of the output values. MCM implementation is recommended in situations where linearization provides an inadequate representation of the measurement functional relationship (i.e., DRE). For linear or linearized models and input quantities for which the probability distribution functions are Gaussian, an MCM yields results consistent with the GUM approach (JCGM, 2008).

7.3 UNCERTAINTY ANALYSIS IMPLEMENTATION

Implementation of the GUM approach to UA typically assumes that the measurement system and experiment are well controlled, all the appropriate calibrations have been executed, and the test objectives, instrument package and data reduction procedures are defined (AIAA, 1995). The basic steps in conducting GUM are illustrated in

Figure 7.3.1. Short descriptions of the practical steps for GUM implementation ensue. The terminology, notations, and procedures used are as they appear in the source references (*i.e.*, GUM, 1993; JCGM, 2008).

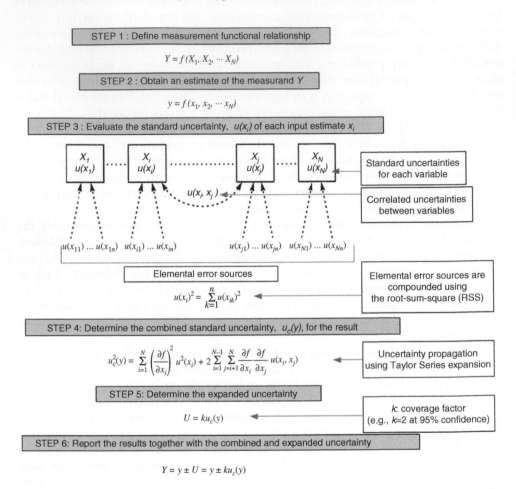

Figure 7.3.1 Flowchart illustrating the steps associated with the GUM approach to implementing UA

7.3.1 Implementation steps

Figure 7.3.1 shows the sequence of steps associated with implementing the GUM approach to UA. The steps are as follow:

1. *Define the measurement functional relationship.* A mathematical relationship of the measurement relates the measurand and input quantities. The measurement process has to provide an estimate (measurement) of each input quantity and the influence quantities involved in the measurement process. If there is no functional relationship for the measurement, such as is the case of reading a length with the scale, the measurement process has to provide the list of sources of uncertainty

affecting that measurement. For the scale measurement case, this list would include: the resolution of the scale, temperature effect, scale condition, etc.

2. *Obtain an estimate of the measurand.* Conduct one or more measurements of the variables in the measurement functional relationship and estimate the results for the measurand. Use the functional relationship to calculate the measurand y in conjunction with the determined input quantities x_i.

3. *Evaluate the standard uncertainty of each input estimate, $u(x_i)$.* Prior to evaluation, the elemental sources of uncertainty need to be identified. At this stage of the analysis, it is necessary to briefly describe the facility, measurement system, environmental conditions, and operator activities to help identify the measurement process errors, including errors not associated with the variable in the modeled measurement. The uncertainty associated with each of the elemental errors (*i.e.*, the standard uncertainty) is the square root of the variance of the measurement error distribution. Standard uncertainties can be evaluated using statistical methods (Type A) or other methods (Type B).

   3a. *Type A evaluation.* The standard uncertainty $u(x_i)$ of an input quantity X_i determined from n independent repeated observations is $u(x_i) = s(\overline{X}_i)$, and is calculated as follows:

$$s^2(\overline{X}_i) = \frac{s^2(x_{ik})}{n} \text{ where } s^2(x_{ik}) = \frac{1}{n-1}\sum_{k=1}^{n}(x_{ik} - \overline{x}_i)^2 \ \overline{x}_i = \frac{1}{n}\sum_{k=1}^{n}x_{ik}$$

$$(7.3.1)$$

X_i is a random variable subjected to n independent observations (*i.e.*, more than 30 are recommended for a significant measurement sample), and x_{ik} is obtained under the same measurement conditions. Based on the available data, several situations can be distinguished: i) small measurement sample and knowledge from one set of previous observations, ii) small measurement sample and knowledge from several sets of previous observations, and, iii) large measurement sample of current measurements (recommended)]. In evaluating Type A uncertainties, the usual assumption is that the Gaussian distribution best describes the measured quantity. However, when uncertainties are determined from a small number of values, the corresponding distribution can be taken as a t-distribution.

   3b. *Type B evaluation.* As evaluation of Type B uncertainties must ensure similar confidence levels as those for Type A, the probability distribution associated with Type B uncertainty and its associated degree of freedom have to be known. Type B uncertainty assumes a dispersion of measurement values and a probability distribution. The dispersion, a_i, is the estimated semi-range of a component of uncertainty associated with an input estimate, x_i, as defined in Figure 7.3.2. The probability distribution can take a variety of forms, but it is generally acceptable to assign well-defined symmetric geometric shapes (*i.e.*, rectangular, Gaussian, triangular) for which the standard uncertainty can be obtained from a single calculation (see Figure 7.3.2). Typical examples of rectangular probability distributions include (ISO, 2005): maximum instrument drift between calibrations; error due to limited resolution of an instrument's display or digitizer; and manufacturers' tolerance limits. A normal probability distribution can also be used in association with calibration certificates quoting a confidence level (or coverage factor) with the

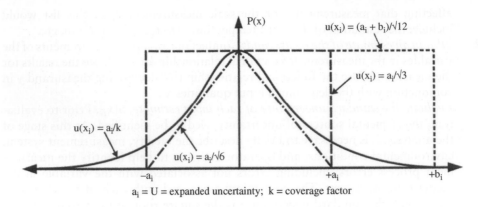

$a_i = U$ = expanded uncertainty; k = coverage factor

Figure 7.3.2 Probability distributions used to evaluate Type B uncertainties

expanded uncertainty. The uniform (rectangular) probability distribution is used when the only information available about a quantity is the maximum bounds within which all values of the quantity are assumed to lie. For intermediate situations between normal and rectangular distributions, triangular distributions can be used. In some measurement situations, the upper and lower bounds for an input quantity are not symmetrical with respect to the best estimate due to, for example, drift in the instrument. For such situations, the asymmetric distribution would be appropriate for estimating the standard uncertainty. The asymmetric distribution can only be applied using MCM.

If an input quantity has multiple sources of errors, the estimated uncertainties (irrespective of their type (i.e., A or B)), are compounded using the root-sum-square (RSS) combination using:

$$u(x_i)^2 = \sum_{j=1}^{K} u(x_i)_j^2 \qquad (7.3.2)$$

where $u(x_i)_j$ is the j-th elemental error associated with the variable x_i.

4. *Determine the combined standard uncertainty, $u_c(y)$.* This step is accomplished by using the variance addition rule. For one input variable, the addition to the variance of various sources of uncertainties of that variable is applied. For a multivariate measurement, in addition to considering the variance of the individual input variables, the possible correlations between the measurement process errors need to be considered. For multivariate analysis, it is also important that the input variable uncertainties are weighted by the appropriate sensitivity coefficients. The degrees of freedom for each uncertainty component, as well as that of the combined uncertainty obtained through the propagation of elemental uncertainties to the final result, are then determined.

The combined standard uncertainty is obtained using the following equation:

$$u_c^2(y) = \sum_{i=1}^{N} \left(\frac{\partial f}{\partial x_i}\right)^2 u^2(x_i) + 2\sum_{i=1}^{N-1}\sum_{j=i+1}^{N} \frac{\partial f}{\partial x_i}\frac{\partial f}{\partial x_j} u(x_i, x_j) \qquad (7.3.3)$$

where f is the functional relationship and each $u(x_i)$ is estimated using either the Type A or B evaluation, or both. x_i and x_j are estimates of X_i and X_j, and $u(x_i, x_j) = u(x_j, x_i)$ is the estimated covariance associated with x_i and x_j. N is the number of input variables. The partial derivatives, called sensitivity coefficients, are evaluated at $X_i = x_i$ using

$$c_i = \partial f / \partial X_i \tag{7.3.4}$$

This step is done using the variance addition rule (a direct method is used in the MCM). For one input variable, the addition to the variance of various sources of uncertainties of that variable is applied. For a multivariate measurement, in addition to considering the variance of the individual input variables, the possible correlations between the measurement process errors need to be considered. For multivariate analysis, it is also important that the input variable uncertainties are weighted by the appropriate sensitivity coefficients. The degrees of freedom for each uncertainty component, as well as that of the combined uncertainty obtained through the propagation of elemental uncertainties to the final result, are then determined.

5. *Determine the expanded uncertainty*, using

$$U = k u_c(y) \tag{7.3.5}$$

where k is the coverage factor. Ideally, uncertainty estimates are based upon reliable Type B and Type A evaluations with a sufficient number of observations such that using a coverage factor of $k = 2$ will ensure a confidence level close to 95 percent (*i.e.*, for a Gaussian distribution, uncertainties are estimated as twice the standard deviation of the average of the measurement sample). If any of these assumptions are not valid, *the effective degrees of freedom* need to be estimated using the Welch-Satterthwaite formula:

$$\nu_{eff} = \frac{u_c^4(y)}{\displaystyle\sum_{i=1}^{N} \frac{u_i^4(y)}{\nu_i}} \tag{7.3.6}$$

where $u_c^2(y) = \displaystyle\sum_{i=1}^{N} u_i^2(y) = \sum_{i=1}^{N} \left(c_i u(x_i) \right)^2$ and c_i is the sensitivity coefficient.

6. *Report the results together with the combined and expanded uncertainty.* The result of a measurement is expressed as $Y = y \pm U = y \pm k u_c(y)$, which is interpreted as the best estimate of the value attributable to the measurand Y, and that $y - U$ to $y + U$ is an interval that may be expected to encompass a large fraction of the distribution of values that could reasonably be attributed to Y. Reported values of uncertainty estimates should present an uncertainty budget containing, at minimum, information such as probability distribution type, standard uncertainty, sensitivity coefficient, degrees of freedom for each component, and applicable cross-correlated uncertainties.

7.3.2 Additional considerations

UA for "timewise"versus "sample-to-sample" experiments. When the measured quantity has a variability unrelated to uncertainty in the experimental process, two types of experiments can be distinguished: timewise and sample-to-sample. A timewise experiment is when the tested entity varies in time not only due to fluctuations inherent in turbulent flows but also due to the longer-time unsteadiness nature of the process, as, for example, is the case for a hydrograph of water flow progressing along a river. For these situations, the measurements are repeated several times during the unsteady flow, and each measurement is treated as being a steady-state measurement with its associated uncertainty. UA is applied for each set of repeated measurements under the assumption that some of the sources of errors are identical for all the repeated measurements (say, environmental factors), while others might change commensurate with the change in variable magnitudes during the propagation (*e.g.*, changes in the operation protocols for different values of the variables). A sample-to-sample experiment is when the process characteristics are determined from samples taken at various locations, as is the case when characterizing the granulometric curve of a sample of river bed sediment. Now, the variability from sample to sample can be significant, and sample identity can be viewed as analogous to time in a timewise experiment.

UA for regression of experimental data. In many experiments, measured data are used in conjunction with regression models to obtain a continuous range of values for the measurand. The measured data are plotted, and a linear regression relationship (based on the method of least squares) is used to determine the best fit of a curve through the experimental data. This relationship, usually a polynomial equation, is then used to represent the relationship between variables over a continuous range. The regression model will incur an uncertainty due to the uncertainty in the direct experimental measurements upon which the regression curves are built (Jeter, 2003). If the uncertainty of the original data used to construct the regression curve is forgotten, the regression model will not predict the true relationship between the variables. Additional uncertainties are introduced if the wrong regression model is adopted to obtain the relationship between variables. Details about the evaluation of uncertainties in regression relationships can be found in Coleman and Steele (1989) and Jeter (2003).

Example 7.3.1 Discharge measurements using velocity-area method. Currently, there are a handful of UA implementation examples for hydraulic engineering applications conducted with standardized methodologies such as GUM. Most of them pertain to hydrometry applications (Joannis & Bertrand-Krajeski, 2009; Gonzalez-Castro & Muste, 2007; Le Coz *et al.*, 2012, 2015; Cohn *et al.*, 2013; Despax *et al.*, 2016). Only a few of these publications, however, demonstrate GUM implementation (Muste *et al.*, 2012; Huang, 2012; Muste & Lee, 2013; Lee *et al.*, 2014). The latter examples involve discharge measurements using conventional instruments (*e.g.*, current meters) as well as other instruments based on more complex principles (*e.g.*, acoustic, magnetic). Extensive efforts have been carried out to assess elemental or lumped uncertainty sources incurred with the various measurement approaches. Le Coz *et al.* (2012)

for example, compare uncertainties associated with simultaneous measurements acquired with multiple instruments in field conditions as illustrated next in Section 7.4, Volume I.

At the outset of this example it is useful to observe that UA is unique for each experiment, and that uncertainty estimates from one experiment can only be partially transferred to other experiments. Attempts at such transfers are a common pitfall in UA (*e.g.*, Muste & Lee, 2013). The UA example presented in this section applies GUM to discharge measured at a cross-section of an open-channel flow with a Price AA mechanical velocimeter, a well-documented instrument extensively used in the past. The example uses surrogate data provided by Boiten (2000) with additional fictitious estimates for some of the error sources for completeness (Muste *et al.*, 2012). It is preferable that uncertainty estimates be obtained from actual measurements or the most recent information available on those uncertainties (see for example, Lee *et al.*, 2014). For the sake of brevity, only the main calculations are presented here.

Measurement process. The discharge measurements discussed herein are acquired with a Price AA current-propeller meter positioned successively at three elevations in several verticals across the channel. The positioning of the current meter at the desired location was made with an A-Pack reel connected to a depth indicator with the finest graduation of tenths of a meter (WMO, 2010). We selected a conventional instrument and measurement protocol for this example as this scenario allows to take advantage of available estimates for most of the elemental error sources involved in the analysis. Velocity measurements were acquired from a bridge spanning the channel at seven verticals across the stream cross section. The locations of the verticals were marked on the bridge, and the distances between markers were subsequently measured with a conventional pocket measuring tape. The water depth in verticals was measured with a sounding reel fitted with a counter (WMO, 2010). It is assumed that the measurements with different instruments follow the usual procedures (*e.g.*, as in Buchanan & Somers, 1969) and that the measurements were not affected by weather or flow unsteadiness. Figure 7.3.3 gives the locations for the measurements.

This current-meter determines flow velocities by measuring the number of propeller revolutions per second, n, for a given measuring period, t, (Boiten, 2000). The velocity

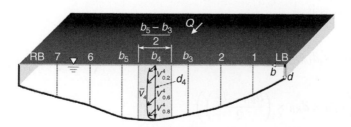

Figure 7.3.3 Sketch of the point velocity measurement locations in one of seven verticals

at a point in the vertical was obtained by determining the rate of propeller revolution Equation (7.3.7) and calibration equations Equation (7.3.8).

$$n = n_1/t \tag{7.3.7}$$

where n is the number of revolutions per second, n_1 is the total number of revolutions during the measuring period;

$$
\begin{aligned}
n &< 0.63 \qquad v = 0.246\,n + 0.017\,(m/s) \\
n &> 0.63 \qquad v = 0.260\,n + 0.008\,(m/s)
\end{aligned}
\tag{7.3.8}
$$

Velocity measurements obtained at three flow elevations (0.2 d, 0.6 d, and 0.8 d; where d is the depth of flow at the location of measurement) were used to estimate the depth-averaged velocity, $\bar{v}$, at each vertical. This three-point method is typically recommended when the flow is uniform and two-dimensional whereby the vertical distribution of the streamwise velocity can be assumed logarithmic WMO (2010). At least 0.75m flow depth is required to use this method. The mean velocity in a vertical is

$$\bar{v} = 0.25v_{0.2} + 0.5v_{0.6} + 0.25v_{0.8} \tag{7.3.9}$$

To illustrate the estimation of Type A uncertainties, we consider a hypothetical scenario whereby the point measurements in vertical 4 were replicated ten times at each depth location (*i.e.*, 0.2 d, 0.6 d, and 0.8 d) for a total duration of 50 seconds at each location. These velocity estimates were made by changing the position of the current-meter reel after each measurement, so as to capture the error associated with the positioning of the probe. The results provide a minimum, but relevant, statistical sample that captures the errors characterizing the measurement environment (flow conditions and operator-related effects).

An estimate of flow discharge is obtained using the mid-section method (Herschy, 2009). In this method it is assumed that the velocity sampled at each vertical represents the mean velocity in a segment of the cross section. The $m = 7$ verticals create seven segments, plus two others at the beginning and end of the cross section, as illustrated in Figure 7.3.3. The mean velocity in each vertical is obtained using $n = 3$ point measurements at specified location on the vertical. Table 7.3.1 summarizes the direct and calculated velocities used for the discharge estimate. Using the notations in Figure 7.3.3, the DRE for the discharge measurement process is:

$$
Q_t = \left(\bar{v}_{LB} \times d_{LB} \times \left(\frac{b_1 - b_{LB}}{2} \right) \right) + \sum_{n=1}^{7} \left(\bar{v}_n \times d_n \times \left(\frac{b_{n+1} - b_{n-1}}{2} \right) \right)
$$

$$
+ \left(\bar{v}_{RB} \times d_{RB} \times \left(\frac{b_{RB} - b_7}{2} \right) \right)
\tag{7.3.10}
$$

Table 7.3.1 Measured depths, widths, velocities, and calculated discharges (Boiten, 2000)

Section	Distance from left bank (b (m))	Depth d (m)	velocity (at 0.2d) v (m/s)	velocity (at 0.6d) v (m/s)	velocity (at 0.8d) v (m/s)	Mean velocity $\bar{v}$ (m/s)	Discharge Q (m³/s)
LB	0	1	0	0	0	0	0
1	10	2.24	0.544	0.523	0.450	0.51	11.4
2	20	3.16	0.741	0.705	0.606	0.69	21.8
3	30	3.72	0.804	0.783	0.679	0.76	28.3
4	40	4.16	0.850	0.830	0.705	0.8*	33.3
5	50	4	0.840	0.809	0.694	0.79	31.6
6	60	3.08	0.726	0.694	0.601	0.68	20.9
7	70	2.44	0.502	0.460	0.382	0.45	11.0
RB	80	1.62	0	0	0	0	0
						Total Discharge	158.30

Note: sampling duration 50 sec.
Note *: Mean and standard uncertainty for operational conditions ($u(\bar{v}_{op})$ in vertical 4 were calculated based on 10 repeated measurements (Type A uncertainty) to account for random errors in the current measurement situation.

where Q_t is the total discharge in the cross section, $\bar{v}_n$ is the mean velocity at the nth vertical (and in the associated panel), d_n is the depths measured at the nth vertical, $b_n = \frac{b_{n+1}-b_{n-1}}{2}$ is the half width between $n + 1^{\text{th}}$ and $n - 1^{\text{th}}$ verticals. In Figure 7.3.3, b_n is centered on the nth vertical (i.e., panel 4 in the figure), and b_0 and b_8 represent b_{LB} and b_{RB}, respectively. Following Boiten (2000), the discharge through the cross-section edges are obtained assuming a parabolic velocity distribution in the verticals close to the banks. Equation (7.3.10) describes a relationship among depth-average velocities (obtained from point velocities and the model described by Equation (7.3.9), depth measurements in each verticals, and the distance measured between verticals combined through the mid-section model used for estimation of the discharge in the cross section. Each of the variables and methods encapsulated in Equation (7.3.10) produce uncertainties that collectively contribute to the total discharge uncertainty.

Assessment of uncertainty sources. Table 7.3.2 provides the estimates of the standard uncertainties for individual sources of uncertainties associated with the independent variables in $\bar{v}_n$, d, and b [i.e., $u(\bar{v}_n)$, $u(d)$, and $u(b)$, respectively]. In addition to the uncertainties associated with the directly measured variables, the table contains uncertainty estimates for the models used to determine the depth-averaged velocity, edge discharge, and the cross-sectional discharge. Finally, the table also include uncertainty sources related to the operational conditions (measurement environment). The provenance of the uncertainties listed in Table 7.3.2 consist of various sources such as previous experiments or expert knowledge. Those sources are cited in the last column of the table. As for some of the uncertainties there are multiple sources of information, a critical judgment of the experimental situations is needed prior to the selection of the uncertainty estimate based on the assessment of the closeness of the measurement situation described in the source and that considered in the current analysis.

Table 7.3.2 Elemental uncertainty sources associated with the stream discharge measurement

Source	Notation	Type*	Standard uncertainty***	Estimation source
Sources associated with the mean velocity in verticals, $\bar{v}_n$				
Instrument accuracy	$u(\bar{v}_{ac})$	B	1.5%	Thibodeaux (2007) for v > 0.2 m/s
Sampling duration	$u(\bar{v}_{st})$	B	3%	ISO 748 (2007), Table E.3
Vertical velocity model	$u(\bar{v}_{vd})$	B	4.8%	ISO 1088 (2007), Table F.1
Operational conditions**	$u(\bar{v}_{op})$	A	0.0066 m/s	10 repeated measurement in vertical 4
Sources associated with the depth in verticals, d				
Instrument accuracy	$u(d_{ac})$	B	0.009 m	Instrument resolution**
Operational conditions	$u(d_{op})$	B	0.02 m	ISO 748 (2007), Table E.2
Sources associated with the distance between verticals, b				
Instrument accuracy	$u(b_{ac})$	B	0.0009 m	Instrument resolution**
Operational conditions	$u(b_{op})$	B	0.1524 m	WMO (2010)
Sources associated with the estimation of discharge, Q_t				
Discharge model	$u(Q_m)$	B	0.5%	Muste et al. (2004)
Number of verticals	$u(Q_{nv})$	B	5.7%****	ISO 1088 (2007), Table G.6
Edge discharge model	$u(Q_{eg})$	B	Not available	
Flow unsteadiness	$u(Q_{us})$	B	0	
Operational conditions	$u(Q_{op})$	B	2.7%	WMO (2010)

* classification compliant with GUM (1993)
** estimated as half of the instrument resolution assumed as a rectangular distribution
*** relative estimates are converted to absolute (dimensional) values in these calculations
**** percentage of the mean velocity in the vertical (not discharge in the subarea)

Total uncertainty. Using Equations (7.3.9) and (7.3.10), the standard and expanded uncertainties can be written as:

$$u_c(Q_t) = \sqrt{\sum_{n=0}^{8} u(\bar{v}_n)^2 \left(\frac{\partial Q_t}{\partial \bar{v}_n}\right)^2 + \sum_{n=0}^{8} u(d_n)^2 \left(\frac{\partial Q_t}{\partial d_n}\right)^2 + \sum_{n=0}^{8} u(b_n)^2 \left(\frac{\partial Q_t}{\partial b_n}\right)^2}$$

$$\sqrt{+2\sum_{n=0}^{7}\sum_{n+1}^{8} \left(\frac{\partial Q_t}{\partial \bar{v}_n}\right)\left(\frac{\partial Q_t}{\partial \bar{v}_{n+1}}\right) u(\bar{v}_n)u(\bar{v}_{n+1})r(\bar{v}_n,\bar{v}_{n+1}) + 2\sum_{n=0}^{7}\sum_{n+1}^{8} \left(\frac{\partial Q_t}{\partial d_n}\right)\left(\frac{\partial Q_t}{\partial d_{n+1}}\right) u(d_n)u(d_{n+1})r(d_n,d_{n+1})}$$

$$\sqrt{+2\sum_{n=0}^{7}\sum_{n+1}^{8}\left(\frac{\partial Q_t}{\partial b_n}\right)\left(\frac{\partial Q_t}{\partial b_{n+1}}\right)u(b_n)u(b_{n+1})r(b_n,b_{n+1})+u(Q_m)^2+u(Q_{nv})^2+u(Q_{eg})^2+u(Q_{us})^2+u(Q_{op})^2}$$

$$\text{(7.3.11)}$$

$$U(Q_t)=ku_c(Q_t) \tag{7.3.12}$$

The first three terms in Equation (7.3.11) are associated with uncertainties in mean vertical velocity, depth and distance between verticals, $u(\bar{v}_n)$, $u(d)$, and $u(b)$, respectively. Each of these uncertainties entail other elemental uncertainty sources that are aggregated through the following relationships (see also Table 7.3.2):

$$u(\bar{v}_n)=\sqrt{u(\bar{v}_{ac})^2+u(\bar{v}_{st})^2+u(\bar{v}_{vd})^2+u(\bar{v}_{op})^2} \tag{7.3.13a}$$

$$u(d_n)=\sqrt{u(d_{ac})^2+u(d_{op})^2} \tag{7.3.13b}$$

$$u(b_n)=\sqrt{u(b_{ac})^2+u(b_{op})^2} \tag{7.3.13c}$$

Table 7.3.3 gives the uncertainty values used in Equations (7.3.13a)–(7.3.13c).

The next three terms in Equation (7.3.11) represent correlated uncertainties for velocity, depth and distances, with $r(x_i, x_i)$ representing the correlation coefficient between the related variables. These additional terms are needed as the total discharge is based on a summation of subareas of the cross sections measured successively with the same instruments. The last five terms are directly related to the estimation of the total discharge, so their corresponding sensitivity coefficients are 1. Note that the uncertainty due to the number of verticals, $u(Q_{nv})$, should include the effect of the limited number of verticals on the accurate replication of the mean velocity field as well as of the cross section geometry. Excepting the work by Le Coz *et al.* (2012) where some preliminary information is noted, the available sources of information do not offer uncertainty estimates for the double effect of this source of error therefore it is neglected in the present analysis.

Propagation of the elemental uncertainties to overall uncertainty associated with the total discharges was conducted using the specialized software QMsys Enterprise (Qualisyst Ltd.) using Equation (7.3.11) as the DRE for the measurement process. The correlated terms were neglected, as there is no information available for their estimates. The effective degrees of freedom for probability distributions associated with the elemental uncertainties and the final result were obtained using Equation (7.3.12). Except for the uncertainty element $u(\bar{v}_{op})$, the elemental uncertainties were assumed to have infinite degrees of freedom therefore the effective degrees of freedom for the measurand is also infinite.

Result reporting and uncertainty budget. The uncertainty budget for the discharge measurement is presented in Table 7.3.3. The total (expanded) uncertainty for the

Table 7.3.3 Uncertainty budget for the discharge estimate

Source	Notation	Probability distribution	Divisor*	Contribution to the total uncertainty (m^3/s)	Relative Contribution to total uncertainty (%)
Instrument accuracy	$u(\bar{v}_{ac})$	normal	2	0.95	4.46
Sampling duration	$u(\bar{v}_{st})$	normal	2	1.91	7.90
Vertical velocity model	$u(\bar{v}_{vd})$	normal	2	2.06	14.32
Operational conditions	$u(\bar{v}_{op})$	t-distribution	1	0.17	0.80
Instrument accuracy	$u(d_{ac})$	rectangular	1.73	0.55	2.61
Operational conditions	$u(d_{op})$	normal	2	0.36	1.65
Instrument accuracy	$u(b_{ac})$	rectangular	1.73	0.07	0.33
Operational conditions	$u(b_{op})$	rectangular	1.73	0.35	1.61
Discharge model	$u(Q_m)$	normal	2	0.78	3.69
Number of verticals	$u(Q_{nv})$	normal	2	10.22	42.71
Edge discharge model	$u(Q_{eg})$			–	
Flow unsteadiness	$u(Q_{us})$			–	
Operational conditions	$u(Q_{op})$	normal	2	4.27	19.92
Combined Uncertainty	$u(Q_t)$	normal		11.22	6.76
Expanded Uncertainty	$U(Q_t)$	normal (k=2)		21.44	13.53

* The divisor is the value by which the standard uncertainty is divided to obtain the standard deviation for the probability distribution assumed for the *j-th* source of uncertainty.

discharge measurement, at a 95-percent confidence level, is $U(Q_t) = \pm 21.44$ m^3/s This uncertainty assessment corresponds to ± 13.53 percent of the total discharge estimate. The uncertainty budget provided by the QMsys Enterprise software illustrates that uncertainties associated with the number of verticals and operational conditions for discharge estimation are the largest contributors to the total uncertainty, followed by the uncertainties associated with the model for determining the depth-averaged velocity. The result of this example can be stated as: the total discharge is $Q_t = 158.44$ m^3/s, with an interval of uncertainty of [137.00 m^3/s, 179.87 m^3/s] estimated at 95-percent confidence level.

7.4 UNCERTAINTY INFERENCES USING INTERCOMPARISON EXPERIMENTS

7.4.1 Overview of intercomparison experiments

The GUM standardized framework presented in the previous section is carried out using a functional expression of the measurement process (*i.e.*, the Data Reduction Equation). Given the complexity of the hydraulic measurements and the often large number of uncertainty sources involved in the measurement process, the GUM approach can be tedious and expensive especially when applied to measurements acquired in field conditions. For example, estimation of a Type A uncertainty would ideally require to "freeze" all sources of errors affecting the measurement process excepting the one that is evaluated and repeat the measurements to obtain a robust sample result population.

Motivated by the continuous improvement of the data quality, some hydrometric communities organize periodically intercomparison experiments whereby multiple teams using multiple instruments from the same category acquire simultaneously measurements in practically identical conditions using well-established measurement protocols. This intercomparison practice is more popular in the user communities operating newer technologies (*e.g.*, acoustic instruments) as means of sharing experience and learning to enhance measurement performance. The practice of intercomparison of experiments originates in metrology and benefits from extensive guidelines (ISO 5725-2, 1994; ISO 21748, 2010; ISO 13528, 2005). The following assumptions are necessary for the implementation of the first two listed standards:

- The experimental results are obtained under conditions in which sources of error are active during the measurements;
- The results can be considered independent samples and the measurement errors follow a Gaussian (or at least unimodal) distribution;
- The measurement environment is the same for all the compared experiments, and,
- Each experiment contains the same number of repeated measurements.

The major output of the intercomparison experiments is the average value of all the repeated measurements obtained by the teams operating multiple instruments. The scatter of the measurements is a result of the combined effect of all the sources of errors active in the measurement process. Customized post-measurement analyses can lead to evaluation of the aggregated effect of the active error sources as will be demonstrated next. If the replication of the experiments is made using a strategy that isolate groups of elemental uncertainty sources, the analysis of the obtained results can provide bulk estimates of the uncertainty produced by that group.

7.4.2 Method implementation

7.4.2.1 Error model

For illustration of the methodology, let's consider the k-th discharge measurement, $Q_{i,k}$, measured by the instrument i in a steady flow:

$$Q_{i,k} = Q_{true} + \delta + B_i + \varepsilon_{i,k} \tag{7.4.1}$$

where Q_{true} is the true discharge value (assumed but not known), δ_{is} the bias associated with the measurement protocol (hypothetically the same for all instruments involved in the intercomparison experiment), and B_i and $\varepsilon_{i,k}$ are the systematic (bias) and random errors related to the k-th discharge measurement of the instrument i, respectively. The reference discharge value for the intercomparison experiment, denoted as $Q_{mean} = Q_{true} + \delta$, is obtained as the average of all discharges measured during the tests. Each of the P instruments involved in the test acquire N measurements.

The measurement error in the discharge acquired with instrument i, Q_i, can be modeled as:

$$Q_{i,k} = Q_{moy} + B_i + \varepsilon_{i,k} \quad \text{with } B_i \sim N(0, \sigma_L) \text{ and } \varepsilon_{i,k} \sim N(0, \sigma_r) \tag{7.4.2}$$

The random error, $\varepsilon_{i,k}$, and the systematic error, B_i, are assumed to follow a Gaussian distribution, hence they both have zero values for the mean and standard deviations σ_r and σ_L, respectively. The above considerations apply to the systematic errors only if all the bias corrections have been applied prior to the actual measurements.

7.4.2.2 Uncertainty estimation using individual repeated measurements

The processing of the intercomparison experiments presented here follows the protocol detailed in ISO 5725-2 (1994). Accordingly, the first processing step after the collection and formatting of the data is the review of the individual measurement values using Mandel's h (for accuracy) and k (for dispersion) criteria to visualize the homogeneity and consistency of the measurements. The implementation of this criteria is primarily intended to identify the instruments and teams exhibiting widely different results. Detecting one or several outliers may lead to statistical tests for outlier detection (e.g., the Cochran and Grubbs tests described in the standard) followed by outlier removal if needed.

The repeatability (s_r), interlaboratory (s_L), and reproducibility (s_R) standard deviations of the comparison test results are subsequently calculated. The repeatability s_r is computed from the experimental standard deviations, s_i, of the n_i repeated discharge measurement, $Q_{i,k}$, provided by each team i, as follows:

$$s_r^2 = \frac{\sum_{i=1}^{p}(n_i - 1)s_i^2}{\sum_{i=1}^{p}(n_i - 1)} \quad \text{with} \quad s_i^2 = \frac{1}{n_i - 1}\sum_{k=1}^{n_i}(Q_{i,k} - \overline{Q_i})^2 \tag{7.4.3}$$

For small samples, say only 2 measurements, the standard deviation is:

$$s_i = |Q_{i,2} - Q_{i,1}|/\sqrt{2} \tag{7.4.4}$$

The intercomparison standard deviation, s_L, is provided by Equation (7.4.5):

$$s_L^2 = \frac{s_d^2 - s_r^2}{N} \tag{7.4.5}$$

with $N = \frac{1}{p-1}\left[\sum_{i=1}^{p} n_i - \frac{\sum_{i=1}^{p} n_i^2}{\sum_{i=1}^{p} n_i}\right]$ being the average number of measurements provided by all teams (note that N may not be an integer if the number of the measurements provided by each team are not equal). The standard deviation of the team averages, s_d, is given by:

$$s_d^2 = \frac{1}{p-1}\sum_{i=1}^{p} n_i(\bar{Q}_i - Q_{moy})^2 \tag{7.4.6}$$

If $s_d < s_r$, s_L is discarded. The reproducibility standard deviation, s_R, is provided by Equation (7.4.7)

$$s_R^2 = s_r^2 + s_L^2 \qquad (7.4.7)$$

For the next processing step, the ISO 21748 (2010) is invoked for obtaining the combined standard uncertainty of the result. Additionally, we introduce the standard uncertainty, $u(\delta)$, related to the bias of the streamgauging technique, δ, which will be discussed hereafter. Using the above approach, the expanded uncertainty in the discharge becomes:

$$u(Q) = k\sqrt{s_R^2 + u^2(\delta)} = k\sqrt{s_r^2 + s_L^2 + u^2(\delta)} \qquad (7.4.8)$$

where k is the coverage factor of the evaluated uncertainty for a prescribed level of confidence. The Hydrometric Uncertainty Guidance (HUG) recommends $k = 2$ for 95% level of confidence as the most appropriate for hydraulic and hydrometric applications (ISO/TS 25377, 2007). Practical methods for estimating δ, and the related uncertainty, $u(\delta)$ are provided in Le Coz et al. (2016). These estimations are not straightforward for measurements where there is no recognized reference method as is the case in stream discharge measurements.

7.4.2.3 Uncertainty estimates for sets of repeated measurements

For some situations the reported result of a measurement is produced from a set of small number of consecutive measurements. For example, some guidelines for best practices in the measurements with moving-boat ADCPs require the acquisition of $n = 4 - 6$ successive transects to report one valid average result. For situation where accuracy at prescribed uncertainty level is required, several ADCPs may be used simultaneously and their results may be averaged altogether. The uncertainty of the discharge acquired with multiple team colleting multiple transects, $U(Q^{n,p})$, can be obtained by applying the procedures described in Section 7.4.2.2, Volume I. For example, the total measurement uncertainty using the ISO 13528 (2005) protocol for n raw ADCP discharge measurements acquired from pairs of successive transects (opposite directions between each transect) using one or several (p) instruments is:

$$U(Q^{n,p}) = k\sqrt{\frac{s_r^2}{np} + \frac{s_L^2}{p} + u^2(\delta)} \qquad (7.4.9)$$

Example 7.4.1: Intercomparison experiment. The example presents the analysis of intercomparison of discharge measurements using moving-boat ADCPs method conducted in the Rhône River downstream of Génissiat dam, France (Le Coz et al., 2016). This intercomparison involved 6 types of ADCP from two manufacturers (Teledyne RDI and Sontek), with different operating frequencies. The two groups of operators were deployed at two different measurement sites as follows. Six boats equipped with 12 ADCPs were deployed just downstream of the Génissiat dam. The site is characterized by deep and uneven cross-section with the river flowing in a developing flow regime (see Figure 7.4.1.a). Other seven boats equipped with 14 ADCPs were deployed at Pyrimont, a site with similar geometry but with the flow fully developed (see 7.4.1.b). Over two days, 6 time windows of constant discharge were provided by controlling the flow at the dam.

Each constant-flow experiment lasted for an hour. The average discharges for the 6 flow scenarios as determined from all ADCP measurements are: 221 m³/s, 335 m³/s, 439 m³/s, 118 m³/s, 227 m³/s, and 333 m³/s, respectively.

The flow at the dam was continuously measured with an acoustic transit-time system installed in the dam penstocks. Previous calibrations displayed uncertainties for the acoustic system in the 1% to 3% range. Consequently, the transit-time discharge measurements are considered as reference, Q_{ref}, for the present analysis. The results of the intercomparison experiments are presented in Table 7.4.1. As can be noted from the data in the table, the standard uncertainty due to the technique bias was estimated to be $u(\delta) = 1.25\%$. The average expanded uncertainties of the successive ADCP transects (estimated with the protocols presented in Section 7.4.2.3, Volume I) are provided in Table 7.4.1.

The total uncertainty for the ADCP discharge at Pyrimont site is $U(Q^{6,1}) = \pm5.4\%$. This values is close to the ±5% value commonly accepted as an indicator of good quality measurement in streamgauging. The total uncertainty for the ADCP discharge at Genissiat site is $U(Q^{6,1}) = \pm9.0\%$, value that exceeds the accepted recommended limit for this type of instrument. The difference between the uncertainties at the two sites acquired with similar instruments and protocols indicate that there is a difference in the measurement environment at the sites. Indeed, a more detailed analysis of the velocity maps produced by the ADCPs confirmed that at Genissiat site the flow was not fully developed with visible three-dimensional, large-scale eddies across the section (Le Coz et al, 2016). The presence of three-dimensional structures in the ADCP measurement volume violate the horizontal homogeneity assumption required for the operation of these instruments. Protocols that do not consider these aspects result in additional uncertainty of the results.

The sensitivity of the total uncertainty of the discharge measurements with the number of transects acquired is displayed in Table 7.4.1 [by the $U(Q^{n,\,p})$ columns] and in more details in Figure 7.4.2. The plots in Figure 7.4.2 allow to observe that the

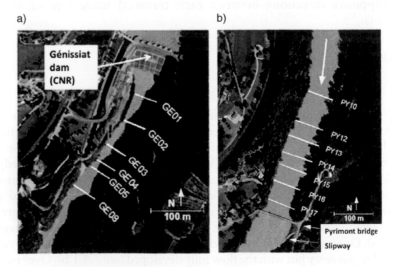

a)   b)

Figure 7.4.1 Aerial view of the ADCP transects used in the intercomparison experiments (Le Coz *et al.*, 2016; with permission from ASCE): a) Genissiat site, and, b) Pyrimont site.

Table 7.4.1 Average results of the 6 interlaboratory comparisons conducted at the two test sites. During the tests, the flow was maintained steady at discharge values of: 221 m³/s, 335 m³/s, 439 m³/s, 118 m³/s, 227 m³/s, and 333 m³/s.

Site	# of ADCPs	s_r	s_L	$u(Q\text{ mean})$	$u(\delta)$ (estim.)	$U(Q^{I,I})$ (Eq. 7.4.9)	$U(Q^{4,I})$ (Eq. 7.4.9)	$U(Q^{6,I})$ (Eq. 7.4.9)
GE	12	4.5%	3.9%	1.5%	1.25%	±12.2%	±9.3%	±9.0%
PY	14	2.4%	2.2%	0.6%	1.25%	±7.0%	±5.6%	±5.4%

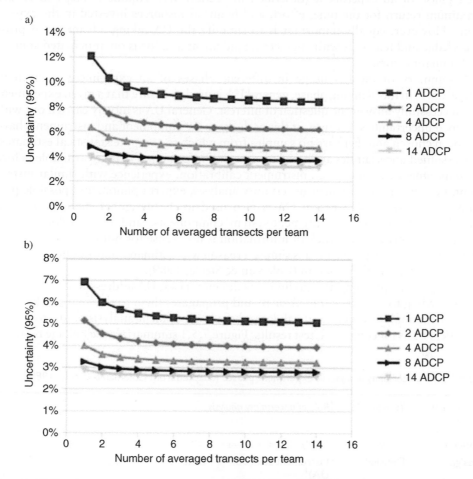

Figure 7.4.2 Total uncertainty for discharges measured with p ADCPs and n transects (Le Coz et al., 2016; with permission from ASCE): a) Génissiat site, and, b) Pyrimont site.

uncertainty reduction levels for more than 6 individual ADCP transects for both measurement sites. If the measurements are averaged in sub-samples of 2 or 4 successive transects, the uncertainty is reduced significantly. Despite being site dependent, these results indicate that that the replication of transects reduces the repeatability uncertainty, for one instrument, while using several instruments and cross-sections allows to reduce both repeatability and intercomparison uncertainties.

7.5 PRACTICAL ISSUES

Data quality assessment should be a key part of any experiment. UA is a useful tool enabling investigators to evaluate a measurement system and method before, during, and after an experimental process is conducted. This important aspect is illustrated in Figure 7.5.1 which shows that UA should be a continuous process, because uncertainty considerations influence both decisions affecting the design of components of an experiment as well as the procedure for conducting an experiment. Use of UA during each phase of an experiment (described in Section 4.1, Volume I) helps to ensure maximum return for the time, effort, and financial resources invested in the experiment. However, equally important is to simplify the UA taking advantage of prior knowledge and temper it with engineering judgment and focus on anticipated sources of dominant errors.

A summary of the UA usage in different phases of an experiment is given in Table 7.5.1. In the planning phase, general UA is used to ensure that a given experiment can successfully answer the question of interest. General UA is mostly carried out with assumed uncertainties, because, at this early stage the equipment and instruments have not been chosen and there are no samples from which to compute statistical estimates. The assumed uncertainties are based on data and information that are available before the experiment execution (i.e., instrument calibration, experience with similar instrumentation, prior measurement uncertainty analysis, expert opinion, and possible preliminary experiments). General UA is also used in the design of the experiments, including selection of measurement-system components, and the development of experimental procedures. Lack of information is no excuse for not doing a UA in the early phases of an experiment; it is simply a reason why estimates may not be as good as they will be later in the program (Coleman & Steele, 1989).

Once past the planning and preliminary design phases, UA addresses the uncertainties according to their type (*i.e.*, A or B) and propagates them to the final result. Post-experiment analysis checks and validates the pre-experiment analysis, and provides a statistical basis for comparing test results. Figure 7.5.1 summarizes the linkage between

Table 7.5.1 Uncertainty analysis in experiment phases (adapted from Coleman & Steele, 1989)

Experiment Phase	Type of UA	Role of uncertainty analysis
Planning	General	Verifying that the experiment addresses the study objectives
Design	Detailed	Instrument selection (0th order UA); finalizing design (N^{th} order UA)[1]
Construction	Detailed	Inform decisions on facility and instrumentation settings
Debugging	Detailed	Verify and qualify operation; 1st order and N^{th} order UA
Execution	Detailed	Capture effect of MS operation and environmental factors
Data Analysis	Detailed	Guide the choice of analysis techniques
Reporting	Detailed	Total uncertainties & confidence levels along with uncertainty budget

Note [1]: The types of experiment replication are discussed in Section 5.2.7 in Volume I.

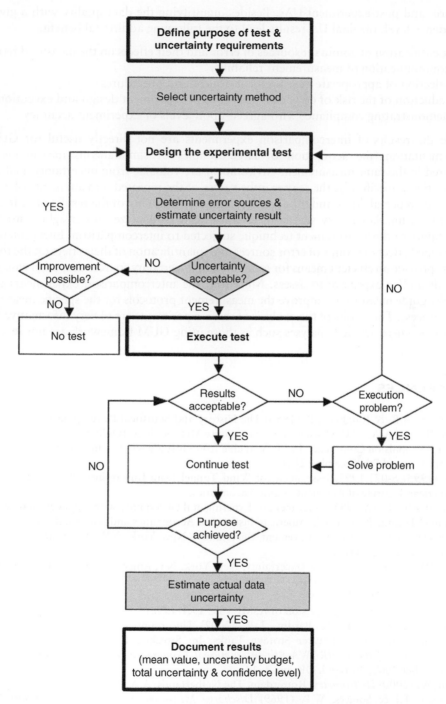

Figure 7.5.1 UA is an integral part of the experimental process (adapted from AIAA, 1995).

the pre- and post-experiment UAs. Besides quantifying the data quality with a given confidence level, the final UA results lead to the following additional benefits:

- identification of dominant sources of error and their effects on the measured result
- communication of measurement reliability
- selection of appropriate measurement devices and procedures
- reduction of the risk of erroneous decisions in experiment design and execution
- demonstrating compliance with agreed-upon levels of experiment accuracy

While the results of intercomparison experiments are not directly useful for GUM implementation, the acquisition of large samples of simultaneous measurements acquired in the same measurement environment are valuable from uncertainty analysis perspectives. Specifically, the scatter inthe values of the repeated measurements of each team (*i.e.*, repeatability standard deviation) and the variability of the results of different teams (*i.e.*, interlaboratory standard deviation) can be analyzed for insights into the uncertainty of the measurement technique subjected to intercomparison. Encapsulating the lumped effect of groups of error sources and quantification of their effect on the total results provide an efficient means for estimation of uncertainty sources that otherwise are difficult and/or expensive to assess. Moreover, the intercomparison experiments can usefully guide measures to improve the measurement protocols for the specific measurement process. The results of the analysis of these experiments are of very informative for situations where detailed analyses such as those using GUM framework are not readily available.

REFERENCES

Abernethy, A.B. & Ringhiser, B. (1985) The History and Statistical Development of the New ASME-SAE-AIAA-ISO Measurement Uncertainty Methodology. [Online] AIAA/SAE/ASME/ ASEE Propulsion Conference 1985. Available from: http://www.barringer1.com/drbob-bio. htm [Accessed 7th December 2016].

AIAA (1995) S-071-1995. Assessment of Wind Tunnel Data Uncertainty. Washington. DC, American Institute of Aeronautics and Astronautics.

AIAA (1999) S-071A-1995. Assessment of Experimental Uncertainty with Application to Wind Tunnel Testing. Reston. VA, American Institute of Aeronautics and Astronautics.

ASME (1986) PTC 19.1. Measurement Uncertainty. New York. NY, American Society of Mechanical Engineering.

ASME (1998) PTC 19.1.1. Test Uncertainty. New York. NY, American Society of Mechanical Engineering (revision of ANSI/ASME (1986).

ASME (2005) PTC 19.1-2005. Test Uncertainty. New York. NY, American Society of Mechanical Engineering (revision of ASME PTC 19.1-1998).

Bertrand-Krajewski J.-L. & Bardin, J.-P. (2002) Uncertainties and Representativity of Measurements in Stormwater Storage Tanks. In: Strecker, E.W. & Huber, W. (eds.) *Proceedings of the IAHR/IWA 9th International Conference on Urban Drainage*, 8–13 September 2002, Portland. OR. pp. 1–14.

Boiten, W. (2000) *Hydrometry*. Rotterdam, The Netherlands, A.A. Balkema.

Buchanan, T.J & Somers, W.P. (1969) *Discharge Measurements at Gaging Stations: U.S. Geological Survey Techniques of Water-Resources Investigations* [Online], U.S. Government Printing Office. Book 3. p.65 Available from: http://pubs.usgs.gov/twri/twri3a8/ [Accessed 7th December 2016].

Cohn, T.A., Kiang, J.E., & Mason, R.R. (2013) Estimating discharge measurement uncertainty using the interpolated variance estimator. *J Hydraul Eng.* 139(5), 502–510.

Coleman, H.W. & Steele, W.G. Jr. (1989) *Experimentation and Uncertainty Analysis for Engineers*, 2nd edition, New York, John Wiley & Sons, Inc.

Coleman, H.W. & Steele, W.G. (1995) Engineering application of experimental uncertainty analysis. *AIAA Journal* 33 (10), 1888–1896.

Despax, A., Perret, C., Garçon, R., Hauet, A., Belleville, A., Le Coz, J., & Favre, A.-C. (2016) Considering sampling strategy and cross-section complexity for estimating the uncertainty of discharge measurements using the velocity-area method. *Journal of Hydrology.* 533, 128–140.

González-Castro, J.A., & Muste, M. (2007) Framework for estimating uncertainty of ADCP measurements from a moving boat by standardized uncertainty analysis. *ASCE J Hydraul Eng.*, [Online] 133, 1390. Available from: doi:10.1061/(ASCE)0733–9429(2007)133:12 (1390) [Accessed 7th December 2016].

GUM (1993) ISBN 92-67-10188-9. Guide to the Expression of Uncertainty in Measurement. Geneva. Switzerland, BIPM. IEC. IFCC. ISO. IUPAC. IUPAP. OIML. International Organization for Standardization.

Herschy, R.W. (2002) The uncertainty in a current meter measurement. *Flow Meas Instrum.* 13 (5–6), 281–284.

Herschy, R.W. (2009) *Streamflow Measurement*. New York, NY, Taylor & Francis.

Huang, H. (2012) Uncertainty Model for In Situ Quality Control of Stationary ADCP Open-Channel Discharge Measurement. *JHydraul Eng.*, 13(1), 4–12.

HUG (2007) Hydrometric Uncertainty Guidance, International Organization for Standardization, Geneva, Switzerland, 51 p.

ISO 5725-2 (1994) Accuracy (trueness and precision) of measurement methods and results. Part 2: Basic method for the determination of repeatability and reproducibility of a standard measurement method. International Organization for Standardization, Geneva, Switzerland, 42 p.

ISO 13528 (2005) Statistical methods for use in proficiency testing by interlaboratory comparisons. International Organization for Standardization, Geneva, Switzerland, 66 p.

ISO 5168 (2005) Measurement of Fluid Flow – Procedures for the Evaluation of Uncertainties. Geneva. Switzerland, International Organization for Standardization.

ISO/TS 25377 (2007) Hydrometric uncertainty guidance (HUG). Geneva, Switzerland, International Organization for Standardization, 51 p.

ISO 748, (2007). Hydrometry – measurement of liquid flow in open channels using current-meters and floats. Geneva, Switzerland, International Organization for Standardization.

ISO 1088, (2007) Hydrometry – velocity-area methods using current-meters collection and processing of data for determination of uncertainties in flow measurement. Geneva, Switzerland, International Organization for Standardization.

ISO 21748 (2010) Guidance for the use of repeatability, reproducibility and trueness estimates in measurement uncertainty estimation. International Organization for Standardization, Geneva, Switzerland, 38 p.

JCGM (2007) 200. International vocabulary of Metrology-Basic and general concepts and associated terms (VIM). Geneva. Switzerland, Joint Committee for Guides in Metrology.

JCGM (2008) 100. Evaluation of measurement data – Guide to the expression of uncertainty in measurement (GUM). JCGM member organizations (BIPM, IEC, IFCC, ILAC, ISO, IUPAC, IUPAP and OIML), 120 p.

JCGM (2008) 101. Supplement 1 to the GUM (Guide to the expression of uncertainty in measurement, – propagation of distributions using a Monte Carlo method). Joint Committee for Guides in Metrology. Available from: http://www.bipm.org/en/publica tions/guides/gum.

JCGM (2009) 104. ISO/IEC Guide 98–1: 2009. Uncertainty of Measurement- Part 1: Introduction to Expression of Uncertainty in Measurement. Geneva. Switzerland, Joint Committee for Guides in Metrology.

Jeter, S.M. (2003) Evaluating the Uncertainty of Polynomial Regression Models Using Excel. American Society for Engineering Education Annual Conference & Exposition, 22–25 June 2003, Nashville, TN, USA.

Joannis, C., & Bertrand-Krajewski, J.-L. (2009) Uncertainties on a measurand defined as an integrated value from a continuous signal temporally discretised – Application to in situ hydrological measurements recorded on the field. *La Houille Blanche*. 3, 82–91.

Kim, D., Muste, M., Gonzalez-Castro, J.A., & Ansar, M. (2005) Graphical User Interface for ADCP Uncertainty Analysis. In: *Proceeding ASCE World Water & Environmental Resources Congress, 15–19 May 2005*, Anchorage. AK. Curran Associates, Inc. pp. 3743–3754.

Kim, Y., Muste, M., Hauet, A., Bradley, A., & Weber, L. (2007) Uncertainty Analysis for LSPIV In-situ Velocity Measurements. In: *Proceedings of the 32nd IAHR Congress*, 1–6 July 2007, Venice. Italy

Kline, S.J., & McClintock, F.A. (1953) Describing uncertainties in single-sample experiments. *Mechanical Engineering*. 75, 3–8.

Le Coz, J., Pobanz, K., Faure, J.-B., Pierrefeu, G., Blanquart, B., & Y. Choquette (2012) *Stage-discharge hysteresis evidenced by multi-ADCP measurements*. RiverFlow 2012 Conference, 1277–1283, 5–7 September 2012, San José, Costa Rica.

Le Coz, J., Camenen, B., Peyrard, X., & Dramais, G. (2012) Uncertainty in open-channel discharges measured with the velocity-area method. *Flow Measurement and Instrumentation*. 26, 18–29.

Le Coz, J., Camenen, B., Peyrard, X., & Dramais, G. (2015) Erratum to "Uncertainty in open-channel discharges measured with the velocity-area method [Flow Meas. Instrum. 26 (2012) 18–29]" Flow Measurement and Instrumentation. Available from: doi:10.1016/j.flowmeasinst.2012.05.001 [Accessed 7th December 2016].

Le Coz, J., Blanquart, B., Pobanz, K., Dramais, G., Pierrefeu, G., Hauet, A., & Despax, A. (2016) Estimating the uncertainty of streamgauging techniques using in situ collaborative interlaboratory experiments. *J Hydraul Eng.*, 142(7), 04016011.

Lee, K., Ho, H-C., Muste, M., & Wu, C-H. (2014) Uncertainty in open-channel discharge measurements acquired with StreamPro ADCP. *J Hydrol.*, 509, 101–114.

Muste, M. & Stern, F. (2000) Proposed uncertainty assessment methodology for hydraulic and water resources engineering. In: Hotchkiss R.H., & Glade, M. (eds.) *Proceedings of the ASCE 2000 Joint Conference on Water Resources Engineering and Water Resources Planning & Management, 30 July – 2 August*, Minneapolis. MN. American Society of Civil Engineers. (CD-ROM).

Gonzalez-Castro, J.A. & Muste, M. (2007) Framework for estimating uncertainty of ADCP measurements from a moving boat by standardized uncertainty analysis. *J Hydraul Eng.*, 133 (12), 1390–1410.

Muste, M. & Lee, K. (2012) Assessment of the Performance of Flow Measurement Instruments and Techniques, Development of the Decision-aid Tool (UADAT). World Metrological Organization. Report number: Phase II, Output 3. "Guidelines for the Assessment of Uncertainty for Hydrometric Measurement" (2012), World Meteorological Organization, WMO-No. 1097, Geneva, Switzerland

Muste, M. & Lee, K. (2013) Uncertainty model for in situ quality control of stationary ADCP open-channel discharge measurement. *J Hydraul Eng.*, 139, 102–104.

Muste, M., Lee, K., & Bertrand-Krajewski, J-L. (2012) Standardized uncertainty analysis frameworks for hydrometry: Review of relevant approaches and implementation examples. *Hydrolog Sci J.*, [Online] Available from: doi:10.1080/0262667.2012.675064 [Accessed 7th December 2016].

Muste, M., Yu, K., González-Castro, J., & Starzmann, E. (2004) Methodology for Estimating ADCP Measurement Uncertainty in Open-Channel Flows. In: Sehlke, G., Hayes, D.F., & Stevens, D.K. (eds.) *Critical transitions in water and environmental resources management: Proceedings of the World Water & Environmental Resources Congress 2004 (EWRI), 27 June – 1 July 2004*, Salt Lake City. UT. American Society of Civil Engineers.

NF ENV (1999) 13005. Guide pour L'Expression de L'incertitude de Mesure. AFNOR. Paris. France (in French)

NIST (2003) E-Handbook of Statistical Methods. NIST/SEMATECH, Available from: http://www.itl.nist.gov/div898/handbook/.

Steele, W.G., Ferguson, R.A., Taylor, R.P., & Coleman, H.W. (1994) Comparison of ANSI/ASME and ISO Models for Calculation of Uncertainty. *ISA Transactions*. 33, 339–352.

Stern, F., Muste, M., Beninati, M.L., & Eichinger, W.E. (1999) Summary of Experimental Uncertainty assessment Methodology with Example. Iowa Institute of Hydraulic Research. The University of Iowa. Iowa City. IA. IIHR. Technical Report: 406.

Taylor, B.N & Kuyatt, C.E. (1994) Guidelines for Evaluating and Expressing the Uncertainty of NIST Measurement Results. NIST Technical Note: 1297.

TCHME (2003) Annotated Bibliography on Uncertainty Analysis. Task Committee on Experimental Uncertainty and Measurement Errors in Hydraulic Engineering. EWRI. ASCE. Available from: http://www.dri.edu/People/Mark.Stone/Tchme/task.html

Thibodeaux, K.G. (2007) Testing and evaluation of inexpensive horizontal-axis mechanical current meters. USCID *Fourth International Conference*, 1295–1308.

Thomas, F. (2002) Open channel flow measurement using international standards: Introducing a standards programme and selecting a standard. *Flow Measurement and Instrumentation*. 13, 303–307.

UKAS (2007) M3003. The expression of uncertainty and confidence in measurement. Feltham, United Kingdom Accreditation Service.

UNESCO (2007) Data Requirements for Integrated Urban Water Management. Fletcher, T.D., & Deletic, A. (eds.) Oxfordshire. UK, Taylor and Francis.

Wahlin, B., Wahl, T., Gonzalez-Castro, J.A., Fulford, J., & Robeson, M. (2005) Task committee on experimental uncertainty and measurement errors in hydraulic engineering. In: *2005 World Water and Environmental Resources Congress: An Update. Impacts of Global Climate Change- Proceedings of the World Water and Environmental Resources Congress*, 15–19 May 2005, Anchorage, Alaska. American Society of Civil Engineers. pp. 1–11.

WMO (2010) Manual on Stream Gauging. Volume I: Fieldwork. World Meteorological Organization. Geneva. Switzerland.

WMO (2017). Guidelines for the assessment of uncertainty for hydrometric measurements, World Meteorological Organization (WMO) report WMO-No 1097, Geneva, Switzerland.

Chapter 8

Hydroinformatics Applied to Hydraulic Experiments

8.1 INTRODUCTION

The advent of digital instruments and non-intrusive measurement technologies based on acoustics and optics principles has resulted in unprecedented opportunities for acquiring high-density temporal measurements at points, along lines and across surfaces and volumes within experimental facilities. The high operational efficiency of the newly-developed technologies has produced a vast amount of laboratory and field data that challenges conventional capabilities for handling, processing, extracting, and visualizing information from the measured data. These challenges have led to the realization that considerable attention needs to be given to the informatics of experimental hydraulics. Fortunately, an interdisciplinary science formed in the hydraulic community in the early 1990s under the label of "hydroinformatics" (Abbott, 1991) addresses these needs. This relatively new discipline is a branch of informatics stemming from aspects of water research seen from a wider perspective. More specifically, hydroinformatics links measurement and modeling technologies to those for data and information management, as well as to methods that make data and information accessible to engineers and scientists (Giordano et al., 2008).

The recent advances in measurement and information technologies (computer and communication hardware and software) now facilitate an "informatics" approach to investigation and management that capitalizes on measurements and their interpretation (Maidment et al., 2009). This *big data* approach differs fundamentally from the earlier "observation-centric" and "model-centric" approaches, whereby hydrologic-cycle components, including river flows, were viewed independently and treated as being uncorrelated. Today, systematic integration of water-related data into an informatics system enables to link a broad set of water-related information (Price et al., 1994; Kapoor, 2006). For example, hydraulic and hydrologic information or coastal information can be connected to contextual analyses concerning economics, political and social sciences, and management practice. Such integrative informatics greatly improves peoples' ability to understand ecosystems and to manage natural and built environments.

As contemporary laboratory and field hydraulic experiments are increasingly using hydroinformatics, it is useful for the present handbook to place hydraulic measurements and experimentation within the larger context of the hydroinfor-matics framework. This chapter first introduces hydroinformatics, explaining the role of this science in organizing the data, irrespective of their source (*i.e.*, experiments (monitoring or modeling). Subsequently, the concept of Digital Environmental Observatories (DEOs) is described as a contemporary approach for assembling various sources of water-related data characterizing large spatio-temporal scales, including river measurements. Finally, the Arc River data model is introduced as a prototype data model for handling and processing multiple databases containing measured or simulated river characteristics. The model allows efficient characterization of rivers and the ancillary processes affecting them. A list of relevant websites leading to further information about hydroinformatics is included at the end of the chapter.

8.2 HYDROINFORMATICS

Arising in the early 1990s as a consequence of the evolution of computers and informa-tion systems, hydroinformatics is a conceptual framework that facilitates informatics-based data management and investigations. Hydroinformatics "integrates knowledge and understanding of both water quantity and quality with the latest developments in information technology to improve technical and business decision making within the water industry. It embraces not only methods of data capture, storage, processing, analysis and graphical display, but also the use of advanced modeling, simulation, optimization and knowledge-based tools and systems infrastructure" (Abbott, 1991). Although initial hydroinformatics efforts have focused on numerical modeling of various components of the hydrologic cycle in natural-scale landscapes, today the discipline focuses on a wider spectrum of information resulting from the integration of laboratory and field, numerical modeling, and theory, with the contextual informa-tion (*e.g.*, social, economic, legal) regarding those landscapes (Abbott, 1993; Abbott, 1994; Price *et al.*, 1994).

Hydroinformatics systems rely on hardware and software, collectively grouped under the label cyberinfrastructure (CI), that are designed to support scientific and practical investigations of the water-related domains (Muste, 2014). CI procedures applied to field conditions typically place datasets in a Geographic Information System (GIS) framework (Maidment, 2002). Several GIS tools are already available to visualize, identify trends and patterns, and interpret and analyze the datasets via customized tools and queries. The assembly of tools and workflows within this digital environment allows a dataset to be seamlessly converted into information and sub-sequently into scientific and practical knowledge through the succession illustrated in Figure 8.2.1. This data-information-knowledge fusion aids investigation of processes and extraction of information in formats that scientists, researchers and managers can use. It also enables the implementation of composite modeling whereby field and laboratory data are merged with data from numerical simulations. Composite mod-eling (presented in Section 5.5.2, Volume I) is currently undergoing rapid develop-ment, becoming an efficient process for solving practical problems in many areas of hydraulics (*e.g.*, design of hydraulic structures, river restoration, and river-related disaster mitigation).

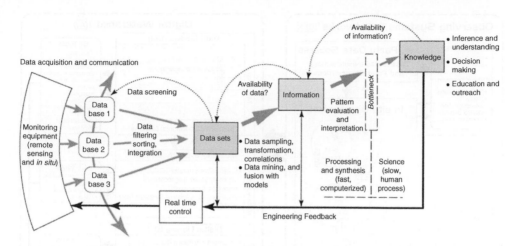

Figure 8.2.1 Data-to-knowledge transformation enabled by a hydroinformatics system (adapted with permission from Fletcher & Deletic, 2006).

8.3 DIGITAL ENVIRONMENTAL OBSERVATORIES

8.3.1 Concepts and terminology

Until recently, most empirically-based investigations for information gathering and analysis regarding watershed processes have been conducted by means of laboratory research, with fieldwork being relatively constrained in its extent and detail. The instrumentation needs for comprehensive fieldwork investigations covering large scales were daunting, due to the potentially large amount of information required to thoroughly quantify the investigated processes. The introduction of new instruments and informatics frameworks has led to significant advances in many areas of science (*e.g.,* health sciences, remote sensing) and society (*e.g.,* social media), therefore it is expected that conceptualization of new information-centric frameworks for the hydro-environment disciplines will have similar positive impacts.

One of these concepts involves Digital Environment Observatories (DEO) described in Muste (2014). The term "observatories" is taken from the field of astronomy, because it refers to massive datasets and information acquired over large spatial scales and durations. DEOs are electronic spatio-temporal representations of landscape units (*e.g.,* a watershed) and ancillary processes as documented with field data acquired at multiple locations, data from simulation models, and layered information from analyses and syntheses of available data. In the case of watersheds, for example, they may include methods for tracking the movement of water, sediments, contaminants, and nutrients. The digital information that describe conditions in the monitored environment include societal infrastructure for water-resources and waste/storm water management. Although the internet has improved access to diverse data sources, gathering the data required for most digital environments requires coordination of multiple observation sites, each with its own access protocols and data-exporting formats.

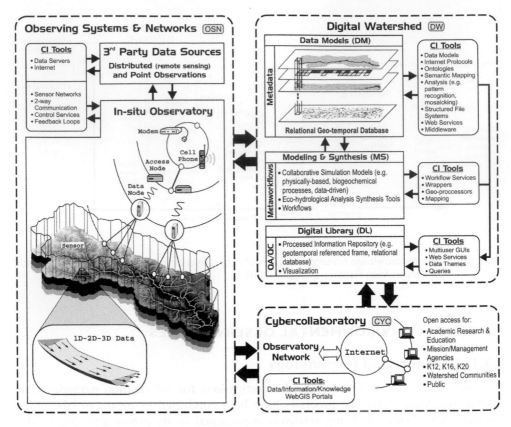

Figure 8.3.1 Conceptual framework for the Digital Environmental Observatory (reprinted from Muste, 2014, with permission from Elsevier).

Given the multitude of users, DEOs are most efficient when they are open-access, web-based, and distributed over a computer network that includes the repositories of data providers, researchers and managers.

DEOs applied to terrestrial landscapes aim at providing a useful framework for organizing, managing, and guiding scientific investigations. While the acquisition of data and the development and application of models were once the sole tasks of expert scientists and engineers, currently the water resources managers increasingly require online access to information for supporting decision-making. In this regard, DEOs can potentially include sophisticated decision-support systems (*e.g.*, Xu *et al.*, 2016). Details about emerging components, workflow, and tools assembled in DEOs for water-centered research are described in Horsburgh *et al.* (2008), Goodall *et al.* (2008), and Muste (2014). The design and operation of a DEO is driven by the needs of the people requiring the DEO. For illustration purposes, the architecture of a scientifically-oriented DEO focused on the characterization of the water cycle over a landscape unit is provided in Figure 8.3.1. The main DEO components and their role is briefly described in Table 8.3.1. As illustrated in the figure, a DEO comprises data-servers and software tools that merge data acquired by various water-related agencies

Table. 8.3.1 Components of a Digital Environmental Observatory and their operational roles (reprinted from Muste, 2014, with permission from Elsevier).

DEO Component	Role
Observing Systems & Networks	Instruments and communication means for data collection in the watershed. (fit with automatic and real-time sampling of the variables)
Digital Watershed	Digital characterization of the watershed eco-hydrologic systems using measured data and simulations models
Data Model	Permanent data & information infrastructure that stores data, metadata on water-cycle fluxes and the related watershed environmental processes
Modeling & Synthesis	Unifying framework for the synthesis of composite modeling results (includes model-based simulators connected to the data and information acquired from field and laboratory experiments)
Digital Library	Storage for historical and new data and information subsequently used for analysis, fusion, visualization, and knowledge dissemination
Cyber Collaboratory	Typically a web portal allowing anytime/anywhere sharing of resources, models, data, and ideas by scientists, managers, and virtual organizations.

with data acquired locally by universities, local mission agencies, and even citizen scientists (through crowdsourcing). The digital description of the watershed is provided by the "Digital Watershed" module that assembles both the field data and modeling results. Previous and future analyses and synthesis of the data are stored in the "Digital library" from which information and knowledge can be accessed with customized tools that match the technical level of the end user.

Example 8.3.1: DEOs. As discussed above, DEOs are being built for various purposes: scientific research (Muste, 2014), water resources management (Demir & Beck, 2009; Xu *et al.*, 2016), sharing of information, public education, and warnings for communities (Demir & Krajewski, 2013). DEOs provide ample visualization opportunities for hydraulic measurements that broaden their context and extends their significance and utility. Consider for illustration purposes the case of the stream stage observing network as measured by conventional gaging stations. Most contemporary gaging systems communicate stream discharges in real-time for immediate use. The stream stage data is of interest for multiple users (from decision makers in various areas of water resources and emergency management to scientists and the general public). While the stage data come from the same source (say, US Geological Survey real-time stream gaging stations), the packaging of the data and of the auxiliary information could be different for various users commensurate with their interest and level of technical ability. For example, the time series of stream stages at multiple downstream locations illustrated in Figure 8.3.2a are useful for hydrologists and scientists. The flood inundation levels for various stages as mapped in Figure 8.3.2b are useful for informing a community about an imminent flood threat. Another view of the stage data is provided in Figure 8.3.2c whereby the real-time stage value is placed directly onto a realistic image of a culvert as obtained from a photogrammetric analysis of the site. This last view does not require any technical knowledge and it is easily understood by a wide variety of public segments.

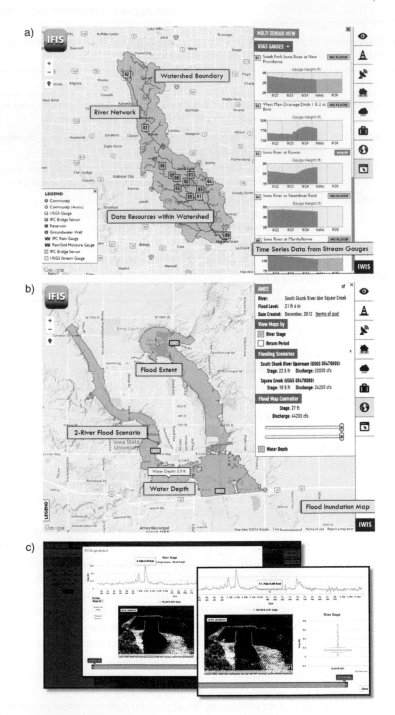

Figure 8.3.2 Illustration of DEO end-user interfaces: a) sample IFIS interface displaying time series of stream gages throughout a watershed; b) sample IFIS illustration displaying areas of flood inundation in local area of a watershed; c) sample IoWaDSS interface illustrating stage within a realistic image of a culvert (lower stage on the left, higher stage on the right). IFIS stands for the Iowa Flood Information System (iowafloodcenter.org); IoWaDSS stands for Iowa Watersheds Decision-support System (iowawatersheds.org/iowadss).

8.3.2 Integrated datasets and models delivered as computer services

Integration of datasets and models is a central task for building a DEO. The datasets associated with scientific investigations in hydrosciences come from a variety of sources and multiple instruments and sensors. Assembling datasets is a difficult task as they are heterogeneous with respect to data type, file format and storage media, accuracy, and their spatial and temporal resolutions. Modeling on the other hand, is accomplished with many different numerical simulations, each one storing input and output data in its own customized format. Converting between data formats can be complicated, and may pose a challenge to using data from diverse sources in models. As the amount and resolution of data increase, so do the difficulties of finding and retrieving the required subset of data, irrespective of their source (*i.e.*, experiments, monitoring or modeling). The above challenges are typical for *big data use cases* and present an impediment for sharing and using information assembled in a DEO (Jackson & Maidment, 2014).

Similar to other areas of sciences and society, the hydroscience community is currently embarking on the emerging trend of developing and delivering a range of cloud-computing services for managing various sources of data, irrespective of their provenance (Tarboton *et al.*, 2014; Rajib *et al.*, 2016). For the experimental hydraulics context discussed herein, the most relevant corresponding term is "data as a service". Typically this service is provided by data providers that host large datasets in remote servers, and provide standardized methods for users to access, search, and download specific datasets. This trend allows to automate the data acquisition as part of a larger workflow procedure. For this purpose, Jackson and Maidment (2014) envision the creation of a single standard "harmonized" transfer language specifically designed for cross-application of water-resources data exchange. Harmonization in this context represents the possibility to combine data from heterogeneous sources into integrated, consistent, and unambiguous information products, in a way that is of no concern to the end-user. Given that water-resources data are acquired on vast areal domains, the GIS emerge as the effective tool for integrating spatial and temporal attributes of in-situ measurements through by their proven capabilities to organize, visualize, and analyze large amounts of data in their geo-spatial context.

While not a representative of the cloud computing services yet, an illustration of the opportunities to assemble large datasets and models in the same DEO is provided next. The example describes the prototype Arc River data model (Kim *et al.*, 2015) that leverages recent advances in CI technology for assembling online data repositories, communication networks, advanced analytical and modeling tools, and computing engines into digital environments. The resulting digital environments enable integrative research and river management to be conducted in one web-GIS platform. The example prototype DEO can be regarded as a computer-based problem-solving environment for rivers seen as systems of integrated processes. The new model can become a key framework for expanding the integration of scientific investigations with management for river systems and the many concerns associated with them (*e.g.*, hydro-, and morphodynamics, water quality, water supply).

Example 8.3.2: Arc River data model. This example describes a river-data model, labeled Arc River, that is linked to a database populated with measured or simulated river characteristics (from geometrical to morpho-hydrodynamics). Arc River expands the one-dimensional river network structure of Arc Hydro (Maidment, 2002) to provide a three-dimensional digital framework for representing rivers. Arc River supports vector as well as scalar information, and can attach metadata to observations. Geometric features in Arc River are assigned a dimensionality based on the type of quantities associated with them, while the network connectivity of higher dimensionality features is determined by their associated lower dimensionality features. Data derived from traditional survey methods (*e.g.*, topo, lidar) as well as datasets generated from Acoustic Doppler Current Profilers (described in Section 3.3, Volume II) can be processed by the Arc River tools to enable visualization of rivers as 3D multi-domain digital objects, rather than the conventional representations accomplished through the linear channel networks (1D) and sets of cross-sections (2D).

The aim of the Arc River data model is to provide an interactive computational environment for investigating the spatio-temporal variation of riverine hydrodynamics and morphodynamics as documented by direct field measurements, hydraulic experiments, and numerical modeling. For this purpose, the following digital representations are enabled (see Figure 8.3.3): (i) river data in a curvilinear coordinate system to

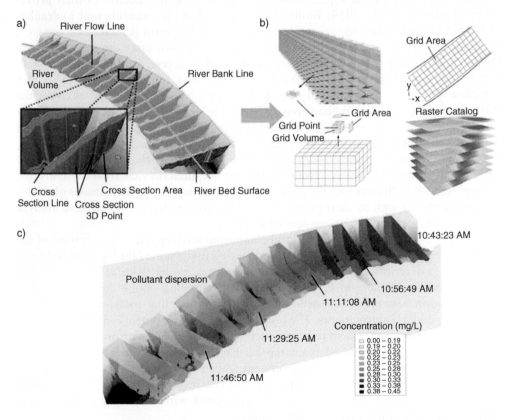

Figure 8.3.3 An illustration of the capabilities of an Arc River data model to: a) 3D representation of the raw ADCP data collected in the Mississippi River; b) representation of the raw ADCP data in various formats; c) use ADCP raw data in conjunction with models for visualizing dispersion of a point pollutant source along the stream within the data model (Reprinted from Kim *et al.*, 2015, with permission from Elsevier).

support river-channel oriented spatial analyses; (ii) multi-dimensional river features through points, lines, polygons, and volumes; (iii) numerical simulation data for river channels that can be efficiently coupled with observed data; and, (iv) spatio-temporal dynamics of moving river objects (such as bedforms) from single or multiple events using Eulerian or Lagrangian frameworks. In this initial stage, the Arc River data model can use customized tools to load datasets generated from ADCPs mounted on a boat. The ADCP data can include three-dimensional hydrodynamic data along any path of a river (*e.g.*, Kim *et al.*, 2015). Figures 8.3.3a and 8.3.3b provide an Arc River representation of a river channel along with modeling results of a hypothetical spill produced by a point source discharging into the river (as shown in Figure 8.3.3c). All of the input data for the model are obtained from ADCP measurements.

8.4 OUTLOOK

The data- and information-reach environment required by many of the laboratory and field experimentation are poised for progression by recent advances in science and engineering research, computing, information and communication technology. As the challenges related to field measurements of multi-disciplinary domains are testing the frontiers of hydroinformatics, the ensuing considerations on DEOs for in-situ experiments characterizing the rivers as systems. Attaining the level of operational science needed for acquiring and handling the data collected at watershed-scale level may appear daunting, as the amount of data from observations and simulations is vast with complex cross-disciplinary workflows and many data producers. Domain scientists lack the skills to effectively manage, curate, and preserve large volumes of digital data and to develop software that connects it through workflows in the digital domain. Such skills are becoming the domain of data professionals with a strong computer programming background, but such specialists rarely have a full understanding of certain specific needs that earth science data entail. Currently, a cross-disciplinary professional category is emerging: the data scientists. The equivalent of this new category of scientists for our area would correspond to hydroinformaticians. These are people who are both specialists in data and workflows management and who also have domain expertise in water-focused areas.

Furthermore, meeting the challenges associated with the building of these new and transformative information systems for investigations at natural scales will require hydroinformaticians to form unconventional and nontraditional research coalitions. In addition to proactively seeking partners and researchers from across the different fields in the mathematical, engineering, physical, life, information, and social sciences, hydroscientists have to acquire knowledge from computer science, information science, and systems theory. Moreover, scientists have to work closer with water resources managers in order to become aware of the interventions made to natural landscapes and subsequently use the vast amount of information that it is gathered by management authorities at all levels (federal, state, local) for monitoring purposes. Only within such a diversified body of participants, representing the broadest spectrum of hydrologic interests - from basic to applied and from regulatory to public interest - will these information systems have the potential of addressing the increasing concern of rivers health and realistically translating the science into practical applications.

REFERENCES

Abbott, M.B. (1991) *Hydroinformatics: Information Technology and the Aquatic Environment*, Aldershot, U.K. and Brookfield, U.S.A, Avebury Technical.

Abbott, M.B. (1993) The electronic encapsulation of knowledge in hydraulics, hydrology, and water resources. *Adv Water Resour.*, 16, 21–39.

Abbott, M.B. (1994) Hydroinformatics: A copernicanrevolution in hydraulics, Special issue on Hydroinformatics in *J Hydraul Eng.*, 32, 3–13.

Demir, I. & Beck, M.B. (2009) GWIS: A Prototype information system for Georgia watersheds, Paper 6.6.4. *Proceedings of Georgia Water Resources Conference: Regional Water Management Opportunities*, April 27–29, Athens, GA, USA, UGA.

Demir, I. & Krajewski, W. (2013) Towards an integrated Flood Information System: Centralized data access, analysis, and visualization. *Environ Modell Softw.*, 50, 77–84, 2013

Fletcher, T. & Deletic, A. (eds.) (2006) *Data Requirements for Integrated Urban Water Management*. Paris, France, UNESCO Publishing.

Giordano, R., Uricchio, V.F., & Vurro, M. (2008) Monitoring information systems to support integrated decision-making. In: Timmerman, J.G., Pahl-Wostl, C., & Moltgen, J. (eds.) *The Adaptiveness of IWRM; Analyzing European IWRM Research*. London, UK, IWA Publishing.

Goodall, J.L., Horsburgh, J.S., Whiteaker, T.L., Maidment, D.R., & Zaslavsky, I. (2008) A first approach to web services for the national water information system. *Environ Modell Softw.*, 23(4), 404–411.

Horsburgh, J.S., Tarboton, D.G., Maidment, D.R., & Zaslavsky, I. (2008) A relational model for environmental and water resources data. *Water Resour Res.*, 44(5), W05406.

Jackson, S.R. & Maidment, D.R. (2014) RiverML: A harmonized transfer language for river hydraulic models [Online] *Report 14–5*, Center for Research in Water Resources, Austin, TX, USA, The University of Texas at Austin (http://www.crwr.utexas.edu/online.shtml – last access November, 10, 2016)

Kapoor, T.R. (2006) Role of information and communication technology in adaptive integrated water resources management. *Proceedings World Environmental & Water Resource*, Omaha, NE, ASCE/EWRI, May 21–26.

Kim, D., Muste, M., & Merwade, V. (2015) A relational data model for representation of multidimensional river hydrodynamics and morphodynamics. *Environ Modell Softw.*, 65, 79–93.

Maidment, D.R. (2002) *Arc Hydro – GIS for Water Resources*. California, USA, ESRI, Redlands.

Maidment, D.R., (ed.) (2009) CUAHSI Hydrologic Information System: 2009 Status Report *Report*for Consortium of Universities for the Advancement of Hydrologic Science, Inc, p. 79 (http://his.cuahsi.org/documents/HISOverview_2009.pdf – last access November 10, 2016)

Merwade, V.M., Maidment, D.R., & Hodges, B.R. (2005) Geospatial representation of river channels. *J Hydrol Eng.*, 10(3), 243–251.

Muste, M. (2014) Information-centric systems for underpinning sustainable watershed resource management, Chapter 13. In: Ahuja, S. (ed.) *Comprehensive Water Quality and Purification.*, 2014, 4, Elsevier, 270–298.

Price, R.K., Samuganathan, K., & Powell, K. (1994) Hydroinformatics and the management of water-based assets. Special Issue on Hydroinformatics in *JHyd Res.*, 32, 65–82.

Rajib, M.A., Nerwade, V., Kim, I.L., Zhao, L., Song, K., & Zhe, S. (2016) SWATShare – A web-platform for collaborative research and education through online sharing, simulation and visualization of SWAT models. *Environ Modell Softw.*[Online] 75, 498–512.Available from: doi: 10.1016/j.envsoft.2015.10.032

Tarboton, D.G., Idaszak, R., Horsburgh, J.S., Heard, J., Ames, D., Goodall, J.L., Band, L., Merwade, V., Couch, A., Arrigo, J., Hooper, R., Valentine, D., & Maidment, D. (2014) HydroShare: Advancing collaboration through hydrologic data and model sharing. In: Ames, D.P., Quinn, N.W.T., & Rizzoli, A.E. (eds.) *Proceedings of the 7th International Congress on Environmental Modelling and Software*, San Diego, California, USA, International Environmental Modelling and Software Society (iEMSs), ISBN: 978–1088-9035-744-2, (http://www.iemss.org/sites/iemss2014/papers/iemss2014_submission_243.pdf – last access on November 10, 2016)

Xu, H., Hameed, H., Muste, M., & Demir, I. (2016) Visualization platform for collaborative modelling. *AWRA Summer Specialty Conference – GIS and Water Resources IX*, Sacramento, CA, American Water Resources Association, July 11–13, 2016.

RELEVANT WEBSITES (LAST ACCESSED NOVEMBER 10, 2016)

- EuroAquae – http://euroaquae.org
- Georgia Watershed Information System (GWIS) – http://www.georgiawis.org
- Global Water Partnership – http://www.gwp.org/
- Group on Earth Observations – http://www.earthobservations.org
- HydroShare – http://hydroshare.cuahsi.org/
- Iowa Flood Information System (IFIS) – http://ifis.iowafloodcenter.org
- Iowa Watershed Decision-Support System (IoWaDSS) – http://iowawatersheds.org/iowadss
- IWA Publishing – Journal of Hydroinformatics - http://iwaponline.com/jh
- UN Water Learning Centre – http://wvlc.uwaterloo.ca
- UNESCO-IHE Institute for Water Education – http://unesco-ihe.org

Subject Index